Matrix Algebra: Exercises and Solutions

Springer
New York
Berlin
Heidelberg
Barcelona
Hong Kong
London
Milan
Paris
Singapore
Tokyo

David A. Harville

Matrix Algebra: Exercises and Solutions

Springer

David A. Harville
Mathematical Sciences Department
IBM T.J. Watson Research Center
Yorktown Heights, NY 10598-0218
USA

Library of Congress Cataloging-in-Publication Data
Harville, David A.
Matrix algebra : exercises and solutions / David A. Harville.
p. cm.
Includes bibliographical references and index.
ISBN 0-387-95318-3 (pbk. : alk. paper)
1. Matrices—Problems, exercises, etc. I. Title.
QA188 .H38 2001
519.9'434—dc21 2001032838

Printed on acid-free paper.

Production managed by Yong-Soon Hwang; manufacturing supervised by Jeffrey Taub.
Photocomposed copy prepared from the author's LaTeX file.
Printed and bound by Maple-Vail Book Manufacturing Group, York, PA.
Printed in the United States of America.

9 8 7 6 5 4 3 2 1

ISBN 0-387-95318-3 SPIN 10841733

Springer-Verlag New York Berlin Heidelberg
A member of BertelsmannSpringer Science+Business Media GmbH

Preface

This book comprises well over three-hundred exercises in matrix algebra and their solutions. The exercises are taken from my earlier book *Matrix Algebra From a Statistician's Perspective*. They have been restated (as necessary) to make them comprehensible independently of their source. To further insure that the restated exercises have this stand-alone property, I have included in the front matter a section on terminology and another on notation. These sections provide definitions, descriptions, comments, or explanatory material pertaining to certain terms and notational symbols and conventions from *Matrix Algebra From a Statistician's Perspective* that may be unfamiliar to a nonreader of that book or that may differ in generality or other respects from those to which he/she is accustomed. For example, the section on terminology includes an entry for scalar and one for matrix. These are standard terms, but their use herein (and in *Matrix Algebra From a Statistician's Perspective*) is restricted to real numbers and to rectangular arrays of real numbers, whereas in various other presentations, a scalar may be a complex number or more generally a member of a field, and a matrix may be a rectangular array of such entities.

It is my intention that *Matrix Algebra: Exercises and Solutions* serve not only as a "solution manual" for the readers of *Matrix Algebra From a Statistician's Perspective*, but also as a resource for anyone with an interest in matrix algebra (including teachers and students of the subject) who may have a need for exercises accompanied by solutions. The early chapters of this volume contain a relatively small number of exercises—in fact, Chapter 7 contains only one exercise and Chapter 3 only two. This is because the corresponding chapters of *Matrix Algebra From a Statistician's Perspective* cover relatively standard material, to which many readers will have had previous exposure, and/or are relatively short. It is the final ten chapters that contain the vast majority of the exercises. The topics of many of these chapters are ones that may not be covered extensively (if at all) in more standard presentations or that may be covered from a different perspective. Consequently, the overlap between the exercises from *Matrix Algebra From a Statistician's Perspective* (and contained herein) and those available from other sources is relatively small.

A considerable number of the exercises consist of verifying or deriving results supplementary to those included in the primary coverage of *Matrix Algebra From a Statistician's Perspective*. Thus, their solutions provide what are in effect proofs. For many of these results, including some of considerable relevance and interest in statistics and related disciplines, proofs have heretofore only been available (if at all) through relatively high-level books or through journal articles.

The exercises are arranged in 22 chapters and within each chapter, are numbered successively (starting with 1). The arrangement, the numbering, and the chapter titles match those in *Matrix Algebra From a Statistician's Perspective*. An exercise from a different chapter is identified by a number obtained by inserting the chapter number (and a decimal point) in front of the exercise number.

A considerable effort was expended in designing the exercises to insure an appropriate level of difficulty—the book *Matrix Algebra From a Statistician's Perspective* is essentially a self-contained treatise on matrix algebra, however it is aimed at a reader who has had at least some previous exposure to the subject (of the kind that might be attained in an introductory course on matrix or linear algebra). This effort included breaking some of the more difficult exercises into relatively palatable parts and/or providing judicious hints.

The solutions presented herein are ones that should be comprehensible to those with exposure to the material presented in the corresponding chapter of *Matrix Algebra From a Statistician's Perspective* (and possibly to that presented in one or more earlier chapters). When deemed helpful in comprehending a solution, references are included to the appropriate results in *Matrix Algebra From a Statistician's Perspective*—unless otherwise indicated a reference to a chapter, section, or subsection or to a numbered result (theorem, lemma, corollary, "equation", etc.) pertains to a chapter, section, or subsection or to a numbered result in *Matrix Algebra From a Statistician's Perspective* (and is made by following the same conventions as in the corresponding chapter of *Matrix Algebra From a Statistician's Perspective*). What constitutes a "legitimate" solution to an exercise depends of course on what one takes to be "given". If additional results are regarded as given, then additional, possibly shorter solutions may become possible.

The ordering of topics in *Matrix Algebra From a Statistician's Perspective* is somewhat nonstandard. In particular, the topic of eigenvalues and eigenvectors is deferred until Chapter 21, which is the next-to-last chapter. Among the key results on that topic is the existence of something called the spectral decomposition. This result if included among those regarded as given, could be used to devise alternative solutions for a number of the exercises in the chapters preceding Chapter 21. However, its use comes at a "price"; the existence of the spectral decomposition can only be established by resort to mathematics considerably deeper than those underlying the results of Chapters 1–20 in *Matrix Algebra From a Statistician's Perspective*.

I am indebted to Emmanuel Yashchin for his support and encouragement in the preparation of the manuscript for *Matrix Algebra: Exercises and Solutions*. I am also indebted to Lorraine Menna, who entered much of the manuscript in LaTeX, and to Barbara White, who participated in the latter stages of the entry process. Finally, I wish to thank John Kimmel, who has been my editor at Springer-Verlag, for his help and advice.

Contents

Some Notation

$\{x_i\}$	A row or (depending on the context) column vector whose ith element is x_i
$\{a_{ij}\}$	A matrix whose ijth element is a_{ij} (and whose dimensions are arbitrary or may be inferred from the context)
$\mathbf{A}'$	The transpose of a matrix $\mathbf{A}$
$\mathbf{A}^p$	The pth (for a positive integer p) power of a square matrix $\mathbf{A}$; i.e., the matrix product $\mathbf{AA}\cdots\mathbf{A}$ defined recursively by setting $\mathbf{A}^0 = \mathbf{I}$ and taking $\mathbf{A}^k = \mathbf{AA}^{k-1}$ $(k = 1, \ldots, p)$
$\mathcal{C}(\mathbf{A})$	Column space of a matrix $\mathbf{A}$
$\mathcal{R}(\mathbf{A})$	Row space of a matrix $\mathbf{A}$
$\mathcal{R}^{m\times n}$	The linear space comprising all $m \times n$ matrices
$\mathcal{R}^n$	The linear space $\mathcal{R}^{n\times 1}$ comprising all n-dimensional column vectors or (depending on the context) the linear space $\mathcal{R}^{1\times n}$ comprising all n-dimensional row vectors
$\mathrm{sp}(S)$	Span of a finite set S of matrices; $\mathrm{sp}(\{\mathbf{A}_1, \ldots, \mathbf{A}_k\})$, which represents the span of the set $\{\mathbf{A}_1, \ldots, \mathbf{A}_k\}$ comprising the k matrices $\mathbf{A}_1, \ldots, \mathbf{A}_k$, is generally abbreviated to $\mathrm{sp}(\mathbf{A}_1, \ldots, \mathbf{A}_k)$
$\subset$	Writing $S \subset T$ (or $T \supset S$) indicates that a set S is a (not necessarily proper) subset of a set T
$\dim(\mathcal{V})$	Dimension of a linear space $\mathcal{V}$
$\mathrm{rank}\,\mathbf{A}$	The rank of a matrix $\mathbf{A}$
$\mathrm{rank}\,T$	The rank of a linear transformation T

$\text{tr}(\mathbf{A})$	The trace of a (square) matrix $\mathbf{A}$
0	The scalar zero (or, depending on the context, the zero transformation from one linear space into another)
$\mathbf{0}$	A null matrix (whose dimensions are arbitrary or may be inferred from the context)
I	An identity transformation
$\mathbf{I}$	An identity matrix (whose order is arbitrary or may be inferred from the context)
$\mathbf{I}_n$	An identity matrix of order n
$\mathbf{A} \bullet \mathbf{B}$	Inner product of a pair of matrices $\mathbf{A}$ and $\mathbf{B}$ (or if so indicated, quasi-inner product of the pair $\mathbf{A}$ and $\mathbf{B}$)
$\Vert \mathbf{A} \Vert$	Norm of a matrix $\mathbf{A}$ (or, in the case of a quasi-inner product, the quasi norm of $\mathbf{A}$)
$\delta(\mathbf{A}, \mathbf{B})$	Distance between two matrices $\mathbf{A}$ and $\mathbf{B}$
T^{-1}	The inverse of an invertible transformation T
$\mathbf{A}^{-1}$	The inverse of an invertible matrix $\mathbf{A}$
$\mathbf{A}^{-}$	An arbitrary generalized inverse of a matrix $\mathbf{A}$
$\mathcal{N}(\mathbf{A})$	Null space of a matrix $\mathbf{A}$
$\mathcal{N}(T)$	Null space of a linear transformation T
$\perp$	A symbol for "is orthogonal to"
$\perp_{\mathbf{W}}$	A symbol used to indicate (by writing $\mathbf{x} \perp_{\mathbf{W}} \mathbf{y}$, $\mathbf{x} \perp_{\mathbf{W}} \mathcal{U}$, or $\mathcal{U} \perp_{\mathbf{W}} \mathcal{V}$) that 2 vectors $\mathbf{x}$ and $\mathbf{y}$, a vector $\mathbf{x}$ and a subspace $\mathcal{U}$, or 2 subspaces $\mathcal{U}$ and $\mathcal{V}$ are orthogonal with respect to a symmetric nonnegative definite matrix $\mathbf{W}$
$\mathbf{P}_{\mathbf{X}}$	The matrix $\mathbf{X}(\mathbf{X}'\mathbf{X})^{-}\mathbf{X}'$ [which is invariant to the choice of the generalized inverse $(\mathbf{X}'\mathbf{X})^{-}$]
$\mathbf{P}_{\mathbf{X},\mathbf{W}}$	The matrix $\mathbf{X}(\mathbf{X}'\mathbf{W}\mathbf{X})^{-}\mathbf{X}'\mathbf{W}$ [which if $\mathbf{W}$ is symmetric and positive definite, is invariant to the choice of the generalized inverse $(\mathbf{X}'\mathbf{W}\mathbf{X})^{-}$]
$\mathcal{U}^{\perp}$	The orthogonal complement of a subspace $\mathcal{U}$ of a linear space $\mathcal{V}$
$\mathcal{C}^{\perp}(\mathbf{X})$	The orthogonal (with respect to the usual inner product) complement of the column space $\mathcal{C}(\mathbf{X})$ of an $n \times p$ matrix $\mathbf{X}$ [when $\mathcal{C}(\mathbf{X})$ is regarded as a subspace of $\mathcal{R}^n$]
$\mathcal{C}^{\perp}_{\mathbf{W}}(\mathbf{X})$	The orthogonal complement of the column space $\mathcal{C}(\mathbf{X})$ of an $n \times p$ matrix $\mathbf{X}$ when, for an $n \times n$ symmetric positive definite matrix $\mathbf{W}$, the inner product is taken to be the bilinear form $\mathbf{x}'\mathbf{W}\mathbf{y}$ [and when $\mathcal{C}(\mathbf{X})$ is regarded as a subspace of $\mathcal{R}^n$]
$\sigma_n(\bullet)$	A function whose value $\sigma_n(i_1, j_1; \ldots; i_n, j_n)$ for any two (not necessarily different) permutations of the first n positive integers is the number of negative pairs among the $\binom{n}{2}$ pairs that can be formed from the $i_1 j_1, \ldots, i_n j_n$th elements of an $n \times n$ matrix
$\phi_n(\bullet)$	A function whose value $\phi_n(i_1, \ldots, i_n)$ for any sequence of n distinct

integers $i_1, \ldots, i_n$ is $p_1 + \cdots + p_{n-1}$, where (for $k = 1, \ldots, n-1$) p_k represents the number of integers in the subsequence $i_{k+1}, \ldots, i_n$ that are smaller than i_k

$|\mathbf{A}|$ The determinant of a square matrix $\mathbf{A}$ — with regard to partitioned matrices, $\left|\begin{pmatrix} \mathbf{A}_{11} & \ldots & \mathbf{A}_{1c} \\ \vdots & & \vdots \\ \mathbf{A}_{r1} & \ldots & \mathbf{A}_{rc} \end{pmatrix}\right|$ may be abbreviated to $\begin{vmatrix} \mathbf{A}_{11} & \ldots & \mathbf{A}_{1c} \\ \vdots & & \vdots \\ \mathbf{A}_{r1} & \ldots & \mathbf{A}_{rc} \end{vmatrix}$

$\det(\mathbf{A})$ The determinant of a square matrix $\mathbf{A}$

$\text{adj}(\mathbf{A})$ The adjoint matrix of a square matrix $\mathbf{A}$

$\mathbf{J}$ A matrix, all of whose elements equal one (and whose dimensions are arbitrary or may be inferred from the context)

$\mathbf{J}_{mn}$ An $m \times n$ matrix, all of whose mn elements equal one

$\mathbf{A} \otimes \mathbf{B}$ The Kronecker product of a matrix $\mathbf{A}$ and a matrix $\mathbf{B}$ — this notation extends in an obvious way to the Kronecker product of 3 or more matrices

vec $\mathbf{A}$ The vec of a matrix $\mathbf{A}$

vech $\mathbf{A}$ The vech of a (square) matrix $\mathbf{A}$

$\mathbf{K}_{mn}$ The $mn \times mn$ vec-permutation (or commutation) matrix

$\mathbf{G}_n$ The $n^2 \times n(n+1)/2$ duplication matrix

$\mathbf{H}_n$ An arbitrary left inverse of $\mathbf{G}_n$, so that $\mathbf{H}_n$ is any $n(n+1)/2 \times n^2$ matrix such that $\mathbf{H}_n\mathbf{G}_n = \mathbf{I}$ or equivalently such that, for every $n \times n$ symmetric matrix $\mathbf{A}$, vech $\mathbf{A} = \mathbf{H}_n$vec $\mathbf{A}$ — one choice for $\mathbf{H}_n$ is $\mathbf{H}_n = (\mathbf{G}_n'\mathbf{G}_n)^{-1}\mathbf{G}_n'$

$D_j f(\mathbf{c})$ The jth (first-order) partial derivative of a function f, with domain S in $\mathcal{R}^{m\times 1}$, at an interior point $\mathbf{c}$ of S — the function whose value at a point $\mathbf{c}$ is $D_j f(\mathbf{c})$ is represented by the symbol $D_j f$

$\dfrac{\partial f}{\partial x_j}$ The jth partial derivative of a function f of an $m \times 1$ vector $\mathbf{x} = (x_1, \ldots, x_m)'$ — an alternative notation to $D_j f$ or $D_j f(\mathbf{x})$

$\mathbf{D}f(\mathbf{c})$ The $1 \times m$ vector $[D_1 f(\mathbf{c}), \ldots, D_m(\mathbf{c})]$ (where f is a function with domain S in $\mathcal{R}^{m\times 1}$ and where $\mathbf{c}$ is an interior point of S)— similarly, $\mathbf{D}f$ is the $1 \times m$ vector $(D_1 f, \ldots, D_m f)$

$\dfrac{\partial f}{\partial \mathbf{x}}$ The $m \times 1$ vector $(\partial f/\partial x_1, \ldots, \partial f/\partial x_m)'$ of partial derivatives of a function f of an $m \times 1$ vector $\mathbf{x} = (x_1, \ldots, x_m)'$ — an alternative [to $(\mathbf{D}f)'$ or $(\mathbf{D}f(\mathbf{x}))'$] notation for the gradient vector

$\dfrac{\partial f}{\partial \mathbf{x}'}$ The $1 \times m$ vector $(\partial f/\partial x_1, \ldots, \partial f/\partial x_m)$ of partial derivatives of a function f of an $m \times 1$ vector $\mathbf{x} = (x_1, \ldots, x_m)'$ — equals $(\partial f/\partial \mathbf{x})'$ and is an alternative notation to $\mathbf{D}f$ or $\mathbf{D}f(\mathbf{x})$

$D_{ij}^2 f(\mathbf{c})$ The ijth second-order partial derivative of a function f, with domain S in $\mathcal{R}^{m\times 1}$, at an interior point $\mathbf{c}$ of S — the function whose value at a point $\mathbf{c}$ is $D_{ij}^2 f(\mathbf{c})$ is represented by the symbol $D_{ij}^2 f$

$\frac{\partial^2 f}{\partial x_i \partial x_j}$ An alternative [to $D_{ij}^2 f$ or $D_{ij}^2 f(\mathbf{x})$] notation for the ijth (second-order) partial derivative of a function f of an $m \times 1$ vector $\mathbf{x} = (x_1, \ldots, x_m)'$—this notation extends in a straightforward way to third- and higher-order partial derivatives

$\mathbf{H}f$ The Hessian matrix of a function f — accordingly, $\mathbf{H}f(\mathbf{c})$ represents the value of $\mathbf{H}f$ at an interior point $\mathbf{c}$ of the domain of f

$D_j\mathbf{f}$ The $p \times 1$ vector $(D_j f_1, \ldots, D_j f_p)'$, whose ith element is the jth partial derivative $D_j f_i$ of the ith element f_i of a $p \times 1$ vector $\mathbf{f} = (f_1, \ldots, f_p)'$ of functions, each of whose domain is a set S in $\mathcal{R}^{m\times 1}$ — similarly, $D_j\mathbf{f}(\mathbf{c}) = [D_j f_1(\mathbf{c}), \ldots, D_j f_p(\mathbf{c})]'$, where $\mathbf{c}$ is an interior point of S

$\frac{\partial \mathbf{F}}{\partial x_j}$ The $p \times q$ matrix whose stth element is the partial derivative $\partial f_{st}/\partial x_j$ of the stth element of a $p \times q$ matrix $\mathbf{F} = \{f_{st}\}$ of functions of a vector $\mathbf{x} = (x_1, \ldots, x_m)'$ of m variables

$\frac{\partial^2 \mathbf{F}}{\partial x_i \partial x_j}$ The $p \times q$ matrix whose stth element is the second-order partial derivative $\partial^2 f_{st}/\partial x_i \partial x_j$ of the stth element of a $p \times q$ matrix $\mathbf{F} = \{f_{st}\}$ of functions of a vector $\mathbf{x} = (x_1, \ldots, x_m)'$ of m variables—this notation extends in a straightforward way to a $p \times q$ matrix whose stth element is one of the third- or higher-order partial derivatives of the stth element of $\mathbf{F}$

$\mathbf{Df}$ The Jacobian matrix $(D_1\mathbf{f}, \ldots, D_m\mathbf{f})$ of a vector $\mathbf{f} = (f_1, \ldots, f_p)'$ of functions, each of whose domain is a set S in $\mathcal{R}^{m\times 1}$ — similarly, $\mathbf{Df}(\mathbf{c}) = [D_1\mathbf{f}(\mathbf{c}), \ldots, D_m\mathbf{f}(\mathbf{c})]$, where $\mathbf{c}$ is an interior point of S

$\frac{\partial \mathbf{f}}{\partial \mathbf{x}'}$ An alternative [to $\mathbf{Df}$ or $\mathbf{Df}(\mathbf{x})$] notation for the Jacobian matrix of a vector $\mathbf{f} = (f_1, \ldots, f_p)'$ of functions of an $m \times 1$ vector $\mathbf{x} = (x_1, \ldots, x_m)'$ — $\partial\mathbf{f}/\partial\mathbf{x}'$ is the $p \times m$ matrix whose ijth element is $\partial f_i/\partial x_j$

$\frac{\partial \mathbf{f}'}{\partial \mathbf{x}}$ An alternative [to $(\mathbf{Df})'$ or $(\mathbf{Df}(\mathbf{x}))'$] notation for the gradient (matrix) of a vector $\mathbf{f} = (f_1, \ldots, f_p)'$ of functions of an $m \times 1$ vector $\mathbf{x} = (x_1, \ldots, x_m)'$— $\partial\mathbf{f}'/\partial\mathbf{x}$ is the $m \times p$ matrix whose jith element is $\partial f_i/\partial x_j$

$\frac{\partial f}{\partial \mathbf{X}}$ The derivative of a function f of an $m \times n$ matrix $\mathbf{X}$ of mn "independent" variables or (depending on the context) of an $n \times n$ symmetric matrix $\mathbf{X}$ — the matrix $\partial f/\partial\mathbf{X}'$ is identical to $(\partial f/\partial\mathbf{X})'$

$\mathcal{U} \cap \mathcal{V}$ The intersection of 2 sets $\mathcal{U}$ and $\mathcal{V}$ of matrices—this notation extends in an obvious way to the intersection of 3 or more sets

$\mathcal{U} \cup \mathcal{V}$ The union of 2 sets $\mathcal{U}$ and $\mathcal{V}$ of matrices (of the same dimensions)—this notation extends in an obvious way to the union of 3 or more sets

$\mathcal{U} + \mathcal{V}$ The sum of 2 nonempty sets $\mathcal{U}$ and $\mathcal{V}$ of matrices (of the same dimensions)—this notation extends in an obvious way to the sum of 3 or more nonempty sets

$\mathcal{U} \oplus \mathcal{V}$ The direct sum of 2 (essentially disjoint) linear spaces $\mathcal{U}$ and $\mathcal{V}$ in $\mathcal{R}^{m\times n}$

— writing $\mathcal{U} \oplus \mathcal{V}$ (rather than $\mathcal{U} + \mathcal{V}$) serves to emphasize, or (in the absence of any previous indication) imply, that $\mathcal{U}$ and $\mathcal{V}$ are essentially disjoint and hence that their sum is a direct sum

$\mathbf{A}^+$ The Moore-Penrose inverse of a matrix $\mathbf{A}$

(kT) The scalar multiple of a scalar k and a transformation T from a linear space $\mathcal{V}$ into a linear space $\mathcal{W}$; in the absence of any ambiguity, the parentheses may be dropped, that is, kT may be written in place of (kT)

$(T + S)$ The sum of two transformations T and S from a linear space $\mathcal{V}$ into a linear space $\mathcal{W}$; in the absence of any ambiguity, the parentheses may be dropped, that is, $T + S$ may be written in place of $(T + S)$ — this notation extends in an obvious way to the sum of three or more transformations

(TS) The product of a transformation T from a linear space $\mathcal{V}$ into a linear space $\mathcal{W}$ and a transformation S from a linear space $\mathcal{U}$ into $\mathcal{V}$; in the absence of any ambiguity, the parentheses may be dropped, that is, TS may be written in place of (TS) — this notation extends in an obvious way to the product of three or more transformations

L_B A transformation defined for any (nonempty) linearly independent set B of matrices (of the same dimensions), say the matrices $\mathbf{Y}_1, \mathbf{Y}_2, \ldots, \mathbf{Y}_n$: it is the transformation from $\mathcal{R}^{n \times 1}$ onto the linear space $\mathcal{W} = \text{sp}(B)$ that assigns to each vector $\mathbf{x} = (x_1, x_2, \ldots, x_n)'$ in $\mathcal{R}^{n \times 1}$ the matrix $x_1\mathbf{Y}_1 + x_2\mathbf{Y}_2 + \cdots + x_n\mathbf{Y}_n$ in $\mathcal{W}$.

Some Terminology

adjoint matrix The adjoint matrix of an $n \times n$ matrix $\mathbf{A} = \{a_{ij}\}$ is the transpose of the cofactor matrix of $\mathbf{A}$ (or equivalently is the $n \times n$ matrix whose ijth element is the cofactor of a_{ji}).

algebraic multiplicity The characteristic polynomial, say $p(\bullet)$, of an $n \times n$ matrix $\mathbf{A}$ has a unique (aside from the order of the factors) representation of the form

$$p(\lambda) = (-1)^n(\lambda - \lambda_1)^{\gamma_1} \cdots (\lambda - \lambda_k)^{\gamma_k} q(\lambda) \qquad (-\infty < \lambda < \infty),$$

where $\{\lambda_1, \ldots, \lambda_k\}$ is the spectrum of $\mathbf{A}$ (comprising the distinct scalars that are eigenvalues of $\mathbf{A}$), $\gamma_1, \ldots, \gamma_k$ are (strictly) positive integers, and q is a polynomial (of degree $n - \sum_{i=1}^{k} \gamma_i$) that has no real roots; for $i = 1, \ldots, k$, γ_i is referred to as the algebraic multiplicity of the eigenvalue λ_i.

basis A basis for a linear space $\mathcal{V}$ is a finite set of linearly independent matrices in $\mathcal{V}$ that spans $\mathcal{V}$.

basis (natural) The natural basis for $\mathcal{R}^{m \times n}$ comprises the mn matrices $\mathbf{U}_{11}, \mathbf{U}_{21}, \ldots, \mathbf{U}_{m1}, \ldots, \mathbf{U}_{1n}, \mathbf{U}_{2n}, \ldots, \mathbf{U}_{mn}$, where (for $i = 1, \ldots, m$ and $j = 1, \ldots n$) $\mathbf{U}_{ij}$ is the $m \times n$ matrix whose ijth element equals 1 and whose remaining $mn - 1$ elements equal 0; the natural (or usual) basis for the linear space of all $n \times n$ symmetric matrices comprises the $n(n + 1)/2$ matrices $\mathbf{U}^*_{11}, \mathbf{U}^*_{21}, \ldots, \mathbf{U}^*_{n1}, \ldots, \mathbf{U}^*_{ii}, \mathbf{U}^*_{i+1,i}, \ldots, \mathbf{U}^*_{ni}, \ldots, \mathbf{U}^*_{nn}$, where (for $i = 1, \ldots, n$) $\mathbf{U}^*_{ii}$ is the $n \times n$ matrix whose ith diagonal element equals 1 and whose remaining $n^2 - 1$ elements equal 0 and (for $j < i = 1, \ldots, n$) $\mathbf{U}^*_{ij}$ is the $n \times n$ matrix whose ijth and jith elements equal 1 and whose remaining $n^2 - 2$ elements equal 0.

bilinear form A bilinear form in an $m \times 1$ vector $\mathbf{x} = (x_1, \ldots, x_m)'$ and an $n \times 1$ vector $\mathbf{y} = (y_1, \ldots, y_n)'$ is a function of $\mathbf{x}$ and $\mathbf{y}$ (defined for $\mathbf{x} \in \mathcal{R}^m$ and

$\mathbf{y} \in \mathcal{R}^n$) that, for some $m \times n$ matrix $\mathbf{A} = \{a_{ij}\}$ (called the matrix of the bilinear form), is expressible as $\mathbf{x}'\mathbf{A}\mathbf{y} = \sum_{i,j} a_{ij}x_i y_j$ — the bilinear form is said to be symmetric if $m = n$ and $\mathbf{x}'\mathbf{A}\mathbf{y} = \mathbf{y}'\mathbf{A}\mathbf{x}$ for all $\mathbf{x}$ and all $\mathbf{y}$ or equivalently if the matrix $\mathbf{A}$ is symmetric.

block-diagonal A partitioned matrix of the form $\begin{pmatrix} \mathbf{A}_{11} & \mathbf{0} & \dots & \mathbf{0} \\ \mathbf{0} & \mathbf{A}_{22} & & \mathbf{0} \\ \vdots & & \ddots & \\ \mathbf{0} & \mathbf{0} & & \mathbf{A}_{rr} \end{pmatrix}$ (all of whose off-diagonal blocks are null matrices) is said to be block-diagonal and may be expressed in abbreviated notation as $\text{diag}(\mathbf{A}_{11}, \mathbf{A}_{22}, \dots, \mathbf{A}_{rr})$.

block-triangular A partitioned matrix of the form $\begin{pmatrix} \mathbf{A}_{11} & \mathbf{A}_{12} & \dots & \mathbf{A}_{1r} \\ \mathbf{0} & \mathbf{A}_{22} & \dots & \mathbf{A}_{2r} \\ \vdots & & \ddots & \vdots \\ \mathbf{0} & \mathbf{0} & & \mathbf{A}_{rr} \end{pmatrix}$ or $\begin{pmatrix} \mathbf{A}_{11} & \mathbf{0} & \dots & \mathbf{0} \\ \mathbf{A}_{21} & \mathbf{A}_{22} & & \mathbf{0} \\ \vdots & \vdots & \ddots & \\ \mathbf{A}_{r1} & \mathbf{A}_{r2} & & \mathbf{A}_{rr} \end{pmatrix}$ is respectively upper or lower block-triangular—to indicate that a partitioned matrix is upper or lower block-triangular (without specifying which), the matrix is referred to simply as block-triangular.

characteristic polynomial (and equation) Corresponding to any $n \times n$ matrix $\mathbf{A}$ is its characteristic polynomial, say $p(\cdot)$, defined (for $-\infty < \lambda < \infty$) by $p(\lambda) = |\mathbf{A} - \lambda\mathbf{I}|$, and its characteristic equation $p(\lambda) = 0$ obtained by setting its characteristic polynomial equal to 0; $p(\lambda)$ is a polynomial in λ of degree n and hence is of the form $p(\lambda) = c_0 + c_1\lambda + \cdots + c_{n-1}\lambda^{n-1} + c_n\lambda^n$, where the coefficients $c_0, c_1, \dots, c_{n-1}, c_n$ depend on the elements of $\mathbf{A}$.

Cholesky decomposition The Cholesky decomposition of a symmetric positive definite matrix, say $\mathbf{A}$, is the unique decomposition of the form $\mathbf{A} = \mathbf{T}'\mathbf{T}$, where $\mathbf{T}$ is an upper triangular matrix with positive diagonal elements. More generally, the Cholesky decomposition of an $n \times n$ symmetric nonnegative definite matrix, say $\mathbf{A}$, of rank r is the unique decomposition of the form $\mathbf{A} = \mathbf{T}'\mathbf{T}$, where $\mathbf{T}$ is an $n \times n$ upper triangular matrix with r positive diagonal elements and $n - r$ null rows.

cofactor (and minor) The cofactor and minor of the ijth element, say a_{ij}, of an $n \times n$ matrix $\mathbf{A}$ are defined in terms of the $(n-1) \times (n-1)$ submatrix, say $\mathbf{A}_{ij}$, of $\mathbf{A}$ obtained by striking out the ith row and jth column (i.e., the row and column containing a_{ij}): the minor of a_{ij} is $|\mathbf{A}_{ij}|$, and the cofactor is the "signed" minor $(-1)^{i+j}|\mathbf{A}_{ij}|$.

cofactor matrix The cofactor matrix (or matrix of cofactors) of an $n \times n$ matrix $\mathbf{A} = \{a_{ij}\}$ is the $n \times n$ matrix whose ijth element is the cofactor of a_{ij}.

column space The column space of an $m \times n$ matrix $\mathbf{A}$ is the set whose elements consist of all m-dimensional column vectors that are expressible as linear combinations of the n columns of $\mathbf{A}$.

commute Two $n \times n$ matrices $\mathbf{A}$ and $\mathbf{B}$ are said to commute if $\mathbf{AB} = \mathbf{BA}$.

commute in pairs $n \times n$ matrices, say $\mathbf{A}_1, \ldots, \mathbf{A}_k$, are said to commute in pairs if $\mathbf{A}_s\mathbf{A}_i = \mathbf{A}_i\mathbf{A}_s$ for $s > i = 1, \ldots, k$.

consistent A linear system is said to be consistent if it has one or more solutions.

continuous A function f, with domain S in $\mathcal{R}^{m\times 1}$, is continuous at an interior point c of S if $\lim_{\mathbf{x}\to\mathbf{c}} f(\mathbf{x}) = f(\mathbf{c})$.

continuously differentiable A function f , with domain S in $\mathcal{R}^{m\times 1}$, is continuously differentiable at an interior point $\mathbf{c}$ of S if $D_1 f(\mathbf{c})$, $D_2 f(\mathbf{c})$, $\ldots$, $D_m f(\mathbf{c})$ exist and are continuous at every point $\mathbf{x}$ in some neighborhood of $\mathbf{c}$ — a vector or matrix of functions is continuously differentiable at $\mathbf{c}$ if all of its elements are continuously differentiable at $\mathbf{c}$.

derivative of a function of a matrix The derivative of a function f of an $m \times n$ matrix $\mathbf{X} = \{x_{ij}\}$ of mn "independent" variables is the $m \times n$ matrix whose ijth element is the partial derivative $\partial f/\partial x_{ij}$ of f with respect to x_{ij} when f is regarded as a function of an mn-dimensional column vector $\mathbf{x}$ formed from $\mathbf{X}$ by rearranging its elements; the derivative of a function f of an $n \times n$ symmetric (but otherwise unrestricted) matrix of variables is the $n \times n$ (symmetric) matrix whose ijth element is the partial derivative $\partial f/\partial x_{ij}$ or $\partial f/\partial x_{ji}$ of f with respect to x_{ij} or x_{ji} when f is regarded as a function of an $n(n+1)/2$-dimensional column vector $\mathbf{x}$ formed from any set of $n(n+1)/2$ nonredundant elements of $\mathbf{X}$.

determinant The determinant of an $n \times n$ matrix $\mathbf{A} = \{a_{ij}\}$ is (by definition) the (scalar-valued) quantity $\sum (-1)^{\phi_n(j_1,\ldots,j_n)} a_{1j_1} \cdots a_{nj_n}$, or equivalently the quantity $\sum (-1)^{\phi_n(i_1,\ldots,i_n)} a_{i_1 1} \cdots a_{i_n n}$, where $j_1, \ldots, j_n$ or $i_1, \ldots, i_n$ is a permutation of the first n positive integers and the summation is over all such permutations.

diagonalization An $n \times n$ matrix, say $\mathbf{A}$, is said to be diagonalizable if there exists an $n \times n$ nonsingular matrix $\mathbf{Q}$ such that $\mathbf{Q}^{-1}\mathbf{AQ}$ is diagonal, in which case $\mathbf{Q}$ is said to diagonalize $\mathbf{A}$ (or $\mathbf{A}$ is said to be diagonalized by $\mathbf{Q}$); a matrix that can be diagonalized by an orthogonal matrix is said to be orthogonally diagonalizable.

diagonalization (simultaneous) k matrices, say $\mathbf{A}_1, \ldots, \mathbf{A}_k$, of dimensions $n \times n$, are said to be simultaneously diagonalizable if all k of them can be diagonalized by the same matrix, that is, if there exists an $n \times n$ nonsingular matrix $\mathbf{Q}$ such that $\mathbf{Q}^{-1}\mathbf{A}_1\mathbf{Q}, \ldots, \mathbf{Q}^{-1}\mathbf{A}_k\mathbf{Q}$ are all diagonal, in which case $\mathbf{Q}$ is said to simultaneously diagonalize $\mathbf{A}_1, \ldots, \mathbf{A}_k$ (or $\mathbf{A}_1, \ldots, \mathbf{A}_k$ are said to be simultaneously diagonalized by $\mathbf{Q}$).

dimension (of a linear space) The dimension of a linear space $\mathcal{V}$ is the number of matrices in a basis for $\mathcal{V}$.

dimension (of a row or column vector) A row or column vector having n elements is said to be of dimension n.

dimensions (of a matrix) A matrix having m rows and n columns is said to be of dimensions $m \times n$.

direct sum If 2 linear spaces in $\mathcal{R}^{m\times n}$ are essentially disjoint, their sum is said to be a direct sum.

distance The distance between two matrices $\mathbf{A}$ and $\mathbf{B}$ in a linear space $\mathcal{V}$ is $\| \mathbf{A} - \mathbf{B} \|$.

dual transformation Corresponding to any linear transformation T from an n-dimensional linear space $\mathcal{V}$ into an m-dimensional linear space $\mathcal{W}$ is a linear transformation from $\mathcal{W}$ into $\mathcal{V}$ called the dual transformation: denoting by $\mathbf{X} \bullet \mathbf{Z}$ the inner product of an arbitrary pair of matrices $\mathbf{X}$ and $\mathbf{Z}$ in $\mathcal{V}$ and by $\mathbf{U} * \mathbf{Y}$ the inner product of an arbitrary pair of matrices $\mathbf{U}$ and $\mathbf{Y}$ in $\mathcal{W}$, the dual transformation is the (unique) linear transformation, say S, from $\mathcal{W}$ into $\mathcal{V}$ such that (for every matrix $\mathbf{X}$ in $\mathcal{V}$ and every matrix $\mathbf{Y}$ in $\mathcal{W}$) $\mathbf{X} \bullet S(\mathbf{Y}) = T(\mathbf{X}) * \mathbf{Y}$; further, for all $\mathbf{Y}$ in $\mathcal{W}$, $S(\mathbf{Y}) = \sum_{j=1}^{n} [\mathbf{Y} * T(\mathbf{X}_j)]\mathbf{X}_j$, where $\mathbf{X}_1, \mathbf{X}_2, \ldots, \mathbf{X}_n$ are any matrices that form an orthonormal basis for $\mathcal{V}$.

duplication matrix The $n^2 \times n(n+1)/2$ duplication matrix is the matrix, denoted by the symbol $\mathbf{G}_n$, such that, for every $n \times n$ symmetric matrix $\mathbf{A}$, $\text{vec}\,\mathbf{A} = \mathbf{G}_n \text{vech}\,\mathbf{A}$.

eigenspace The eigenspace of an eigenvalue, say λ, of an $n \times n$ matrix $\mathbf{A}$ is the linear space $\mathcal{N}(\mathbf{A} - \lambda\mathbf{I})$ — with the exception of the $n \times 1$ null vector, every member of this space is an eigenvector (of $\mathbf{A}$) corresponding to λ.

eigenvalues and eigenvectors An eigenvalue of an $n \times n$ matrix $\mathbf{A}$ is (by definition) a scalar (real number), say λ, for which there exists an $n \times 1$ vector, say $\mathbf{x}$, such that $\mathbf{Ax} = \lambda\mathbf{x}$, or equivalently such that $(\mathbf{A} - \lambda\mathbf{I})\mathbf{x} = \mathbf{0}$; any such vector $\mathbf{x}$ is referred to as an eigenvector (of $\mathbf{A}$) and is said to belong to (or correspond to) the eigenvalue λ — eigenvalues (and eigenvectors), as defined herein, are restricted to real numbers (and vectors of real numbers).

eigenvalues (not necessarily distinct) The characteristic polynomial, say $p(\bullet)$, of an $n \times n$ matrix $\mathbf{A}$ is expressible as

$$p(\lambda) = (-1)^n(\lambda - d_1)(\lambda - d_2)\cdots(\lambda - d_m)q(\lambda) \qquad (-\infty < \lambda < \infty),$$

where $d_1, d_2, \ldots, d_m$ are not-necessarily-distinct scalars and $q(\bullet)$ is a polynomial (of degree $n-m$) that has no real roots; $d_1, d_2, \ldots, d_m$ are referred to as the not-necessarily-distinct eigenvalues of $\mathbf{A}$ or (at the possible risk of confusion) simply as the eigenvalues of $\mathbf{A}$ — if the spectrum of $\mathbf{A}$ has k members, say $\lambda_1, \ldots, \lambda_k$, with algebraic multiplicities of $\gamma_1, \ldots, \gamma_k$, respectively, then $m = \sum_{i=1}^{k} \gamma_i$, and (for $i = 1, \ldots, k$) γ_i of the m not-necessarily-distinct eigenvalues equal λ_i.

essentially disjoint Two subspaces, say $\mathcal{U}$ and $\mathcal{V}$, of $\mathcal{R}^{m\times n}$ are (by definition) essentially disjoint if $\mathcal{U} \cap \mathcal{V} = \{\mathbf{0}\}$, i.e., if the only matrix they have in common is the ($m \times n$) null matrix—note that every subspace of $\mathcal{R}^{m\times n}$ contains the ($m \times n$) null matrix, so that no two subspaces can be entirely disjoint.

full column rank An $m \times n$ matrix $\mathbf{A}$ is said to have full column rank if rank$(\mathbf{A}) = n$.

full row rank An $m \times n$ matrix $\mathbf{A}$ is said to have full row rank if rank$(\mathbf{A}) = m$.

generalized eigenvalue problem The generalized eigenvalue problem consists of finding, for a symmetric matrix $\mathbf{A}$ and a symmetric positive definite matrix $\mathbf{B}$, the roots of the polynomial $|\mathbf{A} - \lambda\mathbf{B}|$ (i.e., the solutions for λ to the equation $|\mathbf{A} - \lambda\mathbf{B}| = 0$).

generalized inverse A generalized inverse of an $m \times n$ matrix $\mathbf{A}$ is any $n \times m$ matrix $\mathbf{G}$ such that $\mathbf{AGA} = \mathbf{A}$ — if $\mathbf{A}$ is nonsingular, its only generalized inverse is $\mathbf{A}^{-1}$; otherwise, it has infinitely many generalized inverses.

geometric multiplicity The geometric multiplicity of an eigenvalue, say λ, of an $n \times n$ matrix $\mathbf{A}$ is (by definition) $\dim[\mathcal{N}(\mathbf{A} - \lambda\mathbf{I})]$ (i.e., the dimension of the eigenspace of λ).

gradient (or gradient matrix) The gradient of a vector $\mathbf{f} = (f_1, \ldots, f_p)'$ of functions, each of whose domain is a set in $\mathcal{R}^{m \times 1}$, is the $m \times p$ matrix $[(\mathbf{D}f_1)', \ldots, (\mathbf{D}f_p)']$, whose jith element is $D_j f_i$—the gradient of $\mathbf{f}$ is the transpose of the Jacobian matrix of $\mathbf{f}$.

gradient vector The gradient vector of a function f, with domain in $\mathcal{R}^{m \times 1}$, is the m-dimensional column vector $(\mathbf{D}f)'$, whose jth element is the partial derivative $D_j f$ of f

Hessian matrix The Hessian matrix of a function f, with domain in $\mathcal{R}^{m \times 1}$, is the $m \times m$ matrix whose ijth element is the ijth partial derivative $D_{ij}^2 f$ of f

homogeneous linear system A linear system (in a matrix $\mathbf{X}$) of the form $\mathbf{AX} = \mathbf{0}$; i.e., a linear system whose right side is a null matrix.

idempotent A (square) matrix $\mathbf{A}$ is idempotent if $\mathbf{A}^2 = \mathbf{A}$.

identity transformation An identity transformation is a transformation from a linear space $\mathcal{V}$ onto $\mathcal{V}$ defined by $T(\mathbf{X}) = \mathbf{X}$.

indefinite A square (symmetric or nonsymmetric) matrix or a quadratic form is (by definition) indefinite if it is neither nonnegative definite nor nonpositive definite—thus, an $n \times n$ matrix $\mathbf{A}$ and the quadratic form $\mathbf{x}'\mathbf{Ax}$ (in an $n \times 1$ vector $\mathbf{x}$) are indefinite if $\mathbf{x}'\mathbf{Ax} < 0$ for some $\mathbf{x}$ and $\mathbf{x}'\mathbf{Ax} > 0$ for some (other) $\mathbf{x}$.

inner product The inner product $\mathbf{A} \bullet \mathbf{B}$ of an arbitrary pair of matrices $\mathbf{A}$ and $\mathbf{B}$ in a linear space $\mathcal{V}$ is the value assigned to $\mathbf{A}$ and $\mathbf{B}$ by a designated function having the following 4 properties: (1) $\mathbf{A} \bullet \mathbf{B} = \mathbf{B} \bullet \mathbf{A}$; (2) $\mathbf{A} \bullet \mathbf{A} \geq 0$, with equality holding if and only if $\mathbf{A} = \mathbf{0}$; (3) $(k\mathbf{A}) \bullet \mathbf{B} = k(\mathbf{A} \bullet \mathbf{B})$ (where k is an arbitrary scalar); (4) $(\mathbf{A} + \mathbf{B}) \bullet \mathbf{C} = (\mathbf{A} \bullet \mathbf{C}) + (\mathbf{B} \bullet \mathbf{C})$ (where $\mathbf{C}$ is an arbitrary matrix in $\mathcal{V}$)—the quasi-inner product $\mathbf{A} \bullet \mathbf{B}$ is defined in the same way as the inner product except that Property (2) is replaced by the weaker property (2′) $\mathbf{A} \bullet \mathbf{A} \geq 0$, with equality holding if $\mathbf{A} = \mathbf{0}$.

inner product (usual) The usual inner product of a pair of matrices $\mathbf{A}$ and $\mathbf{B}$ in a linear space is $\mathrm{tr}(\mathbf{A}'\mathbf{B})$ (which in the special case of a pair of column vectors

a and **b** reduces to $\mathbf{a}'\mathbf{b}$).

interior point A matrix, say $\mathbf{X}$, in a set S of $m \times n$ matrices is an interior point of S if there exists a neighborhood, say N, of $\mathbf{X}$ such that $N \subset S$.

intersection The intersection of 2 sets, say $\mathcal{U}$ and $\mathcal{V}$, of $m \times n$ matrices is the set comprising all matrices that are contained in both $\mathcal{U}$ and $\mathcal{V}$; more generally, the intersection of k sets, say $\mathcal{U}_1, \ldots, \mathcal{U}_k$, of $m \times n$ matrices is the set comprising all matrices that are contained in every one of $\mathcal{U}_1, \ldots, \mathcal{U}_k$.

invariant subspace A subspace $\mathcal{U}$ of the linear space $\mathcal{R}^{n\times 1}$ is said to be invariant relative to an $n \times n$ matrix $\mathbf{A}$ if, for every vector $\mathbf{x}$ in $\mathcal{U}$, the vector $\mathbf{Ax}$ is also in $\mathcal{U}$; a subspace $\mathcal{U}$ of an n-dimensional linear space $\mathcal{V}$ is said to be invariant relative to a linear transformation T from $\mathcal{V}$ into $\mathcal{V}$ if $T(\mathcal{U}) \subset \mathcal{U}$, that is, if the image $T(\mathcal{U})$ of $\mathcal{U}$ is a subspace of $\mathcal{U}$ itself.

inverse (matrix) A matrix $\mathbf{B}$ that is both a right and left inverse of a matrix $\mathbf{A}$ (so that $\mathbf{AB} = \mathbf{I}$ and $\mathbf{BA} = \mathbf{I}$) is called an inverse of $\mathbf{A}$.

inverse (transformation) The inverse of an invertible transformation T from a linear space $\mathcal{V}$ into a linear space $\mathcal{W}$ is the transformation from $\mathcal{W}$ into $\mathcal{V}$ that assigns to each matrix $\mathbf{Y}$ in $\mathcal{W}$ the (unique) matrix $\mathbf{X}$ (in $\mathcal{V}$) such that $T(\mathbf{X}) = \mathbf{Y}$.

invertible (matrix) A matrix that has an inverse is said to be invertible—a matrix is invertible if and only if it is nonsingular.

invertible (transformation) A transformation from a linear space $\mathcal{V}$ into a linear space $\mathcal{W}$ is (by definition) invertible if it is both 1-1 and onto.

involutory A (square) matrix $\mathbf{A}$ is involutory if $\mathbf{A}^2 = \mathbf{I}$, i.e., if it is invertible and is its own inverse.

isomorphic If there exists a 1-1 linear transformation, say T, from a linear space $\mathcal{V}$ onto a linear space $\mathcal{W}$, then $\mathcal{V}$ and $\mathcal{W}$ are said to be isomorphic, and T is said to be an isomorphism of $\mathcal{V}$ onto $\mathcal{W}$.

Jacobian matrix The Jacobian matrix of a p-dimensional vector $\mathbf{f} = (f_1, \ldots, f_p)'$ of functions, each of whose domain is a set in $\mathcal{R}^{m\times 1}$, is the $p \times m$ matrix $(D_1\mathbf{f}, \ldots, D_m\mathbf{f})$, whose ijth element is $D_j f_i$ — in the special case where $p = m$, the determinant of this matrix is referred to as the Jacobian (or Jacobian determinant) of $\mathbf{f}$.

Kronecker product The Kronecker product of two matrices, say an $m \times n$ matrix $\mathbf{A} = \{a_{ij}\}$ and a $p \times q$ matrix $\mathbf{B}$, is the $mp \times nq$ matrix

$$\begin{pmatrix} a_{11}\mathbf{B} & a_{12}\mathbf{B} & \ldots & a_{1n}\mathbf{B} \\ a_{21}\mathbf{B} & a_{22}\mathbf{B} & \ldots & a_{2n}\mathbf{B} \\ \vdots & \vdots & & \vdots \\ a_{m1}\mathbf{B} & a_{m2}\mathbf{B} & \ldots & a_{mn}\mathbf{B} \end{pmatrix}$$

obtained by replacing each element a_{ij} of $\mathbf{A}$ with the $p \times q$ matrix $a_{ij}\mathbf{B}$ — the Kronecker-product operation is associative [for any 3 matrices $\mathbf{A}$, $\mathbf{B}$, and $\mathbf{C}$, $\mathbf{A} \otimes (\mathbf{B} \otimes \mathbf{C}) = (\mathbf{A} \otimes \mathbf{B}) \otimes \mathbf{C}$], so that the notion of a Kronecker product extends in an unambiguous way to 3 or more matrices.

k times continuously differentiable A function f, with domain S in $\mathcal{R}^{m\times 1}$, is k times continuously differentiable at an interior point $\mathbf{c}$ of S if it and all of its first- through $(k-1)$th-order partial derivatives are continuously differentiable at $\mathbf{c}$ or, equivalently, if all of the first- through kth-order partial derivatives of f exist and are continuous at every point in some neighborhood of $\mathbf{c}$ — a vector or matrix of functions is k times continuously differentiable at $\mathbf{c}$ if all of its elements are k times continuously differentiable at $\mathbf{c}$.

LDU decomposition An LDU decomposition of a square matrix, say $\mathbf{A}$, is a decomposition of the form $\mathbf{A} = \mathbf{LDU}$, where $\mathbf{L}$ is a unit lower triangular matrix, $\mathbf{D}$ a diagonal matrix, and $\mathbf{U}$ an upper triangular matrix.

least squares generalized inverse A generalized inverse, say $\mathbf{G}$, of an $m \times n$ matrix $\mathbf{A}$ is said to be a least squares generalized inverse (of $\mathbf{A}$) if $(\mathbf{AG})' = \mathbf{AG}$; or, equivalently, an $n \times m$ matrix is a least squares generalized inverse of $\mathbf{A}$ if it satisfies Moore-Penrose Conditions (1) and (3).

left inverse A left inverse of an $m \times n$ matrix $\mathbf{A}$ is an $n \times m$ matrix $\mathbf{L}$ such that $\mathbf{LA} = \mathbf{I}_n$ — a matrix has a left inverse if and only if it has full column rank.

linear dependence or independence A nonempty (but finite) set of matrices (of the same dimensions), say $\mathbf{A}_1, \mathbf{A}_2, \ldots, \mathbf{A}_k$, is (by definition) linearly dependent if there exist scalars $x_1, x_2, \ldots, x_k$, not all 0, such that $\sum_{i=1}^{k} x_i \mathbf{A}_i = \mathbf{0}$; otherwise (if no such scalars exist), the set is linearly independent—by convention, the empty set is linearly independent.

linear space The use of this term is confined (herein) to sets of matrices (all of which have the same dimensions). A nonempty set, say $\mathcal{V}$, is called a linear space if: (1) for every matrix $\mathbf{A}$ in $\mathcal{V}$ and every matrix $\mathbf{B}$ in $\mathcal{V}$, the sum $\mathbf{A}+\mathbf{B}$ is in $\mathcal{V}$; and (2) for every matrix $\mathbf{A}$ in $\mathcal{V}$ and every scalar k, the product $k\mathbf{A}$ is in $\mathcal{V}$.

linear system A linear system is (for some positive integers m, n, and p) a set of mp simultaneous equations expressible in nonmatrix form as $\sum_{j=1}^{n} a_{ij}x_{jk} = b_{ik}$ $(i = i, \ldots, m;\ k = 1, \ldots, p)$, or in matrix form as $\mathbf{AX} = \mathbf{B}$, where $\mathbf{A} = \{a_{ij}\}$ is an $m \times n$ matrix comprising the "coefficients", $\mathbf{X} = \{x_{jk}\}$ is an $n \times p$ matrix comprising the "unknowns", and $\mathbf{B} = \{b_{ik}\}$ is an $m \times p$ matrix comprising the "right (hand) sides"—$\mathbf{A}$ is referred to as the coefficient matrix and $\mathbf{B}$ as the right side of $\mathbf{AX} = \mathbf{B}$; and to emphasize that $\mathbf{X}$ comprises the unknowns, $\mathbf{AX} = \mathbf{B}$ is referred to as a linear system in $\mathbf{X}$.

linear transformation A transformation, say T, from a linear space $\mathcal{V}$ (of $m \times n$ matrices) into a linear space $\mathcal{W}$ (of $p \times q$ matrices) is said to be linear if it satisfies the following two conditions: (1) for all $\mathbf{X}$ and $\mathbf{Z}$ in $\mathcal{V}$, $T(\mathbf{X}+\mathbf{Z}) = T(\mathbf{X}) + T(\mathbf{Z})$; and (2) for every scalar c and for all $\mathbf{X}$ in $\mathcal{V}$, $T(c\mathbf{X}) = c\,T(\mathbf{X})$ — in the special case where $\mathcal{W} = \mathcal{R}$, it is customary to refer to a linear transformation from $\mathcal{V}$ into $\mathcal{W}$ as a linear functional on $\mathcal{V}$.

matrix The use of the term matrix is confined (herein) to real matrices, i.e., to rectangular arrarys of real numbers.

matrix representation The matrix representation of a linear transformation from an n-dimensional linear space $\mathcal{V}$, with a basis B comprising matrices $\mathbf{V}_1$, $\mathbf{V}_2, \ldots, \mathbf{V}_n$, into a linear space $\mathcal{W}$, with a basis C comprising matrices $\mathbf{W}_1$, $\mathbf{W}_2, \ldots, \mathbf{W}_m$, is the $m \times n$ matrix $\mathbf{A} = \{a_{ij}\}$ whose jth column is (for $j = 1, 2, \ldots, n$) uniquely determined by the equality

$$T(\mathbf{V}_j) = a_{1j}\mathbf{W}_1 + a_{2j}\mathbf{W}_2 + \cdots + a_{mj}\mathbf{W}_m ;$$

this matrix (which depends on the choice of B and C) is such that if $\mathbf{x} = \{x_j\}$ is the $n \times 1$ vector that comprises the coordinates of a matrix $\mathbf{V}$ (in $\mathcal{V}$) in terms of the basis B (i.e., $\mathbf{V} = \sum_j x_j \mathbf{V}_j$), then the $m \times 1$ vector $\mathbf{y} = \{y_i\}$ given by the formula $\mathbf{y} = \mathbf{A}\mathbf{x}$ comprises the coordinates of $T(\mathbf{V})$ in terms of the basis C [i.e., $T(\mathbf{V}) = \sum_i y_i \mathbf{W}_i$].

minimum norm generalized inverse A generalized inverse, say $\mathbf{G}$, of an $m \times n$ matrix $\mathbf{A}$ is said to be a minimum norm generalized inverse (of $\mathbf{A}$) if $(\mathbf{GA})' = \mathbf{GA}$; or, equivalently, an $n \times m$ matrix is a minimum norm generalized inverse of $\mathbf{A}$ if it satisfies Moore-Penrose Conditions (1) and (4).

Moore-Penrose inverse (and conditions) Corresponding to any $m \times n$ matrix $\mathbf{A}$, there is a unique $n \times m$ matrix, say $\mathbf{G}$, such that (1) $\mathbf{AGA} = \mathbf{A}$ (i.e., $\mathbf{G}$ is a generalized inverse of $\mathbf{A}$), (2) $\mathbf{GAG} = \mathbf{G}$ (i.e., $\mathbf{A}$ is a generalized inverse of $\mathbf{G}$), (3) $(\mathbf{AG})' = \mathbf{AG}$ (i.e., $\mathbf{AG}$ is symmetric), and (4) $(\mathbf{GA})' = \mathbf{GA}$ (i.e., $\mathbf{GA}$ is symmetric). This matrix is called the Moore-Penrose inverse (or pseudoinverse) of $\mathbf{A}$, and the four conditions that (in combination) define this matrix are referred to as Moore-Penrose (or Penrose) Conditions (1) – (4).

negative definite An $n \times n$ (symmetric or nonsymmetric) matrix $\mathbf{A}$ and the quadratic form $\mathbf{x}'\mathbf{Ax}$ (in an $n \times 1$ vector $\mathbf{x}$) are (by definition) negative definite if $-\mathbf{x}'\mathbf{Ax}$ is a positive definite quadratic form (or equivalently if $-\mathbf{A}$ is a positive definite matrix)—thus, $\mathbf{A}$ and $\mathbf{x}'\mathbf{Ax}$ are negative definite if $\mathbf{x}'\mathbf{Ax} < 0$ for every nonnull $\mathbf{x}$ in $\mathcal{R}^n$.

negative or positive pair Any pair of elements of an $n \times n$ matrix $\mathbf{A} = \{a_{ij}\}$ that do not lie either in the same row or the same column, say a_{ij} and $a_{i'j'}$ (where $i' \neq i$ and $j' \neq j$) is (by definition) either a negative pair or a positive pair: it is a negative pair if one of the elements is located above and to the right of the other, or equivalently if either $i' > i$ and $j' < j$ or $i' < i$ and $j' > j$; otherwise (if one of the elements is located above and to the left of the other, or equivalently if either $i' > i$ and $j' > j$ or $i' < i$ and $j' < j$), it is a positive pair—note that whether a pair of elements is a negative pair or a positive pair is completely determined by the elements' relative locations and has nothing to do with whether the numerical values of the elements are positive or negative.

negative semidefinite An $n \times n$ (symmetric or nonsymmetric) matrix $\mathbf{A}$ and the quadratic form $\mathbf{x}'\mathbf{Ax}$ (in an $n \times 1$ vector $\mathbf{x}$) are (by definition) negative semidefinite if $-\mathbf{x}'\mathbf{Ax}$ is a positive semidefinite quadratic form (or equivalently if $-\mathbf{A}$ is a positive semidefinite matrix)—thus, $\mathbf{A}$ and $\mathbf{x}'\mathbf{Ax}$ are negative semidefinite if they are nonpositive definite but not negative definite, or equivalently if $\mathbf{x}'\mathbf{Ax} \leq 0$ for every $\mathbf{x}$ in $\mathcal{R}^n$ with equality holding for some nonnull $\mathbf{x}$.

neighborhood A neighborhood of an $m \times n$ matrix $\mathbf{C}$ is a set of the general form $\{\mathbf{X} \in \mathcal{R}^{m \times n} : \|\mathbf{X} - \mathbf{C}\| < r\}$, where r is a positive number called the radius of the neighborhood (and where the norm is the usual norm).

nonhomogeneous linear system A linear system whose right side (which is a column vector or more generally a matrix) is nonnull.

nonnegative definite An $n \times n$ (symmetric or nonsymmetric) matrix $\mathbf{A}$ and the quadratic form $\mathbf{x}'\mathbf{A}\mathbf{x}$ (in an $n \times 1$ vector $\mathbf{x}$) are (by definition) nonnegative definite if $\mathbf{x}'\mathbf{A}\mathbf{x} \geq 0$ for every $\mathbf{x}$ in $\mathcal{R}^n$.

nonpositive definite An $n \times n$ (symmetric or nonsymmetric) matrix $\mathbf{A}$ and the quadratic form $\mathbf{x}'\mathbf{A}\mathbf{x}$ (in an $n \times 1$ vector $\mathbf{x}$) are (by definition) nonpositive definite if $-\mathbf{x}'\mathbf{A}\mathbf{x}$ is a nonnegative definite quadratic form (or equivalently if $-\mathbf{A}$ is a nonnegative definite matrix)—thus, $\mathbf{A}$ and $\mathbf{x}'\mathbf{A}\mathbf{x}$ are nonpositive definite if $\mathbf{x}'\mathbf{A}\mathbf{x} \leq 0$ for every $\mathbf{x}$ in $\mathcal{R}^n$.

nonnull matrix A matrix having 1 or more nonzero elements.

nonsingular A matrix is nonsingular if it has both full row rank and full column rank or equivalently if it is square and its rank equals its order.

norm The norm of a matrix $\mathbf{A}$ in a linear space $\mathcal{V}$ is $(\mathbf{A} \bullet \mathbf{A})^{1/2}$—the use of this term is limited herein to norms defined in terms of an inner product; in the case of a quasi-inner product, $(\mathbf{A} \bullet \mathbf{A})^{1/2}$ is referred to as the quasi norm.

normal equations A linear system (or the equations comprising the linear system) of the form $\mathbf{X}'\mathbf{X}\mathbf{b} = \mathbf{X}'\mathbf{y}$ (in a $p \times 1$ vector $\mathbf{b}$), where $\mathbf{X}$ is an $n \times p$ matrix and $\mathbf{y}$ an $n \times 1$ vector.

null matrix A matrix all of whose elements are 0.

null space (of a matrix) The null space of an $m \times n$ matrix $\mathbf{A}$ is the solution space of the homogeneous linear system $\mathbf{A}\mathbf{x} = \mathbf{0}$ (in an n-dimensional column vector $\mathbf{x}$), or equivalently is the set $\{\mathbf{x} \in \mathcal{R}^{n \times 1} : \mathbf{A}\mathbf{x} = \mathbf{0}\}$.

null space (of a transformation) The null space—also known as the kernel— of a linear transformation T from a linear space $\mathcal{V}$ into a linear space $\mathcal{W}$ is the set $\{\mathbf{X} \in \mathcal{V} : T(\mathbf{X}) = \mathbf{0}\}$, which is a subspace of $\mathcal{V}$.

one to one A transformation T from a set $\mathcal{V}$ into a set $\mathcal{W}$ is said to be 1-1 (one to one) if each member of the range of T is the image of only one member of $\mathcal{V}$.

onto A transformation T from a set $\mathcal{V}$ into a set $\mathcal{W}$ is said to be onto if $T(\mathcal{V}) = \mathcal{W}$ (i.e., if the range of T is all of $\mathcal{W}$), in which case T may be referred to as a transformation from $\mathcal{V}$ onto $\mathcal{W}$.

open set A set S of $m \times n$ matrices is an open set if every matrix in S is an interior point of S.

order A (square) matrix of dimensions $n \times n$ is said to be of order n.

orthogonal complement The orthogonal complement of a subspace $\mathcal{U}$ of a linear space $\mathcal{V}$ is the set comprising all matrices in $\mathcal{V}$ that are orthogonal to $\mathcal{U}$ — note that the orthogonal complement of $\mathcal{U}$ depends on $\mathcal{V}$ as well as $\mathcal{U}$ (and also on the choice of inner product).

orthogonality of a matrix and a subspace A matrix $\mathbf{Y}$ in a linear space $\mathcal{V}$ is orthogonal to a subspace $\mathcal{U}$ (of $\mathcal{V}$) if $\mathbf{Y}$ is orthogonal to every matrix in $\mathcal{U}$.

orthogonality of two subspaces A subspace $\mathcal{U}$ of a linear space $\mathcal{V}$ is orthogonal to a subspace $\mathcal{W}$ (of $\mathcal{V}$) if every matrix in $\mathcal{U}$ is orthogonal to every matrix in $\mathcal{W}$.

orthogonality with respect to a matrix For any $n \times n$ symmetric nonnegative definite matrix $\mathbf{W}$, two $n \times 1$ vectors, say $\mathbf{x}$ and $\mathbf{y}$, are said to be orthogonal with respect to $\mathbf{W}$ if $\mathbf{x}'\mathbf{W}\mathbf{y} = 0$; an $n \times 1$ vector, say $\mathbf{x}$, and a subspace, say $\mathcal{U}$, of $\mathcal{R}^{n\times 1}$ are said to be orthogonal with respect to $\mathbf{W}$ if $\mathbf{x}'\mathbf{W}\mathbf{y} = 0$ for every $\mathbf{y}$ in $\mathcal{U}$; and two subspaces, say $\mathcal{U}$ and $\mathcal{V}$, of $\mathcal{R}^{n\times 1}$ are said to be orthogonal with respect to $\mathbf{W}$ if $\mathbf{x}'\mathbf{W}\mathbf{y} = 0$ for every $\mathbf{x}$ in $\mathcal{U}$ and every $\mathbf{y}$ in $\mathcal{V}$.

orthogonal matrix A (square) matrix $\mathbf{A}$ is orthogonal if $\mathbf{A}'\mathbf{A} = \mathbf{A}\mathbf{A}' = \mathbf{I}$.

orthogonal set A finite set of matrices in a linear space $\mathcal{V}$ is orthogonal if the inner product of every pair of matrices in the set equals 0.

orthonormal set A finite set of matrices in a linear space $\mathcal{V}$ is orthonormal if it is orthogonal and if the norm of every matrix in the set equals 1.

partitioned matrix A partitioned matrix, say $\begin{pmatrix} \mathbf{A}_{11} & \mathbf{A}_{12} & \dots & \mathbf{A}_{1c} \\ \mathbf{A}_{21} & \mathbf{A}_{22} & \dots & \mathbf{A}_{2c} \\ \vdots & \vdots & & \vdots \\ \mathbf{A}_{r1} & \mathbf{A}_{r2} & \dots & \mathbf{A}_{rc} \end{pmatrix}$, is a matrix that has (for some positive integers r and c) been subdivided into rc submatrices $\mathbf{A}_{ij}$ ($i = 1, 2, \dots, r$; $j = 1, 2, \dots, c$), called blocks, by implicitly superimposing on the matrix $r-1$ horizontal lines and $c-1$ vertical lines (so that all of the blocks in the same "row" of blocks have the same number of rows and all of those in the same "column" of blocks have the same number of columns)—in the special case where $c = r$, the blocks $\mathbf{A}_{11}, \mathbf{A}_{22}, \dots, \mathbf{A}_{rr}$ are referred to as the diagonal blocks (and the other blocks are referred to as the off-diagonal blocks).

permutation matrix An $n \times n$ permutation matrix is a matrix that is obtainable from the $n \times n$ identity matrix by permuting its columns; i.e., a matrix of the form $(u_{k_1}, u_{k_2}, \dots, u_{k_n})$, where $u_1, u_2, \dots, u_n$ are respectively the first, second, ..., nth columns of $\mathbf{I}_n$ and where $k_1, k_2, \dots, k_n$ is a permutation of the first n positive integers.

positive definite An $n \times n$ (symmetric or nonsymmetric) matrix $\mathbf{A}$ and the quadratic form $\mathbf{x}'\mathbf{A}\mathbf{x}$ (in an $n \times 1$ vector $\mathbf{x}$) are (by definition) positive definite if $\mathbf{x}'\mathbf{A}\mathbf{x} > 0$ for every nonnull $\mathbf{x}$ in $\mathcal{R}^n$.

positive semidefinite An $n \times n$ (symmetric or nonsymmetric) matrix $\mathbf{A}$ and the quadratic form $\mathbf{x}'\mathbf{A}\mathbf{x}$ (in an $n \times 1$ vector $\mathbf{x}$) are (by definition) positive semidefinite if they are nonnegative definite but not positive definite, or equivalently if $\mathbf{x}'\mathbf{A}\mathbf{x} \geq 0$ for every $\mathbf{x}$ in $\mathcal{R}^n$ with equality holding for some nonnull $\mathbf{x}$.

principal submatrix A submatrix of a square matrix is a principal submatrix if it can be obtained by striking out the same rows as columns (so that the ith row is struck out whenever the ith column is struck out, and vice versa); the $r \times r$ (principal) submatrix of an $n \times n$ matrix obtained by striking out its last $n - r$ rows and columns is referred to as a leading principal submatrix $(r = 1, \ldots, n)$.

product (of transformations) The product (or composition) of a transformation, say T, from a linear space $\mathcal{V}$ into a linear space $\mathcal{W}$ and a transformation, say S, from a linear space $\mathcal{U}$ into $\mathcal{V}$ is the transformation from $\mathcal{U}$ into $\mathcal{W}$ that assigns to each matrix $\mathbf{X}$ in $\mathcal{U}$ the matrix $T[S(\mathbf{X})]$ (in $\mathcal{W}$)—the definition of the term product (or composition) extends in a straightforward way to three or more transformations.

projection (orthogonal) The projection—also known as the orthogonal projection—of a matrix $\mathbf{Y}$ in a linear space $\mathcal{V}$ on a subspace $\mathcal{U}$ (of $\mathcal{V}$) is the unique matrix, say $\mathbf{Z}$, in $\mathcal{U}$ such that $\mathbf{Y} - \mathbf{Z}$ is orthogonal to $\mathcal{U}$; in the special case where (for some positive integer n and for some symmetric positive definite matrix $\mathbf{W}$) $\mathcal{V} = \mathcal{R}^{n \times 1}$ and the inner product is the bilinear form $\mathbf{x}'\mathbf{W}\mathbf{y}$, the projection of $\mathbf{y}$ (an $n \times 1$ vector) on $\mathcal{U}$ is referred to as the projection of $\mathbf{y}$ on $\mathcal{U}$ with respect to $\mathbf{W}$— this terminology can be extended to a symmetric nonnegative definite matrix $\mathbf{W}$ by defining a projection of $\mathbf{y}$ on $\mathcal{U}$ with respect to $\mathbf{W}$ to be any vector $\mathbf{z}$ in $\mathcal{U}$ such that $(\mathbf{y} - \mathbf{z}) \perp_{\mathbf{W}} \mathcal{U}$.

projection along a subspace For a linear space $\mathcal{V}$ of $m \times n$ matrices and for subspaces $\mathcal{U}$ and $\mathcal{W}$ such that $\mathcal{U} \oplus \mathcal{W} = \mathcal{V}$ (essentially disjoint subspaces whose sum is $\mathcal{V}$), the projection of a matrix in $\mathcal{V}$, say the matrix $\mathbf{Y}$, on $\mathcal{U}$ along $\mathcal{W}$ is (by definition) the (unique) matrix $\mathbf{Z}$ in $\mathcal{U}$ such that $\mathbf{Y} - \mathbf{Z} \in \mathcal{W}$.

projection matrix (orthogonal) The projection matrix—also known as the orthogonal projection matrix—for a subspace $\mathcal{U}$ of $\mathcal{R}^{n \times 1}$ is the unique $(n \times n)$ matrix, say $\mathbf{A}$, such that, for every $n \times 1$ vector $\mathbf{y}$, $\mathbf{Ay}$ is the projection (with respect to the usual inner product) of $\mathbf{y}$ on $\mathcal{U}$ — simply saying that a matrix is a projection matrix means that there is some subspace of $\mathcal{R}^{n \times 1}$ for which it is the projection matrix.

projection matrix (general orthogonal) The (orthogonal) projection matrix for a subspace $\mathcal{U}$ of $\mathcal{R}^{n \times 1}$ with respect to an $n \times n$ symmetric positive definite matrix $\mathbf{W}$ is the unique $(n \times n)$ matrix, say $\mathbf{A}$, such that, for every $n \times 1$ vector $\mathbf{y}$, $\mathbf{Ay}$ is the projection of $\mathbf{y}$ on $\mathcal{U}$ with respect to $\mathbf{W}$ — simply saying that a matrix is a projection matrix with respect to $\mathbf{W}$ means that there is some subspace of $\mathcal{R}^{n \times 1}$ for which it is the projection matrix with respect to $\mathbf{W}$— more generally, a projection matrix for $\mathcal{U}$ with respect to an $n \times n$ symmetric nonnegative definite matrix $\mathbf{W}$ is an $(n \times n)$ matrix, say $\mathbf{A}$, such that, for every $n \times 1$ vector $\mathbf{y}$, $\mathbf{Ay}$ is a projection of $\mathbf{y}$ on $\mathcal{U}$ with respect to $\mathbf{W}$.

projection matrix for one subspace along another For subspaces $\mathcal{U}$ and $\mathcal{W}$ (of $\mathcal{R}^{n \times 1}$) such that $\mathcal{U} \oplus \mathcal{W} = \mathcal{R}^{n \times 1}$ (essentially disjoint subspaces whose sum is $\mathcal{R}^{n \times 1}$), the projection matrix for $\mathcal{U}$ along $\mathcal{W}$ is the (unique) $n \times n$ matrix, say $\mathbf{A}$, such that for every $n \times 1$ vector $\mathbf{y}$, $\mathbf{Ay}$ is the projection of $\mathbf{y}$ on $\mathcal{U}$ along $\mathcal{W}$.

QR decomposition The QR decomposition of a matrix of full column rank, say an $m \times k$ matrix $\mathbf{A}$ of rank k, is the unique decomposition of the form $\mathbf{A} = \mathbf{QR}$, where $\mathbf{Q}$ is an $m \times k$ matrix whose columns are orthonormal (with respect to the usual inner product) and $\mathbf{R}$ is a $k \times k$ upper triangular matrix with positive diagonal elements.

quadratic form A quadratic form in an $n \times 1$ vector $\mathbf{x} = (x_1, \ldots, x_n)'$ is a function of $\mathbf{x}$ (defined for $\mathbf{x} \in \mathcal{R}^n$) that, for some $n \times n$ matrix $\mathbf{A} = \{a_{ij}\}$, is expressible as $\mathbf{x}'\mathbf{Ax} = \sum_{i,j} a_{ij}x_ix_j$ — the matrix $\mathbf{A}$ is called the matrix of the quadratic form and, unless $n = 1$ or the choice for $\mathbf{A}$ is restricted (e.g., to symmetric matrices), is nonunique.

range The range of a transformation T from a set $\mathcal{V}$ into a set $\mathcal{W}$ is the set $T(\mathcal{V})$ (i.e., the image of the domain of T)—in the special case of a linear transformation from a linear space $\mathcal{V}$ into a linear space $\mathcal{W}$, the range $T(\mathcal{V})$ of T is a linear space and is referred to as the range space of T.

rank (of a linear transformation) The rank of a linear transformation T from a linear space $\mathcal{V}$ into a linear space $\mathcal{W}$ is (by definition) the dimension $\dim[T(\mathcal{V})]$ of the range space $T(\mathcal{V})$ of T.

rank (of a matrix) The rank of a matrix $\mathbf{A}$ is the dimension of $\mathcal{C}(\mathbf{A})$ or equivalently of $\mathcal{R}(\mathbf{A})$.

rank additivity Two matrices $\mathbf{A}$ and $\mathbf{B}$ (of the same size) are said to be rank additive if $\text{rank}(\mathbf{A} + \mathbf{B}) = \text{rank}(\mathbf{A}) + \text{rank}(\mathbf{B})$; more generally, k matrices $\mathbf{A}_1, \mathbf{A}_2, \ldots, \mathbf{A}_k$ (of the same size) are said to be rank additive if $\text{rank}(\sum_{i=1}^k \mathbf{A}_i) = \sum_{i=1}^k \text{rank}(\mathbf{A}_i)$ (i.e., if the rank of their sum equals the sum of their ranks).

reflexive generalized inverse A generalized inverse, say $\mathbf{G}$, of an $m \times n$ matrix $\mathbf{A}$ is said to be reflexive if $\mathbf{GAG} = \mathbf{G}$; or, equivalently, an $n \times m$ matrix is a reflexive generalized inverse of $\mathbf{A}$ if it satisfies Moore-Penrose Conditions (1) and (2).

restriction If T is a linear transformation from a linear space $\mathcal{V}$ into a linear space $\mathcal{W}$ and if $\mathcal{U}$ is a subspace of $\mathcal{V}$, then the transformation, say R, from $\mathcal{U}$ into $\mathcal{W}$ defined by $R(\mathbf{X}) = T(\mathbf{X})$ (which assigns to each matrix in $\mathcal{U}$ the same matrix in $\mathcal{W}$ assigned by T) is called the restriction of T to $\mathcal{U}$.

right inverse A right inverse of an $m \times n$ matrix $\mathbf{A}$ is an $n \times m$ matrix $\mathbf{R}$ such that $\mathbf{AR} = \mathbf{I}_m$ — a matrix has a right inverse if and only if it has full row rank.

row space The row space of an $m \times n$ matrix $\mathbf{A}$ is the set whose elements consist of all n-dimensional row vectors that are expressible as linear combinations of the m rows of $\mathbf{A}$.

scalar The term scalar is (herein) used interchangeably with real number.

scalar multiple (of a transformation) The scalar multiple of a scalar k and a transformation, say T, from a linear space $\mathcal{V}$ into a linear space $\mathcal{W}$ is the transformation from $\mathcal{V}$ into $\mathcal{W}$ that assigns to each matrix $\mathbf{X}$ in $\mathcal{V}$ the matrix $kT(\mathbf{X})$ (in $\mathcal{W}$).

Schur complement In connection with a partitioned matrix $\mathbf{A}$ of the form $\mathbf{A} = \begin{pmatrix} \mathbf{T} & \mathbf{U} \\ \mathbf{V} & \mathbf{W} \end{pmatrix}$ or $\mathbf{A} = \begin{pmatrix} \mathbf{W} & \mathbf{V} \\ \mathbf{U} & \mathbf{T} \end{pmatrix}$, the matrix $\mathbf{Q} = \mathbf{W} - \mathbf{V}\mathbf{T}^{-}\mathbf{U}$ is referred to as the Schur complement of $\mathbf{T}$ in $\mathbf{A}$ relative to $\mathbf{T}^{-}$ or (especially in a case where $\mathbf{Q}$ is invariant to the choice of the generalized inverse $\mathbf{T}^{-}$) simply as the Schur complement of $\mathbf{T}$ in $\mathbf{A}$ or (in the absence of any ambiguity) even more simply as the Schur complement of $\mathbf{T}$.

second-degree polynomial A second-degree polynomial in an $n \times 1$ vector $\mathbf{x} = (x_1, \ldots, x_n)'$ is a function, say $f(\mathbf{x})$, of $\mathbf{x}$ that is defined for all $\mathbf{x}$ in $\mathcal{R}^n$ and that, for some scalar c, some $n \times 1$ vector $\mathbf{b} = \{b_i\}$, and some $n \times n$ matrix $\mathbf{V} = \{v_{ij}\}$, is expressible as $f(\mathbf{x}) = c - 2\mathbf{b}'\mathbf{x} + \mathbf{x}'\mathbf{V}\mathbf{x}$, or in nonmatrix notation as $f(\mathbf{x}) = c - 2\sum_{i=1}^{n} b_i x_i + \sum_{i=1}^{n}\sum_{j=1}^{n} v_{ij} x_i x_j$ — in the special case where $c = 0$ and $\mathbf{V} = \mathbf{0}$, $f(\mathbf{x}) = -2\mathbf{b}'\mathbf{x}$, which is a linear form (in $\mathbf{x}$), and in the special case where $c = 0$ and $\mathbf{b} = \mathbf{0}$, $f(\mathbf{x}) = \mathbf{x}'\mathbf{V}\mathbf{x}$, which is a quadratic form (in $\mathbf{x}$).

similar An $n \times n$ matrix $\mathbf{B}$ is said to be similar to an $n \times n$ matrix $\mathbf{A}$ if there exists an $n \times n$ nonsingular matrix $\mathbf{C}$ such that $\mathbf{B} = \mathbf{C}^{-1}\mathbf{A}\mathbf{C}$ or, equivalently, such that $\mathbf{C}\mathbf{B} = \mathbf{A}\mathbf{C}$ — if $\mathbf{B}$ is similar to $\mathbf{A}$, then $\mathbf{A}$ is similar to $\mathbf{B}$.

singular A square matrix is singular if its rank is less than its order.

singular value decomposition An $m \times n$ matrix $\mathbf{A}$ of rank r is expressible as

$$\mathbf{A} = \mathbf{P}\begin{pmatrix} \mathbf{D}_1 & \mathbf{0} \\ \mathbf{0} & \mathbf{0} \end{pmatrix}\mathbf{Q}' = \mathbf{P}_1\mathbf{D}_1\mathbf{Q}_1' = \sum_{i=1}^{r} s_i \mathbf{p}_i \mathbf{q}_i' = \sum_{j=1}^{k} \alpha_j \mathbf{U}_j \,,$$

where $\mathbf{Q} = (\mathbf{q}_1, \ldots, \mathbf{q}_n)$ is an $n \times n$ orthogonal matrix and $\mathbf{D}_1 = \mathrm{diag}(s_1, \ldots, s_r)$ an $r \times r$ diagonal matrix such that $\mathbf{Q}'\mathbf{A}'\mathbf{A}\mathbf{Q} = \begin{pmatrix} \mathbf{D}_1^2 & \mathbf{0} \\ \mathbf{0} & \mathbf{0} \end{pmatrix}$, where $s_1, \ldots, s_r$ are (strictly) positive, where $\mathbf{Q}_1 = (\mathbf{q}_1, \ldots, \mathbf{q}_r)$, $\mathbf{P}_1 = (\mathbf{p}_1, \ldots, \mathbf{p}_r) = \mathbf{A}\mathbf{Q}_1\mathbf{D}_1^{-1}$, and, for any $m \times (m - r)$ matrix $\mathbf{P}_2$ such that $\mathbf{P}_1'\mathbf{P}_2 = \mathbf{0}$, $\mathbf{P} = (\mathbf{P}_1, \mathbf{P}_2)$, where $\alpha_1, \ldots, \alpha_k$ are the distinct values represented among $s_1, \ldots, s_r$, and where (for $j = 1, \ldots, k$) $\mathbf{U}_j = \sum_{\{i\,:\,s_i = \alpha_j\}} \mathbf{p}_i \mathbf{q}_i'$; any of these four representations may be referred to as the singular value decomposition of $\mathbf{A}$, and $s_1, \ldots, s_r$ are referred to as the singular values of $\mathbf{A}$ — $s_1, \ldots, s_r$ are the positive square roots of the nonzero eigenvalues of $\mathbf{A}'\mathbf{A}$ (or equivalently $\mathbf{A}\mathbf{A}'$), $\mathbf{q}_1, \ldots, \mathbf{q}_n$ are eigenvectors of $\mathbf{A}'\mathbf{A}$, and the columns of $\mathbf{P}$ are eigenvectors of $\mathbf{A}\mathbf{A}'$.

skew-symmetric An $n \times n$ matrix, say $\mathbf{A} = \{a_{ij}\}$, is (by definition) skew-symmetric if $\mathbf{A}' = -\mathbf{A}$; that is, if $a_{ji} = -a_{ij}$ for all i and j (or equivalently if $a_{ii} = 0$ for $i = 1, \ldots, n$ and $a_{ji} = -a_{ij}$ for $j \neq i = 1, \ldots, n$).

solution A matrix, say $\mathbf{X}_*$, is said to be a solution to a linear system $\mathbf{A}\mathbf{X} = \mathbf{B}$ (in $\mathbf{X}$) if $\mathbf{A}\mathbf{X}_* = \mathbf{B}$.

solution set or space The collection of all solutions to a linear system $\mathbf{A}\mathbf{X} = \mathbf{B}$ (in $\mathbf{X}$) is called the solution set of the linear system; in the special case of a homogeneous linear system $\mathbf{AX} = \mathbf{0}$, the solution set may be called the solution space.

span The span of a finite set of matrices (having the same dimensions) is defined as follows: the span of a finite nonempty set $\{\mathbf{A}_1, \ldots, \mathbf{A}_k\}$ is the set consisting of all matrices that are expressible as linear combinations of $\mathbf{A}_1, \ldots, \mathbf{A}_k$, and the span of the empty set is the set $\{\mathbf{0}\}$, whose only element is the null matrix. And, a finite set S of matrices in a linear space $\mathcal{V}$ is said to span $\mathcal{V}$ if $\text{sp}(S) = \mathcal{V}$.

spectral decomposition An $n \times n$ symmetric matrix $\mathbf{A}$ is expressible as

$$\mathbf{A} = \mathbf{QDQ}' = \sum_{i=1}^{n} d_i \mathbf{q}_i \mathbf{q}_i' = \sum_{j=1}^{k} \lambda_j \mathbf{E}_j \,,$$

where $d_1, \ldots, d_n$ are the not-necessarily-distinct eigenvalues of $\mathbf{A}$, $\mathbf{q}_1, \ldots, \mathbf{q}_n$ are orthonormal eigenvectors corresponding to $d_1, \ldots, d_n$, respectively, $\mathbf{Q} = (\mathbf{q}_1, \ldots, \mathbf{q}_n)$, $\mathbf{D} = \text{diag}(d_1, \ldots, d_n)$, $\{\lambda_1, \ldots, \lambda_k\}$ is the spectrum of $\mathbf{A}$, and (for $j = 1, \ldots, k$) $\mathbf{E}_j = \sum_{\{i \,:\, d_i = \lambda_j\}} \mathbf{q}_i \mathbf{q}_i'$; any of these three representations may be referred to as the spectral decomposition of $\mathbf{A}$.

spectrum The spectrum of an $n \times n$ matrix $\mathbf{A}$ is the set whose members are the distinct (different) scalars that are eigenvalues of $\mathbf{A}$.

subspace A subspace of a linear space $\mathcal{V}$ is a subset of $\mathcal{V}$ that is itself a linear space.

sum (of sets) The sum of 2 nonempty sets, say $\mathcal{U}$ and $\mathcal{V}$, of $m \times n$ matrices is the set $\{\mathbf{A} + \mathbf{B} \,:\, \mathbf{A} \in \mathcal{U}, \mathbf{B} \in \mathcal{V}\}$ comprising every ($m \times n$) matrix that is expressible as the sum of a matrix in $\mathcal{U}$ and a matrix in $\mathcal{V}$; more generally, the sum of k sets, say $\mathcal{U}_1, \ldots, \mathcal{U}_k$, of $m \times n$ matrices is the set $\{\sum_{i=1}^{k} \mathbf{A}_i \,:\, \mathbf{A}_1 \in \mathcal{U}_1, \ldots, \mathbf{A}_k \in \mathcal{U}_k\}$.

sum (of transformations) The sum of two transformations, say T and S, from a linear space $\mathcal{V}$ into a linear space $\mathcal{W}$ is the transformation from $\mathcal{V}$ into $\mathcal{W}$ that assigns to each matrix $\mathbf{X}$ in $\mathcal{V}$ the matrix $T(\mathbf{X}) + S(\mathbf{X})$ (in $\mathcal{W}$)—since the addition of transformations is associative, the definition of the term sum extends in an unambiguous way to three or more transformations.

symmetric A matrix, say $\mathbf{A}$, is symmetric if $\mathbf{A}' = \mathbf{A}$, or equivalently if it is square and (for every i and j) its ijth element equals its jith element.

trace The trace of a (square) matrix is the sum of its diagonal elements.

transformation A transformation (also known as a function, operator, map, or mapping), say T, from a set $\mathcal{V}$, called the domain, into a set $\mathcal{W}$ is a correspondence that assigns to each member X of $\mathcal{V}$ a unique member of $\mathcal{W}$; the member of $\mathcal{W}$ assigned to X is denoted by the symbol $T(X)$ and is referred to as the image of X, and, for any subset $\mathcal{U}$ of $\mathcal{V}$, the set of all members of $\mathcal{W}$ that are the images of one or more members of $\mathcal{U}$ is denoted by the symbol $T(\mathcal{U})$ and is referred to as the image of $\mathcal{U}$ — $\mathcal{V}$ and $\mathcal{W}$ consist of scalars, row or column vectors, matrices, or other "objects".

transpose The transpose of an $m \times n$ matrix $\mathbf{A}$ is the $n \times m$ matrix whose ijth element is the jith element of **A**.

union The union of 2 sets, say $\mathcal{U}$ and $\mathcal{V}$, of $m \times n$ matrices is the set comprising all matrices that belong to either or both of $\mathcal{U}$ and $\mathcal{V}$; more generally, the union of k sets, say $\mathcal{U}_1, \ldots, \mathcal{U}_k$, of $m \times n$ matrices comprises all matrices that belong to at least one of $\mathcal{U}_1, \ldots, \mathcal{U}_k$.

unit (upper or lower) triangular matrix A unit triangular matrix is a triangular matrix all of whose diagonal elements equal one.

U′DU decomposition A U′DU decomposition of a symmetric matrix, say **A**, is a decomposition of the form $\mathbf{A} = \mathbf{U}'\mathbf{D}\mathbf{U}$, where **U** is a unit upper triangular matrix and **D** is a diagonal matrix.

Vandermonde matrix A Vandermonde matrix is a matrix of the general form
$$\begin{pmatrix} 1 & x_1 & x_1^2 & \ldots & x_1^{n-1} \\ 1 & x_2 & x_2^2 & \ldots & x_2^{n-1} \\ \vdots & \vdots & \vdots & & \vdots \\ 1 & x_n & x_n^2 & \ldots & x_n^{n-1} \end{pmatrix}$$
(where $x_1, x_2, \ldots, x_n$ are arbitrary scalars)

vec The vec of an $m \times n$ matrix $\mathbf{A} = (\mathbf{a}_1, \mathbf{a}_2, \ldots, \mathbf{a}_n)$ is the mn-dimensional (column) vector $\begin{pmatrix} \mathbf{a}_1 \\ \mathbf{a}_2 \\ \vdots \\ \mathbf{a}_n \end{pmatrix}$ obtained by successively stacking the first, second, $\ldots$, nth columns of **A** one under the other.

vech The vech of an $n \times n$ matrix $\mathbf{A} = \{a_{ij}\}$ is the $n(n+1)/2$-dimensional (column) vector $\begin{pmatrix} \mathbf{a}_1^* \\ \mathbf{a}_2^* \\ \vdots \\ \mathbf{a}_n^* \end{pmatrix}$, where (for $i = 1, 2, \ldots, n$) $\mathbf{a}_i^* = (a_{ii}, a_{i+1,i}, \ldots, a_{ni})'$ is the subvector of the ith column of **A** obtained by striking out its first $i - 1$ elements.

vec-permutation matrix The $mn \times mn$ vec-permutation matrix is the unique permutation matrix, denoted by the symbol $\mathbf{K}_{mn}$, such that, for every $m \times n$ matrix **A**, $\text{vec}(\mathbf{A}') = \mathbf{K}_{mn}\text{vec}(\mathbf{A})$ — the vec-permutation matrix is also known as the commutation matrix.

zero transformation The linear transformation from a linear space $\mathcal{V}$ into a linear space $\mathcal{W}$ that assigns to every matrix in $\mathcal{V}$ the null matrix (in $\mathcal{W}$) is called the zero transformation.

1

Matrices

EXERCISE 1. Show that, for any matrices **A**, **B**, and **C** (of the same dimensions),

$$(\mathbf{A} + \mathbf{B}) + \mathbf{C} = (\mathbf{C} + \mathbf{A}) + \mathbf{B}.$$

Solution. Since matrix addition is commutative and associative,

$$(\mathbf{A} + \mathbf{B}) + \mathbf{C} = \mathbf{C} + (\mathbf{A} + \mathbf{B}) = (\mathbf{C} + \mathbf{A}) + \mathbf{B}.$$

EXERCISE 2. For any scalars c and k and any matrix **A**,

$$c(k\mathbf{A}) = (ck)\mathbf{A} = (kc)\mathbf{A} = k(c\mathbf{A}), \qquad (*)$$

and, for any scalar c, $m \times n$ matrix **A**, and $n \times p$ matrix **B**,

$$c\mathbf{AB} = (c\mathbf{A})\mathbf{B} = \mathbf{A}(c\mathbf{B}). \qquad (**)$$

Using results $(*)$ and $(**)$ (or other means), show that, for any $m \times n$ matrix **A** and $n \times p$ matrix **B** and for arbitrary scalars c and k,

$$(c\mathbf{A})(k\mathbf{B}) = (ck)\mathbf{AB}.$$

Solution. Making use of results $(**)$ and $(*)$, we find that

$$(c\mathbf{A})(k\mathbf{B}) = k(c\mathbf{A})\mathbf{B} = k(c\mathbf{AB}) = (ck)\mathbf{AB}.$$

EXERCISE 3. (a) Verify the associativeness of matrix multiplication; that is, show that, for any $m \times n$ matrix $\mathbf{A} = \{a_{ij}\}$, $n \times q$ matrix $\mathbf{B} = \{b_{jk}\}$, and $q \times r$ matrix $\mathbf{C} = \{c_{ks}\}$, $\mathbf{A}(\mathbf{BC}) = (\mathbf{AB})\mathbf{C}$.

(b) Verify the distributiveness with respect to addition of matrix multiplication; that is, show that, for any $m \times n$ matrix $\mathbf{A} = \{a_{ij}\}$ and $n \times q$ matrices $\mathbf{B} = \{b_{jk}\}$ and $\mathbf{C} = \{c_{jk}\}$, $\mathbf{A}(\mathbf{B} + \mathbf{C}) = \mathbf{AB} + \mathbf{AC}$.

Solution. (a) The jsth element of $\mathbf{BC}$ equals $\sum_k b_{jk}c_{ks}$, and similarly the ikth element of $\mathbf{AB}$ equals $\sum_j a_{ij}b_{jk}$. Thus, the isth element of $\mathbf{A}(\mathbf{BC})$ equals

$$\sum_j a_{ij}\left(\sum_k b_{jk}c_{ks}\right) = \sum_j \left(\sum_k a_{ij}b_{jk}c_{ks}\right)$$

$$= \sum_k \left(\sum_j a_{ij}b_{jk}c_{ks}\right) = \sum_k \left(\sum_j a_{ij}b_{jk}\right)c_{ks},$$

and $\sum_k(\sum_j a_{ij}b_{jk})c_{ks}$ equals the isth element of $(\mathbf{AB})\mathbf{C}$. Since each element of $\mathbf{A}(\mathbf{BC})$ equals the corresponding element of $(\mathbf{AB})\mathbf{C}$, we conclude that $\mathbf{A}(\mathbf{BC}) = (\mathbf{AB})\mathbf{C}$.

(b) Observing that the jkth element of $\mathbf{B} + \mathbf{C}$ equals $b_{jk} + c_{jk}$, we find that the ikth element of $\mathbf{A}(\mathbf{B} + \mathbf{C})$ equals

$$\sum_j a_{ij}\left(b_{jk} + c_{jk}\right) = \sum_j \left(a_{ij}b_{jk} + a_{ij}c_{jk}\right) = \sum_j a_{ij}b_{jk} + \sum_j a_{ij}c_{jk}.$$

Further, observing that $\sum_j a_{ij}b_{jk}$ is the ikth element of $\mathbf{AB}$ and that $\sum_j a_{ij}c_{jk}$ is the ikth element of $\mathbf{AC}$, we find that $\sum_j a_{ij}b_{jk} + \sum_j a_{ij}c_{jk}$ equals the ikth element of $\mathbf{AB} + \mathbf{AC}$. Since each element of $\mathbf{A}(\mathbf{B} + \mathbf{C})$ equals the corresponding element of $\mathbf{AB} + \mathbf{BC}$, we conclude that $\mathbf{A}(\mathbf{B} + \mathbf{C}) = \mathbf{AB} + \mathbf{BC}$.

EXERCISE 4. Let $\mathbf{A} = \{a_{ij}\}$ represent an $m \times n$ matrix and $\mathbf{B} = \{b_{ij}\}$ a $p \times m$ matrix.

(a) Let $\mathbf{x} = \{x_i\}$ represent an n-dimensional column vector. Show that the ith element of the p-dimensional column vector $\mathbf{BAx}$ is

$$\sum_{j=1}^{m} b_{ij} \sum_{k=1}^{n} a_{jk}x_k. \tag{E.1}$$

(b) Let $\mathbf{X} = \{x_{ij}\}$ represent an $n \times q$ matrix. Generalize formula (E.1) by expressing the irth element of the $p \times q$ matrix $\mathbf{BAX}$ in terms of the elements of $\mathbf{A}$, $\mathbf{B}$, and $\mathbf{X}$.

(c) Let $\mathbf{x} = \{x_i\}$ represent an n-dimensional column vector and $\mathbf{C} = \{c_{ij}\}$ a $q \times p$ matrix. Generalize formula (E.1) by expressing the ith element of the q-dimensional column vector $\mathbf{CBAx}$ in terms of the elements of $\mathbf{A}$, $\mathbf{B}$, $\mathbf{C}$, and $\mathbf{x}$.

(d) Let $\mathbf{y} = \{y_i\}$ represent a p-dimensional column vector. Express the ith element of the n-dimensional row vector $\mathbf{y}'\mathbf{BA}$ in terms of the elements of $\mathbf{A}$, $\mathbf{B}$, and $\mathbf{y}$.

Solution. (a) The jth element of the vector **Ax** is $\sum_{k=1}^{n} a_{jk}x_k$. Thus, upon regarding **BAx** as the product of **B** and **Ax**, we find that the ith element of **BAx** is $\sum_{j=1}^{m} b_{ij} \sum_{k=1}^{n} a_{jk}x_k$.

(b) The irth element of **BAX** is

$$\sum_{j=1}^{m} b_{ij} \sum_{k=1}^{n} a_{jk}x_{kr} ,$$

as is evident from Part (a) upon regarding the irth element of **BAX** as the ith element of the product of **BA** and the rth column of **X**.

(c) According to Part (a), the sth element of the vector **BAx** is

$$\sum_{j=1}^{m} b_{sj} \sum_{k=1}^{n} a_{jk}x_k .$$

Thus, upon regarding **CBAx** as the product of **C** and **BAx**, we find that the ith element of **CBAx** is

$$\sum_{s=1}^{p} c_{is} \sum_{j=1}^{m} b_{sj} \sum_{k=1}^{n} a_{jk}x_k .$$

(d) The ith element of the row vector $\mathbf{y}'\mathbf{BA}$ is the same as the ith element of the column vector $(\mathbf{y}'\mathbf{BA})' = \mathbf{A}'\mathbf{B}'\mathbf{y}$. Thus, according to Part (a), the ith element of $\mathbf{y}'\mathbf{BA}$ is

$$\sum_{j=1}^{m} a_{ji} \sum_{k=1}^{p} b_{kj}y_k .$$

EXERCISE 5. Let **A** and **B** represent $n \times n$ matrices. Show that

$$(\mathbf{A} + \mathbf{B})(\mathbf{A} - \mathbf{B}) = \mathbf{A}^2 - \mathbf{B}^2$$

if and only if **A** and **B** commute.

Solution. Clearly,

$$(\mathbf{A} + \mathbf{B})(\mathbf{A} - \mathbf{B}) = \mathbf{A}(\mathbf{A} - \mathbf{B}) + \mathbf{B}(\mathbf{A} - \mathbf{B}) = \mathbf{A}^2 - \mathbf{AB} + \mathbf{BA} - \mathbf{B}^2.$$

Thus,

$$(\mathbf{A} + \mathbf{B})(\mathbf{A} - \mathbf{B}) = \mathbf{A}^2 - \mathbf{B}^2$$

if and only if $-\mathbf{AB} + \mathbf{BA} = \mathbf{0}$ or equivalently if and only if $\mathbf{AB} = \mathbf{BA}$ (i.e., if and only if **A** and **B** commute).

EXERCISE 6. (a) Show that the product **AB** of two $n \times n$ symmetric matrices **A** and **B** is itself symmetric if and only if **A** and **B** commute.

(b) Give an example of two symmetric matrices (of the same order) whose product is not symmetric.

Solution. (a) Since **A** and **B** are symmetric, $(\mathbf{AB})' = \mathbf{B}'\mathbf{A}' = \mathbf{BA}$. Thus, if $\mathbf{AB}$ is symmetric, that is, if $\mathbf{AB} = (\mathbf{AB})'$, then $\mathbf{AB} = \mathbf{BA}$, that is, **A** and **B** commute. Conversely, if $\mathbf{AB} = \mathbf{BA}$, then $\mathbf{AB} = (\mathbf{AB})'$.

(b) Take $\mathbf{A} = \begin{pmatrix} 1 & 0 \\ 0 & 2 \end{pmatrix}$ and $\mathbf{B} = \begin{pmatrix} 0 & 1 \\ 1 & 0 \end{pmatrix}$. Then,

$$\mathbf{AB} = \begin{pmatrix} 0 & 1 \\ 2 & 0 \end{pmatrix} \neq \begin{pmatrix} 0 & 2 \\ 1 & 0 \end{pmatrix} = \mathbf{BA}.$$

EXERCISE 7. Verify (a) that the transpose of an upper triangular matrix is lower triangular and (b) that the sum of two upper triangular matrices (of the same order) is upper triangular.

Solution. Let $\mathbf{A} = \{a_{ij}\}$ represent an upper triangular matrix of order n. Then, by definition, the ijth element of $\mathbf{A}'$ is a_{ji}. Since **A** is upper triangular, $a_{ji} = 0$ for $i < j = 1, \ldots, n$ or equivalently for $j > i = 1, \ldots, n$. Thus, $\mathbf{A}'$ is lower triangular, which verifies Part (a).

Let $\mathbf{B} = \{b_{ij}\}$ represent another upper triangular matrix of order n. Then, by definition, the ijth element of $\mathbf{A} + \mathbf{B}$ is $a_{ij} + b_{ij}$. Since both **A** and **B** are upper triangular, $a_{ij} = 0$ and $b_{ij} = 0$ for $j < i = 1, \ldots, n$, and hence $a_{ij} + b_{ij} = 0$ for $j < i = 1, \ldots, n$. Thus, $\mathbf{A} + \mathbf{B}$ is upper triangular, which verifies Part (b).

EXERCISE 8. Let $\mathbf{A} = \{a_{ij}\}$ represent an $n \times n$ upper triangular matrix, and suppose that the diagonal elements of **A** equal zero (i.e., that $a_{11} = a_{22} = \cdots = a_{nn} = 0$). Further, let p represent an arbitrary positive integer.

(a) Show that, for $i = 1, \ldots, n$ and $j = 1, \ldots, \min(n, i + p - 1)$, the ijth element of $\mathbf{A}^p$ equals zero.

(b) Show that, for $i \geq n - p + 1$, the ith row of $\mathbf{A}^p$ is null.

(c) Show that, for $p \geq n$, $\mathbf{A}^p = \mathbf{0}$.

Solution. For $i, k = 1, \ldots, n$, let b_{ik} represent that ikth element of $\mathbf{A}^p$.

(a) The proof is by mathematical induction. Clearly, for $i = 1, \ldots, n$ and $j = 1, \ldots, \min(n, i+1-1)$, the ijth element of $\mathbf{A}^1$ equals zero. Now, suppose that, for $i = 1, \ldots, n$ and $j = 1, \ldots, \min(n, i+p-1)$, the ijth element of $\mathbf{A}^p$ equals zero. Then, to complete the induction argument, it suffices to show that, for $i = 1, \ldots, n$ and $j = 1, \ldots, \min(n, i + p)$, the ijth element of $\mathbf{A}^{p+1}$ equals zero. Observing that $\mathbf{A}^{p+1} = \mathbf{A}^p\mathbf{A}$, we find that, for $i = 1, \ldots, n$ and $j = 1, \ldots, \min(n, i + p)$, the ijth element of $\mathbf{A}^{p+1}$ equals

$$\sum_{k=1}^{n} b_{ik}a_{kj} = \sum_{k=1}^{\min(n,i+p-1)} 0\, a_{kj} + \sum_{k=i+p}^{n} b_{ik}a_{kj}$$

(where, if $i > n - p$, the sum $\sum_{k=i+p}^{n} b_{ik}a_{kj}$ is degenerate and is to be interpreted as 0)

$$= 0$$

(since, for $k \geq j$, $a_{kj} = 0$).

(b) For $i \geq n-p+1$, $\min(n, i+p-1) = n$ (since $i \geq n-p+1 \Leftrightarrow i+p-1 \geq n$). Thus, for $i \geq n-p+1$, it follows from Part (a) that all n elements of the ith row of $\mathbf{A}^p$ equal zero and hence that the ith row of $\mathbf{A}^p$ is null.

(c) Clearly, for $p \geq n$, $n-p+1 \leq 1$. Thus, for $p \geq n$, it follows from Part (b) that all n rows of $\mathbf{A}^p$ are null and hence that $\mathbf{A}^p = \mathbf{0}$.

2

Submatrices and Partitioned Matrices

EXERCISE 1. Let $\mathbf{A}_*$ represent an $r \times s$ submatrix of an $m \times n$ matrix $\mathbf{A}$ obtained by striking out the $i_1, \ldots, i_{m-r}$th rows and $j_1, \ldots, j_{n-s}$th columns (of $\mathbf{A}$), and let $\mathbf{B}_*$ represent the $s \times r$ submatrix of $\mathbf{A}'$ obtained by striking out the $j_1, \ldots, j_{n-s}$th rows and $i_1, \ldots, i_{m-r}$th columns (of $\mathbf{A}'$). Verify that

$$\mathbf{B}_* = \mathbf{A}'_* .$$

Solution. Let $i_1^*, \ldots, i_r^*$ ($i_1^* < \cdots < i_r^*$) represent those r of the first m positive integers that are not represented in the sequence $i_1, \ldots, i_{m-r}$. Similarly, let $j_1^*, \ldots, j_s^*$ ($j_1^* < \cdots < j_s^*$) represent those s of the first n positive integers that are not represented in the sequence $j_1, \ldots, j_{n-s}$. Denote by a_{ij} and b_{ij} the ijth elements of $\mathbf{A}$ and $\mathbf{A}'$, respectively. Then,

$$\mathbf{A}'_* = \begin{pmatrix} a_{i_1^* j_1^*} & \cdots & a_{i_1^* j_s^*} \\ \vdots & \ddots & \vdots \\ a_{i_r^* j_1^*} & \cdots & a_{i_r^* j_s^*} \end{pmatrix}' = \begin{pmatrix} a_{i_1^* j_1^*} & \cdots & a_{i_r^* j_1^*} \\ \vdots & \ddots & \vdots \\ a_{i_1^* j_s^*} & \cdots & a_{i_r^* j_s^*} \end{pmatrix}$$

$$= \begin{pmatrix} b_{j_1^* i_1^*} & \cdots & b_{j_1^* i_r^*} \\ \vdots & \ddots & \vdots \\ b_{j_s^* i_1^*} & \cdots & b_{j_s^* i_r^*} \end{pmatrix} = \mathbf{B}_* .$$

EXERCISE 2. Verify (a) that a principal submatrix of a symmetric matrix is symmetric, (b) that a principal submatrix of a diagonal matrix is diagonal, and (c) that a principal submatrix of an upper triangular matrix is upper triangular.

Solution. Let $\mathbf{B} = \{b_{ij}\}$ represent the $r \times r$ principal submatrix of an $n \times n$ matrix $\mathbf{A} = \{a_{ij}\}$ obtained by striking out all of the rows and columns except the k_1, k_2, ..., k_rth rows and columns (where $k_1 < k_2 < \cdots < k_r$). Then, $b_{ij} = a_{k_i k_j}$ $(i, j = 1, \ldots, r)$.

(a) Suppose that $\mathbf{A}$ is symmetric. Then, for $i, j = 1, \ldots, r$, $b_{ij} = a_{k_i k_j} = a_{k_j k_i} = b_{ji}$.

(b) Suppose that $\mathbf{A}$ is diagonal. Then, for $j \neq i = 1, \ldots, r$, $b_{ij} = a_{k_i k_j} = 0$.

(c) Suppose that $\mathbf{A}$ is upper triangular. Then, for $j < i = 1, \ldots, r$, $b_{ij} = a_{k_i k_j} = 0$.

EXERCISE 3. Let

$$\begin{pmatrix} \mathbf{A}_{11} & \mathbf{A}_{12} & \ldots & \mathbf{A}_{1r} \\ \mathbf{0} & \mathbf{A}_{22} & \ldots & \mathbf{A}_{2r} \\ \vdots & & \ddots & \vdots \\ \mathbf{0} & \mathbf{0} & \ldots & \mathbf{A}_{rr} \end{pmatrix}$$

represent an $n \times n$ upper block-triangular matrix whose ijth block $\mathbf{A}_{ij}$ is of dimensions $n_i \times n_j$ $(j \geq i = 1, \ldots, r)$. Show that $\mathbf{A}$ is upper triangular if and only if each of its diagonal blocks $\mathbf{A}_{11}, \mathbf{A}_{22}, \ldots, \mathbf{A}_{rr}$ is upper triangular.

Solution. Let a_{ts} represent the tsth element of $\mathbf{A}$ $(t, s = 1, \ldots, n)$. Then,

$$\mathbf{A}_{ij} = \begin{pmatrix} a_{n_1+\cdots+n_{i-1}+1,n_1+\cdots+n_{j-1}+1} & \cdots & a_{n_1+\cdots+n_{i-1}+1,n_1+\cdots+n_{j-1}+n_j} \\ \vdots & & \vdots \\ a_{n_1+\cdots+n_{i-1}+n_i,n_1+\cdots+n_{j-1}+1} & \cdots & a_{n_1+\cdots+n_{i-1}+n_i,n_1+\cdots+n_{j-1}+n_j} \end{pmatrix}$$

$(j \geq i = 1, \ldots, r)$.

Suppose that $\mathbf{A}$ is upper triangular. Then, by definition, $a_{ts} = 0$ for $s < t = 1, \ldots, n$. Thus, $a_{n_1+\cdots+n_{i-1}+k,n_1+\cdots+n_{i-1}+l}$ (which is the klth element of the ith diagonal block $\mathbf{A}_{ii}$) equals zero for $l < k = 1, \ldots, n_i$, implying that $\mathbf{A}_{ii}$ is upper block-triangular $(i = 1, \ldots, r)$.

Conversely, suppose that $\mathbf{A}_{11}, \mathbf{A}_{22}, \ldots, \mathbf{A}_{rr}$ are upper triangular. Let t and s represent any integers (between 1 and n, inclusive) such that $a_{ts} \neq 0$. Then, clearly, for some integers i and $j \geq i$, a_{ts} is an element of the submatrix $\mathbf{A}_{ij}$, say the klth element, in which case $t = n_1 + \cdots + n_{i-1} + k$ and $s = n_1 + \cdots + n_{j-1} + l$. If $j > i$, then (since $k \leq n_i$) $t < s$. Moreover, if $j = i$, then (since $\mathbf{A}_{ii}$ is upper triangular) $k \leq l$, implying that $t \leq s$. Thus, in either case, $t \leq s$. We conclude that $\mathbf{A}$ is upper triangular.

EXERCISE 4. Let

$$\mathbf{A} = \begin{pmatrix} \mathbf{A}_{11} & \mathbf{A}_{12} & \ldots & \mathbf{A}_{1c} \\ \mathbf{A}_{21} & \mathbf{A}_{22} & \ldots & \mathbf{A}_{2c} \\ \vdots & \vdots & & \vdots \\ \mathbf{A}_{r1} & \mathbf{A}_{r2} & \ldots & \mathbf{A}_{rc} \end{pmatrix}$$

represent a partitioned $m \times n$ matrix whose ijth block $\mathbf{A}_{ij}$ is of dimensions $m_i \times n_j$. Verify that

$$\mathbf{A}' = \begin{pmatrix} \mathbf{A}'_{11} & \mathbf{A}'_{21} & \dots & \mathbf{A}'_{r1} \\ \mathbf{A}'_{12} & \mathbf{A}'_{22} & \dots & \mathbf{A}'_{r2} \\ \vdots & \vdots & & \vdots \\ \mathbf{A}'_{1c} & \mathbf{A}'_{2c} & \dots & \mathbf{A}'_{rc} \end{pmatrix};$$

in other words, verify that $\mathbf{A}'$ can be expressed as a partitioned matrix, comprising c rows and r columns of blocks, the ijth of which is the transpose $\mathbf{A}'_{ji}$ of the jith block $\mathbf{A}_{ji}$ of $\mathbf{A}$. And, letting

$$\mathbf{B} = \begin{pmatrix} \mathbf{B}_{11} & \mathbf{B}_{12} & \dots & \mathbf{B}_{1v} \\ \mathbf{B}_{21} & \mathbf{B}_{22} & \dots & \mathbf{B}_{2v} \\ \vdots & \vdots & & \vdots \\ \mathbf{B}_{u1} & \mathbf{B}_{u2} & \dots & \mathbf{B}_{uv} \end{pmatrix},$$

represent a partitioned $p \times q$ matrix whose ijth block $\mathbf{B}_{ij}$ is of dimensions $p_i \times q_j$, verify also that if $c = u$ and $n_k = p_k$ $(k = 1, \dots, c)$ [in which case all of the products $\mathbf{A}_{ik}\mathbf{B}_{kj}$ $(i = 1, \dots, r;\ j = 1, \dots, v;\ k = 1, \dots, c)$, as well as the product $\mathbf{AB}$, exist], then

$$\mathbf{AB} = \begin{pmatrix} \mathbf{F}_{11} & \mathbf{F}_{12} & \dots & \mathbf{F}_{1v} \\ \mathbf{F}_{21} & \mathbf{F}_{22} & \dots & \mathbf{F}_{2v} \\ \vdots & \vdots & & \vdots \\ \mathbf{F}_{r1} & \mathbf{F}_{r2} & \dots & \mathbf{F}_{rv} \end{pmatrix},$$

where $\mathbf{F}_{ij} = \sum_{k=1}^{c} \mathbf{A}_{ik}\mathbf{B}_{kj} = \mathbf{A}_{i1}\mathbf{B}_{1j} + \mathbf{A}_{i2}\mathbf{B}_{2j} + \cdots + \mathbf{A}_{ic}\mathbf{B}_{cj}$.

Solution. Let a_{ij}, b_{ij}, h_{ij}, and s_{ij} represent the ijth elements of $\mathbf{A}$, $\mathbf{B}$, $\mathbf{A}'$, and $\mathbf{AB}$, respectively. Define $\mathbf{H}_{ij}$ to be a matrix of dimensions $n_i \times m_j$ $(i = l, \dots, c;$ $j = 1, \dots, r)$ such that

$$\mathbf{A}' = \begin{pmatrix} \mathbf{H}_{11} & \mathbf{H}_{12} & \dots & \mathbf{H}_{1r} \\ \mathbf{H}_{21} & \mathbf{H}_{22} & \dots & \mathbf{H}_{2r} \\ \vdots & \vdots & & \vdots \\ \mathbf{H}_{c1} & \mathbf{H}_{c2} & \dots & \mathbf{H}_{cr} \end{pmatrix}.$$

Clearly, $\mathbf{H}_{ij}$ is the submatrix of $\mathbf{A}'$ obtained by striking out the first $n_1 + \cdots + n_{i-1}$ and last $n_{i+1} + \cdots + n_c$ rows of $\mathbf{A}'$ and the first $m_1 + \cdots + m_{j-1}$ and last $m_{j+1} + \cdots + m_r$ columns of $\mathbf{A}'$; and $\mathbf{A}_{ji}$ is the submatrix of $\mathbf{A}$ obtained by striking out the first $m_1 + \cdots + m_{j-1}$ and last $m_{j+1} + \cdots + m_r$ rows of $\mathbf{A}$ and the first $n_1 + \cdots + n_{i-1}$ and last $n_{i+1} + \cdots + n_c$ columns of $\mathbf{A}$. Thus, it follows from result (1.1) that

$$\mathbf{H}_{ij} = \mathbf{A}'_{ji}.$$

Further, define $\mathbf{S}_{ij}$ to be a matrix of dimensions $m_i \times q_j$ ($i = 1, \ldots, r$; $j = 1, \ldots, v$) such that

$$\mathbf{AB} = \begin{pmatrix} \mathbf{S}_{11} & \mathbf{S}_{12} & \ldots & \mathbf{S}_{1v} \\ \mathbf{S}_{21} & \mathbf{S}_{22} & \ldots & \mathbf{S}_{2v} \\ \vdots & \vdots & & \vdots \\ \mathbf{S}_{r1} & \mathbf{S}_{r2} & \ldots & \mathbf{S}_{rv} \end{pmatrix}.$$

Then, for $w = 1, \ldots, m_i$ and $z = 1, \ldots, q_j$, the wzth element of $\mathbf{S}_{ij}$ is

$$\begin{aligned} & s_{m_1+\cdots+m_{i-1}+w,\, q_1+\cdots+q_{j-1}+z} \\ &= \sum_{\ell=1}^{n_1+\cdots+n_c} a_{m_1+\cdots+m_{i-1}+w,\,\ell}\; b_{\ell,\, q_1+\cdots+q_{j-1}+z} \\ &= \sum_{k=1}^{c} \sum_{\ell=n_1+\cdots+n_{k-1}+1}^{n_1+\cdots+n_{k-1}+n_k} a_{m_1+\cdots+m_{i-1}+w,\,\ell}\; b_{\ell,\, q_1+\cdots+q_{j-1}+z} \\ &= \sum_{k=1}^{c} \sum_{t=1}^{n_k} a_{m_1+\cdots+m_{i-1}+w,\, n_1+\cdots+n_{k-1}+t}\; b_{n_1+\cdots+n_{k-1}+t,\, q_1+\cdots+q_{j-1}+z}. \end{aligned}$$

And, upon observing that $a_{m_1+\cdots+m_{i-1}+w,\, n_1+\cdots+n_{k-1}+t}$ is the wtth element of $\mathbf{A}_{ik}$ and that $b_{n_1+\cdots+n_{k-1}+t,\, q_1+\cdots+q_{j-1}+z}$ is the tzth element of $\mathbf{B}_{kj}$, it is clear that

$$\sum_{t=1}^{n_k} a_{m_1+\cdots+m_{i-1}+w,\, n_1+\cdots+n_{k-1}+t}\; b_{n_1+\cdots+n_{k-1}+t,\, q_1+\cdots+q_{j-1}+z}$$

is the wzth element of $\mathbf{A}_{ik}\mathbf{B}_{kj}$ and hence that $s_{m_1+\cdots+m_{i-1}+w,\, q_1+\cdots+q_{j-1}+z}$ is the wzth element of $\mathbf{F}_{ij}$. Thus,

$$\mathbf{S}_{ij} = \mathbf{F}_{ij}.$$

3

Linear Dependence and Independence

EXERCISE 1. For what values of the scalar k are the three row vectors $(k, 1, 0)$, $(1, k, 1)$, and $(0, 1, k)$ linearly dependent, and for what values are they linearly independent? Describe your reasoning.

Solution. Let x_1, x_2, and x_3 represent any scalars such that

$$x_1(k, 1, 0) + x_2(1, k, 1) + x_3(0, 1, k) = \mathbf{0},$$

or equivalently such that

$$\begin{aligned} x_1 k + x_2 &= 0, \\ x_1 + x_2 k + x_3 &= 0, \\ x_2 + x_3 k &= 0, \end{aligned}$$

or also equivalently such that

$$x_2 = -kx_3 = -kx_1, \tag{S.1}$$

$$kx_2 = -x_1 - x_3. \tag{S.2}$$

Suppose that $k = 0$. Then, conditions (S.1) and (S.2) are equivalent to the conditions $x_2 = 0$ and $x_3 = -x_1$.

Alternatively, suppose that $k \neq 0$. Then, conditions (S.1) and (S.2) imply that $x_3 = x_1$ and $-k^2 x_1 = kx_2 = -2x_1$ and hence that $k^2 = 2$ or $x_3 = x_2 = x_1 = 0$. Moreover, if $k^2 = 2$, then either $k = \sqrt{2}$, in which case conditions (S.1) and (S.2) are equivalent to the conditions $x_3 = x_1$ and $x_2 = -\sqrt{2}x_1$, or $k = -\sqrt{2}$, in which case conditions (S.1) and (S.2) are equivalent to the conditions $x_3 = x_1$ and $x_2 = \sqrt{2}x_1$.

Thus, there exist values of x_1, x_2, and x_3 other than $x_1 = x_2 = x_3 = 0$ if and only if $k = 0$ or $k = \pm\sqrt{2}$. And, we conclude that the three vectors $(k, 1, 0)$, $(1, k, 1)$, and $(0, 1, k)$ are linearly dependent if $k = 0$ or $k = \pm\sqrt{2}$, and linearly independent, otherwise.

EXERCISE 2. Let $\mathbf{A}$, $\mathbf{B}$, and $\mathbf{C}$ represent three linearly independent $m \times n$ matrices. Determine whether or not the three pairwise sums $\mathbf{A} + \mathbf{B}$, $\mathbf{A} + \mathbf{C}$, and $\mathbf{B} + \mathbf{C}$ are linearly independent. [*Hint.* Take advantage of the following general result on the linear dependence or independence of linear combinations: Letting $\mathbf{A}_1, \mathbf{A}_2, \ldots, \mathbf{A}_k$ represent $m \times n$ matrices and for $j = 1, \ldots, r$, taking $\mathbf{C}_j = x_{1j}\mathbf{A}_1 + x_{2j}\mathbf{A}_2 + \cdots + x_{kj}\mathbf{A}_k$ (where $x_{1j}, x_{2j}, \ldots, x_{kj}$ are scalars) and letting $\mathbf{x}_j = (x_{1j}, x_{2j}, \ldots, x_{kj})'$, the linear combinations $\mathbf{C}_1, \mathbf{C}_2, \ldots, \mathbf{C}_r$ are linearly independent if $\mathbf{A}_1, \mathbf{A}_2, \ldots, \mathbf{A}_k$ are linearly independent and $\mathbf{x}_1, \mathbf{x}_2, \ldots, \mathbf{x}_r$ are linearly independent, and they are linearly dependent if $\mathbf{x}_1, \mathbf{x}_2, \ldots, \mathbf{x}_r$ are linearly dependent.]

Solution. It follows from the result cited in the hint that $\mathbf{A}+\mathbf{B}$, $\mathbf{A}+\mathbf{C}$, and $\mathbf{B}+\mathbf{C}$ are linearly independent if (and only if) the three vectors $(1, 1, 0)'$, $(1, 0, 1)'$, and $(0, 1, 1,)'$ are linearly independent. Moreover, for any scalars x_1, x_2, and x_3 such that

$$x_1(1, 1, 0)' + x_2(1, 0, 1)' + x_3(0, 1, 1)' = \mathbf{0},$$

we have that $x_1 + x_2 = x_1 + x_3 = x_2 + x_3 = 0$, implying that

$$2x_3 = 0 + 2x_3 = (x_1 + x_2) + 2x_3 = (x_1 + x_3) + (x_2 + x_3) = 0 + 0 = 0$$

and $x_1 = x_2 = -x_3$ and hence that $x_3 = 0$ and $x_1 = x_2 = 0$. Thus, $(1, 1, 0)'$, $(1, 0, 1)'$, and $(0, 1, 1)'$ are linearly independent. And, we conclude that $\mathbf{A}+\mathbf{B}$, $\mathbf{A}+\mathbf{C}$, and $\mathbf{B}+\mathbf{C}$ are linearly independent.

4

Linear Spaces: Row and Column Spaces

EXERCISE 1. Which of the following two sets are linear spaces: (a) the set of all $n \times n$ upper triangular matrices; (b) the set of all $n \times n$ nonsymmetric matrices?

Solution. Clearly, the sum of two $n \times n$ upper triangular matrices is upper triangular. And, the matrix obtained by multiplying any $n \times n$ upper triangular matrix by any scalar is upper triangular. However, the sum of two $n \times n$ nonsymmetric matrices is not necessarily nonsymmetric. For example, if $\mathbf{A}$ is an $n \times n$ nonsymmetric matrix, then $-\mathbf{A}$ and $\mathbf{A}'$ are nonsymmetric, yet the sums $\mathbf{A} + (-\mathbf{A}) = \mathbf{0}$ and $\mathbf{A} + \mathbf{A}'$ are symmetric. Also, the product of the scalar 0 and any $n \times n$ matrix is the null matrix, which is symmetric. Thus, the set of all $n \times n$ upper triangular matrices is a linear space, but the set of all $n \times n$ nonsymmetric matrices is not.

EXERCISE 2. Letting $\mathbf{A}$ represent an $m \times n$ matrix and $\mathbf{B}$ an $m \times p$ matrix, verify that (1) $\mathcal{C}(\mathbf{A}) \subset \mathcal{C}(\mathbf{B})$ if and only if $\mathcal{R}(\mathbf{A}') \subset \mathcal{R}(\mathbf{B}')$, and (2) $\mathcal{C}(\mathbf{A}) = \mathcal{C}(\mathbf{B})$ if and only if $\mathcal{R}(\mathbf{A}') = \mathcal{R}(\mathbf{B}')$.

Solution. (1) Suppose that $\mathcal{R}(\mathbf{A}') \subset \mathcal{R}(\mathbf{B}')$. Then, for any vector $\mathbf{x}$ in $\mathcal{C}(\mathbf{A})$, we have (in light of Lemma 4.1.1) that $\mathbf{x}' \in \mathcal{R}(\mathbf{A}')$, implying that $\mathbf{x}' \in \mathcal{R}(\mathbf{B}')$ and hence (in light of Lemma 4.1.1) that $\mathbf{x} \in \mathcal{C}(\mathbf{B})$. Thus, $\mathcal{C}(\mathbf{A}) \subset \mathcal{C}(\mathbf{B})$.

Conversely, suppose that $\mathcal{C}(\mathbf{A}) \subset \mathcal{C}(\mathbf{B})$. Then, for any m-dimensional column vector $\mathbf{x}$ such that $\mathbf{x}' \in \mathcal{R}(\mathbf{A}')$, we have that $\mathbf{x} \in \mathcal{C}(\mathbf{A})$, implying that $\mathbf{x} \in \mathcal{C}(\mathbf{B})$ and hence that $\mathbf{x}' \in \mathcal{R}(\mathbf{B}')$. Thus, $\mathcal{R}(\mathbf{A}') \subset \mathcal{R}(\mathbf{B}')$.

We conclude that $\mathcal{C}(\mathbf{A}) \subset \mathcal{C}(\mathbf{B})$ if and only if $\mathcal{R}(\mathbf{A}') \subset \mathcal{R}(\mathbf{B}')$.

An alternative verification of Part (1) is obtained by taking advantage of Lemma

4.2.2. We have that

$$\begin{aligned}\mathcal{C}(\mathbf{A}) \subset \mathcal{C}(\mathbf{B}) &\Leftrightarrow \mathbf{A} = \mathbf{BK} \text{ for some matrix } \mathbf{K} \\ &\Leftrightarrow \mathbf{A}' = \mathbf{K}'\mathbf{B}' \text{ for some matrix } \mathbf{K} \\ &\Leftrightarrow \mathcal{R}(\mathbf{A}') \subset \mathcal{R}(\mathbf{B}').\end{aligned}$$

(2) If $\mathcal{R}(\mathbf{A}') = \mathcal{R}(\mathbf{B}')$, then $\mathcal{R}(\mathbf{A}') \subset \mathcal{R}(\mathbf{B}')$ and $\mathcal{R}(\mathbf{B}') \subset \mathcal{R}(\mathbf{A}')$, implying [in light of Part (1)] that $\mathcal{C}(\mathbf{A}) \subset \mathcal{C}(\mathbf{B})$ and $\mathcal{C}(\mathbf{B}) \subset \mathcal{C}(\mathbf{A})$ and hence that $\mathcal{C}(\mathbf{A}) = \mathcal{C}(\mathbf{B})$. Similarly, if $\mathcal{C}(\mathbf{A}) = \mathcal{C}(\mathbf{B})$, then $\mathcal{C}(\mathbf{A}) \subset \mathcal{C}(\mathbf{B})$ and $\mathcal{C}(\mathbf{B}) \subset \mathcal{C}(\mathbf{A})$, implying that $\mathcal{R}(\mathbf{A}') \subset \mathcal{R}(\mathbf{B}')$ and $\mathcal{R}(\mathbf{B}') \subset \mathcal{R}(\mathbf{A}')$ and hence that $\mathcal{R}(\mathbf{A}') = \mathcal{R}(\mathbf{B}')$. Thus, $\mathcal{C}(\mathbf{A}) = \mathcal{C}(\mathbf{B})$ if and only if $\mathcal{R}(\mathbf{A}') = \mathcal{R}(\mathbf{B}')$.

EXERCISE 3. Let $\mathcal{U}$ and $\mathcal{W}$ represent subspaces of a linear space $\mathcal{V}$. Show that if every matrix in $\mathcal{V}$ belongs to $\mathcal{U}$ or $\mathcal{W}$, then $\mathcal{U} = \mathcal{V}$ or $\mathcal{W} = \mathcal{V}$.

Solution. Suppose that every matrix in $\mathcal{V}$ belongs to $\mathcal{U}$ or $\mathcal{W}$. And, assume (for purposes of establishing a contradiction) that neither $\mathcal{U} = \mathcal{V}$ nor $\mathcal{W} = \mathcal{V}$. Then, there exist matrices $\mathbf{A}$ and $\mathbf{B}$ in $\mathcal{V}$ such that $\mathbf{A} \notin \mathcal{U}$ and $\mathbf{B} \notin \mathcal{W}$. And, since $\mathbf{A}$ and $\mathbf{B}$ each belong to $\mathcal{U}$ or $\mathcal{W}$, $\mathbf{A} \in \mathcal{W}$ and $\mathbf{B} \in \mathcal{U}$.

Clearly, $\mathbf{A} = \mathbf{B} - (\mathbf{B} - \mathbf{A})$ and $\mathbf{B} = \mathbf{A} + (\mathbf{B} - \mathbf{A})$, and $\mathbf{B} - \mathbf{A} \in \mathcal{U}$ or $\mathbf{B} - \mathbf{A} \in \mathcal{W}$. If $\mathbf{B} - \mathbf{A} \in \mathcal{U}$, then $\mathbf{B} - (\mathbf{B} - \mathbf{A}) \in \mathcal{U}$ and hence $\mathbf{A} \in \mathcal{U}$. If $\mathbf{B} - \mathbf{A} \in \mathcal{W}$, then $\mathbf{A} + (\mathbf{B} - \mathbf{A}) \in \mathcal{W}$ and hence $\mathbf{B} \in \mathcal{W}$. In either case, we arrive at a contradiction.

We conclude that $\mathcal{U} = \mathcal{V}$ or $\mathcal{W} = \mathcal{V}$.

EXERCISE 4. Let $\mathbf{A}$, $\mathbf{B}$, and $\mathbf{C}$ represent three matrices (having the same dimensions) such that $\mathbf{A} + \mathbf{B} + \mathbf{C} = \mathbf{0}$. Show that $\mathrm{sp}(\mathbf{A}, \mathbf{B}) = \mathrm{sp}(\mathbf{A}, \mathbf{C})$.

Solution. Let $\mathbf{E}$ represent an arbitrary matrix in $\mathrm{sp}(\mathbf{A}, \mathbf{B})$. Then, $\mathbf{E} = d\mathbf{A} + k\mathbf{B}$ for some scalars d and k, implying (since $\mathbf{B} = -\mathbf{A} - \mathbf{C}$) that

$$\mathbf{E} = d\mathbf{A} + k(-\mathbf{A} - \mathbf{C}) = (d - k)\mathbf{A} + (-k)\mathbf{C} \in \mathrm{sp}(\mathbf{A}, \mathbf{C}).$$

Thus, $\mathrm{sp}(\mathbf{A}, \mathbf{B}) \subset \mathrm{sp}(\mathbf{A}, \mathbf{C})$. And, it follows from an analogous argument that $\mathrm{sp}(\mathbf{A}, \mathbf{C}) \subset \mathrm{sp}(\mathbf{A}, \mathbf{B})$. We conclude that $\mathrm{sp}(\mathbf{A}, \mathbf{B}) = \mathrm{sp}(\mathbf{A}, \mathbf{C})$.

EXERCISE 5. Let $\mathbf{A}_1, \ldots, \mathbf{A}_k$ represent any matrices in a linear space $\mathcal{V}$. Show that $\mathrm{sp}(\mathbf{A}_1, \ldots, \mathbf{A}_k)$ is a subspace of $\mathcal{V}$ and that, among all subspaces of $\mathcal{V}$ that contain $\mathbf{A}_1, \ldots \mathbf{A}_k$, it is the smallest [in the sense that, for any subspace $\mathcal{U}$ (of $\mathcal{V}$) that contains $\mathbf{A}_1, \ldots, \mathbf{A}_k$, $\mathrm{sp}(\mathbf{A}_1, \ldots, \mathbf{A}_k) \subset \mathcal{U}$].

Solution. Let $\mathcal{U}$ represent any subspace of $\mathcal{V}$ that contains $\mathbf{A}_1, \ldots, \mathbf{A}_k$. It suffices (since $\mathcal{V}$ itself is a subspace of $\mathcal{V}$) to show that $\mathrm{sp}(\mathbf{A}_1, \ldots, \mathbf{A}_k)$ is a subspace of $\mathcal{U}$.

Let $\mathbf{A}$ represent an arbitrary matrix in $\mathrm{sp}(\mathbf{A}_1, \ldots, \mathbf{A}_k)$. Then, $\mathbf{A} = x_1\mathbf{A}_1 + \cdots + x_k\mathbf{A}_k$ for some scalars $x_1, \ldots, x_k$, implying that $\mathbf{A} \in \mathcal{U}$. Thus, $\mathrm{sp}(\mathbf{A}_1, \ldots, \mathbf{A}_k)$ is a subset of $\mathcal{U}$, and, since $\mathrm{sp}(\mathbf{A}_1, \ldots, \mathbf{A}_k)$ is a linear space, it follows that $\mathrm{sp}(\mathbf{A}_1, \ldots, \mathbf{A}_k)$ is a subspace of $\mathcal{U}$.

EXERCISE 6. Let $\mathbf{A}_1, \ldots, \mathbf{A}_p$ and $\mathbf{B}_1, \ldots, \mathbf{B}_q$ represent matrices in a linear space $\mathcal{V}$. Show that if the set $\{\mathbf{A}_1, \ldots, \mathbf{A}_p\}$ spans $\mathcal{V}$, then so does the set $\{\mathbf{A}_1, \ldots, \mathbf{A}_p, \mathbf{B}_1, \ldots, \mathbf{B}_q\}$. Show also that if the set $\{\mathbf{A}_1, \ldots, \mathbf{A}_p, \mathbf{B}_1, \ldots, \mathbf{B}_q\}$ spans $\mathcal{V}$ and if $\mathbf{B}_1, \ldots, \mathbf{B}_q$ are expressible as linear combinations of $\mathbf{A}_1, \ldots, \mathbf{A}_p$, then the set $\{\mathbf{A}_1, \ldots, \mathbf{A}_p\}$ spans $\mathcal{V}$.

Solution. It suffices (as observed in Section 4.3c) to show that if $\mathbf{B}_1, \ldots, \mathbf{B}_q$ are expressible as linear combinations of $\mathbf{A}_1, \ldots, \mathbf{A}_p$, then any linear combination of the matrices $\mathbf{A}_1, \ldots, \mathbf{A}_p, \mathbf{B}_1, \ldots, \mathbf{B}_q$ is expressible as a linear combination of $\mathbf{A}_1, \ldots, \mathbf{A}_p$ and vice versa. Suppose then that there exist scalars $k_{1j}, \ldots, k_{pj}$ such that $\mathbf{B}_j = \sum_i k_{ij}\mathbf{A}_i$ $(j = 1, \ldots, q)$. Then, for any scalars $x_1, \ldots, x_p, y_1, \ldots, y_q$,

$$\sum_i x_i\mathbf{A}_i + \sum_j y_j\mathbf{B}_j = \sum_i (x_i + \sum_j y_j k_{ij})\mathbf{A}_i,$$

which verifies that any linear combination of $\mathbf{A}_1, \ldots, \mathbf{A}_p, \mathbf{B}_1, \ldots, \mathbf{B}_q$ is expressible as a linear combination of $\mathbf{A}_1, \ldots, \mathbf{A}_p$. That any linear combination of $\mathbf{A}_1, \ldots, \mathbf{A}_p$ is expressible as a linear combination of $\mathbf{A}_1, \ldots, \mathbf{A}_p, \mathbf{B}_1, \ldots, \mathbf{B}_q$ is obvious.

EXERCISE 7. Suppose that $\{\mathbf{A}_1, \ldots, \mathbf{A}_k\}$ is a set of matrices that spans a linear space $\mathcal{V}$ but is not a basis for $\mathcal{V}$. Show that, for any matrix $\mathbf{A}$ in $\mathcal{V}$, the representation of $\mathbf{A}$ in terms of $\mathbf{A}_1, \ldots, \mathbf{A}_k$ is nonunique.

Solution. Let $x_1, \ldots, x_k$ represent any scalars such that $\mathbf{A} = \sum_{i=1}^k x_i\mathbf{A}_i$. [Since $\text{sp}(\mathbf{A}_1, \ldots, \mathbf{A}_k) = \mathcal{V}$, such scalars necessarily exist.] Since the set $\{\mathbf{A}_1, \ldots, \mathbf{A}_k\}$ spans $\mathcal{V}$ but is not a basis for $\mathcal{V}$, it is linearly dependent and hence there exist scalars $z_1, \ldots, z_k$, not all zero, such that $\sum_{i=1}^k z_i\mathbf{A}_i = \mathbf{0}$. Letting $y_i = x_i + z_i$ $(i = 1, \ldots, k)$, we obtain a representation $\mathbf{A} = \sum_{i=1}^k y_i\mathbf{A}_i$ different from the representation $\mathbf{A} = \sum_{i=1}^k x_i\mathbf{A}_i$.

EXERCISE 8. Let

$$\mathbf{A} = \begin{pmatrix} 0 & 1 & 0 & -3 & 2 \\ 0 & -2 & 0 & 6 & 2 \\ 0 & 2 & 2 & 5 & 2 \\ 0 & -4 & -2 & 1 & 0 \end{pmatrix}.$$

(a) Show that each of the two column vectors $(2, -1, 3, -4)'$ and $(0, 9, -3, 12)'$ is expressible as a linear combination of the columns of $\mathbf{A}$ [and hence is in $\mathcal{C}(\mathbf{A})$].

(b) A basis, say S^*, for a linear space $\mathcal{V}$ can be obtained from any finite set S that spans $\mathcal{V}$ by successively applying to each of the matrices in S the following algorithm: include the matrix in S^* if it is nonnull and if it is not expressible as a linear combination of the matrices already included in S^*. Use this algorithm to find a basis for $\mathcal{C}(\mathbf{A})$. (In applying the algorithm, take the spanning set S to be the set consisting of the columns of $\mathbf{A}$.)

(c) What is the value of $\text{rank}(\mathbf{A})$? Explain your reasoning.

(d) A basis for a linear space $\mathcal{V}$ that includes a specified set, say T, of r linearly independent matrices in $\mathcal{V}$ can be obtained by applying the algorithm described in Part (b) to the set S whose first r elements are the elements of T and whose remaining elements are the elements of any finite set U that spans $\mathcal{V}$. Use this generalization of the procedure from Part (b) to find a basis for $\mathcal{C}(\mathbf{A})$ that includes the two column vectors from Part (a). (In applying the generalized procedure, take the spanning set U to be the set consisting of the columns of $\mathbf{A}$.)

Solution. (a) Clearly,

$$\begin{pmatrix} 2 \\ -1 \\ 3 \\ -4 \end{pmatrix} = \begin{pmatrix} 1 \\ -2 \\ 2 \\ -4 \end{pmatrix} + (1/2) \begin{pmatrix} 2 \\ 2 \\ 2 \\ 0 \end{pmatrix}$$

and

$$\begin{pmatrix} 0 \\ 9 \\ -3 \\ 12 \end{pmatrix} = (-3) \begin{pmatrix} 1 \\ -2 \\ 2 \\ -4 \end{pmatrix} + (3/2) \begin{pmatrix} 2 \\ 2 \\ 2 \\ 0 \end{pmatrix}.$$

(b) The basis obtained by applying the algorithm comprises the following 3 vectors:

$$\begin{pmatrix} 1 \\ -2 \\ 2 \\ -4 \end{pmatrix}, \quad \begin{pmatrix} 0 \\ 0 \\ 2 \\ -2 \end{pmatrix}, \quad \begin{pmatrix} 2 \\ 2 \\ 2 \\ 0 \end{pmatrix}.$$

(c) Rank $\mathbf{A} = 3$. The number of vectors in a basis for $\mathcal{C}(\mathbf{A})$ equals 3 [as is evident from Part (b)], implying that the column rank of $\mathbf{A}$ equals 3.

(d) The basis obtained by applying the generalized procedure comprises the following 3 vectors:

$$\begin{pmatrix} 2 \\ -1 \\ 3 \\ -4 \end{pmatrix}, \quad \begin{pmatrix} 0 \\ 9 \\ -3 \\ 12 \end{pmatrix}, \quad \begin{pmatrix} 0 \\ 0 \\ 2 \\ -2 \end{pmatrix}.$$

EXERCISE 9. Let $\mathbf{A}$ represent a $q \times p$ matrix, $\mathbf{B}$ a $p \times n$ matrix, and $\mathbf{C}$ an $m \times q$ matrix. Show that (a) if $\text{rank}(\mathbf{CAB}) = \text{rank}(\mathbf{C})$, then $\text{rank}(\mathbf{CA}) = \text{rank}(\mathbf{C})$ and (b) if $\text{rank}(\mathbf{CAB}) = \text{rank}(\mathbf{B})$, then $\text{rank}(\mathbf{AB}) = \text{rank}(\mathbf{B})$.

Solution. (a) Suppose that $\text{rank}(\mathbf{CAB}) = \text{rank}(\mathbf{C})$. Then, it follows from Corollary 4.4.5 that

$$\text{rank}(\mathbf{C}) \geq \text{rank}(\mathbf{CA}) \geq \text{rank}(\mathbf{CAB}) = \text{rank}(\mathbf{C})$$

and hence that $\text{rank}(\mathbf{CA}) = \text{rank}(\mathbf{C})$.

(b) Similarly, suppose that $\text{rank}(\mathbf{CAB}) = \text{rank}(\mathbf{B})$. Then, it follows from Corollary 4.4.5 that

$$\text{rank}(\mathbf{B}) \geq \text{rank}(\mathbf{AB}) \geq \text{rank}(\mathbf{CAB}) = \text{rank}(\mathbf{B})$$

and hence that $\text{rank}(\mathbf{AB}) = \text{rank}(\mathbf{B})$.

EXERCISE 10. Let $\mathbf{A}$ represent an $m \times n$ matrix of rank r. Show that $\mathbf{A}$ can be expressed as the sum of r matrices of rank 1.

Solution. According to Theorem 4.4.8, there exist an $m \times r$ matrix $\mathbf{B}$ and an $r \times n$ matrix $\mathbf{T}$ such that $\mathbf{A} = \mathbf{BT}$. Let $\mathbf{b}_1, \ldots, \mathbf{b}_r$ represent the first, ..., rth columns of $\mathbf{B}$ and $\mathbf{t}'_1, \ldots, \mathbf{t}'_r$ the first, ..., rth rows of $\mathbf{T}$. Then, applying formula (2.2.9), we find that

$$\mathbf{A} = \sum_{j=1}^{r} \mathbf{A}_j,$$

where (for $j = 1, \ldots, r$) $\mathbf{A}_j = \mathbf{b}_j \mathbf{t}'_j$. Moreover, according to Theorem 4.4.8, $\text{rank}(\mathbf{B}) = \text{rank}(\mathbf{T}) = r$, and it follows that $\mathbf{b}_1, \ldots, \mathbf{b}_r$ and $\mathbf{t}'_1, \ldots, \mathbf{t}'_r$ are nonnull and hence that $\mathbf{A}_1, \ldots, \mathbf{A}_r$ are nonnull. And, upon observing (in light of Corollary 4.4.5 and Lemma 4.4.3) that $\text{rank}(\mathbf{A}_j) \leq \text{rank}(\mathbf{b}_j) \leq 1$, it is clear that $\text{rank}(\mathbf{A}_j) = 1$ $(j = 1, \ldots, r)$.

EXERCISE 11. Let $\mathbf{A}$ represent an $m \times n$ matrix and $\mathbf{C}$ a $q \times n$ matrix.

(a) Confirm that

$$\mathcal{R}(\mathbf{C}) = \mathcal{R}\begin{pmatrix} \mathbf{A} \\ \mathbf{C} \end{pmatrix} \Leftrightarrow \mathcal{R}(\mathbf{A}) \subset \mathcal{R}(\mathbf{C}) .$$

(b) Confirm that $\text{rank}(\mathbf{C}) \leq \text{rank}\begin{pmatrix} \mathbf{A} \\ \mathbf{C} \end{pmatrix}$, with equality holding if and only if $\mathcal{R}(\mathbf{A}) \subset \mathcal{R}(\mathbf{C})$.

Solution. (a) Suppose that $\mathcal{R}(\mathbf{A}) \subset \mathcal{R}(\mathbf{C})$. Then, according to Lemma 4.2.2, there exists an $m \times q$ matrix $\mathbf{L}$ such that $\mathbf{A} = \mathbf{LC}$ and hence such that $\begin{pmatrix} \mathbf{A} \\ \mathbf{C} \end{pmatrix} = \begin{pmatrix} \mathbf{L} \\ \mathbf{I} \end{pmatrix}\mathbf{C}$. Thus, $\mathcal{R}\begin{pmatrix} \mathbf{A} \\ \mathbf{C} \end{pmatrix} \subset \mathcal{R}(\mathbf{C})$, implying [since $\mathcal{R}(\mathbf{C}) \subset \mathcal{R}\begin{pmatrix} \mathbf{A} \\ \mathbf{C} \end{pmatrix}$] that $\mathcal{R}(\mathbf{C}) = \mathcal{R}\begin{pmatrix} \mathbf{A} \\ \mathbf{C} \end{pmatrix}$.

Conversely, suppose that $\mathcal{R}(\mathbf{C}) = \mathcal{R}\begin{pmatrix} \mathbf{A} \\ \mathbf{C} \end{pmatrix}$. Then, since $\mathcal{R}(\mathbf{A}) \subset \mathcal{R}\begin{pmatrix} \mathbf{A} \\ \mathbf{C} \end{pmatrix}$, $\mathcal{R}(\mathbf{A}) \subset \mathcal{R}(\mathbf{C})$. Thus, we have established that $\mathcal{R}(\mathbf{C}) = \mathcal{R}\begin{pmatrix} \mathbf{A} \\ \mathbf{C} \end{pmatrix} \Leftrightarrow \mathcal{R}(\mathbf{A}) \subset \mathcal{R}(\mathbf{C})$.

(b) Since (according to Lemma 4.5.1) $\mathcal{R}(\mathbf{C}) \subset \mathcal{R}\begin{pmatrix} \mathbf{A} \\ \mathbf{C} \end{pmatrix}$, it follows from Theorem 4.4.4 that $\text{rank}(\mathbf{C}) \leq \text{rank}\begin{pmatrix} \mathbf{A} \\ \mathbf{C} \end{pmatrix}$. Moreover, if $\mathcal{R}(\mathbf{A}) \subset \mathcal{R}(\mathbf{C})$, then [according to Part (a) or to Lemma 4.5.1] $\mathcal{R}(\mathbf{C}) = \mathcal{R}\begin{pmatrix}\mathbf{A}\\ \mathbf{C}\end{pmatrix}$ and consequently $\text{rank}(\mathbf{C}) = \text{rank}\begin{pmatrix}\mathbf{A}\\ \mathbf{C}\end{pmatrix}$. And, conversely, if $\text{rank}(\mathbf{C}) = \text{rank}\begin{pmatrix}\mathbf{A}\\ \mathbf{C}\end{pmatrix}$, then since $\mathcal{R}(\mathbf{C}) \subset \mathcal{R}\begin{pmatrix}\mathbf{A}\\ \mathbf{C}\end{pmatrix}$, it follows from Theorem 4.4.6 that $\mathcal{R}(\mathbf{C}) = \mathcal{R}\begin{pmatrix}\mathbf{A}\\ \mathbf{C}\end{pmatrix}$ and hence [in light of Part (a) or of Lemma 4.5.1] that $\mathcal{R}(\mathbf{A}) \subset \mathcal{R}(\mathbf{C})$. Thus, $\text{rank}(\mathbf{C}) \leq \text{rank}\begin{pmatrix}\mathbf{A}\\ \mathbf{C}\end{pmatrix}$, with equality holding if and only if $\mathcal{R}(\mathbf{A}) \subset \mathcal{R}(\mathbf{C})$.

5

Trace of a (Square) Matrix

EXERCISE 1. Show that for any $m \times n$ matrix $\mathbf{A}$, $n \times p$ matrix $\mathbf{B}$, and $p \times q$ matrix $\mathbf{C}$,

$$\text{tr}(\mathbf{ABC}) = \text{tr}(\mathbf{B'A'C'}) = \text{tr}(\mathbf{A'C'B'}).$$

Solution. Making use of results (2.9) and (1.5), we find that

$$\text{tr}(\mathbf{ABC}) = \text{tr}(\mathbf{CAB}) = \text{tr}[(\mathbf{CAB})'] = \text{tr}(\mathbf{B'A'C'}) = \text{tr}(\mathbf{A'C'B'}).$$

EXERCISE 2. Let $\mathbf{A}$, $\mathbf{B}$, and $\mathbf{C}$ represent $n \times n$ matrices.

(a) Using the result of Exercise 1 (or otherwise), show that if $\mathbf{A}$, $\mathbf{B}$, and $\mathbf{C}$ are symmetric, then $\text{tr}(\mathbf{ABC}) = \text{tr}(\mathbf{BAC})$.

(b) Show that [aside from special cases like that considered in Part (a)] $\text{tr}(\mathbf{BAC})$ is not necessarily equal to $\text{tr}(\mathbf{ABC})$.

Solution. (a) If $\mathbf{A}$, $\mathbf{B}$, and $\mathbf{C}$ are symmetric, then $\mathbf{B'A'C'} = \mathbf{BAC}$ and it follows from the result of Exercise 1 that $\text{tr}(\mathbf{ABC}) = \text{tr}(\mathbf{BAC})$.

(b) Let $\mathbf{A} = \text{diag}(\mathbf{A}_*, \mathbf{0})$, $\mathbf{B} = \text{diag}(\mathbf{B}_*, \mathbf{0})$, and $\mathbf{C} = \text{diag}(\mathbf{C}_*, \mathbf{0})$, where

$$\mathbf{A}_* = \begin{pmatrix} 1 & 1 \\ 0 & 0 \end{pmatrix}, \quad \mathbf{B}_* = \begin{pmatrix} 1 & 0 \\ -1 & 0 \end{pmatrix}, \quad \mathbf{C}_* = \begin{pmatrix} 1 & 0 \\ 0 & -1 \end{pmatrix}.$$

Then, $\mathbf{A}_*\mathbf{B}_*\mathbf{C}_* = \mathbf{0}$ and $\mathbf{B}_*\mathbf{A}_*\mathbf{C}_* = \begin{pmatrix} 1 & -1 \\ -1 & 1 \end{pmatrix}$, and, observing that $\mathbf{ABC} = \text{diag}(\mathbf{A}_*\mathbf{B}_*\mathbf{C}_*, \mathbf{0})$ and $\mathbf{BAC} = \text{diag}(\mathbf{B}_*\mathbf{A}_*\mathbf{C}_*, \mathbf{0})$ and making use of result (1.7), we find that

$$\operatorname{tr}(\mathbf{BAC}) = \operatorname{tr}(\mathbf{B}_*\mathbf{A}_*\mathbf{C}_*) = 2 \neq 0 = \operatorname{tr}(\mathbf{A}_*\mathbf{B}_*\mathbf{C}_*) = \operatorname{tr}(\mathbf{ABC}).$$

EXERCISE 3. Let $\mathbf{A}$ represent an $n \times n$ matrix such that $\mathbf{A}'\mathbf{A} = \mathbf{A}^2$.

(a) Show that $\operatorname{tr}[(\mathbf{A} - \mathbf{A}')'(\mathbf{A} - \mathbf{A}')] = 0$.

(b) Show that $\mathbf{A}$ is symmetric.

Solution. (a) Making use of results (2.3) and (1.5), we find that

$$\begin{aligned}
\operatorname{tr}[(\mathbf{A} - \mathbf{A}')'(\mathbf{A} - \mathbf{A}')] &= \operatorname{tr}[\mathbf{A}'\mathbf{A} - \mathbf{A}'\mathbf{A}' - \mathbf{AA} + \mathbf{AA}'] \\
&= \operatorname{tr}(\mathbf{A}'\mathbf{A}) - \operatorname{tr}[(\mathbf{AA})'] - \operatorname{tr}(\mathbf{A}^2) + \operatorname{tr}(\mathbf{AA}') \\
&= \operatorname{tr}(\mathbf{AA}') - \operatorname{tr}[(\mathbf{AA})'] \\
&= \operatorname{tr}(\mathbf{A}'\mathbf{A}) - \operatorname{tr}[(\mathbf{AA})'] \\
&= \operatorname{tr}(\mathbf{A}'\mathbf{A}) - \operatorname{tr}(\mathbf{AA}) = 0.
\end{aligned}$$

(b) In light of Lemma 5.3.1, it follows from Part (a) that $\mathbf{A} - \mathbf{A}' = \mathbf{0}$ or equivalently that $\mathbf{A}' = \mathbf{A}$.

6

Geometrical Considerations

EXERCISE 1. Use the Schwarz inequality to show that, for any two matrices **A** and **B** in a linear space $\mathcal{V}$,

$$\|\mathbf{A}+\mathbf{B}\| \leq \|\mathbf{A}\| + \|\mathbf{B}\|,$$

with equality holding if and only if $\mathbf{B}=\mathbf{0}$ or $\mathbf{A}=k\mathbf{B}$ for some nonnegative scalar k. (This inequality is known as the triangle inequality.)

Solution. We have that

$$\begin{aligned}
\|\mathbf{A}+\mathbf{B}\|^2 &= (\mathbf{A}+\mathbf{B})\bullet(\mathbf{A}+\mathbf{B}) \\
&= \|\mathbf{A}\|^2 + 2\,(\mathbf{A}\bullet\mathbf{B}) + \|\mathbf{B}\|^2 \\
&\leq \|\mathbf{A}\|^2 + 2|\mathbf{A}\bullet\mathbf{B}| + \|\mathbf{B}\|^2 && \text{(S.1)} \\
&\leq \|\mathbf{A}\|^2 + 2\,\|\mathbf{A}\|\,\|\mathbf{B}\| + \|\mathbf{B}\|^2 && \text{(S.2)} \\
&\qquad \text{(using the Schwarz inequality)} \\
&= (\|\mathbf{A}\| + \|\mathbf{B}\|)^2
\end{aligned}$$

or equivalently that

$$\|\mathbf{A}+\mathbf{B}\| \leq \|\mathbf{A}\| + \|\mathbf{B}\|.$$

For this inequality to hold as an equality, it is necessary and sufficient that both of inequalities (S.1) and (S.2) hold as equalities. Recalling (from, for instance, Theorem 6.3.1) the conditions under which the Schwarz inequality holds as an equality, we find that inequalities (S.1) and (S.2) both hold as equalities if and only if $\mathbf{B}=\mathbf{0}$ or $\mathbf{A}=k\mathbf{B}$ with $k \geq 0$.

EXERCISE 2. Letting **A**, **B**, and **C** represent arbitrary matrices in a linear space $\mathcal{V}$, show that

(a) $\delta(\mathbf{B}, \mathbf{A}) = \delta(\mathbf{A}, \mathbf{B})$, that is the distance between **B** and **A** is the same as that between **A** and **B**;

(b)

$$\begin{aligned} \delta(\mathbf{A}, \mathbf{B}) &> 0, \quad \text{if } \mathbf{A} \neq \mathbf{B}, \\ &= 0, \quad \text{if } \mathbf{A} = \mathbf{B}, \end{aligned}$$

that is, the distance between any two matrices is greater than zero, unless the two matrices are identical, in which case the distance between them is zero;

(c) $\delta(\mathbf{A}, \mathbf{B}) \leq \delta(\mathbf{A}, \mathbf{C}) + \delta(\mathbf{C}, \mathbf{B})$, that is, the distance between **A** and **B** is less than or equal to the sum of the distances between **A** and **C** and between **C** and **B**;

(d) $\delta(\mathbf{A}, \mathbf{B}) = \delta(\mathbf{A} + \mathbf{C}, \mathbf{B} + \mathbf{C})$, that is, distance is unaffected by a translation of "axes."

[For Part (c), use the result of Exercise 1, i.e., the triangle inequality.]

Solution. (a)

$$\begin{aligned} \delta(\mathbf{B}, \mathbf{A}) &= \|\mathbf{B} - \mathbf{A}\| \\ &= \|(-1)(\mathbf{A} - \mathbf{B})\| = |-1|\,\|\mathbf{A} - \mathbf{B}\| = \|\mathbf{A} - \mathbf{B}\| = \delta(\mathbf{A}, \mathbf{B}). \end{aligned}$$

(b)

$$\begin{aligned} \delta(\mathbf{A}, \mathbf{B}) = \|\mathbf{A} - \mathbf{B}\| &> 0, \quad \text{if } \mathbf{A} - \mathbf{B} \neq \mathbf{0} \text{ or equivalently if } \mathbf{A} \neq \mathbf{B}, \\ &= 0, \quad \text{if } \mathbf{A} - \mathbf{B} = \mathbf{0} \text{ or equivalently if } \mathbf{A} = \mathbf{B}. \end{aligned}$$

(c)

$$\begin{aligned} \delta(\mathbf{A}, \mathbf{B}) &= \|\mathbf{A} - \mathbf{B}\| \\ &= \|(\mathbf{A} - \mathbf{C}) + (\mathbf{C} - \mathbf{B})\| \\ &\leq \|\mathbf{A} - \mathbf{C}\| + \|\mathbf{C} - \mathbf{B}\| = \delta(\mathbf{A}, \mathbf{C}) + \delta(\mathbf{C}, \mathbf{B}). \end{aligned}$$

(d)

$$\delta(\mathbf{A} + \mathbf{C}, \mathbf{B} + \mathbf{C}) = \|(\mathbf{A} + \mathbf{C}) - (\mathbf{B} + \mathbf{C})\| = \|\mathbf{A} - \mathbf{B}\| = \delta(\mathbf{A}, \mathbf{B}).$$

EXERCISE 3. Let $\mathbf{w}_1'$, $\mathbf{w}_2'$, and $\mathbf{w}_3'$ represent the three linearly independent 4-dimensional row vectors (6, 0, –2, 3), (–2, 4, 4, 2), and (0, 5, –1, 2), respectively, in the linear space $\mathcal{R}^4$, and adopt the usual definition of inner product.

(a) Use Gram-Schmidt orthogonalization to find an orthonormal basis for the linear space $\text{sp}(\mathbf{w}_1', \mathbf{w}_2', \mathbf{w}_3')$.

(b) Find an orthonormal basis for $\mathcal{R}^4$ that includes the three orthonormal vectors from Part (a). Do so by extending the results of the Gram-Schmidt orthogonalization [from Part (a)] to a fourth linearly independent row vector such as (0, 1, 0, 0).

Solution. (a) The 3 orthogonal vectors obtained by applying the formulas (for Gram-Schmidt orthogonalization) of Theorem 6.4.1 are:

$$\mathbf{y}_1' = \mathbf{w}_1' = (6, 0, -2, 3),$$
$$\mathbf{y}_2' = \mathbf{w}_2' - (-2/7)\mathbf{y}_1' = (1/7)(-2, 28, 24, 20),$$
$$\mathbf{y}_3' = \mathbf{w}_3' - (13/21)\mathbf{y}_2' - (8/49)\mathbf{y}_1' = (1/147)(-118, 371, -411, -38).$$

By normalizing $\mathbf{y}_1'$, $\mathbf{y}_2'$, and $\mathbf{y}_3'$, we obtain a basis for $\text{sp}(\mathbf{w}_1', \mathbf{w}_2', \mathbf{w}_3')$ consisting of the following 3 vectors:

$$\mathbf{z}_1' = (1/7)\mathbf{y}_1' = (1/7)(6, 0, -2, 3),$$
$$\mathbf{z}_2' = (1/6)\mathbf{y}_2' = (1/42)(-2, 28, 24, 20),$$
$$\mathbf{z}_3' = (21609/321930)^{1/2}\mathbf{y}_3' = (321930)^{-1/2}(-118, 371, -411, -38).$$

(b) An orthonormal basis for $\mathcal{R}^4$ can be obtained by extending the results of the Gram-Schmidt orthogonalization to a fourth linearly independent vector $\mathbf{w}_4'$. Taking $\mathbf{w}_4' = (0, 1, 0, 0)$ and applying the formulas of Theorem 6.4.1, we obtain the following vector, which is orthogonal to $\mathbf{y}_1'$, $\mathbf{y}_2'$, and $\mathbf{y}_3'$:

$$\begin{aligned}\mathbf{y}_4' &= \mathbf{w}_4' - (371/2190)\mathbf{y}_3' - (1/9)\mathbf{y}_2' - (0)\mathbf{y}_1' \\ &= (1/321930)(53998, 41209, 29841, -88102).\end{aligned}$$

The set consisting of the normalized vector

$$\begin{aligned}\mathbf{z}_4' &= [321930/(13266413370)^{1/2}]\mathbf{y}_4' \\ &= (13266413370)^{-1/2}(53998, 41209, 29841, -88102),\end{aligned}$$

together with $\mathbf{z}_1'$, $\mathbf{z}_2'$, and $\mathbf{z}_3'$, is an orthonormal basis for $\mathcal{R}^4$.

EXERCISE 4. Let $\{\mathbf{A}_1, \ldots, \mathbf{A}_k\}$ represent a nonempty (possibly linearly dependent) set of matrices in a linear space $\mathcal{V}$.

(a) Generalize the results underlying Gram-Schmidt orthogonalization (which are for the special case where the set $\{\mathbf{A}_1, \ldots, \mathbf{A}_k\}$ is linearly independent) by showing (1) that there exist scalars x_{ij} $(i < j = 1, \ldots, k)$ such that the set comprising the k matrices

$$\begin{aligned}\mathbf{B}_1 &= \mathbf{A}_1, \\ \mathbf{B}_2 &= \mathbf{A}_2 - x_{12}\mathbf{B}_1, \\ &\vdots \\ \mathbf{B}_j &= \mathbf{A}_j - x_{j-1,j}\mathbf{B}_{j-1} - \cdots - x_{1j}\mathbf{B}_1, \\ &\vdots \\ \mathbf{B}_k &= \mathbf{A}_k - x_{k-1,k}\mathbf{B}_{k-1} - \cdots - x_{1k}\mathbf{B}_1\end{aligned}$$

is orthogonal; (2) that, for $j = 1, \ldots, k$ and for those $i < j$ such that $\mathbf{B}_i$ is nonnull, x_{ij} is given uniquely by

$$x_{ij} = \frac{\mathbf{A}_j \bullet \mathbf{B}_i}{\mathbf{B}_i \bullet \mathbf{B}_i};$$

and (3) that the number of nonnull matrices among $\mathbf{B}_1, \ldots, \mathbf{B}_k$ equals $\dim[\mathrm{sp}(\mathbf{A}_1, \ldots, \mathbf{A}_k)]$.

(b) Describe a procedure for constructing an orthonormal basis for $\mathrm{sp}(\mathbf{A}_1, \ldots, \mathbf{A}_k)$.

Solution. (a) The proof of (1) and (2) is by mathematical induction. Assertions (1) and (2) are clearly true for $k = 1$. Suppose now that they are true for a set of $k - 1$ matrices. Then, there exist scalars x_{ij} $(i < j = 1, \ldots, k - 1)$ such that the set comprising the $k - 1$ matrices $\mathbf{B}_1, \ldots, \mathbf{B}_{k-1}$ is orthogonal, and, for $j = 1, \ldots, k - 1$ and for those $i < j$ such that $\mathbf{B}_i$ is nonnull, x_{ij} is given uniquely by

$$x_{ij} = \frac{\mathbf{A}_j \bullet \mathbf{B}_i}{\mathbf{B}_i \bullet \mathbf{B}_i}.$$

Moreover, for $i = 1, \ldots, k-1$, we find (as in the proof of the results underlying Gram-Schmidt orthogonalization in Theorem 6.4.1) that $\mathbf{B}_k \bullet \mathbf{B}_i = 0$ if and only if

$$\mathbf{A}_k \bullet \mathbf{B}_i - x_{ik}(\mathbf{B}_i \bullet \mathbf{B}_i) = 0.$$

For those i (between 1 and $k - 1$) such that $\mathbf{B}_i = \mathbf{0}$, this equation is satisfied by any x_{ik}, and, for those i such that $\mathbf{B}_i \neq \mathbf{0}$, it has the unique solution

$$x_{ik} = \frac{\mathbf{A}_k \bullet \mathbf{B}_i}{\mathbf{B}_i \bullet \mathbf{B}_i}.$$

This completes the induction argument, thereby establishing (1) and (2).

Consider now Assertion (3). Each of the matrices $\mathbf{B}_1, \ldots, \mathbf{B}_k$ can (by repeated substitution) be expressed as a linear combination of $\mathbf{A}_1, \ldots, \mathbf{A}_k$. Conversely, each of the matrices $\mathbf{A}_1, \ldots, \mathbf{A}_k$ can be expressed as a linear combination of $\mathbf{B}_1, \ldots, \mathbf{B}_k$. Thus, $\mathrm{sp}(\mathbf{B}_1, \ldots, \mathbf{B}_k) = \mathrm{sp}(\mathbf{A}_1, \ldots, \mathbf{A}_k)$. Since the set $\{\mathbf{B}_1, \ldots, \mathbf{B}_k\}$ is orthogonal, we conclude — in light of Lemma 6.2.1 and Theorem 4.3.2 — that the nonnull matrices among $\mathbf{B}_1, \ldots, \mathbf{B}_k$ form a basis for $\mathrm{sp}(\mathbf{A}_1, \ldots, \mathbf{A}_k)$ and hence that the number of such matrices equals $\dim[\mathrm{sp}(\mathbf{A}_1, \ldots, \mathbf{A}_k)]$.

(b) An orthonormal basis for $\mathrm{sp}(\mathbf{A}_1, \ldots, \mathbf{A}_k)$ can be constructed by making use of the formulas for $\mathbf{B}_1, \ldots, \mathbf{B}_k$ from Part (a). The basis consists of those matrices obtained by normalizing the nonnull matrices among $\mathbf{B}_1, \ldots, \mathbf{B}_k$.

EXERCISE 5. Let $\mathbf{A}$ represent an $m \times k$ matrix of rank r (where r is possibly less than k). Generalize the so-called QR decomposition of $\mathbf{A}$, which is for the special case where $r = k$ and is obtainable through the application of Gram-Schmidt orthogonalization to the columns of $\mathbf{A}$. Do so by using the results of Exercise 4 to obtain a decomposition of the form $\mathbf{A} = \mathbf{Q}\mathbf{R}_1$, where $\mathbf{Q}$ is an $m \times r$ matrix with orthonormal columns and $\mathbf{R}_1$ is an $r \times k$ submatrix whose rows are the r nonnull rows of a $k \times k$ upper triangular matrix $\mathbf{R}$ having r positive diagonal elements and $k - r$ null rows.

Solution. Denote the first, ..., kth columns of $\mathbf{A}$ by $\mathbf{a}_1, \ldots, \mathbf{a}_k$, respectively. Then, according to the results of Exercise 4, there exist scalars x_{ij} $(i < j = 1, \ldots, k)$ such that the k column vectors $\mathbf{b}_1, \ldots, \mathbf{b}_k$ defined recursively by the equalities

$$\begin{aligned}
\mathbf{b}_1 &= \mathbf{a}_1, \\
\mathbf{b}_2 &= \mathbf{a}_2 - x_{12}\mathbf{b}_1, \\
&\vdots \\
\mathbf{b}_j &= \mathbf{a}_j - x_{j-1,j}\mathbf{b}_{j-1} - \cdots - x_{1j}\mathbf{b}_1, \\
&\vdots \\
\mathbf{b}_k &= \mathbf{a}_k - x_{k-1,k}\mathbf{b}_{k-1} - \cdots - x_{1k}\mathbf{b}_1,
\end{aligned}$$

or equivalently by the equalities

$$\begin{aligned}
\mathbf{a}_1 &= \mathbf{b}_1 \\
\mathbf{a}_2 &= \mathbf{b}_2 + x_{12}\mathbf{b}_1, \\
&\vdots \\
\mathbf{a}_j &= \mathbf{b}_j + x_{j-1,j}\mathbf{b}_{j-1} + \cdots + x_{1j}\mathbf{b}_1, \\
&\vdots \\
\mathbf{a}_k &= \mathbf{b}_k + x_{k-1,k}\mathbf{b}_{k-1} + \cdots + x_{1k}\mathbf{b}_1,
\end{aligned}$$

form an orthogonal set. Further, r of the vectors $\mathbf{b}_1, \ldots, \mathbf{b}_k$, say the s_1th, ..., s_rth of them, are nonnull, and, for $j = 1, \ldots, k$ and for those $i < j$ such that $\mathbf{b}_i$ is nonnull, x_{ij} is given uniquely by

$$x_{ij} = \frac{\mathbf{a}_j \cdot \mathbf{b}_i}{\mathbf{b}_i \cdot \mathbf{b}_i}.$$

Now, let $\mathbf{B}$ represent the $m \times k$ matrix whose first, ..., kth columns are $\mathbf{b}_1, \ldots, \mathbf{b}_k$, respectively, and let $\mathbf{X}$ represent the $k \times k$ unit upper triangular matrix whose ijth element is (for $i < j = 1, \ldots, k$) x_{ij}. Then, observing that the first column of $\mathbf{BX}$ is $\mathbf{b}_1$ and that (for $j = 2, \ldots, k$) the jth column of $\mathbf{BX}$ is $\mathbf{b}_j + x_{j-1,j}\mathbf{b}_{j-1} + \cdots + x_{1j}\mathbf{b}_1$ and recalling result (2.2.9), we find that

$$\mathbf{A} = \mathbf{BX} = \mathbf{B}_1\mathbf{X}_1,$$

where $\mathbf{B}_1$ is the $m \times r$ submatrix (of $\mathbf{B}$) whose columns are the s_1th, ..., s_rth columns of $\mathbf{B}$ and $\mathbf{X}_1$ is the $r \times k$ submatrix (of $\mathbf{X}$) whose rows are the s_1th, ..., s_rth rows of $\mathbf{X}$.

And, the decomposition $\mathbf{A} = \mathbf{B}_1\mathbf{X}_1$ can be reexpressed as

$$\mathbf{A} = \mathbf{QR}_1,$$

where $\mathbf{Q} = \mathbf{B}_1\mathbf{D}$, with $\mathbf{D} = \text{diag}(\| \mathbf{b}_{s_1} \|^{-1}, \ldots, \| \mathbf{b}_{s_r} \|^{-1})$, and $\mathbf{R}_1 = \mathbf{E}\mathbf{X}_1$, with $\mathbf{E} = \text{diag}(\| \mathbf{b}_{s_1} \|, \ldots, \| \mathbf{b}_{s_r} \|)$, or equivalently where $\mathbf{Q}$ is the $m \times r$ matrix with jth column $\| \mathbf{b}_{s_j} \|^{-1} \mathbf{b}_{s_j}$ and $\mathbf{R}_1 = \{r_{ij}\}$ is the $r \times k$ matrix with

$$r_{ij} = \begin{cases} \| \mathbf{b}_{s_i} \| x_{s_i j}, & \text{for } j > s_i\,, \\ \| \mathbf{b}_{s_i} \|, & \text{for } j = s_i\,, \\ 0, & \text{for } j < s_i\,. \end{cases}$$

Moreover, the columns of $\mathbf{Q}$ are orthonormal, and $\mathbf{R}_1$ is an $r \times k$ submatrix whose rows are the r nonnull rows of a $k \times k$ upper triangular matrix $\mathbf{R}$ having r positive diagonal elements and $n - r$ null rows — the s_1th, ..., s_rth rows of $\mathbf{R}$ (which are the nonnull rows) are respectively the first, ..., rth rows of $\mathbf{R}_1$.

7

Linear Systems: Consistency and Compatibility

EXERCISE 1. (a) Let $\mathbf{A}$ represent an $m \times n$ matrix, $\mathbf{C}$ an $n \times q$ matrix, and $\mathbf{B}$ a $q \times p$ matrix. Show that if $\text{rank}(\mathbf{AC}) = \text{rank}(\mathbf{C})$, then

$$\mathcal{R}(\mathbf{ACB}) = \mathcal{R}(\mathbf{CB}) \quad \text{and} \quad \text{rank}(\mathbf{ACB}) = \text{rank}(\mathbf{CB})$$

and that if $\text{rank}(\mathbf{CB}) = \text{rank}(\mathbf{C})$, then

$$\mathcal{C}(\mathbf{ACB}) = \mathcal{C}(\mathbf{AC}) \quad \text{and} \quad \text{rank}(\mathbf{ACB}) = \text{rank}(\mathbf{AC})\ .$$

(b) Let $\mathbf{A}$ and $\mathbf{B}$ represent $m \times n$ matrices. (1) Show that if $\mathbf{C}$ is an $r \times q$ matrix and $\mathbf{D}$ a $q \times m$ matrix such that $\text{rank}(\mathbf{CD}) = \text{rank}(\mathbf{D})$, then $\mathbf{CDA} = \mathbf{CDB}$ implies $\mathbf{DA} = \mathbf{DB}$. {*Hint*. To show that $\mathbf{DA} = \mathbf{DB}$, it suffices to show that $\text{rank}[\mathbf{D}(\mathbf{A} - \mathbf{B})] = 0$.} (2) Similarly, show that if $\mathbf{C}$ is an $n \times q$ matrix and $\mathbf{D}$ a $q \times p$ matrix such that $\text{rank}(\mathbf{CD}) = \text{rank}(\mathbf{C})$, then $\mathbf{ACD} = \mathbf{BCD}$ implies $\mathbf{AC} = \mathbf{BC}$.

Solution. (a) It is clear from Corollary 4.2.3 that $\mathcal{R}(\mathbf{ACB}) \subset \mathcal{R}(\mathbf{CB})$ and $\mathcal{C}(\mathbf{ACB}) \subset \mathcal{C}(\mathbf{AC})$.

Now, suppose that $\text{rank}(\mathbf{AC}) = \text{rank}(\mathbf{C})$. Then, according to Corollary 4.4.7, $\mathcal{R}(\mathbf{AC}) = \mathcal{R}(\mathbf{C})$, and it follows from Lemma 4.2.2 that $\mathbf{C} = \mathbf{LAC}$ for some matrix $\mathbf{L}$. Thus,

$$\mathcal{R}(\mathbf{CB}) = \mathcal{R}(\mathbf{LACB}) \subset \mathcal{R}(\mathbf{ACB}),$$

implying that $\mathcal{R}(\mathbf{ACB}) = \mathcal{R}(\mathbf{CB})$ [which implies, in turn, that $\text{rank}(\mathbf{ACB}) = \text{rank}(\mathbf{CB})$].

Similarly, if $\text{rank}(\mathbf{CB}) = \text{rank}(\mathbf{C})$, then $\mathcal{C}(\mathbf{CB}) = \mathcal{C}(\mathbf{C})$, in which case $\mathbf{C} = \mathbf{CBR}$ for some matrix $\mathbf{R}$, implying that $\mathcal{C}(\mathbf{AC}) = \mathcal{C}(\mathbf{ACBR}) \subset \mathcal{C}(\mathbf{ACB})$ and hence that $\mathcal{C}(\mathbf{ACB}) = \mathcal{C}(\mathbf{AC})$ [and $\text{rank}(\mathbf{ACB}) = \text{rank}(\mathbf{AC})$].

(b) Let $\mathbf{F} = \mathbf{A} - \mathbf{B}$. (1) Suppose that $\text{rank}(\mathbf{CD}) = \text{rank}(\mathbf{D})$. Then, if $\mathbf{CDA} = \mathbf{CDB}$, we find, in light of Part (a), that

$$\text{rank}(\mathbf{DF}) = \text{rank}(\mathbf{CDF}) = \text{rank}(\mathbf{CDA} - \mathbf{CDB}) = \text{rank}(\mathbf{0}) = 0,$$

implying that $\mathbf{DF} = \mathbf{0}$ or equivalently that $\mathbf{DA} = \mathbf{DB}$.

(2) Similarly, suppose that $\text{rank}(\mathbf{CD}) = \text{rank}(\mathbf{C})$. Then, if $\mathbf{ACD} = \mathbf{BCD}$, we find, in light of Part (a), that

$$\text{rank}(\mathbf{FC}) = \text{rank}(\mathbf{FCD}) = \text{rank}(\mathbf{ACD} - \mathbf{BCD}) = \text{rank}(\mathbf{0}) = 0,$$

implying that $\mathbf{FC} = \mathbf{0}$ or equivalently that $\mathbf{AC} = \mathbf{BC}$.

8

Inverse Matrices

EXERCISE 1. Let $\mathbf{A}$ represent an $m \times n$ matrix. Show that (a) if $\mathbf{A}$ has a right inverse, then $n \geq m$ and (b) if $\mathbf{A}$ has a left inverse, then $m \geq n$.

Solution. (a) If $\mathbf{A}$ has a right inverse, then, according to Lemma 8.1.1, $\text{rank}(\mathbf{A}) = m$, and, since (according to Lemma 4.4.3) $n \geq \text{rank}(\mathbf{A})$, it follows that $n \geq m$. (b) Similarly, if $\mathbf{A}$ has a left inverse, then according to Lemma 8.1.1, $\text{rank}(\mathbf{A}) = n$, and, since (according to Lemma 4.4.3) $m \geq \text{rank}(\mathbf{A})$, it follows that $m \geq n$.

EXERCISE 2. An $n \times n$ matrix $\mathbf{A}$ is said to be *involutory* if $\mathbf{A}^2 = \mathbf{I}$, that is, if $\mathbf{A}$ is invertible and is its own inverse.

(a) Show that an $n \times n$ matrix $\mathbf{A}$ is involutory if and only if $(\mathbf{I} - \mathbf{A})(\mathbf{I} + \mathbf{A}) = \mathbf{0}$.

(b) Show that a 2×2 matrix $\mathbf{A} = \begin{pmatrix} a & b \\ c & d \end{pmatrix}$ is involutory if and only if (1) $a^2 + bc = 1$ and $d = -a$ or (2) $b = c = 0$ and $d = a = \pm 1$.

Solution. (a) Clearly,

$$(\mathbf{I} - \mathbf{A})(\mathbf{I} + \mathbf{A}) = \mathbf{I} - \mathbf{A} + (\mathbf{I} - \mathbf{A})\mathbf{A} = \mathbf{I} - \mathbf{A} + \mathbf{A} - \mathbf{A}^2 = \mathbf{I} - \mathbf{A}^2.$$

Thus,

$$(\mathbf{I} - \mathbf{A})(\mathbf{I} + \mathbf{A}) = \mathbf{0} \;\Leftrightarrow\; \mathbf{I} - \mathbf{A}^2 = \mathbf{0} \;\Leftrightarrow\; \mathbf{A}^2 = \mathbf{I}.$$

(b) Clearly,

$$\mathbf{A}^2 = \begin{pmatrix} a^2 + bc & ab + bd \\ ac + cd & bc + d^2 \end{pmatrix}.$$

And, if Condition (1) or (2) is satisfied, it is easy to see that **A** is involutory.

Conversely, suppose that **A** is involutory. Then, $ab = -db$ and $ac = -dc$, implying that $d = -a$ or $b = c = 0$. Moreover, $a^2 + bc = 1$ and $d^2 + bc = 1$. Consequently, if $d = -a$, Condition (1) is satisfied. Alternatively, if $b = c = 0$, then $d^2 = a^2 = 1$, implying that $d = a = \pm 1$ (in which case Condition (2) is satisfied) or that $d = -a = \pm 1$ (in which case Condition (1) is satisfied).

EXERCISE 3. Let **A** represent an $n \times n$ nonnull symmetric matrix, and let **B** represent an $n \times r$ matrix of full column rank r and **T** an $r \times n$ matrix of full row rank r such that $\mathbf{A} = \mathbf{BT}$. Show that the $r \times r$ matrix **TB** is nonsingular. (*Hint.* Observe that $\mathbf{A}'\mathbf{A} = \mathbf{A}^2 = \mathbf{BTBT}$.)

Solution. Since $\mathbf{A}'\mathbf{A} = \mathbf{A}^2 = \mathbf{BTBT}$, we have (in light of Corollaries 7.4.5 and 8.3.4) that

$$\text{rank}(\mathbf{BTBT}) = \text{rank}(\mathbf{A}'\mathbf{A}) = \text{rank}(\mathbf{A}) = r.$$

And, making use of Lemma 8.3.2, we find that

$$\text{rank}(\mathbf{BTBT}) = \text{rank}(\mathbf{TBT}) = \text{rank}(\mathbf{TB}).$$

Thus, $\text{rank}(\mathbf{TB}) = r$.

EXERCISE 4. Let **A** represent an $n \times n$ matrix, and partition **A** as $\mathbf{A} = (\mathbf{A}_1, \mathbf{A}_2)$.

(a) Show that if **A** is invertible and $\mathbf{A}^{-1}$ is partitioned as $\mathbf{A}^{-1} = \begin{pmatrix} \mathbf{B}_1 \\ \mathbf{B}_2 \end{pmatrix}$ (where $\mathbf{B}_1$ has the same number of rows as $\mathbf{A}_1$ has columns), then

$$\mathbf{B}_1\mathbf{A}_1 = \mathbf{I}, \quad \mathbf{B}_1\mathbf{A}_2 = \mathbf{0}, \quad \mathbf{B}_2\mathbf{A}_1 = \mathbf{0}, \quad \mathbf{B}_2\mathbf{A}_2 = \mathbf{I}, \tag{E.1}$$

$$\mathbf{A}_1\mathbf{B}_1 = \mathbf{I} - \mathbf{A}_2\mathbf{B}_2, \quad \mathbf{A}_2\mathbf{B}_2 = \mathbf{I} - \mathbf{A}_1\mathbf{B}_1 . \tag{E.2}$$

(b) Show that if **A** is orthogonal, then

$$\mathbf{A}_1'\mathbf{A}_1 = \mathbf{I}, \quad \mathbf{A}_1'\mathbf{A}_2 = \mathbf{0}, \quad \mathbf{A}_2'\mathbf{A}_1 = \mathbf{0} \quad \mathbf{A}_2'\mathbf{A}_2 = \mathbf{I}, \tag{E.3}$$

$$\mathbf{A}_1\mathbf{A}_1' = \mathbf{I} - \mathbf{A}_2\mathbf{A}_2', \quad \mathbf{A}_2\mathbf{A}_2' = \mathbf{I} - \mathbf{A}_1\mathbf{A}_1' \tag{E.4}$$

Solution. (a) To establish results (E.1) and (E.2), it suffices to observe that if **A** is invertible and $\mathbf{A}^{-1}$ is partitioned as $\mathbf{A}^{-1} = \begin{pmatrix} \mathbf{B}_1 \\ \mathbf{B}_2 \end{pmatrix}$, then

$$\begin{pmatrix} \mathbf{B}_1\mathbf{A}_1 & \mathbf{B}_1\mathbf{A}_2 \\ \mathbf{B}_2\mathbf{A}_1 & \mathbf{B}_2\mathbf{A}_2 \end{pmatrix} = \begin{pmatrix} \mathbf{B}_1 \\ \mathbf{B}_2 \end{pmatrix} (\mathbf{A}_1, \mathbf{A}_2) = \mathbf{A}^{-1}\mathbf{A} = \mathbf{I} = \begin{pmatrix} \mathbf{I} & \mathbf{0} \\ \mathbf{0} & \mathbf{I} \end{pmatrix}$$

and

$$\mathbf{A}_1\mathbf{B}_1 + \mathbf{A}_2\mathbf{B}_2 = (\mathbf{A}_1, \mathbf{A}_2) \begin{pmatrix} \mathbf{B}_1 \\ \mathbf{B}_2 \end{pmatrix} = \mathbf{A}\mathbf{A}^{-1} = \mathbf{I}.$$

(b) Results (E.3) and (E.4) can be obtained as a special case of results (E.1) and (E.2) by observing that if $\mathbf{A}$ is orthogonal, then $\mathbf{A}$ is invertible and $\mathbf{A}^{-1} = \mathbf{A}' = \begin{pmatrix} \mathbf{A}_1' \\ \mathbf{A}_2' \end{pmatrix}$.

EXERCISE 5. Let $\mathbf{A}$ represent an $m \times n$ nonnull matrix of rank r. Show that there exists an $m \times m$ orthogonal matrix whose first r columns span $\mathcal{C}(\mathbf{A})$.

Solution. According to Theorem 6.4.3, there exist r m-dimensional vectors that are orthonormal with respect to the usual inner product for $\mathcal{R}^{m\times 1}$ and form a basis for $\mathcal{C}(\mathbf{A})$. And, according to Theorem 6.4.5, there exist $m - r$ additional m-dimensional vectors, say $\mathbf{b}_{r+1}, \ldots, \mathbf{b}_m$, such that $\mathbf{b}_1, \ldots, \mathbf{b}_r, \mathbf{b}_{r+1}, \ldots, \mathbf{b}_m$ are orthonormal with respect to the usual inner product for $\mathcal{R}^{m\times 1}$ and form a basis for $\mathcal{R}^{m\times 1}$. Clearly, the $m \times m$ matrix whose first, $\ldots$, rth, $(r+1)$th, $\ldots$, mth columns are respectively $\mathbf{b}_1, \ldots, \mathbf{b}_r, \mathbf{b}_{r+1}, \ldots, \mathbf{b}_m$ is orthogonal, and its first r columns span $\mathcal{C}(\mathbf{A})$.

EXERCISE 6. Let $\mathbf{T}$ represent an $n \times n$ triangular matrix. Show that rank$(\mathbf{T})$ is greater than or equal to the number of nonzero diagonal elements in $\mathbf{T}$.

Solution. Suppose that $\mathbf{T}$ has m nonzero diagonal elements and that they are located in the i_1th, i_2th, $\ldots$, i_mth rows of $\mathbf{T}$. Let $\mathbf{T}_*$ represent the $m \times m$ submatrix obtained by striking out all of the rows and columns of $\mathbf{T}$ except the i_1th, i_2th, $\ldots$, i_mth rows and columns. Then, $\mathbf{T}_*$ is triangular, and the diagonal elements of $\mathbf{T}_*$, which are identical to the i_1th, i_2th, $\ldots$, i_mth diagonal elements of $\mathbf{T}$, are all nonzero. Thus, it follows from Corollary 8.5.6 that rank$(\mathbf{T}_*) = m$. We conclude, on the basis of Theorem 4.4.10, that rank$(\mathbf{T}) \geq m$.

EXERCISE 7. Let

$$\mathbf{A} = \begin{pmatrix} \mathbf{A}_{11} & \mathbf{A}_{12} & \ldots & \mathbf{A}_{1r} \\ \mathbf{0} & \mathbf{A}_{22} & \ldots & \mathbf{A}_{2r} \\ \vdots & & \ddots & \vdots \\ \mathbf{0} & \mathbf{0} & & \mathbf{A}_{rr} \end{pmatrix}, \quad \mathbf{B} = \begin{pmatrix} \mathbf{B}_{11} & \mathbf{0} & \ldots & \mathbf{0} \\ \mathbf{B}_{21} & \mathbf{B}_{22} & & \mathbf{0} \\ \vdots & \vdots & \ddots & \\ \mathbf{B}_{r1} & \mathbf{B}_{r2} & \ldots & \mathbf{B}_{rr} \end{pmatrix}$$

represent respectively an $n \times n$ upper block-triangular matrix whose ijth block $\mathbf{A}_{ij}$ is of dimensions $n_i \times n_j$ $(j \geq i = 1, \ldots, r)$ and an $n \times n$ lower block-triangular matrix whose ijth block $\mathbf{B}_{ij}$ is of dimensions $n_i \times n_j$ $(j \leq i = 1, \ldots, r)$.

(a) Assume that $\mathbf{A}$ and $\mathbf{B}$ are invertible, and "recall" that

$$\mathbf{A}^{-1} = \begin{pmatrix} \mathbf{F}_{11} & \mathbf{F}_{12} & \ldots & \mathbf{F}_{1r} \\ \mathbf{0} & \mathbf{F}_{22} & \ldots & \mathbf{F}_{2r} \\ \vdots & & \ddots & \vdots \\ \mathbf{0} & \mathbf{0} & & \mathbf{F}_{rr} \end{pmatrix}, \quad \mathbf{B}^{-1} = \begin{pmatrix} \mathbf{G}_{11} & \mathbf{0} & \ldots & \mathbf{0} \\ \mathbf{G}_{21} & \mathbf{G}_{22} & & \mathbf{0} \\ \vdots & \vdots & \ddots & \\ \mathbf{G}_{r1} & \mathbf{G}_{r2} & \ldots & \mathbf{G}_{rr} \end{pmatrix},$$

where

$$\mathbf{F}_{ii} = \mathbf{A}_{ii}^{-1}, \quad \mathbf{F}_{ij} = -\mathbf{A}_{ii}^{-1} \sum_{k=i+1}^{j} \mathbf{A}_{ik}\mathbf{F}_{kj} \quad (j > i = 1, \ldots, r), \qquad (*)$$

$$\mathbf{G}_{ii} = \mathbf{B}_{ii}^{-1}, \quad \mathbf{G}_{ij} = -\mathbf{B}_{ii}^{-1} \sum_{k=j} \mathbf{B}_{ik}\mathbf{G}_{kj} \quad (j < i = 1, \ldots, r). \qquad (**)$$

Show that the submatrices $\mathbf{F}_{ij}$ $(j \geq i = 1, \ldots, r)$ and $\mathbf{G}_{ij}$ $(j \leq i = 1, \ldots, r)$ are also expressible as

$$\mathbf{F}_{jj} = \mathbf{A}_{jj}^{-1}, \quad \mathbf{F}_{ij} = -(\sum_{k=i}^{j-1} \mathbf{F}_{ik}\mathbf{A}_{kj})\mathbf{A}_{jj}^{-1} \quad (i < j = 1, \ldots, r), \qquad \text{(E.5)}$$

$$\mathbf{G}_{jj} = \mathbf{B}_{jj}^{-1}, \quad \mathbf{G}_{ij} = -(\sum_{k=j+1} \mathbf{G}_{ik}\mathbf{B}_{kj})\mathbf{B}_{jj}^{-1} \quad (i > j = 1, \ldots, r). \qquad \text{(E.6)}$$

Do so by applying results (∗∗) and (∗) to $\mathbf{A}'$ and $\mathbf{B}'$, respectively.

(b) Formulas (∗) form the basis for an algorithm for computing $\mathbf{A}^{-1}$ in r steps: the first step is to compute the matrix $\mathbf{F}_{rr} = \mathbf{A}_{rr}^{-1}$; the $(r - i + 1)$th step is to compute the matrices $\mathbf{F}_{ii}, \mathbf{F}_{i,i+1}, \ldots, \mathbf{F}_{ir}$ from formulas (∗) $(i = r - 1, r - 2, \ldots, 1)$. Similarly, formulas (∗∗) form the basis for an algorithm for computing $\mathbf{B}^{-1}$ in r steps: the first step is to compute the matrix $\mathbf{G}_{11} = \mathbf{B}_{11}^{-1}$; the ith step is to compute the matrices $\mathbf{G}_{i1}, \mathbf{G}_{i2}, \ldots, \mathbf{G}_{ii}$ from formulas (∗∗) $(i = 2, \ldots, r)$. Describe how formulas (E.5) and (E.6) in Part (a) can be used to devise r-step algorithms for computing $\mathbf{A}^{-1}$ and $\mathbf{B}^{-1}$, and indicate how these algorithms differ from those based on formulas (∗) and (∗∗).

Solution. (a) Clearly, it suffices to show that

$$(\mathbf{A}')^{-1} = \begin{pmatrix} \mathbf{F}'_{11} & \mathbf{0} & \ldots & \mathbf{0} \\ \mathbf{F}'_{12} & \mathbf{F}'_{22} & & \mathbf{0} \\ \vdots & \vdots & \ddots & \\ \mathbf{F}'_{1r} & \mathbf{F}'_{2r} & \ldots & \mathbf{F}'_{rr} \end{pmatrix}, \quad (\mathbf{B}')^{-1} = \begin{pmatrix} \mathbf{G}'_{11} & \mathbf{G}'_{21} & \ldots & \mathbf{G}'_{r1} \\ \mathbf{0} & \mathbf{G}'_{22} & \ldots & \mathbf{G}'_{r2} \\ \vdots & & \ddots & \vdots \\ \mathbf{0} & \mathbf{0} & & \mathbf{G}'_{rr} \end{pmatrix},$$

where

$$\mathbf{F}'_{jj} = (\mathbf{A}'_{jj})^{-1}, \quad \mathbf{F}'_{ij} = -(\mathbf{A}'_{jj})^{-1}(\sum_{k=i}^{j-1} \mathbf{A}'_{kj}\mathbf{F}'_{ik}) \quad (i < j = 1, \ldots, r),$$

$$\mathbf{G}'_{jj} = (\mathbf{B}'_{jj})^{-1}, \quad \mathbf{G}'_{ij} = -(\mathbf{B}'_{jj})^{-1}(\sum_{k=j+1} \mathbf{B}'_{kj}\mathbf{G}'_{ik}) \quad (i > j = 1, \ldots, r),$$

or equivalently (after relabeling the i and j subscripts) where

$$\mathbf{F}'_{ii} = (\mathbf{A}'_{ii})^{-1}, \quad \mathbf{F}'_{ji} = -(\mathbf{A}'_{ii})^{-1}(\sum_{k=j}^{i-1} \mathbf{A}'_{ki}\mathbf{F}'_{jk}) \quad (j < i = 1, \ldots, r),$$

$$\mathbf{G}'_{ii} = (\mathbf{B}'_{ii})^{-1}, \quad \mathbf{G}'_{ji} = -(\mathbf{B}'_{ii})^{-1}\Big(\sum_{k=i+1}^{j} \mathbf{B}'_{ki}\mathbf{G}'_{jk}\Big) \quad (j > i = 1, \ldots, r)\,.$$

Upon observing that

$$\mathbf{A}' = \begin{pmatrix} \mathbf{A}'_{11} & \mathbf{0} & \ldots & \mathbf{0} \\ \mathbf{A}'_{12} & \mathbf{A}'_{22} & & \mathbf{0} \\ \vdots & \vdots & \ddots & \\ \mathbf{A}'_{1r} & \mathbf{A}'_{2r} & \ldots & \mathbf{A}'_{rr} \end{pmatrix}, \quad \mathbf{B}' = \begin{pmatrix} \mathbf{B}'_{11} & \mathbf{B}'_{21} & \ldots & \mathbf{B}'_{r1} \\ \mathbf{0} & \mathbf{B}'_{22} & \ldots & \mathbf{B}'_{r2} \\ \vdots & & \ddots & \vdots \\ \mathbf{0} & \mathbf{0} & & \mathbf{B}'_{rr} \end{pmatrix},$$

the validity of these formulas for $(\mathbf{A}')^{-1}$ and $(\mathbf{B}')^{-1}$ is seen to be an immediate consequence of formulas (∗∗) and (∗), respectively.

(b) To compute $\mathbf{A}^{-1}$, we can employ an r-step algorithm, whose first step is to compute $\mathbf{F}_{11} = \mathbf{A}_{11}^{-1}$ and whose jth step is to compute the matrices $\mathbf{F}_{1j}, \mathbf{F}_{2j}, \ldots, \mathbf{F}_{jj}$ from formulas (E.5) ($j = 2, \ldots, r$). To compute $\mathbf{B}^{-1}$, we can employ an r-step algorithm, whose first step is to compute $\mathbf{G}_{rr} = \mathbf{B}_{rr}^{-1}$ and whose $(r - j + 1)$th step is to compute the matrices $\mathbf{G}_{jj}, \mathbf{G}_{j+1,j}, \ldots, \mathbf{G}_{rj}$ from formulas (E.6) ($j = r-1, r-2, \ldots, 1$). These algorithms differ from those based on formulas (∗) and (∗∗) in that they generate $\mathbf{A}^{-1}$ and $\mathbf{B}^{-1}$ one "column" of blocks at a time, rather than one "row" at a time.

9

Generalized Inverses

EXERCISE 1. Let $\mathbf{A}$ represent any $m \times n$ matrix and $\mathbf{B}$ any $m \times p$ matrix. Show that if $\mathbf{AHB} = \mathbf{B}$ for some $n \times m$ matrix $\mathbf{H}$, then $\mathbf{AGB} = \mathbf{B}$ for every generalized inverse $\mathbf{G}$ of $\mathbf{A}$.

Solution. Suppose that $\mathbf{AHB} = \mathbf{B}$ for some $n \times m$ matrix $\mathbf{H}$, and let $\mathbf{G}$ represent an arbitrary generalized inverse of $\mathbf{A}$. Then,

$$\mathbf{AGB} = \mathbf{AGAHB} = \mathbf{AHB} = \mathbf{B}.$$

[Or, alternatively, observe that $\mathbf{HB}$ is a solution to the linear system $\mathbf{AX} = \mathbf{B}$ (in $\mathbf{X}$), so that this linear system is consistent and it follows from Theorem 9.1.2 that $\mathbf{GB}$ is a solution to $\mathbf{AX} = \mathbf{B}$ or equivalently that $\mathbf{AGB} = \mathbf{B}$.]

EXERCISE 2. (a) Let $\mathbf{A}$ represent an $m \times n$ matrix. Show that any $n \times m$ matrix $\mathbf{X}$ such that $\mathbf{A}'\mathbf{AX} = \mathbf{A}'$ is a generalized inverse of $\mathbf{A}$ and similarly that any $n \times m$ matrix $\mathbf{Y}$ such that $\mathbf{AA}'\mathbf{Y}' = \mathbf{A}$ is a generalized inverse of $\mathbf{A}$.

(b) Use Part (a), together with the result that (for any matrix $\mathbf{A}$) the linear system $\mathbf{A}'\mathbf{AX} = \mathbf{A}'$ (in $\mathbf{X}$) is consistent, to conclude that every matrix has at least one generalized inverse.

Solution. (a) Suppose that $\mathbf{X}$ is such that $\mathbf{A}'\mathbf{AX} = \mathbf{A}'$. Then,

$$\mathbf{A}'\mathbf{AXA} = \mathbf{A}'\mathbf{A} = \mathbf{A}'\mathbf{AI},$$

and it follows from Corollary 5.3.3 that

$$\mathbf{AXA} = \mathbf{AI} = \mathbf{A}$$

(i.e., that $\mathbf{X}$ is a generalized inverse of $\mathbf{A}$). Similarly, if $\mathbf{Y}$ is such that $\mathbf{AA'Y'} = \mathbf{A}$, then

$$\mathbf{AA'Y'A'} = \mathbf{AA'} = \mathbf{AA'I},$$

implying that $\mathbf{A'Y'A'} = \mathbf{A'I} = \mathbf{A'}$ and hence that

$$\mathbf{AYA} = (\mathbf{A'Y'A'})' = (\mathbf{A'})' = \mathbf{A}.$$

(b) The consistency (for any matrix $\mathbf{A}$) of the linear system $\mathbf{A'AX} = \mathbf{A'}$ implies that corresponding to any matrix $\mathbf{A}$, there exists a matrix $\mathbf{X}$ such that $\mathbf{A'AX} = \mathbf{A'}$ (and a matrix $\mathbf{Y}$ such that $\mathbf{AA'Y'} = \mathbf{A}$). Thus, it follows from Part (a) that every matrix has at least one generalized inverse.

EXERCISE 3. Let $\mathbf{A}$ represent an $m \times n$ nonnull matrix, let $\mathbf{B}$ represent a matrix of full column rank and $\mathbf{T}$ a matrix of full row rank such that $\mathbf{A} = \mathbf{BT}$, and let $\mathbf{L}$ represent a left inverse of $\mathbf{B}$ and $\mathbf{R}$ a right inverse of $\mathbf{T}$.

(a) Show that the matrix $\mathbf{R}(\mathbf{B'B})^{-1}\mathbf{R'}$ is a generalized inverse of the matrix $\mathbf{A'A}$ and that the matrix $\mathbf{L'}(\mathbf{TT'})^{-1}\mathbf{L}$ is a generalized inverse of the matrix $\mathbf{AA'}$.

(b) Show that if $\mathbf{A}$ is symmetric, then the matrix $\mathbf{R}(\mathbf{TB})^{-1}\mathbf{L}$ is a generalized inverse of the matrix $\mathbf{A}^2$. (If $\mathbf{A}$ is symmetric, then it follows from the result of Exercise 8.3 that $\mathbf{TB}$ is nonsingular.)

Solution. (a) Clearly,

$$\begin{aligned}\mathbf{A'A}[\mathbf{R}(\mathbf{B'B})^{-1}\mathbf{R'}]\mathbf{A'A} &= \mathbf{T'B'BTR}(\mathbf{B'B})^{-1}\mathbf{R'T'B'BT}\\ &= \mathbf{T'B'BI}(\mathbf{B'B})^{-1}(\mathbf{TR})'\mathbf{B'BT}\\ &= \mathbf{T'I'B'BT} = \mathbf{T'B'BT} = \mathbf{A'A},\end{aligned}$$

and similarly

$$\begin{aligned}\mathbf{AA'}[\mathbf{L'}(\mathbf{TT'})^{-1}\mathbf{L}]\mathbf{AA'} &= \mathbf{BTT'B'L'}(\mathbf{TT'})^{-1}\mathbf{LBTT'B'}\\ &= \mathbf{BTT'}(\mathbf{LB})'(\mathbf{TT'})^{-1}\mathbf{ITT'B'}\\ &= \mathbf{BTT'I'B'} = \mathbf{BTT'B'} = \mathbf{AA'}.\end{aligned}$$

(b) Clearly,

$$\begin{aligned}\mathbf{A}^2[\mathbf{R}(\mathbf{TB})^{-1}\mathbf{L}]\mathbf{A}^2 &= \mathbf{BTBTR}(\mathbf{TB})^{-1}\mathbf{LBTBT}\\ &= \mathbf{BTBI}(\mathbf{TB})^{-1}\mathbf{ITBT} = \mathbf{BTBT} = \mathbf{A}^2.\end{aligned}$$

EXERCISE 4. A generalized inverse, say $\mathbf{G}$, of an $m \times n$ matrix $\mathbf{A}$ of rank r can be obtained by an approach consisting of (1) finding r linearly independent rows (of $\mathbf{A}$), say rows $i_1, i_2, \ldots, i_r$ (where $i_1 < i_2 < \cdots < i_r$), and r linearly independent columns, say $j_1, j_2, \ldots, j_r$ (where $j_1 < j_2 < \cdots < j_r$), (2) inverting the submatrix, say $\mathbf{B}_{11}$, of $\mathbf{A}$ obtained by striking out all of the rows and columns (of $\mathbf{A}$) save rows $i_1, i_2, \ldots, i_r$ and columns $j_1, j_2, \ldots, j_r$, and (3) taking (for $s = 1, 2, \ldots, r$ and $t = 1, 2, \ldots, r$) the $j_s i_t$th element of $\mathbf{G}$ to be the stth element of $\mathbf{B}_{11}^{-1}$ and taking its other $(n-r)(m-r)$ elements to be 0. Use this approach to find a generalized inverse of the matrix

$$\mathbf{A} = \begin{pmatrix} 0 & 0 & 0 \\ 0 & 4 & 2 \\ 0 & -2 & -1 \\ 0 & 3 & 3 \\ 0 & 8 & 4 \end{pmatrix}.$$

Solution. The second and third columns of $\mathbf{A}$ are linearly independent (as can be easily verified), implying (since the first column is null) that $r = 2$. Choose, for example, the linearly independent rows and linearly independent columns so that $i_1 = 2$ and $i_2 = 4$ — clearly, the second and fourth rows of $\mathbf{A}$ are linearly independent— and $j_1 = 2$ and $j_2 = 3$. Then,

$$\mathbf{B}_{11} = \begin{pmatrix} 4 & 2 \\ 3 & 3 \end{pmatrix}.$$

Applying formula (8.1.2) for the inverse of a 2×2 nonsingular matrix, we find that

$$\mathbf{B}_{11}^{-1} = (1/6)\begin{pmatrix} 3 & -2 \\ -3 & 4 \end{pmatrix}.$$

Thus, one generalized inverse of $\mathbf{A}$ is

$$\mathbf{G} = (1/6)\begin{pmatrix} 0 & 0 & 0 & 0 & 0 \\ 0 & 3 & 0 & -2 & 0 \\ 0 & -3 & 0 & 4 & 0 \end{pmatrix}.$$

EXERCISE 5. Let $\mathbf{A}$ represent an $m \times n$ nonnull matrix of rank r. Take $\mathbf{B}$ and $\mathbf{K}$ to be nonsingular matrices (of orders m and n, respectively) such that

$$\mathbf{A} = \mathbf{B}\begin{pmatrix} \mathbf{I}_r & \mathbf{0} \\ \mathbf{0} & \mathbf{0} \end{pmatrix}\mathbf{K}$$

(the existence of which is guaranteed). Show (a) that an $n \times m$ matrix $\mathbf{G}$ is a generalized inverse of $\mathbf{A}$ if and only if $\mathbf{G}$ is expressible in the form

$$\mathbf{G} = \mathbf{K}^{-1}\begin{pmatrix} \mathbf{I}_r & \mathbf{U} \\ \mathbf{V} & \mathbf{W} \end{pmatrix}\mathbf{B}^{-1} \tag{E.1}$$

for some $r \times (m-r)$ matrix $\mathbf{U}$, $(n-r) \times r$ matrix $\mathbf{V}$, and $(n-r) \times (m-r)$ matrix $\mathbf{W}$, and (b) that distinct choices for $\mathbf{U}$, $\mathbf{V}$, and/or $\mathbf{W}$ lead to distinct generalized inverses.

Solution. (a) Let $\mathbf{H} = \mathbf{KGB}$, and partition $\mathbf{H}$ as

$$\mathbf{H} = \begin{pmatrix} \mathbf{H}_{11} & \mathbf{H}_{12} \\ \mathbf{H}_{21} & \mathbf{H}_{22} \end{pmatrix},$$

where $\mathbf{H}_{11}$ is of dimensions $r \times r$.

Clearly, $\mathbf{G}$ is a generalized inverse of $\mathbf{A}$ if and only if

$$\mathbf{B}\begin{pmatrix} \mathbf{I}_r & \mathbf{0} \\ \mathbf{0} & \mathbf{0} \end{pmatrix}\mathbf{KGB}\begin{pmatrix} \mathbf{I}_r & \mathbf{0} \\ \mathbf{0} & \mathbf{0} \end{pmatrix}\mathbf{K} = \mathbf{B}\begin{pmatrix} \mathbf{I}_r & \mathbf{0} \\ \mathbf{0} & \mathbf{0} \end{pmatrix}\mathbf{K},$$

or equivalently (since $\mathbf{B}$ and $\mathbf{K}$ are nonsingular) if and only if

$$\begin{pmatrix} \mathbf{I}_r & \mathbf{0} \\ \mathbf{0} & \mathbf{0} \end{pmatrix}\mathbf{H}\begin{pmatrix} \mathbf{I}_r & \mathbf{0} \\ \mathbf{0} & \mathbf{0} \end{pmatrix} = \begin{pmatrix} \mathbf{I}_r & \mathbf{0} \\ \mathbf{0} & \mathbf{0} \end{pmatrix},$$

and hence if and only if $\mathbf{H}_{11} = \mathbf{I}$.

Moreover, if $\mathbf{G}$ is expressible in the form (E.1), then $\mathbf{H} = \begin{pmatrix} \mathbf{I}_r & \mathbf{U} \\ \mathbf{V} & \mathbf{W} \end{pmatrix}$, so that $\mathbf{H}_{11} = \mathbf{I}$. Conversely, if $\mathbf{H}_{11} = \mathbf{I}$, then

$$\mathbf{G} = \mathbf{K}^{-1}\mathbf{H}\mathbf{B}^{-1} = \mathbf{K}^{-1}\begin{pmatrix} \mathbf{I} & \mathbf{H}_{12} \\ \mathbf{H}_{21} & \mathbf{H}_{22} \end{pmatrix}\mathbf{B}^{-1} = \mathbf{K}^{-1}\begin{pmatrix} \mathbf{I} & \mathbf{U} \\ \mathbf{V} & \mathbf{W} \end{pmatrix}\mathbf{B}^{-1},$$

with $\mathbf{U} = \mathbf{H}_{12}$, $\mathbf{V} = \mathbf{H}_{21}$, and $\mathbf{W} = \mathbf{H}_{22}$, so that $\mathbf{G}$ is expressible in the form (E.1). We conclude that $\mathbf{G}$ is a generalized inverse of $\mathbf{A}$ if and only if $\mathbf{G}$ is expressible in the form (E.1).

(b) Let $\mathbf{G}_1 = \mathbf{K}^{-1}\begin{pmatrix} \mathbf{I}_r & \mathbf{U}_1 \\ \mathbf{V}_1 & \mathbf{W}_1 \end{pmatrix}\mathbf{B}^{-1}$ and $\mathbf{G}_2 = \mathbf{K}^{-1}\begin{pmatrix} \mathbf{I}_r & \mathbf{U}_2 \\ \mathbf{V}_2 & \mathbf{W}_2 \end{pmatrix}\mathbf{B}^{-1}$, where $\mathbf{U}_1$ and $\mathbf{U}_2$ are $r \times (m-r)$ matrices, $\mathbf{V}_1$ and $\mathbf{V}_2$ are $(n-r) \times r$ matrices, and $\mathbf{W}_1$ and $\mathbf{W}_2$ are $(n-r) \times (m-r)$ matrices. Then, $\mathbf{G}_1 = \mathbf{G}_2$ only if $\mathbf{KG}_1\mathbf{B} = \mathbf{KG}_2\mathbf{B}$, or equivalently only if

$$\begin{pmatrix} \mathbf{I}_r & \mathbf{U}_1 \\ \mathbf{V}_1 & \mathbf{W}_1 \end{pmatrix} = \begin{pmatrix} \mathbf{I}_r & \mathbf{U}_2 \\ \mathbf{V}_2 & \mathbf{W}_2 \end{pmatrix},$$

that is, only if $\mathbf{U}_2 = \mathbf{U}_1$, $\mathbf{V}_2 = \mathbf{V}_1$, and $\mathbf{W}_2 = \mathbf{W}_1$.

EXERCISE 6. Let k represent a nonzero scalar. For any matrix $\mathbf{A}$, $(1/k)\mathbf{A}^-$ is a generalized inverse of the matrix $k\mathbf{A}$. Generalize this result to partitioned matrices of the form $(\mathbf{A}, k\mathbf{B})$ and $\begin{pmatrix} \mathbf{A} \\ k\mathbf{C} \end{pmatrix}$, where $\mathbf{A}$ is an $m \times n$ matrix, $\mathbf{B}$ an $m \times p$ matrix, and $\mathbf{C}$ a $q \times n$ matrix. Do so by showing (1) that, for any generalized inverse $\begin{pmatrix} \mathbf{G}_1 \\ \mathbf{G}_2 \end{pmatrix}$ of the partitioned matrix $(\mathbf{A}, \mathbf{B})$ (where $\mathbf{G}_{11}$ is of dimensions $n \times m$), $\begin{pmatrix} \mathbf{G}_1 \\ k^{-1}\mathbf{G}_2 \end{pmatrix}$ is a generalized inverse of $(\mathbf{A}, k\mathbf{B})$ and (2) that, for any generalized inverse $(\mathbf{H}_1, \mathbf{H}_2)$

of the partitioned matrix $\begin{pmatrix} \mathbf{A} \\ \mathbf{C} \end{pmatrix}$ (where $\mathbf{H}_1$ is of dimensions $n \times m$) $(\mathbf{H}_1, k^{-1}\mathbf{H}_2)$ is a generalized inverse of $\begin{pmatrix} \mathbf{A} \\ k\mathbf{C} \end{pmatrix}$.

Solution. (1) Clearly,

$$(\mathbf{A}, k\mathbf{B}) = (\mathbf{A}, \mathbf{B}) \begin{pmatrix} \mathbf{I}_n & \mathbf{0} \\ \mathbf{0} & k\mathbf{I}_p \end{pmatrix}.$$

Thus, it follows from Part (2) of Lemma 9.2.4 that the matrix

$$\begin{pmatrix} \mathbf{I}_n & \mathbf{0} \\ \mathbf{0} & k\mathbf{I}_p \end{pmatrix}^{-1} \begin{pmatrix} \mathbf{G}_1 \\ \mathbf{G}_2 \end{pmatrix} = \begin{pmatrix} \mathbf{I}_n & \mathbf{0} \\ \mathbf{0} & k^{-1}\mathbf{I}_p \end{pmatrix} \begin{pmatrix} \mathbf{G}_1 \\ \mathbf{G}_2 \end{pmatrix} = \begin{pmatrix} \mathbf{G}_1 \\ k^{-1}\mathbf{G}_2 \end{pmatrix}$$

is a generalized inverse of $(\mathbf{A}, k\mathbf{B})$.

(2) Similarly,

$$\begin{pmatrix} \mathbf{A} \\ k\mathbf{C} \end{pmatrix} = \begin{pmatrix} \mathbf{I}_m & \mathbf{0} \\ \mathbf{0} & k\mathbf{I}_q \end{pmatrix} \begin{pmatrix} \mathbf{A} \\ \mathbf{C} \end{pmatrix}.$$

Thus, it follows from Part (1) of Lemma 9.2.4 that the matrix

$$(\mathbf{H}_1, \mathbf{H}_2) \begin{pmatrix} \mathbf{I}_m & \mathbf{0} \\ \mathbf{0} & k\mathbf{I}_q \end{pmatrix}^{-1} = (\mathbf{H}_1, \mathbf{H}_2) \begin{pmatrix} \mathbf{I}_m & \mathbf{0} \\ \mathbf{0} & k^{-1}\mathbf{I}_q \end{pmatrix} = (\mathbf{H}_1, k^{-1}\mathbf{H}_2)$$

is a generalized inverse of $\begin{pmatrix} \mathbf{A} \\ k\mathbf{C} \end{pmatrix}$.

EXERCISE 7. Let $\mathbf{T}$ represent an $m \times p$ matrix and $\mathbf{W}$ an $n \times q$ matrix.

(a) Show that, unless $\mathbf{T}$ and $\mathbf{W}$ are both nonsingular, there exist generalized inverses of $\begin{pmatrix} \mathbf{T} & \mathbf{0} \\ \mathbf{0} & \mathbf{W} \end{pmatrix}$ that are not of the form $\begin{pmatrix} \mathbf{T}^- & \mathbf{0} \\ \mathbf{0} & \mathbf{W}^- \end{pmatrix}$. [*Hint*. Make use of the result that, for any $m \times n$ matrix $\mathbf{A}$ and for any particular generalized inverse $\mathbf{G}$ of $\mathbf{A}$, an $n \times m$ matrix $\mathbf{G}^*$ is a generalized inverse of $\mathbf{A}$ if and only if $\mathbf{G}^* = \mathbf{G} + (\mathbf{I} - \mathbf{GA})\mathbf{T} + \mathbf{S}(\mathbf{I} - \mathbf{AG})$ for some $n \times m$ matrices $\mathbf{T}$ and $\mathbf{S}$.]

(b) Take $\mathbf{U}$ to be an $m \times q$ matrix and $\mathbf{V}$ an $n \times p$ matrix such that $\mathcal{C}(\mathbf{U}) \subset \mathcal{C}(\mathbf{T})$ and $\mathcal{R}(\mathbf{V}) \subset \mathcal{R}(\mathbf{T})$, define $\mathbf{Q} = \mathbf{W} - \mathbf{V}\mathbf{T}^-\mathbf{U}$, and "recall" that the partitioned matrix

$$\begin{pmatrix} \mathbf{T}^- + \mathbf{T}^-\mathbf{U}\mathbf{Q}^-\mathbf{V}\mathbf{T}^- & -\mathbf{T}^-\mathbf{U}\mathbf{Q}^- \\ -\mathbf{Q}^-\mathbf{V}\mathbf{T}^- & \mathbf{Q}^- \end{pmatrix} \qquad (*)$$

is a generalized inverse of the matrix $\begin{pmatrix} \mathbf{T} & \mathbf{U} \\ \mathbf{V} & \mathbf{W} \end{pmatrix}$. Generalize the result of Part (a) by showing that, unless $\mathbf{T}$ and $\mathbf{Q}$ are both nonsingular, there exist generalized inverses of $\begin{pmatrix} \mathbf{T} & \mathbf{U} \\ \mathbf{V} & \mathbf{W} \end{pmatrix}$ that are not of the form (∗). [*Hint*. Use Part (a), together with the result that, for any $r \times s$ matrix $\mathbf{B}$, any $r \times r$ nonsingular matrix $\mathbf{A}$, and any $s \times s$

nonsingular matrix $\mathbf{C}$, a matrix $\mathbf{G}$ is a generalized inverse of $\mathbf{ABC}$ if and only if $\mathbf{G} = \mathbf{C}^{-1}\mathbf{H}\mathbf{A}^{-1}$ for some generalized inverse $\mathbf{H}$ of $\mathbf{B}$.]

Solution. (a) Making use of the result cited in the hint, we find that, for any $p \times n$ matrix $\mathbf{X}$ and $q \times m$ matrix $\mathbf{Y}$, the partitioned matrix

$$\begin{pmatrix} \mathbf{T}^- & \mathbf{0} \\ \mathbf{0} & \mathbf{W}^- \end{pmatrix} + \left[\begin{pmatrix} \mathbf{I}_p & \mathbf{0} \\ \mathbf{0} & \mathbf{I}_q \end{pmatrix} - \begin{pmatrix} \mathbf{T}^- & \mathbf{0} \\ \mathbf{0} & \mathbf{W}^- \end{pmatrix}\begin{pmatrix} \mathbf{T} & \mathbf{0} \\ \mathbf{0} & \mathbf{W} \end{pmatrix}\right]\begin{pmatrix} \mathbf{0} & \mathbf{X} \\ \mathbf{0} & \mathbf{0} \end{pmatrix}$$
$$+ \begin{pmatrix} \mathbf{0} & \mathbf{0} \\ \mathbf{Y} & \mathbf{0} \end{pmatrix}\left[\begin{pmatrix} \mathbf{I}_m & \mathbf{0} \\ \mathbf{0} & \mathbf{I}_n \end{pmatrix} - \begin{pmatrix} \mathbf{T} & \mathbf{0} \\ \mathbf{0} & \mathbf{W} \end{pmatrix}\begin{pmatrix} \mathbf{T}^- & \mathbf{0} \\ \mathbf{0} & \mathbf{W}^- \end{pmatrix}\right]$$
$$= \begin{pmatrix} \mathbf{T}^- & (\mathbf{I}_p - \mathbf{T}^-\mathbf{T})\mathbf{X} \\ \mathbf{Y}(\mathbf{I}_m - \mathbf{T}\mathbf{T}^-) & \mathbf{W}^- \end{pmatrix}$$

is a generalized inverse of $\begin{pmatrix} \mathbf{T} & \mathbf{0} \\ \mathbf{0} & \mathbf{W} \end{pmatrix}$. If $\mathbf{T}$ is not nonsingular, then either $\mathbf{I} - \mathbf{T}^-\mathbf{T} \neq \mathbf{0}$ or $\mathbf{I} - \mathbf{T}\mathbf{T}^- \neq \mathbf{0}$, as is evident from Corollary 8.1.2. Moreover, if $\mathbf{I} - \mathbf{T}^-\mathbf{T} \neq \mathbf{0}$, then $\mathbf{X}$ can be chosen so that $(\mathbf{I} - \mathbf{T}^-\mathbf{T})\,\mathbf{X} \neq \mathbf{0}$, and similarly if $\mathbf{I} - \mathbf{T}\mathbf{T}^- \neq \mathbf{0}$, then $\mathbf{Y}$ can be chosen so that $\mathbf{Y}(\mathbf{I} - \mathbf{T}\mathbf{T}^-) \neq \mathbf{0}$. We conclude that if $\mathbf{T}$ is not nonsingular, then there exists a generalized inverse of $\begin{pmatrix} \mathbf{T} & \mathbf{0} \\ \mathbf{0} & \mathbf{W} \end{pmatrix}$ that is not of the form $\begin{pmatrix} \mathbf{T}^- & \mathbf{0} \\ \mathbf{0} & \mathbf{W}^- \end{pmatrix}$. It follows from an analogous argument that if $\mathbf{W}$ is not nonsingular, then again there exists a generalized inverse of $\begin{pmatrix} \mathbf{T} & \mathbf{0} \\ \mathbf{0} & \mathbf{W} \end{pmatrix}$ that is not of the form $\begin{pmatrix} \mathbf{T}^- & \mathbf{0} \\ \mathbf{0} & \mathbf{W}^- \end{pmatrix}$.

(b) Suppose that either $\mathbf{T}$ or $\mathbf{Q}$ is not nonsingular, and assume (for purposes of establishing a contradiction) that every generalized inverse of $\begin{pmatrix} \mathbf{T} & \mathbf{U} \\ \mathbf{V} & \mathbf{W} \end{pmatrix}$ is of the form $(*)$. Upon observing that (in light of Lemma 9.3.5)

$$\begin{pmatrix} \mathbf{T} & \mathbf{0} \\ \mathbf{0} & \mathbf{Q} \end{pmatrix} = \begin{pmatrix} \mathbf{I} & \mathbf{0} \\ -\mathbf{V}\mathbf{T}^- & \mathbf{I} \end{pmatrix}\begin{pmatrix} \mathbf{T} & \mathbf{U} \\ \mathbf{V} & \mathbf{W} \end{pmatrix}\begin{pmatrix} \mathbf{I} & -\mathbf{T}^-\mathbf{U} \\ \mathbf{0} & \mathbf{I} \end{pmatrix}$$

and that (according to Lemma 8.5.2) the matrices $\begin{pmatrix} \mathbf{I} & \mathbf{0} \\ \mathbf{V}\mathbf{T}^- & \mathbf{I} \end{pmatrix}$ and $\begin{pmatrix} \mathbf{I} & -\mathbf{T}^-\mathbf{U} \\ \mathbf{0} & \mathbf{I} \end{pmatrix}$ are nonsingular and upon applying the result cited in the hint and making use of formulas (8.5.6), we find that every generalized inverse of $\begin{pmatrix} \mathbf{T} & \mathbf{0} \\ \mathbf{0} & \mathbf{Q} \end{pmatrix}$ is of the form

$$\begin{pmatrix} \mathbf{I} & -\mathbf{T}^-\mathbf{U} \\ \mathbf{0} & \mathbf{I} \end{pmatrix}^{-1}\begin{pmatrix} \mathbf{T}^- + \mathbf{T}^-\mathbf{U}\mathbf{Q}^-\mathbf{V}\mathbf{T}^- & -\mathbf{T}^-\mathbf{U}\mathbf{Q}^- \\ -\mathbf{Q}^-\mathbf{V}\mathbf{T}^- & \mathbf{Q}^- \end{pmatrix}\begin{pmatrix} \mathbf{I} & \mathbf{0} \\ -\mathbf{V}\mathbf{T}^- & \mathbf{I} \end{pmatrix}^{-1}$$
$$= \begin{pmatrix} \mathbf{I} & \mathbf{T}^-\mathbf{U} \\ \mathbf{0} & \mathbf{I} \end{pmatrix}\begin{pmatrix} \mathbf{T}^- + \mathbf{T}^-\mathbf{U}\mathbf{Q}^-\mathbf{V}\mathbf{T}^- & -\mathbf{T}^-\mathbf{U}\mathbf{Q}^- \\ -\mathbf{Q}^-\mathbf{V}\mathbf{T}^- & \mathbf{Q}^- \end{pmatrix}\begin{pmatrix} \mathbf{I} & \mathbf{0} \\ \mathbf{V}\mathbf{T}^- & \mathbf{I} \end{pmatrix}$$
$$= \begin{pmatrix} \mathbf{T}^- & \mathbf{0} \\ \mathbf{0} & \mathbf{Q}^- \end{pmatrix},$$

which contradicts Part (a). We conclude that, unless $\mathbf{T}$ and $\mathbf{Q}$ are both nonsingular, there exist generalized inverses of $\begin{pmatrix} \mathbf{T} & \mathbf{U} \\ \mathbf{V} & \mathbf{W} \end{pmatrix}$ that are not of the form $(*)$.

EXERCISE 8. Let $\mathbf{T}$ represent an $m \times p$ matrix, $\mathbf{U}$ an $m \times q$ matrix, $\mathbf{V}$ an $n \times p$ matrix, and $\mathbf{W}$ an $n \times q$ matrix, take $\mathbf{A} = \begin{pmatrix} \mathbf{T} & \mathbf{U} \\ \mathbf{V} & \mathbf{W} \end{pmatrix}$, and define $\mathbf{Q} = \mathbf{W} - \mathbf{V}\mathbf{T}^-\mathbf{U}$.

(a) Show that the matrix

$$\mathbf{G} = \begin{pmatrix} \mathbf{T}^- + \mathbf{T}^-\mathbf{U}\mathbf{Q}^-\mathbf{V}\mathbf{T}^- & -\mathbf{T}^-\mathbf{U}\mathbf{Q}^- \\ -\mathbf{Q}^-\mathbf{V}\mathbf{T}^- & \mathbf{Q}^- \end{pmatrix} \qquad (*)$$

is a generalized inverse of the matrix $\mathbf{A}$ if and only if

(1) $(\mathbf{I} - \mathbf{T}\mathbf{T}^-)\mathbf{U}(\mathbf{I} - \mathbf{Q}^-\mathbf{Q}) = \mathbf{0}$,

(2) $(\mathbf{I} - \mathbf{Q}\mathbf{Q}^-)\mathbf{V}(\mathbf{I} - \mathbf{T}^-\mathbf{T}) = \mathbf{0}$, and

(3) $(\mathbf{I} - \mathbf{T}\mathbf{T}^-)\mathbf{U}\mathbf{Q}^-\mathbf{V}(\mathbf{I} - \mathbf{T}^-\mathbf{T}) = \mathbf{0}$.

(b) Verify that (together) the two conditions $\mathcal{C}(\mathbf{U}) \subset \mathcal{C}(\mathbf{T})$ and $\mathcal{R}(\mathbf{V}) \subset \mathcal{R}(\mathbf{T})$ imply Conditions (1) – (3) of Part (a).

(c) Exhibit matrices $\mathbf{T}$, $\mathbf{U}$, $\mathbf{V}$, and $\mathbf{W}$ that (regardless of how the generalized inverses $\mathbf{T}^-$ and $\mathbf{Q}^-$ are chosen) satisfy Conditions (1) – (3) of Part (a) but do not satisfy (both of) the conditions $\mathcal{C}(\mathbf{U}) \subset \mathcal{C}(\mathbf{T})$ and $\mathcal{R}(\mathbf{V}) \subset \mathcal{R}(\mathbf{T})$.

Solution. (a) It is a straightforward exercise to show that

$$\mathbf{AGA} = \begin{pmatrix} \mathbf{T} + (\mathbf{I} - \mathbf{T}\mathbf{T}^-)\mathbf{U}\mathbf{Q}^-\mathbf{V}(\mathbf{I} - \mathbf{T}^-\mathbf{T}) & \mathbf{U} - (\mathbf{I} - \mathbf{T}\mathbf{T}^-)\mathbf{U}(\mathbf{I} - \mathbf{Q}^-\mathbf{Q}) \\ \mathbf{V} - (\mathbf{I} - \mathbf{Q}\mathbf{Q}^-)\mathbf{V}(\mathbf{I} - \mathbf{T}^-\mathbf{T}) & \mathbf{W} \end{pmatrix}.$$

Thus, $\mathbf{AGA} = \mathbf{A}$ if and only if Conditions (1) – (3) are satisfied.

(b) Suppose that $\mathcal{C}(\mathbf{U}) \subset \mathcal{C}(\mathbf{T})$ and $\mathcal{R}(\mathbf{V}) \subset \mathcal{R}(\mathbf{T})$. Then, it follows from Lemma 9.3.5 that $(\mathbf{I} - \mathbf{T}\mathbf{T}^-)\mathbf{U} = \mathbf{0}$ and $\mathbf{V}(\mathbf{I} - \mathbf{T}^-\mathbf{T}) = \mathbf{0}$, and hence that Conditions (1) – (3) are satisfied.

(c) Take $\mathbf{T} = \mathbf{0}$ and $\mathbf{U} = \mathbf{0}$, take $\mathbf{W}$ to be an arbitrary nonnull matrix, and take $\mathbf{V}$ to be any nonnull matrix such that $\mathcal{C}(\mathbf{V}) \subset \mathcal{C}(\mathbf{W})$. Conditions (1) and (3) are clearly satisfied. Moreover, $\mathbf{Q} = \mathbf{W}$, and (in light of Lemma 9.3.5) $(\mathbf{I} - \mathbf{Q}\mathbf{Q}^-)\mathbf{V} = \mathbf{0}$, so that condition (2) is also satisfied. On the other hand, the condition $\mathcal{R}(\mathbf{V}) \subset \mathcal{R}(\mathbf{T})$ is obviously not satisfied.

EXERCISE 9. Suppose that a matrix $\mathbf{A}$ is partitioned as

$$\mathbf{A} = \begin{pmatrix} \mathbf{A}_{11} & \mathbf{A}_{12} & \mathbf{A}_{13} \\ \mathbf{A}_{21} & \mathbf{A}_{22} & \mathbf{A}_{23} \\ \mathbf{A}_{31} & \mathbf{A}_{32} & \mathbf{A}_{33} \end{pmatrix}$$

and that $\mathcal{C}(\mathbf{A}_{12}) \subset \mathcal{C}(\mathbf{A}_{11})$ and $\mathcal{R}(\mathbf{A}_{21}) \subset \mathcal{R}(\mathbf{A}_{11})$. Take $\mathbf{Q}$ to be the Schur complement of $\mathbf{A}_{11}$ in $\mathbf{A}$ relative to $\mathbf{A}_{11}^-$, and partition $\mathbf{Q}$ as

$$\mathbf{Q} = \begin{pmatrix} \mathbf{Q}_{11} & \mathbf{Q}_{12} \\ \mathbf{Q}_{21} & \mathbf{Q}_{22} \end{pmatrix}$$

(where $\mathbf{Q}_{11}$, $\mathbf{Q}_{12}$, $\mathbf{Q}_{21}$, and $\mathbf{Q}_{22}$ are of the same dimensions as $\mathbf{A}_{22}$, $\mathbf{A}_{23}$, $\mathbf{A}_{32}$, and $\mathbf{A}_{33}$, respectively), so that $\mathbf{Q}_{11} = \mathbf{A}_{22} - \mathbf{A}_{21}\mathbf{A}_{11}^-\mathbf{A}_{12}$, $\mathbf{Q}_{12} = \mathbf{A}_{23} - \mathbf{A}_{21}\mathbf{A}_{11}^-\mathbf{A}_{13}$, $\mathbf{Q}_{21} = \mathbf{A}_{32} - \mathbf{A}_{31}\mathbf{A}_{11}^-\mathbf{A}_{12}$, and $\mathbf{Q}_{22} = \mathbf{A}_{33} - \mathbf{A}_{31}\mathbf{A}_{11}^-\mathbf{A}_{13}$. Let

$$\mathbf{G} = \begin{pmatrix} \mathbf{A}_{11}^- + \mathbf{A}_{11}^-\mathbf{A}_{12}\mathbf{Q}_{11}^-\mathbf{A}_{21}\mathbf{A}_{11}^- & -\mathbf{A}_{11}^-\mathbf{A}_{12}\mathbf{Q}_{11}^- \\ -\mathbf{Q}_{11}^-\mathbf{A}_{21}\mathbf{A}_{11}^- & \mathbf{Q}_{11}^- \end{pmatrix}.$$

Define $\mathbf{T} = \begin{pmatrix} \mathbf{A}_{11} & \mathbf{A}_{12} \\ \mathbf{A}_{21} & \mathbf{A}_{22} \end{pmatrix}$, $\mathbf{U} = \begin{pmatrix} \mathbf{A}_{13} \\ \mathbf{A}_{23} \end{pmatrix}$, and $\mathbf{V} = (\mathbf{A}_{31}, \mathbf{A}_{32})$, or equivalently define $\mathbf{T}$, $\mathbf{U}$, and $\mathbf{V}$ to satisfy

$$\mathbf{A} = \begin{pmatrix} \mathbf{T} & \mathbf{U} \\ \mathbf{V} & \mathbf{A}_{33} \end{pmatrix}.$$

Show that (1) $\mathbf{G}$ is a generalized inverse of $\mathbf{T}$; (2) the Schur complement $\mathbf{Q}_{22} - \mathbf{Q}_{21}\mathbf{Q}_{11}^-\mathbf{Q}_{12}$ of $\mathbf{Q}_{11}$ in $\mathbf{Q}$ relative to $\mathbf{Q}_{11}^-$ equals the Schur complement $\mathbf{A}_{33} - \mathbf{VGU}$ of $\mathbf{T}$ in $\mathbf{A}$ relative to $\mathbf{G}$; and (3)

$$\mathbf{GU} = \begin{pmatrix} \mathbf{A}_{11}^-\mathbf{A}_{13} - \mathbf{A}_{11}^-\mathbf{A}_{12}\mathbf{Q}_{11}^-\mathbf{Q}_{12} \\ \mathbf{Q}_{11}^-\mathbf{Q}_{12} \end{pmatrix},$$
$$\mathbf{VG} = (\mathbf{A}_{31}\mathbf{A}_{11}^- - \mathbf{Q}_{21}\mathbf{Q}_{11}^-\mathbf{A}_{21}\mathbf{A}_{11}^-,\ \mathbf{Q}_{21}\mathbf{Q}_{11}^-).$$

(b) Let $\mathbf{A}$ represent an $n \times n$ matrix (where $n \geq 2$), let $n_1, \ldots, n_k$ represent positive integers such that $n_1 + \cdots + n_k = n$ (where $k \geq 2$), and (for $i = 1, \ldots, k$) let $n_i^* = n_1 + \cdots + n_i$. Define (for $i = 1, \ldots, k$) $\mathbf{A}_i$ to be the leading principal submatrix of $\mathbf{A}$ of order n_i^* and define (for $i = 1, \ldots, k-1$) $\mathbf{U}_i$ to be the $n_i^* \times (n - n_i^*)$ matrix obtained by striking out all of the rows and columns of $\mathbf{A}$ except the first n_i^* rows and the last $n - n_i^*$ columns, $\mathbf{V}_i$ to be the $(n - n_i^*) \times n_i^*$ matrix obtained by striking out all of the rows and columns of $\mathbf{A}$ except the last $n - n_i^*$ rows and first n_i^* columns, and $\mathbf{W}_i$ to be the $(n - n_i^*) \times (n - n_i^*)$ submatrix obtained by striking out all of the rows and columns of $\mathbf{A}$ except the last $n - n_i^*$ rows and columns, so that (for $i = 1, \ldots, k - 1$)

$$\mathbf{A} = \begin{pmatrix} \mathbf{A}_i & \mathbf{U}_i \\ \mathbf{V}_i & \mathbf{W}_i \end{pmatrix}.$$

Suppose that (for $i = 1, \ldots, k - 1$) $\mathcal{C}(\mathbf{U}_i) \subset \mathcal{C}(\mathbf{A}_i)$ and $\mathcal{R}(\mathbf{V}_i) \subset \mathcal{R}(\mathbf{A}_i)$. Let

$$\mathbf{B}^{(i)} = \begin{pmatrix} \mathbf{B}_{11}^{(i)} & \mathbf{B}_{12}^{(i)} \\ \mathbf{B}_{21}^{(i)} & \mathbf{B}_{22}^{(i)} \end{pmatrix}$$

$(i = 1, \ldots, k-1)$ and $\mathbf{B}^{(k)} = \mathbf{B}_{11}^{(k)}$, where $\mathbf{B}_{11}^{(1)} = \mathbf{A}_1^-$, $\mathbf{B}_{12}^{(1)} = \mathbf{A}_1^-\mathbf{U}_1$, $\mathbf{B}_{21}^{(1)} = \mathbf{V}_1\mathbf{A}_1^-$, and $\mathbf{B}_{22}^{(1)} = \mathbf{W}_1 - \mathbf{V}_1\mathbf{A}_1^-\mathbf{U}_1$ and where (for $i \geq 2$) $\mathbf{B}_{11}^{(i)}$, $\mathbf{B}_{12}^{(i)}$, $\mathbf{B}_{21}^{(i)}$, and $\mathbf{B}_{22}^{(i)}$ are defined recursively by partitioning $\mathbf{B}_{12}^{(i-1)}$, $\mathbf{B}_{21}^{(i-1)}$, and $\mathbf{B}_{22}^{(i-1)}$ as

$$\mathbf{B}_{12}^{(i-1)} = (\mathbf{X}_1^{(i-1)}, \mathbf{X}_2^{(i-1)}),$$

$$\mathbf{B}_{21}^{(i-1)} = \begin{pmatrix} \mathbf{Y}_1^{(i-1)} \\ \mathbf{Y}_2^{(i-1)} \end{pmatrix}, \quad \mathbf{B}_{22}^{(i-1)} = \begin{pmatrix} \mathbf{Q}_{11}^{(i-1)} & \mathbf{Q}_{12}^{(i-1)} \\ \mathbf{Q}_{21}^{(i-1)} & \mathbf{Q}_{22}^{(i-1)} \end{pmatrix}$$

(in such a way that $\mathbf{X}_1^{(i-1)}$ has n_i columns, $\mathbf{Y}_1^{(i-1)}$ has n_i rows, and $\mathbf{Q}_{11}^{(i-1)}$ is of dimensions $n_i \times n_i$) and (using $\mathbf{Q}_{11}^{-(i-1)}$ to represent a generalized inverse of $\mathbf{Q}_{11}^{(i-1)}$) by taking

$$\mathbf{B}_{11}^{(i)} = \begin{pmatrix} \mathbf{B}_{11}^{(i-1)} + \mathbf{X}_1^{(i-1)}\mathbf{Q}_{11}^{-(i-1)}\mathbf{Y}_1^{(i-1)} & -\mathbf{X}_1^{(i-1)}\mathbf{Q}_{11}^{-(i-1)} \\ -\mathbf{Q}_{11}^{-(i-1)}\mathbf{Y}_1^{(i-1)} & \mathbf{Q}_{11}^{-(i-1)} \end{pmatrix},$$

$$\mathbf{B}_{12}^{(i)} = \begin{pmatrix} \mathbf{X}_2^{(i-1)} - \mathbf{X}_1^{(i-1)}\mathbf{Q}_{11}^{-(i-1)}\mathbf{Q}_{12}^{(i-1)} \\ \mathbf{Q}_{11}^{-(i-1)}\mathbf{Q}_{12}^{(i-1)} \end{pmatrix},$$

$$\mathbf{B}_{21}^{(i)} = (\mathbf{Y}_2^{(i-1)} - \mathbf{Q}_{21}^{(i-1)}\mathbf{Q}_{11}^{-(i-1)}\mathbf{Y}_1^{(i-1)}, \ \mathbf{Q}_{21}^{(i-1)}\mathbf{Q}_1^{-(i-1)}),$$

$$\mathbf{B}_{22}^{(i)} = \mathbf{Q}_{22}^{(i-1)} - \mathbf{Q}_{21}^{(i-1)}\mathbf{Q}_{11}^{-(i-1)}\mathbf{Q}_{12}^{(i-1)}.$$

Show that (1) $\mathbf{B}_{11}^{(i)}$ is a generalized inverse of $\mathbf{A}_i$ $(i = 1, \ldots, k)$; (2) $\mathbf{B}_{22}^{(i)}$ is the Schur complement of $\mathbf{A}_i$ in $\mathbf{A}$ relative to $\mathbf{B}_{11}^{(i)}$ $(i = 1, \ldots, k-1)$; (3) $\mathbf{B}_{12}^{(i)} = \mathbf{B}_{11}^{(i)}\mathbf{U}_i$ and $\mathbf{B}_{21}^{(i)} = \mathbf{V}_i\mathbf{B}_{11}^{(i)}$ $(i = 1, \ldots, k-1)$.

[*Note.* The recursive formulas given in Part (b) for the sequence of matrices $\mathbf{B}^{(1)}$, ..., $\mathbf{B}^{(k-1)}$, $\mathbf{B}^{(k)}$ can be used to generate $\mathbf{B}^{(k-1)}$ in $k-1$ steps or to generate $\mathbf{B}^{(k)}$ in k steps — the formula for generating $\mathbf{B}^{(i)}$ from $\mathbf{B}^{(i-1)}$ involves a generalized inverse of the $n_i \times n_i$ matrix $\mathbf{Q}_{11}^{(i-1)}$. The various parts of $\mathbf{B}^{(k-1)}$ consist of a generalized inverse $\mathbf{B}_{11}^{(k-1)}$ of $\mathbf{A}_{k-1}$, the Schur complement $\mathbf{B}_{22}^{(k-1)}$ of $\mathbf{A}_{k-1}$ in $\mathbf{A}$ relative to $\mathbf{B}_{11}^{(k-1)}$, a solution $\mathbf{B}_{12}^{(k-1)}$ of the linear system $\mathbf{A}_{k-1}\mathbf{X} = \mathbf{U}_{k-1}$ (in $\mathbf{X}$), and a solution $\mathbf{B}_{21}^{(k-1)}$ of the linear system $\mathbf{Y}\mathbf{A}_{k-1} = \mathbf{V}_{k-1}$ (in $\mathbf{Y}$). The matrix $\mathbf{B}^{(k)}$ is a generalized inverse of $\mathbf{A}$. In the special case where $n_i = 1$, the process of generating the elements of the $n \times n$ matrix $\mathbf{B}^{(i)}$ from those of the $n \times n$ matrix $\mathbf{B}^{(i-1)}$ is called a *sweep operation* — see, e.g., Goodnight (1979).]

Solution. (a) (1) That $\mathbf{G}$ is a generalized inverse of $\mathbf{T}$ is evident upon setting $\mathbf{T} = \mathbf{A}_{11}, \mathbf{U} = \mathbf{A}_{12}, \mathbf{V} = \mathbf{A}_{21}$, and $\mathbf{W} = \mathbf{A}_{22}$ in formula (6.2a) of Theorem 9.6.1 [or equivalently in formula $(*)$ of Exercise 7 or 8]—the conditions $\mathcal{C}(\mathbf{A}_{12}) \subset \mathcal{C}(\mathbf{A}_{11})$ and $\mathcal{R}(\mathbf{A}_{21}) \subset \mathcal{R}(\mathbf{A}_{11})$ insure that this formula is applicable.

(2)

$\mathbf{Q}_{22} - \mathbf{Q}_{21}\mathbf{Q}_{11}^-\mathbf{Q}_{12}$

$$= \mathbf{A}_{33} - \mathbf{A}_{31}\mathbf{A}_{11}^{-}\mathbf{A}_{13} - (\mathbf{A}_{32} - \mathbf{A}_{31}\mathbf{A}_{11}^{-}\mathbf{A}_{12})\mathbf{Q}_{11}^{-}(\mathbf{A}_{23} - \mathbf{A}_{21}\mathbf{A}_{11}^{-}\mathbf{A}_{13})$$
$$= \mathbf{A}_{33} - \mathbf{A}_{31}(\mathbf{A}_{11}^{-} + \mathbf{A}_{11}^{-}\mathbf{A}_{12}\mathbf{Q}_{11}^{-}\mathbf{A}_{21}\mathbf{A}_{11}^{-})\mathbf{A}_{13}$$
$$-\mathbf{A}_{31}(-\mathbf{A}_{11}^{-}\mathbf{A}_{12}\mathbf{Q}_{11}^{-})\mathbf{A}_{23} - \mathbf{A}_{32}(-\mathbf{Q}_{11}^{-}\mathbf{A}_{21}\mathbf{A}_{11}^{-})\mathbf{A}_{13} - \mathbf{A}_{32}\mathbf{Q}_{11}^{-}\mathbf{A}_{23}$$
$$= \mathbf{A}_{33} - \mathbf{VGU}.$$

(3) Partition $\mathbf{GU}$ as $\mathbf{GU} = \begin{pmatrix}\mathbf{X}_1\\ \mathbf{X}_2\end{pmatrix}$ and $\mathbf{VG}$ as $\mathbf{VG} = (\mathbf{Y}_1, \mathbf{Y}_2)$ (where $\mathbf{X}_1$, $\mathbf{X}_2$, $\mathbf{Y}_1$, and $\mathbf{Y}_2$ are of the same dimensions as $\mathbf{A}_{13}$, $\mathbf{A}_{23}$, $\mathbf{A}_{31}$, and $\mathbf{A}_{32}$, respectively). Then,

$$\begin{aligned}\mathbf{X}_1 &= (\mathbf{A}_{11}^{-} + \mathbf{A}_{11}^{-}\mathbf{A}_{12}\mathbf{Q}_{11}^{-}\mathbf{A}_{21}\mathbf{A}_{11}^{-})\mathbf{A}_{13} + (-\mathbf{A}_{11}^{-}\mathbf{A}_{12}\mathbf{Q}_{11}^{-})\mathbf{A}_{23}\\ &= \mathbf{A}_{11}^{-}\mathbf{A}_{13} - \mathbf{A}_{11}^{-}\mathbf{A}_{12}\mathbf{Q}_{11}^{-}(\mathbf{A}_{23} - \mathbf{A}_{21}\mathbf{A}_{11}^{-}\mathbf{A}_{13})\\ &= \mathbf{A}_{11}^{-}\mathbf{A}_{13} - \mathbf{A}_{11}^{-}\mathbf{A}_{12}\mathbf{Q}_{11}^{-}\mathbf{Q}_{12}\end{aligned}$$

and

$$\mathbf{X}_2 = (-\mathbf{Q}_{11}^{-}\mathbf{A}_{21}\mathbf{A}_{11}^{-})\mathbf{A}_{13} + \mathbf{Q}_{11}^{-}\mathbf{A}_{23} = \mathbf{Q}_{11}^{-}(\mathbf{A}_{23} - \mathbf{A}_{21}\mathbf{A}_{11}^{-}\mathbf{A}_{13}) = \mathbf{Q}_{11}^{-}\mathbf{A}_{12}.$$

It can be established in similar fashion that $\mathbf{Y}_1 = \mathbf{A}_{31}\mathbf{A}_{11}^{-} - \mathbf{Q}_{21}\mathbf{Q}_{11}^{-}\mathbf{A}_{21}\mathbf{A}_{11}^{-}$ and $\mathbf{Y}_2 = \mathbf{Q}_{21}\mathbf{Q}_{11}^{-}$.

(b) The proof of results (1), (2), and (3) is by mathematical induction. By definition, $\mathbf{B}_{11}^{(1)}$ is a generalized inverse of $\mathbf{A}_1$, $\mathbf{B}_{22}^{(1)}$ is the Schur complement of $\mathbf{A}_1$ in $\mathbf{A}$ relative to $\mathbf{B}_{11}^{(1)}$, and $\mathbf{B}_{12}^{(1)} = \mathbf{B}_{11}^{(1)}\mathbf{U}_1$ and $\mathbf{B}_{21}^{(1)} = \mathbf{V}_1\mathbf{B}_{11}^{(1)}$.

Suppose now that $\mathbf{B}_{11}^{(i-1)}$ is a generalized inverse of $\mathbf{A}_{i-1}$, that $\mathbf{B}_{22}^{(i-1)}$ is the Schur complement of $\mathbf{A}_{i-1}$ in $\mathbf{A}$ relative to $\mathbf{B}_{11}^{(i-1)}$, and that $\mathbf{B}_{12}^{(i-1)} = \mathbf{B}_{11}^{(i-1)}\mathbf{U}_{i-1}$ and $\mathbf{B}_{21}^{(i-1)} = \mathbf{V}_{i-1}\mathbf{B}_{11}^{(i-1)}$ (where $2 \le i \le k-1$). Partition $\mathbf{A}_i$, $\mathbf{U}_i$, and $\mathbf{V}_i$ as

$$\mathbf{A}_i = \begin{pmatrix}\mathbf{A}_{i-1} & \mathbf{A}_{12}^{(i-1)}\\ \mathbf{A}_{21}^{(i-1)} & \mathbf{A}_{22}^{(i-1)}\end{pmatrix}, \quad \mathbf{U}_i = \begin{pmatrix}\mathbf{A}_{13}^{(i-1)}\\ \mathbf{A}_{23}^{(i-1)}\end{pmatrix}, \quad \text{and } \mathbf{V}_i = (\mathbf{A}_{31}^{(i-1)}, \mathbf{A}_{32}^{(i-1)})$$

(where $\mathbf{A}_{13}^{(i-1)}$ has n_{i-1}^* rows and $\mathbf{A}_{31}^{(i-1)}$ has n_{i-1}^* columns). Then, clearly,

$$\mathbf{U}_{i-1} = (\mathbf{A}_{12}^{(i-1)}, \mathbf{A}_{13}^{(i-1)}) \quad \text{and} \quad \mathbf{V}_{i-1} = \begin{pmatrix}\mathbf{A}_{21}^{(i-1)}\\ \mathbf{A}_{31}^{(i-1)}\end{pmatrix},$$

so that $\mathbf{X}_1^{(i-1)} = \mathbf{B}_{11}^{(i-1)}\mathbf{A}_{12}^{(i-1)}$, $\mathbf{X}_2^{(i-1)} = \mathbf{B}_{11}^{(i-1)}\mathbf{A}_{13}^{(i-1)}$, $\mathbf{Y}_1^{(i-1)} = \mathbf{A}_{21}^{(i-1)}\mathbf{B}_{11}^{(i-1)}$, and $\mathbf{Y}_2^{(i-1)} = \mathbf{A}_{31}^{(i-1)}\mathbf{B}_{11}^{(i-1)}$. Thus, it follows from Part (a) that $\mathbf{B}_{11}^{(i)}$ is a generalized inverse of $\mathbf{A}_i$, that $\mathbf{B}_{22}^{(i)}$ is the Schur complement of $\mathbf{A}_i$ in $\mathbf{A}$ relative to $\mathbf{B}_{11}^{(i)}$, and that $\mathbf{B}_{12}^{(i)} = \mathbf{B}_{11}^{(i)}\mathbf{U}_i$ and $\mathbf{B}_{21}^{(i)} = \mathbf{V}_i\mathbf{B}_{11}^{(i)}$.

We conclude (based on mathematical induction) that (for $i = 1, \ldots, k-1$) $\mathbf{B}_{11}^{(i)}$ is a generalized inverse of $\mathbf{A}_i$, $\mathbf{B}_{22}^{(i)}$ is the Schur complement of $\mathbf{A}_i$ in $\mathbf{A}$ relative to

$\mathbf{B}_{11}^{(i)}$, and $\mathbf{B}_{12}^{(i)} = \mathbf{B}_{11}^{(i)}\mathbf{U}_i$ and $\mathbf{B}_{21}^{(i)} = \mathbf{V}_i\mathbf{B}_{11}^{(i)}$. Moreover, since $\mathbf{B}_{11}^{(k-1)}$ is a generalized inverse of $\mathbf{A}_{k-1}$, since $\mathbf{Q}_{11}^{(k-1)} = \mathbf{B}_{22}^{(k-1)}$ and $\mathbf{B}_{22}^{(k-1)}$ is the Schur complement of $\mathbf{A}_{k-1}$ in $\mathbf{A}$ relative to $\mathbf{B}_{11}^{(k-1)}$, since $\mathbf{X}_1^{(k-1)} = \mathbf{B}_{12}^{(k-1)} = \mathbf{B}_{11}^{(k-1)}\mathbf{U}_{k-1}$ and $\mathbf{Y}_1^{(k-1)} = \mathbf{B}_{21}^{(k-1)} = \mathbf{V}_{k-1}\mathbf{B}_{11}^{(k-1)}$, and since $\mathbf{A}_k = \mathbf{A}$, it is evident upon setting $\mathbf{T} = \mathbf{A}_{k-1}$, $\mathbf{U} = \mathbf{U}_{k-1}$, $\mathbf{V} = \mathbf{V}_{k-1}$, and $\mathbf{W} = \mathbf{W}_{k-1}$ in formula (6.2a) of Theorem 9.6.1 [or equivalently in formula (*) of Exercise (7) or (8)] that $\mathbf{B}_{11}^{(k)}$ is a generalized inverse of $\mathbf{A}_k$.

EXERCISE 10. Let $\mathbf{T}$ represent an $m \times p$ matrix and $\mathbf{W}$ an $n \times q$ matrix, and let $\mathbf{G} = \begin{pmatrix} \mathbf{G}_{11} & \mathbf{G}_{12} \\ \mathbf{G}_{21} & \mathbf{G}_{22} \end{pmatrix}$ (where $\mathbf{G}_{11}$ is of dimensions $p \times m$) represent an arbitrary generalized inverse of the $(m+n) \times (p+q)$ block-diagonal matrix $\mathbf{A} = \begin{pmatrix} \mathbf{T} & \mathbf{0} \\ \mathbf{0} & \mathbf{W} \end{pmatrix}$. Show that $\mathbf{G}_{11}$ is a generalized inverse of $\mathbf{T}$ and $\mathbf{G}_{22}$ a generalized inverse of $\mathbf{W}$. Show also that $\mathbf{T}\mathbf{G}_{12}\mathbf{W} = \mathbf{0}$ and $\mathbf{W}\mathbf{G}_{21}\mathbf{T} = \mathbf{0}$.

Solution. Clearly,

$$\begin{pmatrix} \mathbf{T} & \mathbf{0} \\ \mathbf{0} & \mathbf{W} \end{pmatrix} = \mathbf{A} = \mathbf{AGA} = \begin{pmatrix} \mathbf{T}\mathbf{G}_{11} & \mathbf{T}\mathbf{G}_{12} \\ \mathbf{W}\mathbf{G}_{21} & \mathbf{W}\mathbf{G}_{22} \end{pmatrix}\mathbf{A} = \begin{pmatrix} \mathbf{T}\mathbf{G}_{11}\mathbf{T} & \mathbf{T}\mathbf{G}_{12}\mathbf{W} \\ \mathbf{W}\mathbf{G}_{21}\mathbf{T} & \mathbf{W}\mathbf{G}_{22}\mathbf{W} \end{pmatrix}.$$

Thus, $\mathbf{T}\mathbf{G}_{11}\mathbf{T} = \mathbf{T}$ (i.e., $\mathbf{G}_{11}$ is a generalized inverse of $\mathbf{T}$), $\mathbf{W}\mathbf{G}_{22}\mathbf{W} = \mathbf{W}$ (i.e., $\mathbf{G}_{22}$ is a generalized inverse of $\mathbf{W}$), $\mathbf{T}\mathbf{G}_{12}\mathbf{W} = \mathbf{0}$, and $\mathbf{W}\mathbf{G}_{21}\mathbf{T} = \mathbf{0}$.

EXERCISE 11. Let $\mathbf{T}$ represent an $m \times p$ matrix, $\mathbf{U}$ an $m \times q$ matrix, $\mathbf{V}$ an $n \times p$ matrix, and $\mathbf{W}$ an $n \times q$ matrix, and define $\mathbf{Q} = \mathbf{W} - \mathbf{V}\mathbf{T}^-\mathbf{U}$. Suppose that $\mathcal{C}(\mathbf{U}) \subset \mathcal{C}(\mathbf{T})$ and $\mathcal{R}(\mathbf{V}) \subset \mathcal{R}(\mathbf{T})$. Prove that for any generalized inverse $\mathbf{G} = \begin{pmatrix} \mathbf{G}_{11} & \mathbf{G}_{12} \\ \mathbf{G}_{21} & \mathbf{G}_{22} \end{pmatrix}$ of the partitioned matrix $\begin{pmatrix} \mathbf{T} & \mathbf{U} \\ \mathbf{V} & \mathbf{W} \end{pmatrix}$, the $(q \times n)$ submatrix $\mathbf{G}_{22}$ is a generalized inverse of $\mathbf{Q}$. Do so via an approach that consists of showing that

$$\begin{pmatrix} \mathbf{I} & \mathbf{0} \\ -\mathbf{V}\mathbf{T}^- & \mathbf{I} \end{pmatrix}\begin{pmatrix} \mathbf{T} & \mathbf{U} \\ \mathbf{V} & \mathbf{W} \end{pmatrix}\begin{pmatrix} \mathbf{I} & -\mathbf{T}^-\mathbf{U} \\ \mathbf{0} & \mathbf{I} \end{pmatrix} = \begin{pmatrix} \mathbf{T} & \mathbf{0} \\ \mathbf{0} & \mathbf{Q} \end{pmatrix}$$

and of then using the result cited in the hint for Part (b) of Exercise 7, along with the result of Exercise 10.

Solution. Observing (in light of Lemma 9.3.5) that $\mathbf{V} - \mathbf{V}\mathbf{T}^-\mathbf{T} = \mathbf{0}$ and that $\mathbf{U} - \mathbf{T}\mathbf{T}^-\mathbf{U} = \mathbf{0}$, we find that

$$\begin{aligned} \begin{pmatrix} \mathbf{I} & \mathbf{0} \\ -\mathbf{V}\mathbf{T}^- & \mathbf{I} \end{pmatrix}\begin{pmatrix} \mathbf{T} & \mathbf{U} \\ \mathbf{V} & \mathbf{W} \end{pmatrix}\begin{pmatrix} \mathbf{I} & -\mathbf{T}^-\mathbf{U} \\ \mathbf{0} & \mathbf{I} \end{pmatrix} &= \begin{pmatrix} \mathbf{T} & \mathbf{U} \\ \mathbf{0} & \mathbf{Q} \end{pmatrix}\begin{pmatrix} \mathbf{I} & -\mathbf{T}^-\mathbf{U} \\ \mathbf{0} & \mathbf{I} \end{pmatrix} \\ &= \begin{pmatrix} \mathbf{T} & \mathbf{0} \\ \mathbf{0} & \mathbf{Q} \end{pmatrix}. \end{aligned}$$

Moreover, recalling Lemma 8.5.2, it follows from the result cited in the hint for Part (b) of Exercise 7, or (equivalently) from Part (3) of Lemma 9.2.4, that the matrix

$$\begin{pmatrix} \mathbf{I} & -\mathbf{T}^-\mathbf{U} \\ \mathbf{0} & \mathbf{I} \end{pmatrix}^{-1} \begin{pmatrix} \mathbf{G}_{11} & \mathbf{G}_{12} \\ \mathbf{G}_{21} & \mathbf{G}_{22} \end{pmatrix} \begin{pmatrix} \mathbf{I} & \mathbf{0} \\ -\mathbf{V}\mathbf{T}^- & \mathbf{I} \end{pmatrix}^{-1}$$
$$= \begin{pmatrix} \mathbf{I} & \mathbf{T}^-\mathbf{U} \\ \mathbf{0} & \mathbf{I} \end{pmatrix} \begin{pmatrix} \mathbf{G}_{11} & \mathbf{G}_{12} \\ \mathbf{G}_{21} & \mathbf{G}_{22} \end{pmatrix} \begin{pmatrix} \mathbf{I} & \mathbf{0} \\ \mathbf{V}\mathbf{T}^- & \mathbf{I} \end{pmatrix}$$
$$= \begin{pmatrix} \mathbf{G}_{11} + \mathbf{T}^-\mathbf{U}\mathbf{G}_{21} + \mathbf{G}_{12}\mathbf{V}\mathbf{T}^- + \mathbf{T}^-\mathbf{U}\mathbf{G}_{22}\mathbf{V}\mathbf{T}^- & \mathbf{G}_{12} + \mathbf{T}^-\mathbf{U}\mathbf{G}_{22} \\ \mathbf{G}_{21} + \mathbf{G}_{22}\mathbf{V}\mathbf{T}^- & \mathbf{G}_{22} \end{pmatrix}$$

is a generalized inverse of the matrix $\begin{pmatrix} \mathbf{T} & \mathbf{0} \\ \mathbf{0} & \mathbf{Q} \end{pmatrix}$. Based on the result of Exercise 10, we conclude that $\mathbf{G}_{22}$ is a generalized inverse of $\mathbf{Q}$..

EXERCISE 12. Let $\mathbf{T}$ represent an $m \times p$ matrix, $\mathbf{U}$ an $m \times q$ matrix, $\mathbf{V}$ an $n \times p$ matrix, and $\mathbf{W}$ an $n \times q$ matrix, and take $\mathbf{A} = \begin{pmatrix} \mathbf{T} & \mathbf{U} \\ \mathbf{V} & \mathbf{W} \end{pmatrix}$.

(a) Define $\mathbf{Q} = \mathbf{W} - \mathbf{V}\mathbf{T}^-\mathbf{U}$, and let $\mathbf{G} = \begin{pmatrix} \mathbf{G}_{11} & \mathbf{G}_{12} \\ \mathbf{G}_{21} & \mathbf{G}_{22} \end{pmatrix}$, where $\mathbf{G}_{11} = \mathbf{T}^- + \mathbf{T}^-\mathbf{U}\mathbf{Q}^-\mathbf{V}\mathbf{T}^-$, $\mathbf{G}_{12} = -\mathbf{T}^-\mathbf{U}\mathbf{Q}^-$, $\mathbf{G}_{21} = -\mathbf{Q}^-\mathbf{V}\mathbf{T}^-$, and $\mathbf{G}_{22} = \mathbf{Q}^-$. Show that the matrix

$$\mathbf{G}_{11} - \mathbf{G}_{12}\mathbf{G}_{22}^-\mathbf{G}_{21}$$

is a generalized inverse of $\mathbf{T}$. {*Note*. If the conditions $\mathcal{C}(\mathbf{U}) \subset \mathcal{C}(\mathbf{T})$ and $\mathcal{R}(\mathbf{V}) \subset \mathcal{R}(\mathbf{T})$ are satisfied or more generally if Conditions (1) – (3) of Part (a) of Exercise 8 are satisfied, then $\mathbf{G}$ is a generalized inverse of $\mathbf{A}$}.

(b) Show by example that, for some values of $\mathbf{T}$, $\mathbf{U}$, $\mathbf{V}$, and $\mathbf{W}$, there exists a generalized inverse $\mathbf{G} = \begin{pmatrix} \mathbf{G}_{11} & \mathbf{G}_{12} \\ \mathbf{G}_{21} & \mathbf{G}_{22} \end{pmatrix}$ of $\mathbf{A}$ (where $\mathbf{G}_{11}$ is of dimensions $p \times m$, $\mathbf{G}_{12}$ of dimensions $p \times n$, $\mathbf{G}_{21}$ of dimensions $q \times m$, and $\mathbf{G}_{22}$ of dimensions $q \times n$) such that the matrix

$$\mathbf{G}_{11} - \mathbf{G}_{12}\mathbf{G}_{22}^-\mathbf{G}_{21}$$

is *not* a generalized inverse of $\mathbf{T}$.

Solution. (a) We find that

$$\begin{aligned} \mathbf{G}_{11} - \mathbf{G}_{12}\mathbf{G}_{22}^-\mathbf{G}_{21} &= \mathbf{T}^- + \mathbf{T}^-\mathbf{U}\mathbf{Q}^-\mathbf{V}\mathbf{T}^- - \mathbf{T}^-\mathbf{U}\mathbf{Q}^-(\mathbf{Q}^-)^-\mathbf{Q}^-\mathbf{V}\mathbf{T}^- \\ &= \mathbf{T}^- + \mathbf{T}^-\mathbf{U}\mathbf{Q}^-\mathbf{V}\mathbf{T}^- - \mathbf{T}^-\mathbf{U}\mathbf{Q}^-\mathbf{V}\mathbf{T}^- \\ &= \mathbf{T}^-. \end{aligned}$$

(b) Take $\mathbf{T} = \begin{pmatrix} \mathbf{I} & \mathbf{0} \\ \mathbf{0} & \mathbf{0} \end{pmatrix}$, $\mathbf{U} = \mathbf{0}$, $\mathbf{V} = \mathbf{0}$, and $\mathbf{W} = \mathbf{0}$, and take

$$\mathbf{G} = \begin{pmatrix} \mathbf{G}_{11} & \mathbf{G}_{12} \\ \mathbf{G}_{21} & \mathbf{G}_{22} \end{pmatrix}.$$

where $\mathbf{G}_{11} = \mathbf{T}^-$, $\mathbf{G}_{22} = \mathbf{0}$, and $\mathbf{G}_{12}$ and $\mathbf{G}_{21}$ are arbitrary. Then, clearly $\mathbf{G}$ is a generalized inverse of $\mathbf{A}$. Further,

$$\mathbf{T}(\mathbf{G}_{11} - \mathbf{G}_{12}\mathbf{G}_{22}^-\mathbf{G}_{21})\mathbf{T} = \mathbf{T} - \mathbf{T}\mathbf{G}_{12}\mathbf{G}_{22}^-\mathbf{G}_{21}\mathbf{T},$$

so that $\mathbf{G}_{11} - \mathbf{G}_{12}\mathbf{G}_{22}^-\mathbf{G}_{21}$ is a generalized inverse of $\mathbf{T}$ if and only if

$$\mathbf{T}\mathbf{G}_{12}\mathbf{G}_{22}^-\mathbf{G}_{21}\mathbf{T} = \mathbf{0}.$$

Suppose, for example, that $\mathbf{G}_{12}$, $\mathbf{G}_{22}^-$, and $\mathbf{G}_{21}$ are chosen so that the $(1, 1)$th element of $\mathbf{G}_{12}\mathbf{G}_{22}^-\mathbf{G}_{21}$ is nonzero (which — since any $n \times q$ matrix is a generalized inverse of $\mathbf{G}_{22}$ — is clearly possible). Then, the $(1, 1)$th element of $\mathbf{T}\mathbf{G}_{12}\mathbf{G}_{22}^-\mathbf{G}_{21}\mathbf{T}$ is nonzero and hence $\mathbf{T}\mathbf{G}_{12}\mathbf{G}_{22}^-\mathbf{G}_{21}\mathbf{T}$ is nonnull. We conclude that $\mathbf{G}_{11} - \mathbf{G}_{12}\mathbf{G}_{22}^-\mathbf{G}_{21}$ is not a generalized inverse of $\mathbf{T}$.

10
Idempotent Matrices

EXERCISE 1. Show that if an $n \times n$ matrix $\mathbf{A}$ is idempotent, then (a) for any $n \times n$ nonsingular matrix $\mathbf{B}$, $\mathbf{B}^{-1}\mathbf{AB}$ is idempotent; and (b) for any integer k greater than or equal to 2, $\mathbf{A}^k = \mathbf{A}$.

Solution. Suppose that $\mathbf{A}$ is idempotent.

(a) $\mathbf{B}^{-1}\mathbf{AB}(\mathbf{B}^{-1}\mathbf{AB}) = \mathbf{B}^{-1}\mathbf{A}^2\mathbf{B} = \mathbf{B}^{-1}\mathbf{AB}$.

(b) The proof is by mathematical induction. By definition, $\mathbf{A}^k = \mathbf{A}$ for $k = 2$. Suppose that $\mathbf{A}^k = \mathbf{A}$ for $k = k^*$. Then,

$$\mathbf{A}^{k^*+1} = \mathbf{A}\mathbf{A}^{k^*} = \mathbf{A}\mathbf{A} = \mathbf{A},$$

that is, $\mathbf{A}^k = \mathbf{A}$ for $k = k^* + 1$. Thus, for any integer $k \geq 2$, $\mathbf{A}^k = \mathbf{A}$.

EXERCISE 2. Let $\mathbf{P}$ represent an $m \times n$ matrix (where $m \geq n$) such that $\mathbf{P}'\mathbf{P} = \mathbf{I}_n$, or equivalently an $m \times n$ matrix whose columns are orthonormal (with respect to the usual inner product). Show that the $m \times m$ symmetric matrix $\mathbf{PP}'$ is idempotent.

Solution. Clearly,

$$(\mathbf{PP}')\mathbf{PP}' = \mathbf{P}(\mathbf{P}'\mathbf{P})\mathbf{P}' = \mathbf{P}\mathbf{I}_n\mathbf{P}' = \mathbf{PP}'.$$

EXERCISE 3. Show that, for any symmetric idempotent matrix $\mathbf{A}$, the matrix $\mathbf{I} - 2\mathbf{A}$ is orthogonal.

Solution. Clearly,

$$(\mathbf{I}-2\mathbf{A})'(\mathbf{I}-2\mathbf{A}) = (\mathbf{I}-2\mathbf{A})(\mathbf{I}-2\mathbf{A}) = \mathbf{I}-2\mathbf{A}-2\mathbf{A}+4\mathbf{A}^2 = \mathbf{I}-2\mathbf{A}-2\mathbf{A}+4\mathbf{A} = \mathbf{I}.$$

EXERCISE 4. Let $\mathbf{A}$ represent an $m \times n$ matrix. Show that if $\mathbf{A}'\mathbf{A}$ is idempotent, then $\mathbf{A}\mathbf{A}'$ is idempotent.

Solution. Suppose that $\mathbf{A}'\mathbf{A}$ is idempotent. Then,

$$\mathbf{A}'\mathbf{A}\mathbf{A}'\mathbf{A} = \mathbf{A}'\mathbf{A} = \mathbf{A}'\mathbf{A}\mathbf{I},$$

implying (in light of Corollary 5.3.3) that

$$\mathbf{A}\mathbf{A}'\mathbf{A} = \mathbf{A}\mathbf{I} = \mathbf{A}$$

and hence that

$$(\mathbf{A}\mathbf{A}')^2 = (\mathbf{A}\mathbf{A}'\mathbf{A})\mathbf{A}' = \mathbf{A}\mathbf{A}'.$$

EXERCISE 5. Let $\mathbf{A}$ represent a symmetric matrix and k an integer greater than or equal to 1. Show that if $\mathbf{A}^{k+1} = \mathbf{A}^k$, then $\mathbf{A}$ is idempotent.

Solution. It suffices to show that, for every integer m between k and 2 inclusive, $\mathbf{A}^{m+1} = \mathbf{A}^m$ implies $\mathbf{A}^m = \mathbf{A}^{m-1}$. (If $\mathbf{A}^{k+1} = \mathbf{A}^k$ but $\mathbf{A}^2$ were not equal to $\mathbf{A}$, then there would exist an integer m between k and 2 inclusive such that $\mathbf{A}^{m+1} = \mathbf{A}^m$ but $\mathbf{A}^m \neq \mathbf{A}^{m-1}$.)

Suppose that $\mathbf{A}^{m+1} = \mathbf{A}^m$. Then, since $\mathbf{A}$ is symmetric,

$$\mathbf{A}'\mathbf{A}\mathbf{A}^{m-1} = \mathbf{A}'\mathbf{A}\mathbf{A}^{m-2}$$

(where $\mathbf{A}^0 = \mathbf{I}$), and it follows from Corollary 5.3.3 that

$$\mathbf{A}\mathbf{A}^{m-1} = \mathbf{A}\mathbf{A}^{m-2}$$

or equivalently that

$$\mathbf{A}^m = \mathbf{A}^{m-1}.$$

EXERCISE 6. Let $\mathbf{A}$ represent an $n \times n$ matrix. Show that $(1/2)(\mathbf{I} + \mathbf{A})$ is idempotent if and only if $\mathbf{A}$ is involutory (where involutory is as defined in Exercise 8.2).

Solution. Clearly,

$$[(1/2)(\mathbf{I} + \mathbf{A})]^2 = (1/4)\mathbf{I} + (1/2)\mathbf{A} + (1/4)\mathbf{A}^2.$$

Thus, $(1/2)(\mathbf{I} + \mathbf{A})$ is idempotent if and only if

$$(1/4)\mathbf{I} + (1/2)\mathbf{A} + (1/4)\mathbf{A}^2 = (1/2)\mathbf{I} + (1/2)\mathbf{A},$$

or equivalently if and only if $(1/4)\mathbf{A}^2 = (1/4)\mathbf{I}$, and hence if and only if $\mathbf{A}^2 = \mathbf{I}$ (i.e., if and only if $\mathbf{A}$ is involutory).

EXERCISE 7. Let $\mathbf{A}$ and $\mathbf{B}$ represent $n \times n$ symmetric idempotent matrices. Show that if $\mathcal{C}(\mathbf{A}) = \mathcal{C}(\mathbf{B})$, then $\mathbf{A} = \mathbf{B}$.

Solution. Suppose that $\mathcal{C}(\mathbf{A}) = \mathcal{C}(\mathbf{B})$. Then, according to Lemma 4.2.2, $\mathbf{A} = \mathbf{BR}$ and $\mathbf{B} = \mathbf{AS}$ for some $n \times n$ matrices $\mathbf{R}$ and $\mathbf{S}$. Further,

$$\mathbf{B} = \mathbf{B}' = (\mathbf{AS})' = \mathbf{S}'\mathbf{A}' = \mathbf{S}'\mathbf{A}.$$

Thus,

$$\mathbf{A} = \mathbf{BBR} = \mathbf{BA} = \mathbf{S}'\mathbf{AA} = \mathbf{S}'\mathbf{A} = \mathbf{B}.$$

EXERCISE 8. Let $\mathbf{A}$ represent an $r \times m$ matrix and $\mathbf{B}$ an $m \times n$ matrix.

(a) Show that $\mathbf{B}^-\mathbf{A}^-$ is a generalized inverse of $\mathbf{AB}$ if and only if $\mathbf{A}^-\mathbf{ABB}^-$ is idempotent.

(b) Show that if $\mathbf{A}$ has full column rank or $\mathbf{B}$ has full row rank, then $\mathbf{B}^-\mathbf{A}^-$ is a generalized inverse of $\mathbf{AB}$.

Solution. (a) In light of the definition of a generalized inverse and the definition of an idempotent matrix, it suffices to show that

$$\mathbf{ABB}^-\mathbf{A}^-\mathbf{AB} = \mathbf{AB}$$

if and only if

$$\mathbf{A}^-\mathbf{ABB}^-\mathbf{A}^-\mathbf{ABB}^- = \mathbf{A}^-\mathbf{ABB}^-.$$

Premultiplication and postmultiplication of both sides of the first of these two equalities by $\mathbf{A}^-$ and $\mathbf{B}^-$, respectively, give the second equality, and premultiplication and postmultiplication of both sides of the second equality by $\mathbf{A}$ and $\mathbf{B}$, respectively, give the first equality. Thus, these two equalities are equivalent.

(b) Suppose that $\mathbf{A}$ has full column rank. Then, according to Lemma 9.2.8, $\mathbf{A}^-$ is a left inverse of $\mathbf{A}$ (i.e., $\mathbf{A}^-\mathbf{A} = \mathbf{I}$). It follows that $\mathbf{A}^-\mathbf{ABB}^- = \mathbf{BB}^-$ and hence, in light of Lemma 10.2.5, that $\mathbf{A}^-\mathbf{ABB}^-$ is idempotent. We conclude, on the basis of Part (a), that $\mathbf{B}^-\mathbf{A}^-$ is a generalized inverse of $\mathbf{AB}$. It follows from an analogous argument that if $\mathbf{B}$ has full row rank, then $\mathbf{B}^-\mathbf{A}^-$ is a generalized inverse of $\mathbf{AB}$.

EXERCISE 9. Let $\mathbf{T}$ represent an $m \times p$ matrix, $\mathbf{U}$ an $m \times q$ matrix, $\mathbf{V}$ an $n \times p$ matrix, and $\mathbf{W}$ an $n \times q$ matrix, and define $\mathbf{Q} = \mathbf{W} - \mathbf{VT}^-\mathbf{U}$. Using the result of Part (a) of Exercise 9.8, together with the result that (for any matrix $\mathbf{B}$) $\text{rank}(\mathbf{B}) = \text{tr}(\mathbf{B}^-\mathbf{B}) = \text{tr}(\mathbf{BB}^-)$, show that if

(1) $(\mathbf{I} - \mathbf{TT}^-)\mathbf{U}(\mathbf{I} - \mathbf{Q}^-\mathbf{Q}) = \mathbf{0}$,

(2) $(\mathbf{I} - \mathbf{QQ}^-)\mathbf{V}(\mathbf{I} - \mathbf{T}^-\mathbf{T}) = \mathbf{0}$, and

(3) $(\mathbf{I} - \mathbf{TT}^-)\mathbf{UQ}^-\mathbf{V}(\mathbf{I} - \mathbf{T}^-\mathbf{T}) = \mathbf{0}$,

then

$$\text{rank}\begin{pmatrix} \mathbf{T} & \mathbf{U} \\ \mathbf{V} & \mathbf{W} \end{pmatrix} = \text{rank}(\mathbf{T}) + \text{rank}(\mathbf{Q}).$$

Solution. Let $\mathbf{A} = \begin{pmatrix} \mathbf{T} & \mathbf{U} \\ \mathbf{V} & \mathbf{W} \end{pmatrix}$, and define $\mathbf{G}$ as in Part (a) of Exercise 9.8. Suppose that Conditions (1) – (3) are satisfied. Then, in light of Exercise 9.8, $\mathbf{G}$ is a generalized inverse of **A**, and, making use of the result that (for any matrix **B**) rank(**B**) = tr (**BB**$^-$) [which is part of result (2.1)], we find that

$$\begin{aligned}\text{rank}(\mathbf{A}) &= \text{tr}\,(\mathbf{AG})\\ &= \text{tr}\begin{pmatrix}\mathbf{TT}^- - (\mathbf{I}-\mathbf{TT}^-)\mathbf{UQ}^-\mathbf{VT}^- & (\mathbf{I}-\mathbf{TT}^-)\mathbf{UQ}^- \\ (\mathbf{I}-\mathbf{QQ}^-)\mathbf{VT}^- & \mathbf{QQ}^-\end{pmatrix}\\ &= \text{tr}\,(\mathbf{TT}^-) - \text{tr}\,[(\mathbf{I}-\mathbf{TT}^-)\mathbf{UQ}^-\mathbf{VT}^-] + \text{tr}\,(\mathbf{QQ}^-)\\ &= \text{rank}(\mathbf{T}) - \text{tr}\,[(\mathbf{I}-\mathbf{TT}^-)\mathbf{UQ}^-\mathbf{VT}^-] + \text{rank}(\mathbf{Q}).\end{aligned}$$

Moreover, it follows from Condition (3) that

$$(\mathbf{I}-\mathbf{TT}^-)\mathbf{UQ}^-\mathbf{V} = (\mathbf{I}-\mathbf{TT}^-)\mathbf{UQ}^-\mathbf{VT}^-\mathbf{T}$$

and hence that

$$\begin{aligned}\text{tr}\,[(\mathbf{I}-\mathbf{TT}^-)\mathbf{UQ}^-\mathbf{VT}^-] &= \text{tr}\,[(\mathbf{I}-\mathbf{TT}^-)\mathbf{UQ}^-\mathbf{VT}^-\mathbf{TT}^-]\\ &= \text{tr}\,[\mathbf{TT}^-(\mathbf{I}-\mathbf{TT}^-)\mathbf{UQ}^-\mathbf{VT}^-]\\ &= \text{tr}\,[(\mathbf{TT}^- - \mathbf{TT}^-)\mathbf{UQ}^-\mathbf{VT}^-]\\ &= \text{tr}\,(\mathbf{0})\\ &= 0.\end{aligned}$$

We conclude that

$$\text{rank}(\mathbf{A}) = \text{rank}(\mathbf{T}) + \text{rank}(\mathbf{Q}).$$

EXERCISE 10. Let **T** represent an $m \times p$ matrix, **U** an $m \times q$ matrix, **V** an $n \times p$ matrix, and **W** an $n \times q$ matrix, take $\mathbf{A} = \begin{pmatrix}\mathbf{T} & \mathbf{U}\\ \mathbf{V} & \mathbf{W}\end{pmatrix}$, and define $\mathbf{Q} = \mathbf{W} - \mathbf{VT}^-\mathbf{U}$. Further, let

$$\mathbf{E}_T = \mathbf{I} - \mathbf{TT}^-,\ \ \mathbf{F}_T = \mathbf{I} - \mathbf{T}^-\mathbf{T},\ \ \mathbf{X} = \mathbf{E}_T\mathbf{U},\ \ \mathbf{Y} = \mathbf{VF}_T\,,\ \ \mathbf{E}_Y = \mathbf{I} - \mathbf{YY}^-,$$
$$\mathbf{F}_X = \mathbf{I} - \mathbf{X}^-\mathbf{X},\ \ \mathbf{Z} = \mathbf{E}_Y\mathbf{QF}_X,\ \text{ and }\ \mathbf{Q}^* = \mathbf{F}_X\mathbf{Z}^-\mathbf{E}_Y\,.$$

(a) (Meyer 1973, Theorem 3.1) Show that the matrix

$$\mathbf{G} = \mathbf{G}_1 + \mathbf{G}_2\,, \tag{E.1}$$

where

$$\mathbf{G}_1 = \left(\begin{array}{c|c}\begin{array}{c}\mathbf{T}^- - \mathbf{T}^-\mathbf{U}(\mathbf{I}-\mathbf{Q}^*\mathbf{Q})\mathbf{X}^-\mathbf{E}_T\\ -\,\mathbf{F}_T\mathbf{Y}^-(\mathbf{I}-\mathbf{QQ}^*)\mathbf{VT}^-\\ -\,\mathbf{F}_T\mathbf{Y}^-(\mathbf{I}-\mathbf{QQ}^*)\mathbf{QX}^-\mathbf{E}_T\end{array} & \mathbf{F}_T\mathbf{Y}^-(\mathbf{I}-\mathbf{QQ}^*)\\ \hline (\mathbf{I}-\mathbf{Q}^*\mathbf{Q})\mathbf{X}^-\mathbf{E}_T & \mathbf{0}\end{array}\right)$$

and

$$\mathbf{G}_2 = \begin{pmatrix}-\mathbf{T}^-\mathbf{U}\\ \mathbf{I}_q\end{pmatrix}\mathbf{Q}^*(-\mathbf{VT}^-,\ \mathbf{I}_n)\,,$$

is a generalized inverse of $\mathbf{A}$.

(b) (Meyer 1973, Theorem 4.1) Show that

$$\text{rank}(\mathbf{A}) = \text{rank}(\mathbf{T}) + \text{rank}(\mathbf{X}) + \text{rank}(\mathbf{Y}) + \text{rank}(\mathbf{Z})\,. \tag{E.2}$$

[*Hint*. Use Part (a), together with the result that (for any matrix $\mathbf{B}$) $\text{rank}(\mathbf{B}) = \text{tr}\,(\mathbf{B}^-\mathbf{B}) = \text{tr}\,(\mathbf{B}\mathbf{B}^-)$.]

(c) Show that if $\mathcal{C}(\mathbf{U}) \subset \mathcal{C}(\mathbf{T})$ and $\mathcal{R}(\mathbf{V}) \subset \mathcal{R}(\mathbf{T})$, then formula (E.2) for $\text{rank}(\mathbf{A})$ reduces to the formula

$$\text{rank}(\mathbf{A}) = \text{rank}(\mathbf{T}) + \text{rank}(\mathbf{Q}),$$

and the formula

$$\begin{pmatrix} \mathbf{T}^- + \mathbf{T}^-\mathbf{U}\mathbf{Q}^-\mathbf{V}\mathbf{T}^- & -\mathbf{T}^-\mathbf{U}\mathbf{Q}^- \\ -\mathbf{Q}^-\mathbf{V}\mathbf{T}^- & \mathbf{Q}^- \end{pmatrix}, \tag{$*$}$$

which is reexpressible as

$$\begin{pmatrix} \mathbf{T}^- & \mathbf{0} \\ \mathbf{0} & \mathbf{0} \end{pmatrix} + \begin{pmatrix} -\mathbf{T}^-\mathbf{U} \\ \mathbf{I}_q \end{pmatrix} \mathbf{Q}^-(-\mathbf{V}\mathbf{T}^-, \mathbf{I}_n), \tag{$**$}$$

can be obtained as a special case of formula (E.1) for a generalized inverse of $\mathbf{A}$.

Solution. (a) It can be shown, via some painstaking algebraic manipulation, that

$$\mathbf{A}\mathbf{G}_1\mathbf{A} = \begin{pmatrix} \mathbf{T} & \mathbf{U} \\ \mathbf{V} & \mathbf{W} - \mathbf{Q}\mathbf{Q}^*\mathbf{Q} \end{pmatrix} \quad \text{and} \quad \mathbf{A}\mathbf{G}_2\mathbf{A} = \begin{pmatrix} \mathbf{0} & \mathbf{0} \\ \mathbf{0} & \mathbf{Q}\mathbf{Q}^*\mathbf{Q} \end{pmatrix}$$

and hence that

$$\mathbf{A}\mathbf{G}\mathbf{A} = \mathbf{A}\mathbf{G}_1\mathbf{A} + \mathbf{A}\mathbf{G}_2\mathbf{A} = \mathbf{A}.$$

(b) Taking $\mathbf{G}$ to be the generalized inverse (E.1), it is easy to show that

$$\mathbf{A}\mathbf{G} = \left(\begin{array}{c|c} \mathbf{T}\mathbf{T}^- + \mathbf{X}\mathbf{X}^-\mathbf{E}_T & \mathbf{0} \\ \text{omitted} & \mathbf{Y}\mathbf{Y}^- + \mathbf{E}_Y\mathbf{Q}\mathbf{Q}^* \end{array}\right).$$

Thus, making use of the result that (for any matrix $\mathbf{B}$) $\text{rank}(\mathbf{B}) = \text{tr}\,(\mathbf{B}\mathbf{B}^-)$ [which is part of result (2.1)], we find that

$$\begin{aligned} \text{rank}(\mathbf{A}) &= \text{tr}\,(\mathbf{A}\mathbf{G}) \\ &= \text{tr}\,(\mathbf{T}\mathbf{T}^-) + \text{tr}\,(\mathbf{X}\mathbf{X}^-\mathbf{E}_T) + \text{tr}\,(\mathbf{Y}\mathbf{Y}^-) + \text{tr}\,(\mathbf{E}_Y\mathbf{Q}\mathbf{Q}^*) \\ &= \text{rank}(\mathbf{T}) + \text{tr}\,(\mathbf{X}\mathbf{X}^-\mathbf{E}_T) + \text{rank}(\mathbf{Y}) + \text{tr}\,(\mathbf{E}_Y\mathbf{Q}\mathbf{Q}^*). \end{aligned}$$

Moreover,

$$\begin{aligned} \text{tr}\,(\mathbf{X}\mathbf{X}^-\mathbf{E}_T) = \text{tr}\,(\mathbf{E}_T\mathbf{X}\mathbf{X}^-) &= \text{tr}\,(\mathbf{E}_T\mathbf{E}_T\mathbf{U}\mathbf{X}^-) \\ &= \text{tr}\,(\mathbf{E}_T\mathbf{U}\mathbf{X}^-) = \text{tr}\,(\mathbf{X}\mathbf{X}^-) = \text{rank}(\mathbf{X}), \end{aligned}$$

and similarly

$$\begin{aligned}\operatorname{tr}(\mathbf{E}_Y\mathbf{Q}\mathbf{Q}^*) = \operatorname{tr}(\mathbf{E}_Y\mathbf{Q}\mathbf{F}_X\mathbf{Z}^-\mathbf{E}_Y) &= \operatorname{tr}(\mathbf{E}_Y\mathbf{E}_Y\mathbf{Q}\mathbf{F}_X\mathbf{Z}^-)\\ &= \operatorname{tr}(\mathbf{E}_Y\mathbf{Q}\mathbf{F}_X\mathbf{Z}^-)\\ &= \operatorname{tr}(\mathbf{Z}\mathbf{Z}^-) = \operatorname{rank}(\mathbf{Z}).\end{aligned}$$

(c) Suppose that $\mathcal{C}(\mathbf{U}) \subset \mathcal{C}(\mathbf{T})$ and $\mathcal{R}(\mathbf{V}) \subset \mathcal{R}(\mathbf{T})$. Then, it follows from Lemma 9.3.5 that $\mathbf{X} = \mathbf{0}$ and $\mathbf{Y} = \mathbf{0}$. Accordingly, $\mathbf{F}_X = \mathbf{I}$ and $\mathbf{E}_Y = \mathbf{I}$, implying that $\mathbf{Z} = \mathbf{Q}$. Thus, formula (E.2) reduces to

$$\operatorname{rank}(\mathbf{A}) = \operatorname{rank}(\mathbf{T}) + \operatorname{rank}(\mathbf{Q}),$$

[which is formula (9.6.1)].

Clearly, $\mathbf{Q}^*$ is an arbitrary generalized inverse of $\mathbf{Q}$, and the $q \times m$ and $p \times n$ null matrices are generalized inverses of $\mathbf{X}$ and $\mathbf{Y}$, respectively. Thus, formula (∗∗) can be obtained as a special case of formula (E.1) by setting $\mathbf{X}^- = \mathbf{0}$ and $\mathbf{Y}^- = \mathbf{0}$ — formula (∗∗) is identical to formula (9.6.2b), and formula (∗) identical to formula (9.6.2a) [and to formula (∗) of Exercise 9.7 or 9.8].

11

Linear Systems: Solutions

EXERCISE 1. Show that, for any matrix $\mathbf{A}$,

$$\mathcal{C}(\mathbf{A}) = \mathcal{N}(\mathbf{I} - \mathbf{AA}^-) .$$

Solution. Letting $\mathbf{x}$ represent a column vector (whose dimension equals the number of rows in $\mathbf{A}$), it follows from Corollary 9.3.6 that $\mathbf{x} \in \mathcal{C}(\mathbf{A})$ if and only if $\mathbf{x} = \mathbf{AA}^-\mathbf{x}$, or equivalently if and only if $(\mathbf{I} - \mathbf{AA}^-)\mathbf{x} = \mathbf{0}$, and hence if and only if $\mathbf{x} \in \mathcal{N}(\mathbf{I} - \mathbf{AA}^-)$. We conclude that $\mathcal{C}(\mathbf{A}) = \mathcal{N}(\mathbf{I} - \mathbf{AA}^-)$.

EXERCISE 2. Show that if $\mathbf{X}_1, \ldots, \mathbf{X}_k$ are solutions to a linear system $\mathbf{AX} = \mathbf{B}$ (in $\mathbf{X}$) and $c_1, \ldots, c_k$ are scalars such that $\sum_{i=1}^k c_i = 1$, then the matrix $\sum_{i=1}^k c_i\mathbf{X}_i$ is a solution to $\mathbf{AX} = \mathbf{B}$.

Solution. If $\mathbf{X}_1, \ldots, \mathbf{X}_k$ are solutions to $\mathbf{AX} = \mathbf{B}$ and $c_1, \ldots, c_k$ are scalars such that $\sum_{i=1}^k c_i = 1$, then

$$\mathbf{A}\left(\sum_{i=1}^k c_i\mathbf{X}_i\right) = \sum_{i=1}^k c_i(\mathbf{AX}_i) = \sum_{i=1}^k c_i\mathbf{B} = \mathbf{B}.$$

EXERCISE 3. Let $\mathbf{A}$ and $\mathbf{Z}$ represent $n \times n$ matrices. Suppose that $\text{rank}(\mathbf{A}) = n - 1$, and let $\mathbf{x}$ and $\mathbf{y}$ represent nonnull n-dimensional column vectors such that $\mathbf{Ax} = \mathbf{0}$ and $\mathbf{A}'\mathbf{y} = \mathbf{0}$.

(a) Show that $\mathbf{AZ} = \mathbf{0}$ if and only if $\mathbf{Z} = \mathbf{xk}'$ for some n-dimensional row vector $\mathbf{k}'$.

(b) Show that $\mathbf{AZ} = \mathbf{ZA} = \mathbf{0}$ if and only if $\mathbf{Z} = c\mathbf{xy}'$ for some scalar c.

Solution. (a) Suppose that $\mathbf{Z} = \mathbf{xk}'$ for some row vector $\mathbf{k}'$. Then,

$$\mathbf{AZ} = (\mathbf{Ax})\mathbf{k}' = \mathbf{0k}' = \mathbf{0}.$$

Conversely, suppose that $\mathbf{AZ} = \mathbf{0}$. Let $\mathbf{z}_j$ represent the jth column of $\mathbf{Z}$. Since (in light of Lemma 11.3.1 and Theorem 4.3.9) $\{\mathbf{x}\}$ is a basis for $\mathcal{N}(\mathbf{A})$, $\mathbf{z}_j = k_j\mathbf{x}$ for some scalar k_j $(j = 1, \ldots, n)$, in which case $\mathbf{Z} = \mathbf{xk}'$, where $\mathbf{k}' = (k_1, \ldots, k_n)$.

(b) Suppose that $\mathbf{Z} = c\mathbf{xy}'$ for some scalar c. Then, $\mathbf{AZ} = \mathbf{0}$ [as is evident from Part (a)], and

$$\mathbf{ZA} = c\mathbf{xy}'\mathbf{A} = c\mathbf{x}(\mathbf{A}'\mathbf{y})' = c\mathbf{x0}' = \mathbf{0}.$$

Conversely, suppose that $\mathbf{AZ} = \mathbf{ZA} = \mathbf{0}$. Then, it follows from Part (a) that $\mathbf{Z} = \mathbf{xk}'$ for some row vector $\mathbf{k}'$. Moreover, $\mathbf{k}' = (\mathbf{x}'\mathbf{x})^{-1}(\mathbf{x}'\mathbf{x})\mathbf{k}' = (\mathbf{x}'\mathbf{x})^{-1}\mathbf{x}'\mathbf{Z}$, so that $\mathbf{k}'\mathbf{A} = (\mathbf{x}'\mathbf{x})^{-1}\mathbf{x}'\mathbf{ZA} = \mathbf{0}$, implying that $\mathbf{A}'\mathbf{k} = (\mathbf{k}'\mathbf{A})' = \mathbf{0}$ and hence that $\mathbf{k} \in \mathcal{N}(\mathbf{A}')$. Since (in light of Lemma 11.3.1 and Theorem 4.3.9) $\{\mathbf{y}\}$ is a basis for $\mathcal{N}(\mathbf{A}')$, $\mathbf{k} = c\mathbf{y}$ for some scalar c. Thus, $\mathbf{Z} = c\mathbf{xy}'$.

EXERCISE 4. Suppose that $\mathbf{AX} = \mathbf{B}$ is a nonhomogeneous linear system (in an $n \times p$ matrix $\mathbf{X}$). Let $s = p[n - \text{rank}(\mathbf{A})]$, and take $\mathbf{Z}_1, \ldots, \mathbf{Z}_s$ to be any s $n \times p$ matrices that form a basis for the solution space of the homogeneous linear system $\mathbf{AZ} = \mathbf{0}$ (in an $n \times p$ matrix $\mathbf{Z}$). Define $\mathbf{X}_0$ to be any particular solution to $\mathbf{AX} = \mathbf{B}$, and let $\mathbf{X}_i = \mathbf{X}_0 + \mathbf{Z}_i$ $(i = 1, \ldots, s)$.

(a) Show that the $s + 1$ matrices $\mathbf{X}_0, \mathbf{X}_1, \ldots, \mathbf{X}_s$ are linearly independent solutions to $\mathbf{AX} = \mathbf{B}$.

(b) Show that every solution to $\mathbf{AX} = \mathbf{B}$ is expressible as a linear combination of $\mathbf{X}_0, \mathbf{X}_1, \ldots, \mathbf{X}_s$.

(c) Show that a linear combination $\sum_{i=0}^{s} k_i\mathbf{X}_i$ of $\mathbf{X}_0, \mathbf{X}_1, \ldots, \mathbf{X}_s$ is a solution to $\mathbf{AX} = \mathbf{B}$ if and only if the scalars $k_0, k_1, \ldots, k_s$ are such that $\sum_{i=0}^{s} k_i = 1$.

(d) Show that the solution set of $\mathbf{AX} = \mathbf{B}$ is a proper subset of the linear space $\text{sp}(\mathbf{X}_0, \mathbf{X}_1, \ldots, \mathbf{X}_s)$.

Solution. (a) It follows from Theorem 11.2.3 that $\mathbf{X}_1, \ldots, \mathbf{X}_s$, like $\mathbf{X}_0$, are solutions to $\mathbf{AX} = \mathbf{B}$.

For purposes of showing that $\mathbf{X}_0, \mathbf{X}_1, \ldots, \mathbf{X}_s$ are linearly independent, suppose that $k_0, k_1, \ldots, k_s$ are scalars such that $\sum_{i=0}^{s} k_i\mathbf{X}_i = \mathbf{0}$. Then,

$$\left(\sum_{i=0}^{s} k_i\right)\mathbf{X}_0 + \sum_{i=1}^{s} k_i\mathbf{Z}_i = \sum_{i=1}^{s} k_i\mathbf{X}_i = \mathbf{0}. \tag{S.1}$$

Consequently,

$$\left(\sum_{i=0}^{s} k_i\right)\mathbf{B} = \mathbf{A}\left[\left(\sum_{i=0}^{s} k_i\right)\mathbf{X}_0 + \sum_{i=1}^{s} k_i\mathbf{Z}_i\right] = \mathbf{0},$$

implying (since $\mathbf{B} \neq \mathbf{0}$) that

$$\sum_{i=0}^{s} k_i = 0 \tag{S.2}$$

and hence [in light of equality (S.1)] that $\sum_{i=1}^{s} k_i \mathbf{Z}_i = \mathbf{0}$. Since $\mathbf{Z}_1, \ldots, \mathbf{Z}_s$ are linearly independent, we have that $k_1 = \ldots = k_s = 0$, which, together with equality (S.2), further implies that $k_0 = 0$. We conclude that $\mathbf{X}_0, \mathbf{X}_1, \ldots, \mathbf{X}_s$ are linearly independent.

(b) Let $\mathbf{X}^*$ represent any solution to $\mathbf{AX} = \mathbf{B}$. Then, according to Theorem 11.2.3, $\mathbf{X}^* = \mathbf{X}_0 + \mathbf{Z}^*$ for some solution $\mathbf{Z}^*$ to $\mathbf{AZ} = \mathbf{0}$. Since $\mathbf{Z}_1, \ldots, \mathbf{Z}_s$ form a basis for the solution space of $\mathbf{AZ} = \mathbf{0}$, $\mathbf{Z}^* = \sum_{i=1}^{s} k_i \mathbf{Z}_i$ for some scalars $k_i, \ldots, k_s$. Thus,

$$\mathbf{X}^* = \mathbf{X}_0 + \sum_{i=1}^{s} k_i \mathbf{Z}_i = \left(1 - \sum_{i=1}^{s} k_i\right)\mathbf{X}_0 + \sum_{i=1}^{s} k_i \mathbf{X}_i.$$

(c) We find that

$$\mathbf{A}\left(\sum_{i=0}^{s} k_i \mathbf{X}_i\right) = \mathbf{A}\left[\left(\sum_{i=0}^{s} k_i\right)\mathbf{X}_0 + \sum_{i=1}^{s} k_i \mathbf{Z}_i\right] = \left(\sum_{i=0}^{s} k_i\right)\mathbf{B}.$$

Thus, if $\sum_{i=0}^{s} k_i = 1$, then $\sum_{i=0}^{s} k_i \mathbf{X}_i$ is a solution to $\mathbf{AX} = \mathbf{B}$. Conversely, if $\sum_{i=0}^{s} k_i \mathbf{X}_i$ is a solution to $\mathbf{AX} = \mathbf{B}$, then clearly $\left(\sum_{i=0}^{s} k_i\right)\mathbf{B} = \mathbf{B}$, implying (since $\mathbf{B} \neq \mathbf{0}$) that $\sum_{i=0}^{s} k_i = 1$.

(d) It is clear from Part (b) that every solution to $\mathbf{AX} = \mathbf{B}$ belongs to sp($\mathbf{X}_0$, $\mathbf{X}_1, \ldots, \mathbf{X}_s$). However, not every matrix in sp($\mathbf{X}_0, \mathbf{X}_1, \ldots, \mathbf{X}_s$) is a solution to $\mathbf{AX} = \mathbf{B}$, as is evident from Part (c). Thus, the solution set of $\mathbf{AX} = \mathbf{B}$ is a proper subset of sp($\mathbf{X}_0, \mathbf{X}_1, \ldots, \mathbf{X}_s$).

EXERCISE 5. Suppose that $\mathbf{AX} = \mathbf{B}$ is a consistent linear system (in an $n \times p$ matrix $\mathbf{X}$). Show that if rank($\mathbf{A}$) $< n$ and rank($\mathbf{B}$) $< p$, then there exists a solution $\mathbf{X}^*$ to $\mathbf{AX} = \mathbf{B}$ that is not expressible as $\mathbf{X}^* = \mathbf{GB}$ for any generalized inverse $\mathbf{G}$ of $\mathbf{A}$.

Solution. Suppose that rank($\mathbf{A}$) $< n$ and rank($\mathbf{B}$) $< p$. Then, since the columns of $\mathbf{B}$ are linearly dependent, there exists a nonnull vector $\mathbf{k}_1$ such that $\mathbf{Bk}_1 = \mathbf{0}$, and, according to Theorem 4.3.12, there exist $p-1$ p-dimensional column vectors $\mathbf{k}_2, \ldots, \mathbf{k}_p$ such that the set $\{\mathbf{k}_1, \mathbf{k}_2, \ldots, \mathbf{k}_p\}$ is a basis for $\mathcal{R}^p$. Define $\mathbf{K} = (\mathbf{k}_1, \mathbf{K}_2)$, where $\mathbf{K}_2$ is the $p \times (p-1)$ matrix whose columns are $\mathbf{k}_2, \ldots, \mathbf{k}_p$. Clearly, the matrix $\mathbf{K}$ is nonsingular.

Since the columns of $\mathbf{A}$ are linearly dependent, there exists a nonnull vector $\mathbf{y}_1^*$ such that $\mathbf{Ay}_1^* = \mathbf{0}$. Let $\mathbf{Y}^* = (\mathbf{y}_1^*, \mathbf{Y}_2^*)$, where $\mathbf{Y}_2^*$ is any solution to the linear system $\mathbf{AY}_2 = \mathbf{BK}_2$ (in $\mathbf{Y}_2$). (Since $\mathbf{AX} = \mathbf{B}$ is consistent, so is $\mathbf{AY}_2 = \mathbf{BK}_2$.) Clearly, $\mathbf{AY}^* = \mathbf{BK}$.

Define $\mathbf{X}^* = \mathbf{Y}^*\mathbf{K}^{-1}$. Then,

$$\mathbf{AX}^* = \mathbf{AY}^*\mathbf{K}^{-1} = \mathbf{BKK}^{-1} = \mathbf{B},$$

so that $\mathbf{X}^*$ is a solution to $\mathbf{AX} = \mathbf{B}$.

To complete the proof, it suffices to show that $\mathbf{X}^*$ is not expressible as $\mathbf{X}^* = \mathbf{GB}$ for any generalized inverse $\mathbf{G}$ of $\mathbf{A}$. Assume the contrary, that is, assume that $\mathbf{X}^* = \mathbf{GB}$ for some generalized inverse $\mathbf{G}$ of $\mathbf{A}$. Then, since

$$(\mathbf{y}_1^*, \mathbf{Y}_2^*) = \mathbf{Y}^* = \mathbf{X}^*\mathbf{K} = (\mathbf{X}^*\mathbf{k}_1, \mathbf{X}^*\mathbf{K}_2),$$

we have that

$$\mathbf{y}_1^* = \mathbf{X}^*\mathbf{k}_1 = \mathbf{GBk}_1 = \mathbf{0},$$

which (since, by definition, $\mathbf{y}_1^*$ is nonnull) establishes a contradiction.

EXERCISE 6. Let $\mathbf{A}$ represent an $m \times n$ matrix and $\mathbf{B}$ an $m \times p$ matrix. If $\mathbf{C}$ is an $r \times m$ matrix of full column rank (i.e., of rank m), then the linear system $\mathbf{CAX} = \mathbf{CB}$ is equivalent to the linear system $\mathbf{AX} = \mathbf{B}$ (in $\mathbf{X}$). Use the result of Part (b) of Exercise 7.1 to generalize this result.

Solution. If $\mathbf{C}$ is an $r \times q$ matrix and $\mathbf{D}$ a $q \times m$ matrix such that $\text{rank}(\mathbf{CD}) = \text{rank}(\mathbf{D})$, then the linear system $\mathbf{CDAX} = \mathbf{CDB}$ (in $\mathbf{X}$) is equivalent to the linear system $\mathbf{DAX} = \mathbf{DB}$ (in $\mathbf{X}$) [as is evident from Part (b) of Exercise 7.1].

EXERCISE 7. Let $\mathbf{A}$ represent an $m \times n$ matrix, $\mathbf{B}$ an $m \times p$ matrix, and $\mathbf{C}$ a $q \times m$ matrix, and suppose that $\mathbf{AX} = \mathbf{B}$ and $\mathbf{CAX} = \mathbf{CB}$ are linear systems (in $\mathbf{X}$).

(a) Show that if $\text{rank}[\mathbf{C}(\mathbf{A}, \mathbf{B})] = \text{rank}(\mathbf{A}, \mathbf{B})$, then $\mathbf{CAX} = \mathbf{CB}$ is equivalent to $\mathbf{AX} = \mathbf{B}$ — this result is a generalization of the result that $\mathbf{CAX} = \mathbf{CB}$ is equivalent to $\mathbf{AX} = \mathbf{B}$ if $\mathbf{C}$ is of full column rank (i.e., of rank m) and also of the result that (for any $n \times s$ matrix $\mathbf{F}$, the linear system $\mathbf{A}'\mathbf{AX} = \mathbf{A}'\mathbf{AF}$ is equivalent to the linear system $\mathbf{AX} = \mathbf{AF}$ (in $\mathbf{X}$).

(b) Show that if $\text{rank}[\mathbf{C}(\mathbf{A}, \mathbf{B})] < \text{rank}(\mathbf{A}, \mathbf{B})$ and if $\mathbf{CAX} = \mathbf{CB}$ is consistent, then the solution set of $\mathbf{AX} = \mathbf{B}$ is a proper subset of that of $\mathbf{CAX} = \mathbf{CB}$ (i.e., there exists a solution to $\mathbf{CAX} = \mathbf{CB}$ that is not a solution to $\mathbf{AX} = \mathbf{B}$).

(c) Show, by example, that if $\text{rank}[\mathbf{C}(\mathbf{A}, \mathbf{B})] < \text{rank}(\mathbf{A}, \mathbf{B})$ and if $\mathbf{AX} = \mathbf{B}$ is inconsistent, then $\mathbf{CAX} = \mathbf{CB}$ can be either consistent or inconsistent.

Solution. (a) Suppose that $\text{rank}[\mathbf{C}(\mathbf{A}, \mathbf{B})] = \text{rank}(\mathbf{A}, \mathbf{B})$. Then, according to Corollary 4.4.7, $\mathcal{R}[\mathbf{C}(\mathbf{A}, \mathbf{B})] = \mathcal{R}(\mathbf{A}, \mathbf{B})$ and hence $\mathcal{R}(\mathbf{A}, \mathbf{B}) \subset \mathcal{R}[\mathbf{C}(\mathbf{A}, \mathbf{B})]$, implying (in light of Lemma 4.2.2) that $(\mathbf{A}, \mathbf{B}) = \mathbf{LC}(\mathbf{A}, \mathbf{B})$ for some matrix $\mathbf{L}$. Therefore,

$$\mathbf{A} = \mathbf{LCA} \quad \text{and} \quad \mathbf{B} = \mathbf{LCB}.$$

For any solution $\mathbf{X}^*$ to $\mathbf{CAX} = \mathbf{CB}$, we find that

$$\mathbf{AX}^* = \mathbf{LCAX}^* = \mathbf{LCB} = \mathbf{B}.$$

Thus, any solution to $\mathbf{CAX} = \mathbf{CB}$ is a solution to $\mathbf{AX} = \mathbf{B}$, and hence (since any solution to $\mathbf{AX} = \mathbf{B}$ is a solution to $\mathbf{CAX} = \mathbf{CB}$) $\mathbf{CAX} = \mathbf{CB}$ is equivalent to $\mathbf{AX} = \mathbf{B}$.

(b) Suppose that $\text{rank}[\mathbf{C}(\mathbf{A}, \mathbf{B})] < \text{rank}(\mathbf{A}, \mathbf{B})$ and that $\mathbf{CAX} = \mathbf{CB}$ is consistent. And, assume that $\mathbf{AX} = \mathbf{B}$ is consistent — if $\mathbf{AX} = \mathbf{B}$ is inconsistent, then clearly the solution set of $\mathbf{AX} = \mathbf{B}$ is a proper subset of that of $\mathbf{CAX} = \mathbf{CB}$. Then, making use of Theorem 7.2.1, we find that

$$\text{rank}(\mathbf{A}) = \text{rank}(\mathbf{A}, \mathbf{B}) > \text{rank}[\mathbf{C}(\mathbf{A}, \mathbf{B})] = \text{rank}(\mathbf{CA}, \mathbf{CB}) = \text{rank}(\mathbf{CA}),$$

implying that

$$n - \text{rank}(\mathbf{A}) < n - \text{rank}(\mathbf{CA}). \tag{S.3}$$

Let $\mathbf{X}_0$ represent any particular solution to $\mathbf{AX} = \mathbf{B}$. According to Theorem 11.2.3, the solution set of $\mathbf{AX} = \mathbf{B}$ is comprised of every $n \times p$ matrix $\mathbf{X}^*$ that is expressible as

$$\mathbf{X}^* = \mathbf{X}_0 + \mathbf{Z}^*$$

for some solution $\mathbf{Z}^*$ to the homogeneous linear system $\mathbf{AZ} = \mathbf{0}$ (in an $n \times p$ matrix $\mathbf{Z}$). Similarly, since $\mathbf{X}_0$ is also a solution to $\mathbf{CAX} = \mathbf{CB}$, the solution set of $\mathbf{CAX} = \mathbf{CB}$ is comprised of every matrix $\mathbf{X}^*$ that is expressible as

$$\mathbf{X}^* = \mathbf{X}_0 + \mathbf{Z}^*$$

for some solution $\mathbf{Z}^*$ to the homogeneous linear system $\mathbf{CAZ} = \mathbf{0}$.

It follows from Lemma 11.3.2 that the dimension of the solution space of $\mathbf{AZ} = \mathbf{0}$ equals $p[n - \text{rank}(\mathbf{A})]$ and the dimension of the solution space of $\mathbf{CAZ} = \mathbf{0}$ equals $p[n - \text{rank}(\mathbf{CA})]$. Clearly, the solution space of $\mathbf{AZ} = \mathbf{0}$ is a subspace of the solution space of $\mathbf{CAZ} = \mathbf{0}$ and hence, in light of inequality (S.3), it is a proper subspace. We conclude that the solution set of $\mathbf{AX} = \mathbf{B}$ is a proper subset of the solution set of $\mathbf{CAX} = \mathbf{CB}$.

(c) Suppose that $\mathbf{AX} = \mathbf{B}$ is any inconsistent linear system and that $\mathbf{C} = \mathbf{0}$, in which case

$$\text{rank}[\mathbf{C}(\mathbf{A}, \mathbf{B})] = 0 < \text{rank}(\mathbf{A}, \mathbf{B}).$$

Then, $\mathbf{CAX} = \mathbf{CB}$ is clearly consistent.

Alternatively, suppose that

$$\mathbf{A} = \begin{pmatrix} 0 & 0 \\ 1 & 0 \end{pmatrix}, \quad \mathbf{B} = \begin{pmatrix} 1 \\ 0 \end{pmatrix}, \quad \text{and} \quad \mathbf{C} = (1, 0),$$

in which case $\mathbf{AX} = \mathbf{B}$ is obviously inconsistent and

$$\text{rank}[\mathbf{C}(\mathbf{A}, \mathbf{B})] = 1 < 2 = \text{rank}(\mathbf{A}, \mathbf{B}).$$

Then, $\mathbf{CAX} = \mathbf{CB}$ is clearly inconsistent.

EXERCISE 8. Let $\mathbf{A}$ represent a $q \times n$ matrix, $\mathbf{B}$ an $m \times p$ matrix, and $\mathbf{C}$ an $m \times q$ matrix; and suppose that the linear system $\mathbf{CAX} = \mathbf{B}$ (in an $n \times p$ matrix $\mathbf{X}$) is consistent. Show that the value of $\mathbf{AX}$ is the same for every solution to $\mathbf{CAX} = \mathbf{B}$ if and only if $\text{rank}(\mathbf{CA}) = \text{rank}(\mathbf{A})$.

Solution. It suffices (in light of Theorem 11.10.1) to show that $\text{rank}(\mathbf{CA}) = \text{rank}(\mathbf{A})$ if and only if $\mathcal{R}(\mathbf{A}) \subset \mathcal{R}(\mathbf{CA})$ or equivalently [since $\mathcal{R}(\mathbf{CA} \subset \mathcal{R}(\mathbf{A})$] if and only if $\mathcal{R}(\mathbf{A}) = \mathcal{R}(\mathbf{CA})$. If $\mathcal{R}(\mathbf{A}) = \mathcal{R}(\mathbf{CA})$, then it follows from the very definition of the rank of a matrix that $\text{rank}(\mathbf{CA}) = \text{rank}(\mathbf{A})$. Conversely, if $\text{rank}(\mathbf{CA}) = \text{rank}(\mathbf{A})$, then it follows from Corollary 4.4.7 that $\mathcal{R}(\mathbf{A}) = \mathcal{R}(\mathbf{CA})$.

EXERCISE 9. Let $\mathbf{A}$ represent an $m \times n$ matrix, $\mathbf{B}$ an $m \times p$ matrix, and $\mathbf{K}$ an $n \times q$ matrix. Verify (1) that if $\mathbf{X}^*$ and $\mathbf{L}^*$ are the first and second parts, respectively, of any solution to the linear system

$$\begin{pmatrix} \mathbf{A} & \mathbf{0} \\ -\mathbf{K}' & \mathbf{I} \end{pmatrix} \begin{pmatrix} \mathbf{X} \\ \mathbf{L} \end{pmatrix} = \begin{pmatrix} \mathbf{B} \\ \mathbf{0} \end{pmatrix} \tag{$*$}$$

(in $\mathbf{X}$ and $\mathbf{L}$), then $\mathbf{X}^*$ is a solution to the linear system $\mathbf{AX} = \mathbf{B}$ (in $\mathbf{X}$), and $\mathbf{L}^* = \mathbf{K}'\mathbf{X}^*$. and, conversely, if $\mathbf{X}^*$ is any solution to $\mathbf{AX} = \mathbf{B}$, then $\mathbf{X}^*$ and $\mathbf{K}'\mathbf{X}^*$ are the first and second parts, respectively, of some solution to linear system ($*$); and (2) (restricting attention to the special case where $m = n$) that If $\mathbf{X}^*$ and $\mathbf{L}^*$ are the first and second parts, respectively, of any solution to the linear system

$$\begin{pmatrix} \mathbf{A} + \mathbf{KK}' & -\mathbf{K} \\ -\mathbf{K}' & \mathbf{I} \end{pmatrix} \begin{pmatrix} \mathbf{X} \\ \mathbf{L} \end{pmatrix} = \begin{pmatrix} \mathbf{B} \\ \mathbf{0} \end{pmatrix} \tag{$**$}$$

(in $\mathbf{X}$ and $\mathbf{L}$), then $\mathbf{X}^*$ is a solution to the linear system $\mathbf{AX} = \mathbf{B}$ (in $\mathbf{X}$) and $\mathbf{L}^* = \mathbf{K}'\mathbf{X}^*$, and, conversely, if $\mathbf{X}^*$ is any solution to $\mathbf{AX} = \mathbf{B}$, then $\mathbf{X}^*$ and $\mathbf{K}'\mathbf{X}^*$ are the first and second parts, respectively, of some solution to linear system ($**$).

Solution. (1) Suppose that $\mathbf{X}^*$ and $\mathbf{L}^*$ are the first and second parts, respectively, of any solution to linear system ($*$). Then, clearly

$$-\mathbf{K}'\mathbf{X}^* + \mathbf{L}^* = \mathbf{0},$$

or equivalently $\mathbf{L}^* = \mathbf{K}'\mathbf{X}^*$, and

$$\mathbf{AX}^* = \mathbf{AX}^* + \mathbf{0L}^* = \mathbf{B},$$

so that $\mathbf{X}^*$ is a solution to $\mathbf{AX} = \mathbf{B}$.

Conversely, suppose that $\mathbf{X}^*$ is a solution to $\mathbf{AX} = \mathbf{B}$. Then, clearly

$$\begin{pmatrix} \mathbf{A} & \mathbf{0} \\ -\mathbf{K}' & \mathbf{I} \end{pmatrix} \begin{pmatrix} \mathbf{X}^* \\ \mathbf{K}'\mathbf{X}^* \end{pmatrix} = \begin{pmatrix} \mathbf{AX}^* \\ -\mathbf{K}'\mathbf{X}^* + \mathbf{K}'\mathbf{X}^* \end{pmatrix} = \begin{pmatrix} \mathbf{B} \\ \mathbf{0} \end{pmatrix},$$

so that $\mathbf{X}^*$ and $\mathbf{K}'\mathbf{X}^*$ are the first and second parts, respectively, of some solution to linear system ($*$).

(2) Suppose that $\mathbf{X}^*$ and $\mathbf{L}^*$ are the first and second parts, respectively, of any solution to linear system ($**$). Then, clearly

$$-\mathbf{K}'\mathbf{X}^* + \mathbf{L}^* = \mathbf{0},$$

or equivalently $\mathbf{L}^* = \mathbf{K}'\mathbf{X}^*$, and

$$\mathbf{AX}^* = (\mathbf{A} + \mathbf{KK}')\mathbf{X}^* - \mathbf{K}(\mathbf{K}'\mathbf{X}^*) = (\mathbf{A} + \mathbf{KK}')\mathbf{X}^* - \mathbf{KL}^* = \mathbf{B},$$

so that $\mathbf{X}^*$ is a solution to $\mathbf{AX} = \mathbf{B}$.

Conversely, suppose that $\mathbf{X}^*$ is a solution to $\mathbf{AX} = \mathbf{B}$. Then, clearly,

$$\begin{pmatrix} \mathbf{A} + \mathbf{KK}' & -\mathbf{K} \\ -\mathbf{K}' & \mathbf{I} \end{pmatrix} \begin{pmatrix} \mathbf{X}^* \\ \mathbf{K}'\mathbf{X}^* \end{pmatrix} = \begin{pmatrix} (\mathbf{A} + \mathbf{KK}')\mathbf{X}^* - \mathbf{KK}'\mathbf{X}^* \\ -\mathbf{K}'\mathbf{X}^* + \mathbf{K}'\mathbf{X}^* \end{pmatrix} = \begin{pmatrix} \mathbf{AX}^* \\ \mathbf{0} \end{pmatrix} = \begin{pmatrix} \mathbf{B} \\ \mathbf{0} \end{pmatrix},$$

so that $\mathbf{X}^*$ and $\mathbf{K}'\mathbf{X}^*$ are the first and second parts, respectively, of some solution to linear system (∗∗).

12

Projections and Projection Matrices

EXERCISE 1. Let $\mathbf{Y}$ represent a matrix in a linear space $\mathcal{V}$, let $\mathcal{U}$ and $\mathcal{W}$ represent subspaces of $\mathcal{V}$, and take $\{\mathbf{X}_1, \ldots, \mathbf{X}_s\}$ to be a set of matrices that spans $\mathcal{U}$ and $\{\mathbf{Z}_1, \ldots, \mathbf{Z}_t\}$ to be a set that spans $\mathcal{W}$. Verify that $\mathbf{Y} \perp \mathcal{U}$ if and only if $\mathbf{Y} \bullet \mathbf{X}_i = 0$ for $i = 1, \ldots, s$ (i.e., that $\mathbf{Y}$ is orthogonal to $\mathcal{U}$ if and only if $\mathbf{Y}$ is orthogonal to each of the matrices $\mathbf{X}_1, \ldots, \mathbf{X}_s$); and, similarly, that $\mathcal{U} \perp \mathcal{W}$ if and only if $\mathbf{X}_i \bullet \mathbf{Z}_j = 0$ for $i = 1, \ldots, s$ and $j = 1, \ldots, t$ (i.e., that $\mathcal{U}$ is orthogonal to $\mathcal{W}$ if and only if each of the matrices $\mathbf{X}_1, \ldots, \mathbf{X}_s$ is orthogonal to each of the matrices $\mathbf{Z}_1, \ldots, \mathbf{Z}_t$).

Solution. Suppose that $\mathbf{Y} \perp \mathcal{U}$. Then, since $\mathbf{X}_i \in \mathcal{U}$, we have that $\mathbf{Y} \bullet \mathbf{X}_i = 0$ $(i = 1, \ldots, s)$.

Conversely, suppose that $\mathbf{Y} \bullet \mathbf{X}_i = 0$ for $i = 1, \ldots, s$. For each matrix $\mathbf{X} \in \mathcal{U}$, there exist scalars $c_1, \ldots, c_s$ such that $\mathbf{X} = c_1\mathbf{X}_1 + \cdots + c_s\mathbf{X}_s$, so that

$$\mathbf{Y} \bullet \mathbf{X} = c_1(\mathbf{Y} \bullet \mathbf{X}_1) + \cdots + c_s(\mathbf{Y} \bullet \mathbf{X}_s) = 0.$$

Thus, $\mathbf{Y}$ is orthogonal to every matrix in $\mathcal{U}$, that is, $\mathbf{Y} \perp \mathcal{U}$.

The verification of the first assertion is now complete. For purposes of verifying the second assertion, suppose that $\mathcal{U} \perp \mathcal{W}$. Then, since $\mathbf{X}_i \in \mathcal{U}$ and $\mathbf{Y}_j \in \mathcal{W}$, we have that $\mathbf{X}_i \bullet \mathbf{Y}_j = 0$ $(i = 1, \ldots, s;\ j = 1, \ldots, t)$.

Conversely, suppose that $\mathbf{X}_i \bullet \mathbf{Z}_j = 0$ for $i = 1, \ldots, s$ and $j = 1, \ldots, t$. For each matrix $\mathbf{X} \in \mathcal{U}$, there exist scalars $c_1, \ldots, c_s$ such that $\mathbf{X} = c_1\mathbf{X}_1 + \cdots + c_s\mathbf{X}_s$ and, for each matrix $\mathbf{Z}$ in $\mathcal{W}$, there exist scalars $d_1, \ldots, d_t$ such that $\mathbf{Z} = d_1\mathbf{Z}_1 +$

$\cdots + d_t\mathbf{Z}_t$, so that

$$\mathbf{X} \bullet \mathbf{Z} = \sum_i c_i \left(\mathbf{X}_i \bullet \sum_j d_j \mathbf{Z}_j \right) = \sum_i c_i \sum_j d_j (\mathbf{X}_i \bullet \mathbf{Z}_j) = 0.$$

Thus, $\mathcal{U} \perp \mathcal{W}$.

EXERCISE 2. Let $\mathcal{U}$ and $\mathcal{V}$ represent subspaces of $\mathcal{R}^{m \times n}$. Show that if $\dim(\mathcal{V}) > \dim(\mathcal{U})$, then $\mathcal{V}$ contains a nonnull matrix that is orthogonal to $\mathcal{U}$.

Solution. Let $r = \dim(\mathcal{U})$ and $s = \dim(\mathcal{V})$. And, let $\{\mathbf{A}_1, \ldots, \mathbf{A}_r\}$ and $\{\mathbf{B}_1, \ldots, \mathbf{B}_s\}$ represent bases for $\mathcal{U}$ and $\mathcal{V}$, respectively. Further, define $\mathbf{H} = \{h_{ij}\}$ to be the $r \times s$ matrix whose ijth element equals $\mathbf{A}_i \bullet \mathbf{B}_j$.

Now, suppose that $s > r$. Then, since $\text{rank}(\mathbf{H}) \le r < s$, there exists an $s \times 1$ nonnull vector $\mathbf{x} = \{x_j\}$ such that $\mathbf{H}\mathbf{x} = \mathbf{0}$.

Let $\mathbf{C} = x_1\mathbf{B}_1 + \cdots + x_s\mathbf{B}_s$. Then, $\mathbf{C}$ is nonnull. Moreover, for $i = 1, \ldots, r$,

$$\mathbf{A}_i \bullet \mathbf{C} = x_1(\mathbf{A}_i \bullet \mathbf{B}_1) + \cdots + x_s(\mathbf{A}_i \bullet \mathbf{B}_s) = \sum_j h_{ij} x_j.$$

Since $\sum_j h_{ij} x_j$ is the ith element of the vector $\mathbf{H}\mathbf{x}$, $\sum_j h_{ij} x_j = 0$, and hence $\mathbf{A}_i \bullet \mathbf{C} = 0$ $(i = 1, \ldots, r)$. Thus, $\mathbf{C}$ is orthogonal to each of the matrices $\mathbf{A}_1, \ldots, \mathbf{A}_r$. We conclude on the basis of Lemma 12.1.1 (or equivalently the result of Exercise 1) that $\mathbf{C}$ is orthogonal to $\mathcal{U}$.

EXERCISE 3. Let $\mathcal{U}$ represent a subspace of the linear space $\mathcal{R}^m$ of all m-dimensional column vectors. Take $\mathcal{M}$ to be the subspace of $\mathcal{R}^{m \times n}$ defined by $\mathbf{W} \in \mathcal{M}$ if and only if $\mathbf{W} = (\mathbf{w}_1, \ldots, \mathbf{w}_n)$ for some vectors $\mathbf{w}_1, \ldots, \mathbf{w}_n$ in $\mathcal{U}$. Let $\mathbf{Z}$ represent the projection (with respect to the usual inner product) of an $m \times n$ matrix $\mathbf{Y}$ on $\mathcal{M}$, and let $\mathbf{X}$ represent any $m \times p$ matrix whose columns span $\mathcal{U}$. Show that $\mathbf{Z} = \mathbf{X}\mathbf{B}^*$ for any solution $\mathbf{B}^*$ to the linear system

$$\mathbf{X}'\mathbf{X}\mathbf{B} = \mathbf{X}'\mathbf{Y} \quad (\text{in } \mathbf{B}).$$

Solution. Let $\mathbf{y}_i$ represent the ith column of $\mathbf{Y}$, and take $\mathbf{v}_i$ to be the projection (with respect to the usual inner product) of $\mathbf{y}_i$ on $\mathcal{U}$ $(i = 1, \ldots, n)$. Define $\mathbf{V} = (\mathbf{v}_1, \ldots, \mathbf{v}_n)$. Then, by definition, $(\mathbf{y}_i - \mathbf{v}_i)'\mathbf{w} = 0$ for every vector $\mathbf{w}$ in $\mathcal{U}$, so that, for every matrix $\mathbf{W} = (\mathbf{w}_1, \ldots, \mathbf{w}_n)$ in $\mathcal{M}$,

$$\text{tr}[(\mathbf{Y} - \mathbf{V})'\mathbf{W}] = \sum_{i=1}^{n} (\mathbf{y}_i - \mathbf{v}_i)'\mathbf{w}_i = 0,$$

implying that $\mathbf{Z} = \mathbf{V}$.

Now, suppose that $\mathbf{B}^*$ is a solution to $\mathbf{X}'\mathbf{X}\mathbf{B} = \mathbf{X}'\mathbf{Y}$. Then, for $i = 1, \ldots, n$, the ith column $\mathbf{b}_i^*$ of $\mathbf{B}^*$ is clearly a solution to the linear system $\mathbf{X}'\mathbf{X}\mathbf{b}_i = \mathbf{X}'\mathbf{y}_i$ (in

$\mathbf{b}_i$). We conclude, on the basis of Theorem 12.2.1, that $\mathbf{v}_i = \mathbf{X}\mathbf{b}_i^*$ $(i = 1, \ldots, n)$ and hence that

$$\mathbf{Z} = \mathbf{V} = (\mathbf{v}_1, \ldots, \mathbf{v}_n) = (\mathbf{X}\mathbf{b}_1^*, \ldots, \mathbf{X}\mathbf{b}_n^*) = \mathbf{X}\mathbf{B}^*.$$

EXERCISE 4. The projection (with respect to the usual inner product) of an n-dimensional column vector $\mathbf{y}$ on a subspace $\mathcal{U}$ of $\mathcal{R}^n$ in the special case where $n = 3$, $\mathbf{y} = (3, -38/5, 74/5)'$ and $\mathcal{U} = \text{sp}\{\mathbf{x}_1, \mathbf{x}_2, \mathbf{x}_3\}$, with

$$\mathbf{x}_1 = \begin{pmatrix} 0 \\ 3 \\ 6 \end{pmatrix}, \quad \mathbf{x}_2 = \begin{pmatrix} -2 \\ 2 \\ 4 \end{pmatrix}, \quad \mathbf{x}_3 = \begin{pmatrix} -2 \\ 1 \\ 2 \end{pmatrix},$$

was determined to be the vector $(3, 22/5, 44/5)'$—and it was observed that $\mathbf{x}_1$ and $\mathbf{x}_2$ are linearly independent and that $\mathbf{x}_3 = \mathbf{x}_2 - (1/3)\mathbf{x}_1$, with the consequence that $\dim(\mathcal{U}) = 2$. Recompute the projection of $\mathbf{y}$ on $\mathcal{U}$ (in this special case) by taking $\mathbf{X}$ to be the 3×2 matrix

$$\begin{pmatrix} 0 & -2 \\ 3 & 2 \\ 6 & 4 \end{pmatrix}$$

and carrying out the following two steps: (1) compute the solution to the normal equations $\mathbf{X}'\mathbf{X}\mathbf{b} = \mathbf{X}'\mathbf{y}$; and (2) postmultiply $\mathbf{X}$ by the solution you computed in Step (1).

Solution. (1) The normal equations are

$$\begin{pmatrix} 45 & 30 \\ 30 & 24 \end{pmatrix} \mathbf{b} = \begin{pmatrix} 66 \\ 38 \end{pmatrix}.$$

They have the unique solution

$$\mathbf{b} = \begin{pmatrix} 45 & 30 \\ 30 & 24 \end{pmatrix}^{-1} \begin{pmatrix} 66 \\ 38 \end{pmatrix} = \begin{pmatrix} 2/15 & -1/6 \\ -1/6 & 1/4 \end{pmatrix} \begin{pmatrix} 66 \\ 38 \end{pmatrix} = \begin{pmatrix} 37/15 \\ -3/2 \end{pmatrix}.$$

(2) The projection of $\mathbf{y}$ on $\mathcal{U}$ is

$$\mathbf{z} = \begin{pmatrix} 0 & -2 \\ 3 & 2 \\ 6 & 4 \end{pmatrix} \begin{pmatrix} 37/15 \\ -3/2 \end{pmatrix} = \begin{pmatrix} 3 \\ 22/5 \\ 44/5 \end{pmatrix}.$$

EXERCISE 5. Let $\mathbf{X}$ represent any $n \times p$ matrix. If a $p \times n$ matrix $\mathbf{B}^*$ is a solution to the linear system $\mathbf{X}'\mathbf{X}\mathbf{B} = \mathbf{X}'$ (in $\mathbf{B}$), then $\mathbf{B}^*$ is a generalized inverse of $\mathbf{X}$ and $\mathbf{X}\mathbf{B}^*$ is symmetric. Show that, conversely, if a $p \times n$ matrix $\mathbf{G}$ is a generalized inverse of $\mathbf{X}$ and if $\mathbf{X}\mathbf{G}$ is symmetric, then $\mathbf{X}'\mathbf{X}\mathbf{G} = \mathbf{X}'$ (i.e., $\mathbf{G}$ is a solution to $\mathbf{X}'\mathbf{X}\mathbf{B} = \mathbf{X}'$).

Solution. Suppose that $\mathbf{G}$ is a generalized inverse of $\mathbf{X}$ and $\mathbf{XG}$ is symmetric. Then,

$$\mathbf{X}'\mathbf{XG} = \mathbf{X}'(\mathbf{XG})' = (\mathbf{XGX})' = \mathbf{X}'.$$

EXERCISE 6. Using the result of Part (b) of Exercise 9.3 (or otherwise), show that, for any nonnull symmetric matrix $\mathbf{A}$,

$$\mathbf{P_A} = \mathbf{B}(\mathbf{TB})^{-1}\mathbf{T},$$

where $\mathbf{B}$ is any matrix of full column rank and $\mathbf{T}$ any matrix of full row rank such that $\mathbf{A} = \mathbf{BT}$. (That $\mathbf{TB}$ is nonsingular follows from the result of Exercise 8.3.)

Solution. Let $\mathbf{L}$ represent a left inverse of $\mathbf{B}$ and $\mathbf{R}$ a right inverse of $\mathbf{T}$. Then, according to Part (b) of Exercise 9.3, the matrix $\mathbf{R}(\mathbf{TB})^{-1}\mathbf{L}$ is a generalized inverse of $\mathbf{A}^2$ or equivalently (since $\mathbf{A}$ is symmetric) of $\mathbf{A}'\mathbf{A}$. Thus,

$$\mathbf{P_A} = \mathbf{AR}(\mathbf{TB})^{-1}\mathbf{LA} = \mathbf{BTR}(\mathbf{TB})^{-1}\mathbf{LBT} = \mathbf{BI}(\mathbf{TB})^{-1}\mathbf{IT} = \mathbf{B}(\mathbf{TB})^{-1}\mathbf{T}.$$

EXERCISE 7. Let $\mathcal{V}$ represent a k-dimensional subspace of the linear space $\mathcal{R}^n$ of all n-dimensional column vectors. Take $\mathbf{X}$ to be any $n \times p$ matrix whose columns span $\mathcal{V}$, let $\mathcal{U}$ represent a subspace of $\mathcal{V}$, and define $\mathbf{A}$ to be the projection matrix for $\mathcal{U}$. Show (1) that a matrix $\mathbf{B}$ (of dimensions $n \times n$) is such that $\mathbf{By}$ is the projection of $\mathbf{y}$ on $\mathcal{U}$ for every $\mathbf{y} \in \mathcal{V}$ if and only if $\mathbf{B} = \mathbf{A} + \mathbf{Z}_*'$ for some solution $\mathbf{Z}_*$ to the homogeneous linear system $\mathbf{X}'\mathbf{Z} = \mathbf{0}$ (in an $n \times n$ matrix $\mathbf{Z}$) and (2) that, unless $k = n$, there is more than one matrix $\mathbf{B}$ such that $\mathbf{By}$ is the projection of $\mathbf{y}$ on $\mathcal{U}$ for every $\mathbf{y} \in \mathcal{V}$.

Solution. (1) The vector $\mathbf{Ay}$ is the projection of $\mathbf{y}$ on $\mathcal{U}$ for every $\mathbf{y} \in \mathcal{R}^n$. Thus, $\mathbf{By}$ is the projection of $\mathbf{y}$ on $\mathcal{U}$ for every $\mathbf{y} \in \mathcal{V}$ if and only if $\mathbf{By} = \mathbf{Ay}$ for every $\mathbf{y} \in \mathcal{V}$, or equivalently if and only if $\mathbf{BXr} = \mathbf{AXr}$ for every $p \times 1$ vector $\mathbf{r}$, and hence (in light of Lemma 2.3.2) if and only if $\mathbf{BX} = \mathbf{AX}$.

Furthermore, $\mathbf{BX} = \mathbf{AX}$ if and only if $\mathbf{X}'(\mathbf{B} - \mathbf{A})' = \mathbf{0}$, or equivalently if and only if $(\mathbf{B} - \mathbf{A})'$ is a solution to the homogeneous linear system $\mathbf{X}'\mathbf{Z} = \mathbf{0}$ (in an $n \times n$ matrix $\mathbf{Z}$), and hence if and only if $\mathbf{B}' = \mathbf{A}' + \mathbf{Z}_*$ for some solution $\mathbf{Z}_*$ to $\mathbf{X}'\mathbf{Z} = \mathbf{0}$, that is, if and only if $\mathbf{B} = \mathbf{A} + \mathbf{Z}_*'$ for some solution $\mathbf{Z}_*$ to $\mathbf{X}'\mathbf{Z} = \mathbf{0}$.

(2) According to Lemma 11.3.2, the solution space of the homogeneous linear system $\mathbf{X}'\mathbf{Z} = \mathbf{0}$ (in an $n \times n$ matrix $\mathbf{Z}$) is of dimension $n[n - \text{rank}(\mathbf{X})] = n(n-k)$. Thus, unless $k = n$, there is more than one solution to $\mathbf{AZ} = \mathbf{0}$, and hence [in light of the result of Part (1)] there is more than one matrix $\mathbf{B}$ such that $\mathbf{By}$ is the projection of $\mathbf{y}$ on $\mathcal{U}$ for every $\mathbf{y} \in \mathcal{V}$.

EXERCISE 8. Let $\{\mathbf{A}_1, \ldots, \mathbf{A}_k\}$ represent a nonempty linearly independent set of matrices in a linear space $\mathcal{V}$. And, define (as in Gram-Schmidt orthogonalization) k

nonnull orthogonal linear combinations, say $\mathbf{B}_1, \ldots, \mathbf{B}_k$, of $\mathbf{A}_1, \ldots, \mathbf{A}_k$ as follows:

$$\begin{aligned}
\mathbf{B}_1 &= \mathbf{A}_1, \\
\mathbf{B}_2 &= \mathbf{A}_2 - x_{12}\mathbf{B}_1, \\
&\vdots \\
\mathbf{B}_j &= \mathbf{A}_j - x_{j-1,j}\mathbf{B}_{j-1} - \cdots - x_{1j}\mathbf{B}_1, \\
&\vdots \\
\mathbf{B}_k &= \mathbf{A}_k - x_{k-1,k}\mathbf{B}_{k-1} - \cdots - x_{1k}\mathbf{B}_1,
\end{aligned}$$

where (for $i < j = 1, \ldots, k$)

$$x_{ij} = \frac{\mathbf{A}_j \cdot \mathbf{B}_i}{\mathbf{B}_i \cdot \mathbf{B}_i}.$$

Show that $\mathbf{B}_j$ is the (orthogonal) projection of $\mathbf{A}_j$ on some subspace $\mathcal{U}_j$ (of $\mathcal{V}$) and describe $\mathcal{U}_j$ $(j = 2, \ldots, k)$.

Solution. For $j = 1, \ldots, k$, define $\mathbf{C}_j = \| \mathbf{B}_j \|^{-1}\mathbf{B}_j$ (as in Corollary 6.4.2). And, define $\mathcal{W}_j = \text{sp}(\mathbf{C}_1, \ldots, \mathbf{C}_j)$. Then, for $j = 2, \ldots, k$,

$$\mathbf{B}_j = \mathbf{A}_j - \sum_{i=1}^{j-1} \frac{\mathbf{A}_j \cdot \mathbf{B}_i}{\mathbf{B}_i \cdot \mathbf{B}_i}\mathbf{B}_i = \mathbf{A}_j - \sum_{i=1}^{j-1} \frac{\mathbf{A}_j \cdot \mathbf{B}_i}{\| \mathbf{B}_i \|}\mathbf{C}_i = \mathbf{A}_j - \sum_{i=1}^{j-1}(\mathbf{A}_j \cdot \mathbf{C}_i)\mathbf{C}_i.$$

Moreover, upon observing that the set $\{\mathbf{C}_1, \ldots, \mathbf{C}_j\}$ is orthonormal and applying result (1.1), we find that $\sum_{i=1}^{j-1}(\mathbf{A}_j \cdot \mathbf{C}_i)\mathbf{C}_i$ is the projection of $\mathbf{A}_j$ on $\mathcal{W}_{j-1}$.

Thus, it follows from Theorem 12.5.8 that $\mathbf{B}_j$ is the projection of $\mathbf{A}_j$ on $\mathcal{W}_{j-1}^{\perp}$. And, since (in light of the discussion of Section 6.4b) $\mathcal{W}_{j-1} = \text{sp}(\mathbf{A}_1, \ldots, \mathbf{A}_{j-1})$, we conclude that $\mathbf{B}_j$ is the projection of $\mathbf{A}_j$ on the orthogonal complement of the subspace (of $\mathcal{V}$) spanned by $\mathbf{A}_1, \ldots, \mathbf{A}_{j-1}$.

13
Determinants

EXERCISE 1. Let

$$\mathbf{A} = \begin{pmatrix} a_{11} & a_{12} & a_{13} & \boxed{a_{14}} \\ \boxed{a_{21}} & a_{22} & a_{23} & a_{24} \\ a_{31} & a_{32} & \boxed{a_{33}} & a_{34} \\ a_{41} & \boxed{a_{42}} & a_{43} & a_{44} \end{pmatrix}.$$

(a) Write out all of the pairs that can be formed from the four boxed elements of **A**.

(b) Indicate which of the pairs from Part (a) are positive and which are negative.

(c) Use the formula

$$\sigma_n(1, i_1; \ldots; n, i_n) = \sigma_n(i_1, 1; \ldots; i_n, n) = \phi_n(i_1, \ldots, i_n)$$

(in which $i_1, \ldots, i_n$ represents an arbitrary permutation of the first n positive integers) to compute the number of pairs from Part (a) that are negative, and check that the result of this computation is consistent with your answer to Part (b).

Solution. (a) and (b)

Pair	"Sign"
$a_{14},\ a_{21}$	$-$
$a_{14},\ a_{33}$	$-$
$a_{14},\ a_{42}$	$-$
$a_{21},\ a_{33}$	$+$
$a_{21},\ a_{42}$	$+$
$a_{33},\ a_{42}$	$-$

(c) $\phi_4(4, 1, 3, 2) = 3+0+1 = 4$ [or alternatively $\phi_4(2, 4, 3, 1) = 1+2+1 = 4$].

EXERCISE 2. Consider the $n \times n$ matrix

$$\mathbf{A} = \begin{pmatrix} x+\lambda & x & \dots & x \\ x & x+\lambda & & x \\ \vdots & & \ddots & \\ x & x & & x+\lambda \end{pmatrix} .$$

"Recall" that

$$|\mathbf{S}| = |\mathbf{R}| \qquad (*)$$

for any $n \times n$ matrix $\mathbf{R}$ and for any matrix $\mathbf{S}$ formed from $\mathbf{R}$ by adding to any one of its rows or columns, scalar multiples of one or more other rows or columns; and use this result to show that

$$|\mathbf{A}| = \lambda^{n-1}(nx + \lambda) .$$

(*Hint.* Add the last $n - 1$ columns of $\mathbf{A}$ to the first column, and then subtract the first row of the resultant matrix from each of the last $n - 1$ rows).

Solution. The matrix obtained from $\mathbf{A}$ by adding the last $n - 1$ columns of $\mathbf{A}$ to the first column is

$$\mathbf{B} = \begin{pmatrix} nx+\lambda & x & \dots & x \\ nx+\lambda & x+\lambda & & x \\ \vdots & & \ddots & \\ nx+\lambda & x & & x+\lambda \end{pmatrix} .$$

The matrix obtained from $\mathbf{B}$ by subtracting the first row of $\mathbf{B}$ from each of the next i rows is

$$\mathbf{C}_i = \begin{pmatrix} nx+\lambda & x & \dots & x & x & \dots & x \\ 0 & \lambda & & 0 & 0 & \dots & 0 \\ \vdots & & \ddots & & \vdots & & \vdots \\ 0 & 0 & & \lambda & 0 & \dots & 0 \\ nx+\lambda & x & \dots & x & x+\lambda & & x \\ \vdots & \vdots & & \vdots & & \ddots & \\ nx+\lambda & x & \dots & x & x & & x+\lambda \end{pmatrix} \begin{matrix} \left.\begin{matrix} \\ \\ \\ \end{matrix}\right\} i \text{ rows} \\ \left.\begin{matrix} \\ \\ \\ \end{matrix}\right\} n-1-i \text{ rows} \end{matrix}$$

Observing that $\mathbf{C}_i$ can be obtained from $\mathbf{C}_{i-1}$ by subtracting the first row of $\mathbf{C}_{i-1}$ from the $(i + 1)$th row and making use of result $(*)$ (or equivalently Theorem 13.2.10) and Lemma 13.1.1, we find that

$$|\mathbf{A}| = |\mathbf{B}| = |\mathbf{C}_1| = |\mathbf{C}_2| = \cdots = |\mathbf{C}_{n-1}| = \lambda^{n-1}(nx + \lambda).$$

EXERCISE 3. Let $\mathbf{A}$ represent an $n \times n$ nonsingular matrix. Show that if the elements of $\mathbf{A}$ and $\mathbf{A}^{-1}$ are all integers, then $|\mathbf{A}| = \pm 1$.

Solution. Suppose that the elements of $\mathbf{A}$ and $\mathbf{A}^{-1}$ are all integers. Then, it follows from the very definition of a determinant that $|\mathbf{A}|$ and $|\mathbf{A}^{-1}|$ are both integers. Thus, since (according to Theorem 13.3.7) $|\mathbf{A}^{-1}| = 1/|\mathbf{A}|$, $|\mathbf{A}|$ and $1/|\mathbf{A}|$ are both integers. We conclude that $|\mathbf{A}| = \pm 1$.

EXERCISE 4. Let $\mathbf{T}$ represent an $m \times m$ matrix, $\mathbf{U}$ an $m \times n$ matrix, $\mathbf{V}$ an $n \times m$ matrix, and $\mathbf{W}$ an $n \times n$ matrix. Show that if $\mathbf{T}$ is nonsingular, then

$$\begin{vmatrix} \mathbf{V} & \mathbf{W} \\ \mathbf{T} & \mathbf{U} \end{vmatrix} = \begin{vmatrix} \mathbf{U} & \mathbf{T} \\ \mathbf{W} & \mathbf{V} \end{vmatrix} = (-1)^{mn} |\mathbf{T}| |\mathbf{W} - \mathbf{V}\mathbf{T}^{-1}\mathbf{U}|.$$

Solution. It follows from Theorem 13.2.7 that

$$\begin{vmatrix} \mathbf{V} & \mathbf{W} \\ \mathbf{T} & \mathbf{U} \end{vmatrix} = (-1)^{mn} \begin{vmatrix} \mathbf{T} & \mathbf{U} \\ \mathbf{V} & \mathbf{W} \end{vmatrix} \quad \text{and} \quad \begin{vmatrix} \mathbf{U} & \mathbf{T} \\ \mathbf{W} & \mathbf{V} \end{vmatrix} = (-1)^{mn} \begin{vmatrix} \mathbf{T} & \mathbf{U} \\ \mathbf{V} & \mathbf{W} \end{vmatrix}.$$

Thus, making use of Theorem 13.3.8, we find that

$$\begin{vmatrix} \mathbf{V} & \mathbf{W} \\ \mathbf{T} & \mathbf{U} \end{vmatrix} = \begin{vmatrix} \mathbf{U} & \mathbf{T} \\ \mathbf{W} & \mathbf{V} \end{vmatrix} = (-1)^{mn} |\mathbf{T}| |\mathbf{W} - \mathbf{V}\mathbf{T}^{-1}\mathbf{U}|.$$

EXERCISE 5. Compute the determinant of the $n \times n$ matrix $\mathbf{A} = \{a_{ij}\}$ in the special case where $n = 4$ and

$$\mathbf{A} = \begin{pmatrix} 0 & 4 & 0 & 5 \\ 1 & 0 & -1 & 2 \\ 0 & 3 & 0 & -2 \\ 0 & 0 & -6 & 0 \end{pmatrix}.$$

Do so in each of the following two ways:

(a) by finding and summing the nonzero terms in the expression

$$\sum (-1)^{\phi_n(j_1,\ldots,j_n)} a_{1j_1} \cdots a_{nj_n} \quad \text{or} \quad \sum (-1)^{\phi_n(i_1,\ldots,i_n)} a_{i_1 1} \cdots a_{i_n n},$$

(where $j_1, \ldots, j_n$ or $i_1, \ldots, i_n$ is a permutation of the first n positive integers and the summation is over all such permutations);

(b) by repeated expansion in terms of cofactors—use the (general) formula

$$|\mathbf{A}| = \sum_{j=1}^{n} a_{ij}\alpha_{ij} \quad \text{or} \quad |\mathbf{A}| = \sum_{i=1}^{n} a_{ij}\alpha_{ij}$$

(where i or j, respectively, is any integer between 1 and n inclusive and where α_{ij} is the cofactor of a_{ij}) to expand $|\mathbf{A}|$ (in the special case) in terms of the determinants of 3×3 matrices, to expand the determinants of the 3×3 matrices in terms of the determinants of 2×2 matrices, and finally to expand the determinants of the 2×2 matrices in terms of the determinants of 1×1 matrices.

Solution. (a)

$$\begin{aligned}|\mathbf{A}| &= (-1)^{\phi_4(2,1,4,3)}4(1)(-2)(-6) + (-1)^{\phi_4(4,1,2,3)}5(1)(3)(-6)\\ &= (-1)^{1+0+1}48 + (-1)^{3+0+0}(-90)\\ &= 48 + 90\\ &= 138.\end{aligned}$$

(b)

$$\begin{aligned}|\mathbf{A}| &= (1)(-1)^{2+1}\begin{vmatrix}4 & 0 & 5\\ 3 & 0 & -2\\ 0 & -6 & 0\end{vmatrix}\\ &= (-1)^3(-6)(-1)^{3+2}\begin{vmatrix}4 & 5\\ 3 & -2\end{vmatrix}\\ &= (-1)^3(-6)(-1)^5[4(-1)^{1+1}(-2) + 5(-1)^{1+2}(3)]\\ &= (-6)(-8-15)\\ &= 138.\end{aligned}$$

EXERCISE 6. ,Let $\mathbf{A} = \{a_{ij}\}$ represent an $n \times n$ matrix. Verify that if $\mathbf{A}$ is symmetric, then the matrix of cofactors (of $\mathbf{A}$) is also symmetric.

Solution. Let α_{ij} represent the cofactor of a_{ij}, let $\mathbf{A}_{ij}$ represent the $(n-1)\times(n-1)$ submatrix of $\mathbf{A}$ obtained by striking out the ith row and the jth column (of $\mathbf{A}$), and let $\mathbf{B}_{ji}$ represent the $(n-1) \times (n-1)$ submatrix of $\mathbf{A}'$ obtained by striking out the jth row and the ith column of $\mathbf{A}'$. Then, making use of Lemma 13.2.1 and result (2.1.1), we find that

$$\alpha_{ij} = (-1)^{i+j}|\mathbf{A}_{ij}| = (-1)^{i+j}|\mathbf{A}'_{ij}| = (-1)^{i+j}|\mathbf{B}_{ji}|.$$

Moreover, if $\mathbf{A}$ is symmetric, then $\mathbf{B}_{ji} = \mathbf{A}_{ji}$, implying that

$$\alpha_{ij} = (-1)^{i+j}|\mathbf{A}_{ji}| = \alpha_{ji}$$

and hence that the matrix of cofactors is symmetric.

EXERCISE 7. Let $\mathbf{A}$ represent an $n \times n$ matrix.

(a) Show that if $\mathbf{A}$ is singular, then $\text{adj}(\mathbf{A})$ is singular.

(b) Show that $\det[\text{adj}(\mathbf{A})] = [\det(\mathbf{A})]^{n-1}$.

Solution. (a) If $\mathbf{A}$ is null, then it is clear that $\text{adj}(\mathbf{A}) = \mathbf{0}$ and hence that $\text{adj}(\mathbf{A})$ is singular.

Suppose now that $\mathbf{A}$ is singular but nonnull, in which case $\mathbf{A}$ contains a nonnull row, say the ith row $\mathbf{a}_i'$. Since $\mathbf{A}$ is singular, $|\mathbf{A}| = 0$, and it follows from Theorem 13.5.3 that $\mathbf{A}$ adj $(\mathbf{A}) = \mathbf{0}$ and hence that

$$\mathbf{a}_i' \operatorname{adj}(\mathbf{A}) = \mathbf{0},$$

implying (since $\mathbf{a}_i'$ is nonnull) that the rows of adj($\mathbf{A}$) are linearly dependent. We conclude that adj($\mathbf{A}$) is singular.

(b) Making use of Theorems 13.3.4 and 13.5.3, Corollary 13.2.4, and result (1.9), we find that

$$|\mathbf{A}|\ |\operatorname{adj}(\mathbf{A})| = |\mathbf{A} \operatorname{adj}(\mathbf{A})| = \det(|\mathbf{A}|\mathbf{I}_n) = |\mathbf{A}|^n |\mathbf{I}_n| = |\mathbf{A}|^n. \qquad \text{(S.1)}$$

If $\mathbf{A}$ is nonsingular, then $|\mathbf{A}| \neq 0$, and it follows from result (S.1) that

$$|\operatorname{adj}(\mathbf{A})| = |\mathbf{A}|^{n-1}.$$

Alternatively, if $\mathbf{A}$ is singular, then it follows from Part (a) that adj($\mathbf{A}$) is singular and hence that

$$|\operatorname{adj}(\mathbf{A})| = 0 = |\mathbf{A}|^{n-1}.$$

EXERCISE 8. For any $n \times n$ nonsingular matrix $\mathbf{A}$,

$$\mathbf{A}^{-1} = (1/|\mathbf{A}|) \operatorname{adj}(\mathbf{A}). \qquad (*)$$

Use formula $(*)$ to verify that, for any 2×2 nonsingular matrix $\mathbf{A} = \begin{pmatrix} a_{11} & a_{12} \\ a_{21} & a_{22} \end{pmatrix}$,

$$\mathbf{A}^{-1} = (1/k) \begin{pmatrix} a_{22} & -a_{12} \\ -a_{21} & a_{11} \end{pmatrix}, \qquad (**)$$

where $k = a_{11}a_{22} - a_{12}a_{21}$.

Solution. Let a_{ij} represent the ijth element of a 2×2 matrix $\mathbf{A}$, and let α_{ij} represent the cofactor of a_{ij} $(i, j = 1, 2)$. Then, as a special case of formula $(*)$ [or equivalently formula (5.7)], we have that

$$\mathbf{A}^{-1} = (1/|\mathbf{A}|) \begin{pmatrix} \alpha_{11} & \alpha_{21} \\ \alpha_{12} & \alpha_{22} \end{pmatrix}. \qquad \text{(S.2)}$$

Moreover, it follows from the very definition of a cofactor and from formulas (1.3) and (1.4) that

$$\alpha_{11} = (-1)^{1+1}a_{22} = a_{22}, \qquad \alpha_{21} = (-1)^{2+1}a_{12} = -a_{12},$$
$$\alpha_{12} = (-1)^{1+2}a_{21} = -a_{21}, \qquad \alpha_{22} = (-1)^{2+2}a_{11} = a_{11}, \quad \text{and}$$
$$|\mathbf{A}| = a_{11}a_{22} - a_{12}a_{21}.$$

Upon substituting these expressions in formula (S.2), we obtain formula $(**)$ [or equivalently formula (8.1.2)].

EXERCISE 9. Let

$$\mathbf{A} = \begin{pmatrix} 2 & 0 & -1 \\ -1 & 3 & 1 \\ 0 & -4 & 5 \end{pmatrix}.$$

(a) Compute the cofactor of each element of **A**.

(b) Compute $|\mathbf{A}|$ by expanding $|\mathbf{A}|$ in terms of the cofactors of the elements of the second row of **A**, and then check your answer by expanding $|\mathbf{A}|$ in terms of the cofactors of the elements of the second column of **A**.

(c) Use formula (∗) of Exercise 8 to compute $\mathbf{A}^{-1}$.

Solution. (a) Let α_{ij} represent the cofactor of the ijth element of **A**. Then,

$$\alpha_{11} = (-1)^{1+1}\begin{vmatrix} 3 & 1 \\ -4 & 5 \end{vmatrix} = 19, \qquad \alpha_{12} = (-1)^{1+2}\begin{vmatrix} -1 & 1 \\ 0 & 5 \end{vmatrix} = 5,$$

$$\alpha_{13} = (-1)^{1+3}\begin{vmatrix} -1 & 3 \\ 0 & -4 \end{vmatrix} = 4, \qquad \alpha_{21} = (-1)^{2+1}\begin{vmatrix} 0 & -1 \\ -4 & 5 \end{vmatrix} = 4,$$

$$\alpha_{22} = (-1)^{2+2}\begin{vmatrix} 2 & -1 \\ 0 & 5 \end{vmatrix} = 10, \qquad \alpha_{23} = (-1)^{2+3}\begin{vmatrix} 2 & 0 \\ 0 & -4 \end{vmatrix} = 8,$$

$$\alpha_{31} = (-1)^{3+1}\begin{vmatrix} 0 & -1 \\ 3 & 1 \end{vmatrix} = 3, \qquad \alpha_{32} = (-1)^{3+2}\begin{vmatrix} 2 & -1 \\ -1 & 1 \end{vmatrix} = -1, \text{ and}$$

$$\alpha_{33} = (-1)^{3+3}\begin{vmatrix} 2 & 0 \\ -1 & 3 \end{vmatrix} = 6.$$

(b) Expanding $|\mathbf{A}|$ in terms of the cofactors of the elements of the second row of **A** gives

$$|\mathbf{A}| = (-1)4 + 3(10) + 1(8) = 34.$$

Expanding $|\mathbf{A}|$ in terms of the cofactors of the elements of the second column of **A** gives

$$|\mathbf{A}| = 0(5) + 3(10) + (-4)(-1) = 34.$$

(c) Substituting from Parts (a) and (b) in formula (∗) of Exercise 8 [or equivalently in formula (5.7)], we find that

$$\mathbf{A}^{-1} = (1/34)\begin{pmatrix} 19 & 4 & 3 \\ 5 & 10 & -1 \\ 4 & 8 & 6 \end{pmatrix}.$$

EXERCISE 10. Let $\mathbf{A} = \{a_{ij}\}$ represent an $n \times n$ matrix (where $n \geq 2$), and let α_{ij} represent the cofactor of a_{ij}.

(a) Show [by for instance, making use of the result of Part (b) of Exercise 11.3] that if rank($\mathbf{A}$) $= n - 1$, then there exists a scalar c such that adj($\mathbf{A}$) $= c\mathbf{x}\mathbf{y}'$, where $\mathbf{x} = \{x_j\}$ and $\mathbf{y} = \{y_i\}$ are any nonnull n-dimensional column vectors such that $\mathbf{A}\mathbf{x} = \mathbf{0}$ and $\mathbf{A}'\mathbf{y} = \mathbf{0}$. Show also that c is nonzero and is expressible as $c = \alpha_{ij}/(y_i x_j)$ for any i and j such that $y_i \neq 0$ and $x_j \neq 0$.

(b) Show that if rank($\mathbf{A}$) $\leq n - 2$, then adj($\mathbf{A}$) $= \mathbf{0}$.

Solution. (a) Suppose that rank($\mathbf{A}$) $= n-1$. Then, det($\mathbf{A}$) $= 0$ and hence (according to Theorem 13.5.3) $\mathbf{A}$ adj($\mathbf{A}$) $=$ adj($\mathbf{A}$)$\mathbf{A} = \mathbf{0}$. Thus, it follows from the result of Part (b) of Exercise 11.3 that there exists a scalar c such that adj($\mathbf{A}$) $= c\mathbf{x}\mathbf{y}'$ [or equivalently such that (adj $\mathbf{A}$)$' = c\mathbf{y}\mathbf{x}'$] and hence such that (for all i and j)

$$\alpha_{ij} = c y_i x_j. \tag{S.3}$$

Moreover, since (according to Theorem 4.4.10) $\mathbf{A}$ contains an $(n-1) \times (n-1)$ nonsingular submatrix, $\alpha_{ij} \neq 0$ for some i and j, implying that $c \neq 0$. And, for any i and j such that $y_i \neq 0$ and $x_j \neq 0$, we have [in light of result (S.3)] that $c = \alpha_{ij}/(y_i x_j)$.

(b) If rank($\mathbf{A}$) $\leq n - 2$, then it follows from Theorem 4.4.10 that every $(n - 1) \times (n - 1)$ submatrix of $\mathbf{A}$ is singular, implying that $\alpha_{ij} = 0$ for all i and j or equivalently that adj($\mathbf{A}$) $= \mathbf{0}$.

EXERCISE 11. Let $\mathbf{A}$ represent an $n \times n$ nonsingular matrix and $\mathbf{b}$ an $n \times 1$ vector. Show that the solution to the linear system $\mathbf{A}\mathbf{x} = \mathbf{b}$ (in $\mathbf{x}$) is the $n \times 1$ vector whose jth component is

$$|\mathbf{A}_j| \,/\, |\mathbf{A}|,$$

where $\mathbf{A}_j$ is a matrix formed from $\mathbf{A}$ by substituting $\mathbf{b}$ for the jth column of $\mathbf{A}$ $(j = 1, \ldots, n)$. [This result is called Cramer's rule, after Gabriel Cramer (1704–1752).]

Solution. The (unique) solution to $\mathbf{A}\mathbf{x} = \mathbf{b}$ is expressible as $\mathbf{A}^{-1}\mathbf{b}$. Let b_i represent the ith element of $\mathbf{b}$ and α_{ij} the cofactor of the ijth element of $\mathbf{A}$ $(i, j = 1, \ldots, n)$. It follows from Corollary 13.5.4 that the jth element of $\mathbf{A}^{-1}\mathbf{b}$ is

$$(1/|\mathbf{A}|)(\alpha_{1j}, \ldots, \alpha_{nj})\begin{pmatrix} b_1 \\ \vdots \\ b_n \end{pmatrix} = (1/|\mathbf{A}|)\sum_{i=1}^{n} b_i \alpha_{ij}.$$

Clearly, the cofactor of the ijth element of $\mathbf{A}_j$ is the same as the cofactor of the ijth element of $\mathbf{A}$ $(i = 1, \ldots, n)$, so that, according to Theorem 13.5.1, the jth element of $\mathbf{A}^{-1}\mathbf{b}$ is $|\mathbf{A}_j| \,/\, |\mathbf{A}|$.

EXERCISE 12. Let c represent a scalar, let $\mathbf{x}$ and $\mathbf{y}$ represent $n \times 1$ vectors, and let $\mathbf{A}$ represent an $n \times n$ matrix.

(a) Show that

$$\begin{vmatrix} \mathbf{A} & \mathbf{y} \\ \mathbf{x}' & c \end{vmatrix} = c|\mathbf{A}| - \mathbf{x}'\text{adj}(\mathbf{A})\mathbf{y}. \tag{E.1}$$

(b) Show that, in the special case where $\mathbf{A}$ is nonsingular, result (E.1) can be reexpressed as

$$\begin{vmatrix} \mathbf{A} & \mathbf{y} \\ \mathbf{x}' & c \end{vmatrix} = |\mathbf{A}|(c - \mathbf{x}'\mathbf{A}^{-1}\mathbf{y}),$$

in agreement with the more general result that, for any $n \times n$ nonsingular matrix $\mathbf{T}$, $n \times m$ matrix $\mathbf{U}$, $m \times n$ matrix $\mathbf{V}$, and $m \times m$ matrix $\mathbf{W}$,

$$\begin{vmatrix} \mathbf{T} & \mathbf{U} \\ \mathbf{V} & \mathbf{W} \end{vmatrix} = \begin{vmatrix} \mathbf{W} & \mathbf{V} \\ \mathbf{U} & \mathbf{T} \end{vmatrix} = |\mathbf{T}||\mathbf{W} - \mathbf{V}\mathbf{T}^{-1}\mathbf{U}|. \tag{$*$}$$

Solution. (a) Denote by x_i the ith element of $\mathbf{x}$, and by y_i the ith element of $\mathbf{y}$. Let $\mathbf{A}_j$ represent the $n \times (n-1)$ submatrix of $\mathbf{A}$ obtained by striking out the jth column, let $\mathbf{A}_{ij}$ represent the $(n-1) \times (n-1)$ submatrix of $\mathbf{A}$ obtained by striking out the ith row and the jth column, and let α_{ij} represent the cofactor of the ijth element of $\mathbf{A}$.

Expanding $\begin{vmatrix} \mathbf{A} & \mathbf{y} \\ \mathbf{x}' & c \end{vmatrix}$ in terms of the cofactors of the last row of $\begin{pmatrix} \mathbf{A} & \mathbf{y} \\ \mathbf{x}' & c \end{pmatrix}$, we obtain

$$\begin{vmatrix} \mathbf{A} & \mathbf{y} \\ \mathbf{x}' & c \end{vmatrix} = \sum_j x_j(-1)^{n+1+j}\det(\mathbf{A}_j, \mathbf{y}) + c(-1)^{2(n+1)}|\mathbf{A}|. \tag{S.4}$$

Further, expanding $\det(\mathbf{A}_j, \mathbf{y})$ in terms of the cofactors of the last column of $(\mathbf{A}_j, \mathbf{y})$, we obtain

$$\det(\mathbf{A}_j, \mathbf{y}) = \sum_i y_i(-1)^{i+n}|\mathbf{A}_{ij}|. \tag{S.5}$$

Substituting expression (S.5) in equality (S.4), we find that

$$\begin{aligned} \begin{vmatrix} \mathbf{A} & \mathbf{y} \\ \mathbf{x}' & c \end{vmatrix} &= \sum_{i,j} y_i x_j(-1)^{2n+1+i+j}|\mathbf{A}_{ij}| + c|\mathbf{A}| \\ &= c|\mathbf{A}| - \sum_{i,j} y_i x_j(-1)^{i+j}|\mathbf{A}_{ij}| \\ &= c|\mathbf{A}| - \sum_{i,j} y_i x_j \alpha_{ij} \\ &= c|\mathbf{A}| - \mathbf{x}'\,\text{adj}(\mathbf{A})\mathbf{y}. \end{aligned}$$

(b) Suppose that $\mathbf{A}$ is nonsingular, in which case $|\mathbf{A}| \neq 0$. Then, using Corollary 13.5.4, result (E.1) can be reexpressed as

$$\begin{aligned} \begin{vmatrix} \mathbf{A} & \mathbf{y} \\ \mathbf{x}' & c \end{vmatrix} &= |\mathbf{A}|\{c - \mathbf{x}'[(1/|\mathbf{A}|)\,\text{adj}(\mathbf{A})]\mathbf{y}\} \\ &= |\mathbf{A}|(c - \mathbf{x}'\mathbf{A}^{-1}\mathbf{y}). \end{aligned}$$

Note that this same expression can be obtained by setting $\mathbf{T} = \mathbf{A}$, $\mathbf{U} = \mathbf{y}$, $\mathbf{V} = \mathbf{x}'$, and $\mathbf{W} = c$ in result (*) [or equivalently result (3.13)].

EXERCISE 13. Let $\mathbf{V}_k$ represent the $(n-1) \times (n-1)$ submatrix of the $n \times n$ Vandermonde matrix

$$\mathbf{V} = \begin{pmatrix} 1 & x_1 & x_1^2 & \dots & x_1^{n-1} \\ 1 & x_2 & x_2^2 & \dots & x_2^{n-1} \\ \vdots & \vdots & \vdots & & \vdots \\ 1 & x_n & x_n^2 & \dots & x_n^{n-1} \end{pmatrix}$$

(where $x_1, x_2, \dots, x_n$ are arbitrary scalars) obtained by striking out the kth row and the nth (last) column (of $\mathbf{V}$). Show that

$$|\mathbf{V}| = |\mathbf{V}_k|(-1)^{n-k} \prod_{i \neq k}(x_k - x_i).$$

Solution. Let $\mathbf{V}^*$ represent the $n \times n$ matrix whose first, $\dots$, $(k-1)$th rows are respectively the first, $\dots$, $(k-1)$th rows of $\mathbf{V}$, whose kth, $\dots$, $(n-1)$th rows are respectively the $(k+1)$th, $\dots$, nth rows of $\mathbf{V}$, and whose nth row is the kth row of $\mathbf{V}$. Then, $\mathbf{V}^*$ (like $\mathbf{V}$) is an $n \times n$ Vandermonde matrix, and $\mathbf{V}_k$ equals the $(n-1) \times (n-1)$ submatrix of $\mathbf{V}^*$ obtained by striking out the last row and the last column (of $\mathbf{V}^*$). Moreover, $\mathbf{V}$ can be obtained from $\mathbf{V}^*$ by $n-k$ successive interchanges of pairs of rows — specifically, $\mathbf{V}$ can be obtained from $\mathbf{V}^*$ by successively interchanging the nth row of $\mathbf{V}^*$ with the $(n-1)$th, $\dots$, kth rows of $\mathbf{V}^*$. Thus, making use of Theorem 13.2.6 and of result (6.4), we find that

$$\begin{aligned} |\mathbf{V}| &= (-1)^{n-k}|\mathbf{V}^*| \\ &= (-1)^{n-k}(x_k - x_1)\cdots(x_k - x_{k-1})(x_k - x_{k+1})\cdots(x_k - x_n)|\mathbf{V}_k| \\ &= |\mathbf{V}_k|\,(-1)^{n-k}\prod_{i \neq k}(x_k - x_i). \end{aligned}$$

EXERCISE 14. Show that, for $n \times n$ matrices $\mathbf{A}$ and $\mathbf{B}$,

$$\operatorname{adj}(\mathbf{AB}) = \operatorname{adj}(\mathbf{B})\operatorname{adj}(\mathbf{A}).$$

(*Hint.* Use the Binet-Cauchy formula to establish that the ijth element of adj($\mathbf{AB}$) equals the ijth element of adj($\mathbf{B}$)adj($\mathbf{A}$).)

Solution. Let $\mathbf{A}_j$ represent the $(n-1) \times n$ submatrix of $\mathbf{A}$ obtained by striking out the jth row of $\mathbf{A}$, and let $\mathbf{B}_i$ represent the $n \times (n-1)$ submatrix of $\mathbf{B}$ obtained by striking out the ith column of $\mathbf{B}$. Further, let $\mathbf{A}_{js}$ represent the $(n-1) \times (n-1)$ submatrix of $\mathbf{A}$ obtained by striking out the jth row and the sth column of $\mathbf{A}$, and let $\mathbf{B}_{si}$ represent the $(n-1) \times (n-1)$ submatrix of $\mathbf{B}$ obtained by striking out the sth row and the ith column of $\mathbf{B}$. Then, application of formula (8.3) (the Binet-Cauchy formula) gives

$$|\mathbf{A}_j\mathbf{B}_i| = \sum_{s=1}^{n} |\mathbf{A}_{js}|\ |\mathbf{B}_{si}|,$$

implying that

$$(-1)^{j+i}|\mathbf{A}_j\mathbf{B}_i| = \sum_{s=1}^{n}(-1)^{s+i}|\mathbf{B}_{si}|\ (-1)^{j+s}\ |\mathbf{A}_{js}|. \tag{S.6}$$

Note that $\mathbf{A}_j\mathbf{B}_i$ equals the $(n-1)\times(n-1)$ submatrix of $\mathbf{AB}$ obtained by striking out the jth row and the ith column of $\mathbf{AB}$, so that the left side of equality (S.6) is the cofactor of the jith element of $\mathbf{AB}$ and hence is the ijth element of $\text{adj}(\mathbf{AB})$. Note also that $(-1)^{s+i}|\mathbf{B}_{si}|$ is the cofactor of the sith element of $\mathbf{B}$ and hence is the isth element of $\text{adj}(\mathbf{B})$ and similarly that $(-1)^{j+s}|\mathbf{A}_{js}|$ is the cofactor of the jsth element of $\mathbf{A}$ and hence is the sjth element of $\text{adj}(\mathbf{A})$. Thus, the right side of equality (S.6) is the ijth element of $\text{adj}(\mathbf{B})\text{adj}(\mathbf{A})$.

We conclude that

$$\text{adj}(\mathbf{AB}) = \text{adj}(\mathbf{B})\text{adj}(\mathbf{A}).$$

14

Linear, Bilinear, and Quadratic Forms

EXERCISE 1. Show that a symmetric bilinear form $\mathbf{x}'\mathbf{A}\mathbf{y}$ (in n-dimensional vectors $\mathbf{x}$ and $\mathbf{y}$) can be expressed in terms of the corresponding quadratic form, that is, the quadratic form whose matrix is $\mathbf{A}$. Do so by verifying that

$$\mathbf{x}'\mathbf{A}\mathbf{y} = (1/2)[(\mathbf{x}+\mathbf{y})'\mathbf{A}(\mathbf{x}+\mathbf{y}) - \mathbf{x}'\mathbf{A}\mathbf{x} - \mathbf{y}'\mathbf{A}\mathbf{y}]\,.$$

Solution. Since the bilinear form $\mathbf{x}'\mathbf{A}\mathbf{y}$ is symmetric, we have that

$$\begin{aligned}(1/2)&[(\mathbf{x}+\mathbf{y})'\mathbf{A}(\mathbf{x}+\mathbf{y}) - \mathbf{x}'\mathbf{A}\mathbf{x} - \mathbf{y}'\mathbf{A}\mathbf{y}] \\ &= (1/2)(\mathbf{x}'\mathbf{A}\mathbf{x} + \mathbf{x}'\mathbf{A}\mathbf{y} + \mathbf{y}'\mathbf{A}\mathbf{x} + \mathbf{y}'\mathbf{A}\mathbf{y} - \mathbf{x}'\mathbf{A}\mathbf{x} - \mathbf{y}'\mathbf{A}\mathbf{y}) \\ &= (1/2)(\mathbf{x}'\mathbf{A}\mathbf{y} + \mathbf{y}'\mathbf{A}\mathbf{x}) = (1/2)(\mathbf{x}'\mathbf{A}\mathbf{y} + \mathbf{x}'\mathbf{A}\mathbf{y}) = \mathbf{x}'\mathbf{A}\mathbf{y}.\end{aligned}$$

EXERCISE 2. Show that corresponding to any quadratic form $\mathbf{x}'\mathbf{A}\mathbf{x}$ (in the n-dimensional vector $\mathbf{x}$) there exists a unique upper triangular matrix $\mathbf{B}$ such that $\mathbf{x}'\mathbf{A}\mathbf{x}$ and $\mathbf{x}'\mathbf{B}\mathbf{x}$ are identically equal, and express the elements of $\mathbf{B}$ in terms of the elements of $\mathbf{A}$.

Solution. Let a_{ij} represent the ijth element of $\mathbf{A}$ $(i, j = 1, \ldots, n)$. When $\mathbf{B} = \{b_{ij}\}$ is upper triangular, the conditions $a_{ii} = b_{ii}$ and $a_{ij} + a_{ji} = b_{ij} + b_{ji}$ $(j \neq i = 1, \ldots, n)$ of Lemma 14.1.1 are equivalent to the conditions $a_{ii} = b_{ii}$ and $a_{ij} + a_{ji} = b_{ij}$ $(j > i = 1, \ldots, n)$. Thus, it follows from the lemma that there exists a unique upper triangular matrix $\mathbf{B}$ such that $\mathbf{x}'\mathbf{A}\mathbf{x}$ and $\mathbf{x}'\mathbf{B}\mathbf{x}$ are identically equal, namely, the upper triangular matrix $\mathbf{B} = \{b_{ij}\}$, where $b_{ii} = a_{ii}$ and $b_{ij} = a_{ij} + a_{ji}$ $(j > i = 1, \ldots, n)$.

EXERCISE 3. Show, by example, that the sum of two positive semidefinite matrices can be positive definite.

Solution. Consider the two $n \times n$ matrices $\begin{pmatrix} 1 & \mathbf{0} \\ \mathbf{0} & \mathbf{0} \end{pmatrix}$ and $\begin{pmatrix} 0 & \mathbf{0} \\ \mathbf{0} & \mathbf{I}_{n-1} \end{pmatrix}$. Clearly, both of these two matrices are positive semidefinite, however, their sum is the $n \times n$ identity matrix $\mathbf{I}_n$, which is positive definite.

EXERCISE 4. Show, via an example, that there exist (nonsymmetric) nonsingular positive semidefinite matrices.

Solution. Consider the $n \times n$ upper triangular matrix

$$\mathbf{A} = \begin{pmatrix} 1 & 2 & 0 & \dots & 0 \\ 0 & 1 & 0 & \dots & 0 \\ 0 & 0 & 1 & \dots & 0 \\ \vdots & \vdots & & \ddots & \\ 0 & 0 & 0 & & 1 \end{pmatrix}.$$

For an arbitrary n-dimensional vector $\mathbf{x} = (x_1, x_2, x_3, \dots, x_n)'$, we find that

$$\mathbf{x}'\mathbf{A}\mathbf{x} = (x_1 + x_2)^2 + x_3^2 + \cdots + x_n^2 \geq 0$$

and that $\mathbf{x}'\mathbf{A}\mathbf{x} = 0$ if $x_1 = -x_2$ and $x_3 = \cdots = x_n = 0$. Thus, $\mathbf{A}$ is positive semidefinite. Moreover, it follows from Corollary 8.5.6 that $\mathbf{A}$ is nonsingular.

EXERCISE 5. Show, by example, that there exist an $n \times n$ positive semidefinite matrix $\mathbf{A}$ and an $n \times m$ matrix $\mathbf{P}$ (where $m < n$) such that $\mathbf{P}'\mathbf{A}\mathbf{P}$ is positive definite.

Solution. Take $\mathbf{A}$ to be the $n \times n$ diagonal matrix $\text{diag}(\mathbf{I}_m, \mathbf{0})$, which is clearly positive semidefinite, and take $\mathbf{P}$ to be the $n \times m$ (partitioned) matrix $\begin{pmatrix} \mathbf{I}_m \\ \mathbf{0} \end{pmatrix}$. Then, $\mathbf{P}'\mathbf{A}\mathbf{P} = \mathbf{I}_m$, which is an $m \times m$ positive definite matrix.

EXERCISE 6. For an $n \times n$ matrix $\mathbf{A}$ and an $n \times m$ matrix $\mathbf{P}$, it is the case that (1) if $\mathbf{A}$ is nonnegative definite, then $\mathbf{P}'\mathbf{A}\mathbf{P}$ is nonnegative definite; (2) if $\mathbf{A}$ is nonnegative definite and $\text{rank}(\mathbf{P}) < m$, then $\mathbf{P}'\mathbf{A}\mathbf{P}$ is positive semidefinite; and (3) if $\mathbf{A}$ is positive definite and $\text{rank}(\mathbf{P}) = m$, then $\mathbf{P}'\mathbf{A}\mathbf{P}$ is positive definite. Convert these results, which are for nonnegative definite (positive definite or positive semidefinite) matrices, into equivalent results for nonpositive definite matrices.

Solution. As in results (1)–(3) (of the exercise or equivalently of Theorem 14.2.9), let $\mathbf{A}$ represent an $n \times n$ matrix and $\mathbf{P}$ an $n \times m$ matrix. Upon applying results (1)–(3) with $-\mathbf{A}$ in place of $\mathbf{A}$, we find that (1′) if $-\mathbf{A}$ is nonnegative definite, then $-\mathbf{P}'\mathbf{A}\mathbf{P}$ is nonnegative definite; (2′) if $-\mathbf{A}$ is nonnegative definite and $\text{rank}(\mathbf{P}) < m$, then $-\mathbf{P}'\mathbf{A}\mathbf{P}$ is positive semidefinite; and (3′) if $-\mathbf{A}$ is positive definite and $\text{rank}(\mathbf{P}) = m$, then $-\mathbf{P}'\mathbf{A}\mathbf{P}$ is positive definite. These three results can be restated as follows:

(1′) if $\mathbf{A}$ is nonpositive definite, then $\mathbf{P}'\mathbf{A}\mathbf{P}$ is nonpositive definite; (2′) if $\mathbf{A}$ is nonpositive definite and $\text{rank}(\mathbf{P}) < m$, then $\mathbf{P}'\mathbf{A}\mathbf{P}$ is negative semidefinite; and (3′) if $\mathbf{A}$ is negative definite and $\text{rank}(\mathbf{P}) = m$, then $\mathbf{P}'\mathbf{A}\mathbf{P}$ is negative definite.

EXERCISE 7. Let $\{\mathbf{X}_1, \ldots, \mathbf{X}_r\}$ represent a set of matrices from a linear space $\mathcal{V}$. And, let $\mathbf{A} = \{a_{ij}\}$ represent the $r \times r$ matrix whose ijth element is $\mathbf{X}_i \bullet \mathbf{X}_j$ — this matrix is referred to as the *Gram matrix* (or the Gramian) of the set $\{\mathbf{X}_1, \ldots, \mathbf{X}_r\}$ and its determinant is referred to as the *Gramian* (or the Gram determinant) of $\{\mathbf{X}_1, \ldots, \mathbf{X}_r\}$.

(a) Show that $\mathbf{A}$ is symmetric and nonnegative definite.

(b) Show that $\mathbf{X}_1, \ldots, \mathbf{X}_r$ are linearly independent if and only if $\mathbf{A}$ is nonsingular.

Solution. Let $\mathbf{Y}_1, \ldots, \mathbf{Y}_n$ represent any matrices that form an orthonormal basis for $\mathcal{V}$. Then, for $j = 1, \ldots, r$, there exist scalars $b_{1j}, \ldots, b_{nj}$ such that

$$\mathbf{X}_j = b_{1j}\mathbf{Y}_1 + \cdots + b_{nj}\mathbf{Y}_n.$$

And, for $i, j = 1, \ldots, r$,

$$\begin{aligned} a_{ij} = \mathbf{X}_i \bullet \mathbf{X}_j &= \left(\sum_{k=1}^{n} b_{ki}\mathbf{Y}_k\right) \bullet \left(\sum_{s=1}^{n} b_{sj}\mathbf{Y}_s\right) \\ &= \sum_{k=1}^{n} b_{ki}\left[\mathbf{Y}_k \bullet \left(\sum_{s=1}^{n} b_{sj}\mathbf{Y}_s\right)\right] \\ &= \sum_{k=1}^{n} b_{ki} \sum_{s=1}^{n} b_{sj}(\mathbf{Y}_s \bullet \mathbf{Y}_k) \\ &= \sum_{k=1}^{n} b_{ki}b_{kj}. \end{aligned}$$

Moreover, $\sum_{k=1}^{n} b_{ki}b_{kj}$ is the ijth element of the $r \times r$ matrix $\mathbf{B}'\mathbf{B}$, where $\mathbf{B}$ is the $n \times r$ matrix whose kjth element is b_{kj} (and hence where $\mathbf{B}'$ is the $r \times n$ matrix whose ikth element is b_{ki}). Thus, $\mathbf{A} = \mathbf{B}'\mathbf{B}$, and since $\mathbf{B}'\mathbf{B}$ is symmetric (and in light of Corollary 14.2.14) nonnegative definite, the solution of Part (a) is complete.

Now, consider Part (b). For $j = 1, \ldots, r$, let $\mathbf{b}_j = (b_{1j}, \ldots, b_{nj})'$. Then, since clearly $\mathbf{Y}_1, \ldots, \mathbf{Y}_n$ are linearly independent, it follows from Lemma 3.2.4 that $\mathbf{X}_1, \ldots, \mathbf{X}_r$ are linearly independent if and only if $\mathbf{b}_1, \ldots, \mathbf{b}_r$ are linearly independent. Thus, since $\mathbf{b}_1, \ldots, \mathbf{b}_r$ are the columns of $\mathbf{B}$, $\mathbf{X}_1, \ldots, \mathbf{X}_r$ are linearly independent if and only if $\text{rank}(\mathbf{B}) = r$ or equivalently (in light of Corollary 7.4.5) if and only if $\text{rank}(\mathbf{B}'\mathbf{B}) = r$. And, since $\mathbf{A} = \mathbf{B}'\mathbf{B}$, we conclude that $\mathbf{X}_1, \ldots, \mathbf{X}_r$ are linearly independent if and only if $\mathbf{A}$ is nonsingular.

EXERCISE 8. Let $\mathbf{A} = \{a_{ij}\}$ represent an $n \times n$ symmetric positive definite matrix, and let $\mathbf{B} = \{b_{ij}\} = \mathbf{A}^{-1}$. Show that, for $i = 1, \ldots, n$,

$$b_{ii} \geq 1/a_{ii} \,,$$

with equality holding if and only if, for all $j \neq i$, $a_{ij} = 0$.

Solution. Let $\mathbf{U} = (\mathbf{u}_1, \mathbf{U}_2)$, where $\mathbf{u}_1$ is the ith column of $\mathbf{I}_n$ and $\mathbf{U}_2$ is the submatrix of $\mathbf{I}_n$ obtained by striking out the ith column, and observe that $\mathbf{U}$ is a permutation matrix.

Define $\mathbf{R} = \mathbf{U}'\mathbf{A}\mathbf{U}$ and $\mathbf{S} = \mathbf{R}^{-1}$. Partition $\mathbf{R}$ and $\mathbf{S}$ as

$$\mathbf{R} = \begin{pmatrix} r_{11} & \mathbf{r}' \\ \mathbf{r} & \mathbf{R}_* \end{pmatrix} \quad \text{and} \quad \mathbf{S} = \begin{pmatrix} s_{11} & \mathbf{s}' \\ \mathbf{s} & \mathbf{S}_* \end{pmatrix}$$

[where the dimensions of both $\mathbf{R}_*$ and $\mathbf{S}_*$ are $(n-1) \times (n-1)$]. Then,

$$r_{11} = \mathbf{u}_1'\mathbf{A}\mathbf{u}_1 = a_{ii}, \tag{S.1}$$

$$r' = \mathbf{u}_1'\mathbf{A}\mathbf{U}_2 = (a_{i1}, a_{i2}, \ldots, a_{i,\,i-1}, a_{i,\,i+1}, \ldots, a_{i,\,n-1}, a_{in}), \tag{S.2}$$

and (since $\mathbf{S} = \mathbf{U}'\mathbf{B}\mathbf{U}$)

$$s_{11} = \mathbf{u}_1'\mathbf{B}\mathbf{u}_1 = b_{ii}. \tag{S.3}$$

It follows from Corollary 14.2.10 that $\mathbf{R}$ is positive definite, implying (in light of Corollary 14.2.12) that $\mathbf{R}_*$ is positive definite and hence (in light of Corollary 14.2.11) that $\mathbf{R}_*$ is invertible and that $\mathbf{R}_*^{-1}$ is positive definite. Thus, making use of Theorem 8.5.11, we find [in light of results (S.1) and (S.3)] that

$$b_{ii} = (a_{ii} - \mathbf{r}'\mathbf{R}_*^{-1}\mathbf{r})^{-1}$$

and also that $\mathbf{r}'\mathbf{R}_*^{-1}\mathbf{r} \geq 0$ with equality holding if and only if $\mathbf{r} = \mathbf{0}$. Since $b_{ii} > 0$ (and hence $a_{ii} - \mathbf{r}'\mathbf{R}_*^{-1}\mathbf{r} > 0$), we conclude that $b_{ii} \geq 1/a_{ii}$ with equality holding if and only if $\mathbf{r} = \mathbf{0}$ or equivalently [in light of result (S.2)] if and only if, for $j \neq i$, $a_{ij} = 0$.

EXERCISE 9. Let $\mathbf{A}$ represent an $m \times n$ matrix and $\mathbf{D}$ a diagonal matrix such that $\mathbf{A} = \mathbf{PDQ}$ for some matrix $\mathbf{P}$ of full column rank and some matrix $\mathbf{Q}$ of full row rank. Show that rank($\mathbf{A}$) equals the number of nonzero diagonal elements in $\mathbf{D}$.

Solution. Making use of Lemma 8.3.2, we find that

$$\text{rank}(\mathbf{A}) = \text{rank}(\mathbf{PDQ}) = \text{rank}(\mathbf{DQ}) = \text{rank}(\mathbf{D}).$$

Moreover, rank($\mathbf{D}$) equals the number of nonzero diagonal elements in $\mathbf{D}$.

EXERCISE 10. Let $\mathbf{A}$ represent an $n \times n$ symmetric idempotent matrix and $\mathbf{V}$ an $n \times n$ symmetric positive definite matrix. Show that rank($\mathbf{AVA}$) = tr($\mathbf{A}$).

Solution. According to Corollary 14.3.13, $\mathbf{V} = \mathbf{P}'\mathbf{P}$ for some nonsingular matrix $\mathbf{P}$. Thus, making use of Corollary 7.4.5, Corollary 8.3.3, and Corollary 10.2.2, we find that

$$\text{rank}(\mathbf{AVA}) = \text{rank}[(\mathbf{PA})'\mathbf{PA}] = \text{rank}(\mathbf{PA}) = \text{rank}(\mathbf{A}) = \text{tr}(\mathbf{A}).$$

EXERCISE 11. Show that if an $n \times n$ matrix $\mathbf{A}$ is such that $\mathbf{x}'\mathbf{A}\mathbf{x} \neq 0$ for every $n \times 1$ nonnull vector $\mathbf{x}$, then $\mathbf{A}$ is either positive definite or negative definite.

Solution. Let $\mathbf{A}$ represent an $n \times n$ matrix such that $\mathbf{x}'\mathbf{A}\mathbf{x} \neq 0$ for every $n \times 1$ nonnull vector $\mathbf{x}$.

Define $\mathbf{B} = (1/2)(\mathbf{A} + \mathbf{A}')$. Then, $\mathbf{x}'\mathbf{B}\mathbf{x} = \mathbf{x}'\mathbf{A}\mathbf{x}$ for every $n \times 1$ vector $\mathbf{x}$. Moreover, $\mathbf{B}$ is symmetric, implying (in light of Corollary 14.3.5) that there exists a nonsingular matrix $\mathbf{P}$ and a diagonal matrix $\mathbf{D} = \text{diag}(d_1, \ldots, d_n)$ such that $\mathbf{B} = \mathbf{P}'\mathbf{D}\mathbf{P}$. Thus, $(\mathbf{P}\mathbf{x})'\mathbf{D}\mathbf{P}\mathbf{x} = \mathbf{x}'\mathbf{A}\mathbf{x}$ for every $n \times 1$ vector $\mathbf{x}$ and hence $(\mathbf{P}\mathbf{x})'\mathbf{D}\mathbf{P}\mathbf{x} \neq 0$ for every $n \times 1$ nonnull vector $\mathbf{x}$.

There exists no i such that $d_i = 0$ [since, if $d_i = 0$, then, taking $\mathbf{x}$ to be the nonnull vector $\mathbf{P}^{-1}\mathbf{e}_i$, where $\mathbf{e}_i$ is the ith column of $\mathbf{I}_n$, we would have that $(\mathbf{P}\mathbf{x})'\mathbf{D}\mathbf{P}\mathbf{x} = \mathbf{e}_i'\mathbf{D}\mathbf{e}_i = d_i = 0$]. Moreover, there exists no i and j such that $d_i > 0$ and $d_j < 0$ [since, if $d_i > 0$ and $d_j < 0$, then, taking $\mathbf{x}$ to be the (nonnull) vector $\mathbf{P}^{-1}\mathbf{y}$, where $\mathbf{y}$ is the $n \times 1$ vector with ith element $1/\sqrt{d_i}$ and jth element $1/\sqrt{-d_j}$, we would have that $(\mathbf{P}\mathbf{x})'\mathbf{D}\mathbf{P}\mathbf{x} = \mathbf{y}'\mathbf{D}\mathbf{y} = 1 - 1 = 0$].

It follows that the n scalars $d_1, \ldots, d_n$ are either all positive, in which case $\mathbf{B}$ is (according to Corollary 14.2.15) positive definite, or all negative, in which case $-\mathbf{B}$ is positive definite and hence $\mathbf{B}$ is negative definite. We conclude (on the basis of Corollary 14.2.7) that $\mathbf{A}$ is either positive definite or negative definite.

EXERCISE 12. (a) Let $\mathbf{A}$ represent an $n \times n$ symmetric matrix of rank r. Take $\mathbf{P}$ to be an $n \times n$ nonsingular matrix and $\mathbf{D}$ an $n \times n$ diagonal matrix such that $\mathbf{A} = \mathbf{P}'\mathbf{D}\mathbf{P}$ — the existence of such matrices is guaranteed. The number, say m, of diagonal elements of $\mathbf{D}$ that are positive is called the *index of inertia* of $\mathbf{A}$ (or of the quadratic form $\mathbf{x}'\mathbf{A}\mathbf{x}$ whose matrix is $\mathbf{A}$). Show that the index of inertia is well-defined in the sense that m does not vary with the choice of $\mathbf{P}$ or $\mathbf{D}$. That is, show that, if $\mathbf{P}_1$ and $\mathbf{P}_2$ are nonsingular matrices and $\mathbf{D}_1$ and $\mathbf{D}_2$ diagonal matrices such that $\mathbf{A} = \mathbf{P}_1'\mathbf{D}_1\mathbf{P}_1 = \mathbf{P}_2'\mathbf{D}_2\mathbf{P}_2$, then $\mathbf{D}_2$ contains the same number of positive diagonal elements as $\mathbf{D}_1$. Show also that the number of diagonal elements of $\mathbf{D}$ that are negative equals $r - m$.

(b) Let $\mathbf{A}$ represent an $n \times n$ symmetric matrix. Show that $\mathbf{A} = \mathbf{P}' \,\text{diag}(\mathbf{I}_m, -\mathbf{I}_{r-m}, \mathbf{0})\mathbf{P}$ for some $n \times n$ nonsingular matrix $\mathbf{P}$ and some nonnegative integers m and r. Show further that m equals the index of inertia of the matrix $\mathbf{A}$ and that $r = \text{rank}(\mathbf{A})$.

(c) An $n \times n$ symmetric matrix $\mathbf{B}$ is said to be *congruent* to an $n \times n$ symmetric matrix $\mathbf{A}$ if there exists an $n \times n$ nonsingular matrix $\mathbf{P}$ such that $\mathbf{B} = \mathbf{P}'\mathbf{A}\mathbf{P}$. (If $\mathbf{B}$ is congruent to $\mathbf{A}$, then clearly $\mathbf{A}$ is congruent to $\mathbf{B}$.) Show that $\mathbf{B}$ is congruent to

A if and only if **B** has the same rank and the same index of inertia as **A**. This result is called *Sylvester's law of inertia*, after James Joseph Sylvester (1814–1897).

(d) Let **A** represent an $n \times n$ symmetric matrix of rank r and with index of inertia m. Show that **A** is nonnegative definite if and only if $m = r$ and is positive definite if and only if $m = r = n$.

Solution. (a) Take $\mathbf{P}_1$ and $\mathbf{P}_2$ to be $n \times n$ nonsingular matrices and $\mathbf{D}_1 = \{d_i^{(1)}\}$ and $\mathbf{D}_2 = \{d_i^{(2)}\}$ to be $n \times n$ diagonal matrices such that $\mathbf{A} = \mathbf{P}_1'\mathbf{D}_1\mathbf{P}_1 = \mathbf{P}_2'\mathbf{D}_2\mathbf{P}_2$. Let m_1 represent the number of diagonal elements of $\mathbf{D}_1$ that are positive and m_2 the number of diagonal elements of $\mathbf{D}_2$ that are positive. Take $i_1, i_2, \ldots, i_n$ to be a permutation of the first n positive integers such that $d_{i_j}^{(1)} > 0$ for $j = 1, 2, \ldots, m_1$, and similarly take $k_1, k_2, \ldots, k_n$ to be a permutation such that $d_{k_j}^{(2)} > 0$ for $j = 1, 2, \ldots, m_2$. Further, take $\mathbf{U}_1$ to be the $n \times n$ permutation matrix whose first, second, ..., nth columns are respectively the i_1th, i_2th, ..., i_nth columns of $\mathbf{I}_n$ and $\mathbf{U}_2$ to be the $n \times n$ permutation matrix whose first, second, ..., nth columns are respectively the k_1th, k_2th, ..., k_nth columns of $\mathbf{I}_n$, and define $\mathbf{D}_1^* = \mathbf{U}_1'\mathbf{D}_1\mathbf{U}_1$ and $\mathbf{D}_2^* = \mathbf{U}_2'\mathbf{D}_2\mathbf{U}_2$. Then, $\mathbf{D}_1^* = \text{diag}(d_{i_1}^{(1)}, d_{i_2}^{(1)}, \ldots, d_{i_n}^{(1)})$ and $\mathbf{D}_2^* = \text{diag}(d_{k_1}^{(2)}, d_{k_2}^{(2)}, \ldots, d_{k_n}^{(2)})$.

Suppose, for purposes of establishing a contradiction, that $m_1 < m_2$, and observe that

$$\begin{aligned}\mathbf{D}_2^* = \mathbf{U}_2'(\mathbf{P}_2^{-1})'\mathbf{A}\mathbf{P}_2^{-1}\mathbf{U}_2 &= \mathbf{U}_2'(\mathbf{P}_2^{-1})'\mathbf{P}_1'\mathbf{D}_1\mathbf{P}_1\mathbf{P}_2^{-1}\mathbf{U}_2 \\ &= \mathbf{U}_2'(\mathbf{P}_2^{-1})'\mathbf{P}_1'\mathbf{U}_1\mathbf{D}_1^*\mathbf{U}_1'\mathbf{P}_1\mathbf{P}_2^{-1}\mathbf{U}_2 = \mathbf{R}'\mathbf{D}_1^*\mathbf{R},\end{aligned}$$

where $\mathbf{R} = \mathbf{U}_1'\mathbf{P}_1\mathbf{P}_2^{-1}\mathbf{U}_2$. Partition the $n \times n$ matrix $\mathbf{R}$ as $\mathbf{R} = \begin{pmatrix} \mathbf{R}_{11} & \mathbf{R}_{12} \\ \mathbf{R}_{21} & \mathbf{R}_{22} \end{pmatrix}$, where $\mathbf{R}_{11}$ is of dimensions $m_1 \times m_2$.

Take $\mathbf{x} = \{x_j\}$ to be an m_2-dimensional nonnull column vector such that $\mathbf{R}_{11}\mathbf{x} = \mathbf{0}$ — since (by supposition) $m_1 < m_2$, such a vector necessarily exists. Letting $y_1, y_2, \ldots, y_{n-m_1}$, represent the elements of the vector $\mathbf{R}_{21}\mathbf{x}$, we find that

$$\begin{aligned}\sum_{j=1}^{m_2} d_{k_j}^{(2)} x_j^2 = \begin{pmatrix}\mathbf{x}\\ \mathbf{0}\end{pmatrix}'\mathbf{D}_2^*\begin{pmatrix}\mathbf{x}\\ \mathbf{0}\end{pmatrix} &= \begin{pmatrix}\mathbf{x}\\ \mathbf{0}\end{pmatrix}'\mathbf{R}'\mathbf{D}_1^*\mathbf{R}\begin{pmatrix}\mathbf{x}\\ \mathbf{0}\end{pmatrix} \\ &= \begin{pmatrix}\mathbf{0}\\ \mathbf{R}_{21}\mathbf{x}\end{pmatrix}'\mathbf{D}_1^*\begin{pmatrix}\mathbf{0}\\ \mathbf{R}_{21}\mathbf{x}\end{pmatrix} \\ &= \sum_{j=m_1+1}^{n} d_{i_j}^{(1)} y_{j-m_1}^2 \qquad \text{(S.4)}\end{aligned}$$

Moreover,

$$\sum_{j=1}^{m_2} d_{k_j}^{(2)} x_j^2 > 0,$$

and, since the last $n - m_1$ diagonal elements of $\mathbf{D}_1^*$ are either zero or negative,

$$\sum_{j=m_1+1}^{n} d_{i_j}^{(1)} y_{j-m_1}^2 \leq 0.$$

These two inequalities, in combination with equality (S.4), establish the sought-after contradiction.

We conclude that $m_1 \geq m_2$. It can be established, via an analogous argument, that $m_1 \leq m_2$. Together, those two inequalities imply that $m_2 = m_1$.

Consider now the number of negative diagonal elements in the diagonal matrix $\mathbf{D}$. According to Lemma 14.3.1, the number of nonzero diagonal elements in $\mathbf{D}$ equals r. Thus, the number of negative diagonal elements in $\mathbf{D}$ equals $r - m$.

(b) According to Corollary 14.3.5, there exists an $n \times n$ nonsingular matrix $\mathbf{P}_*$ and an $n \times n$ diagonal matrix $\mathbf{D} = \{d_i\}$ such that $\mathbf{A} = \mathbf{P}_*'\mathbf{D}\mathbf{P}_*$. Take $i_1, i_2, \ldots, i_n$ to be any permutation of the first n positive integers such that — for some integers m and r $(0 \leq m \leq r \leq n)$ — $d_{i_j} > 0$, for $j = 1, \ldots, m$, $d_{i_j} < 0$, for $j = m+1, \ldots, r$, and $d_{i_j} = 0$, for $j = r+1, \ldots, n$. Further, take $\mathbf{U}$ to be the $n \times n$ permutation matrix whose first, second, ..., nth columns are respectively the i_1th, i_2th, ..., i_nth columns of $\mathbf{I}_n$, and define $\mathbf{D}_* = \mathbf{U}'\mathbf{D}\mathbf{U}$. Then,

$$\mathbf{D}_* = \text{diag}(d_{i_1}, d_{i_2}, \ldots, d_{i_n}).$$

We find that

$$\mathbf{A} = \mathbf{P}_*'\mathbf{U}\mathbf{U}'\mathbf{D}\mathbf{U}\mathbf{U}'\mathbf{P}_* = (\mathbf{U}'\mathbf{P}_*)'\mathbf{D}_*\mathbf{U}'\mathbf{P}_*.$$

And, taking Δ to be the diagonal matrix whose first m diagonal elements are $\sqrt{d_{i_1}}, \sqrt{d_{i_2}}, \ldots, \sqrt{d_{i_m}}$, whose $(m+1)$th, $(m+2)$th, ..., rth diagonal elements are $\sqrt{-d_{i_{m+1}}}, \sqrt{-d_{i_{m+2}}}, \ldots, \sqrt{-d_{i_r}}$, and whose last $n-r$ diagonal elements equal one, we have that

$$\Delta^{-1}\mathbf{D}_*\Delta^{-1} = \text{diag}(\mathbf{I}_m, -\mathbf{I}_{r-m}, \mathbf{0})$$

and hence that

$$\begin{aligned}\mathbf{A} = (\mathbf{U}'\mathbf{P}_*)'\mathbf{D}_*\mathbf{U}'\mathbf{P}_* &= (\Delta\mathbf{U}'\mathbf{P}_*)'\Delta^{-1}\mathbf{D}_*\Delta^{-1}\Delta\mathbf{U}'\mathbf{P}_* \\ &= \mathbf{P}'\text{diag}(\mathbf{I}_m, -\mathbf{I}_{r-m}, \mathbf{0})\mathbf{P},\end{aligned}$$

where $\mathbf{P} = \Delta\mathbf{U}'\mathbf{P}_*$. Clearly, $\mathbf{P}$ is nonsingular.

That m equals the index of inertia and that $r = \text{rank}(\mathbf{A})$ are immediate consequences of the results of Part (a).

(c) Suppose that $\mathbf{B}$ is congruent to $\mathbf{A}$. Then, by definition, $\mathbf{B} = \mathbf{P}'\mathbf{A}\mathbf{P}$ for some $n \times n$ nonsingular matrix $\mathbf{P}$. Moreover, according to Corollary 14.3.5, $\mathbf{A} = \mathbf{Q}'\mathbf{D}\mathbf{Q}$ for some $n \times n$ nonsingular matrix $\mathbf{Q}$ and some $n \times n$ diagonal matrix $\mathbf{D}$. Thus,

$$\mathbf{B} = \mathbf{P}'\mathbf{Q}'\mathbf{D}\mathbf{Q}\mathbf{P} = \mathbf{P}_*'\mathbf{D}\mathbf{P}_*,$$

where $\mathbf{P}_* = \mathbf{Q}\mathbf{P}$. Clearly, $\mathbf{P}_*$ is nonsingular. And, in light of Part (a), we conclude that $\mathbf{B}$ has the same rank and the same index of inertia as $\mathbf{A}$.

Conversely, suppose that $\mathbf{A}$ and $\mathbf{B}$ have the same rank, say r, and the same index of inertia, say m. Then, according to Part (b), $\mathbf{A} = \mathbf{P}' \operatorname{diag}(\mathbf{I}_m, -\mathbf{I}_{r-m}, \mathbf{0})\, \mathbf{P}$ and $\mathbf{B} = \mathbf{Q}' \operatorname{diag}(\mathbf{I}_m, -\mathbf{I}_{r-m}, \mathbf{0})\, \mathbf{Q}$ for some $n \times n$ nonsingular matrices $\mathbf{P}$ and $\mathbf{Q}$. Thus,

$$(\mathbf{Q}')^{-1}\mathbf{B}\mathbf{Q}^{-1} = \operatorname{diag}(\mathbf{I}_m, -\mathbf{I}_{r-m}, \mathbf{0}) = (\mathbf{P}^{-1})'\mathbf{A}\mathbf{P}^{-1},$$

and consequently

$$\mathbf{B} = \mathbf{Q}'(\mathbf{P}^{-1})'\mathbf{A}\mathbf{P}^{-1}\mathbf{Q} = \mathbf{P}_*'\mathbf{A}\mathbf{P}_*,$$

where $\mathbf{P}_* = \mathbf{P}^{-1}\mathbf{Q}$. Clearly, $\mathbf{P}_*$ is nonsingular. We conclude that $\mathbf{B}$ is congruent to $\mathbf{A}$.

(d) According to Part (b),

$$\mathbf{A} = \mathbf{P}'\operatorname{diag}(\mathbf{I}_m, -\mathbf{I}_{r-m}, \mathbf{0})\, \mathbf{P}$$

for some $n \times n$ nonsingular matrix $\mathbf{P}$. Thus, we have as an immediate consequence of Corollary 14.2.15 that $\mathbf{A}$ is nonnegative definite if and only if $m = r$ and is positive definite if and only if $m = r = n$.

EXERCISE 13. Let $\mathbf{A}$ represent an $n \times n$ symmetric nonnegative definite matrix of rank r. Then, there exists an $n \times r$ matrix $\mathbf{B}$ (of rank r) such that $\mathbf{A} = \mathbf{B}\mathbf{B}'$. Let $\mathbf{X}$ represent any $n \times m$ matrix (where $m \geq r$) such that $\mathbf{A} = \mathbf{X}\mathbf{X}'$.

(a) Show that $\mathbf{X} = \mathbf{P}_{\mathbf{B}}\mathbf{X}$.

(b) Show that $\mathbf{X} = (\mathbf{B},\ \mathbf{0})\mathbf{Q}$ for some orthogonal matrix $\mathbf{Q}$.

Solution. (a) It follows from Corollary 7.4.5 that $\mathcal{C}(\mathbf{X}) = \mathcal{C}(\mathbf{A}) = \mathcal{C}(\mathbf{B})$, implying (in light of Corollary 12.3.6) that $\mathbf{P}_{\mathbf{X}} = \mathbf{P}_{\mathbf{B}}$. Thus, making use of Part (1) of Theorem 12.3.4, we find that

$$\mathbf{X} = \mathbf{P}_{\mathbf{X}}\mathbf{X} = \mathbf{P}_{\mathbf{B}}\mathbf{X}.$$

(b) Since $\operatorname{rank}(\mathbf{B}'\mathbf{B}) = \operatorname{rank}(\mathbf{B}) = r$, $\mathbf{B}'\mathbf{B}$ (which is of dimensions $r \times r$) is invertible. Thus, it follows from Part (a) that

$$\mathbf{X} = \mathbf{P}_{\mathbf{B}}\mathbf{X} = \mathbf{B}(\mathbf{B}'\mathbf{B})^{-1}\mathbf{B}'\mathbf{X} = \mathbf{B}\mathbf{Q}_1, \tag{S.5}$$

where $\mathbf{Q}_1 = (\mathbf{B}'\mathbf{B})^{-1}\mathbf{B}'\mathbf{X}$. Moreover,

$$\begin{aligned} \mathbf{Q}_1\mathbf{Q}_1' &= (\mathbf{B}'\mathbf{B})^{-1}\mathbf{B}'\mathbf{X}\mathbf{X}'\mathbf{B}(\mathbf{B}'\mathbf{B})^{-1} \\ &= (\mathbf{B}'\mathbf{B})^{-1}\mathbf{B}'\mathbf{A}\mathbf{B}(\mathbf{B}'\mathbf{B})^{-1} \\ &= (\mathbf{B}'\mathbf{B})^{-1}\mathbf{B}'\mathbf{B}\mathbf{B}'\mathbf{B}(\mathbf{B}'\mathbf{B})^{-1} = \mathbf{I}, \end{aligned}$$

so that the rows of the $r \times m$ matrix $\mathbf{Q}_1$ are orthonormal (with respect to the usual inner product).

It follows from Theorem 6.4.5 that there exists an $(m - r) \times m$ matrix $\mathbf{Q}_2$ whose rows, together with the rows of $\mathbf{Q}_1$, form an orthonormal (with respect to the usual inner product) basis for $\mathcal{R}^m$. Take $\mathbf{Q} = \begin{pmatrix} \mathbf{Q}_1 \\ \mathbf{Q}_2 \end{pmatrix}$. Then, clearly, $\mathbf{Q}$ is orthogonal. Further, $(\mathbf{B}, \mathbf{0})\mathbf{Q} = \mathbf{B}\mathbf{Q}_1$, implying, in light of result (S.5), that $\mathbf{X} = (\mathbf{B}, \mathbf{0})\mathbf{Q}$.

EXERCISE 14. Show that if a symmetric matrix $\mathbf{A}$ has a nonnegative definite generalized inverse, then $\mathbf{A}$ is nonnegative definite.

Solution. Suppose that the symmetric matrix $\mathbf{A}$ has a nonnegative definite generalized inverse, say $\mathbf{G}$. Then, $\mathbf{A} = \mathbf{AGA} = \mathbf{A}'\mathbf{GA}$, implying (in light of Theorem 14.2.9) that $\mathbf{A}$ is nonnegative definite.

EXERCISE 15. Suppose that an $n \times n$ matrix $\mathbf{A}$ has an LDU decomposition, say $\mathbf{A} = \mathbf{LDU}$, and let $d_1, d_2, \ldots, d_n$ represent the diagonal elements of the diagonal matrix $\mathbf{D}$. Show that

$$|\mathbf{A}| = d_1 d_2 \cdots d_n .$$

Solution. Making use of Theorem 13.2.11 and of Corollary 13.1.2, we find that

$$|\mathbf{A}| = |\mathbf{LDU}| = |\mathbf{LD}| = |\mathbf{D}| = d_1 d_2 \cdots d_n .$$

EXERCISE 16. (a) Suppose that an $n \times n$ matrix $\mathbf{A}$ (where $n \geq 2$) has a unique LDU decomposition, say $\mathbf{A} = \mathbf{LDU}$, and let $d_1, d_2, \ldots, d_n$ represent the first, second, ..., nth diagonal elements of $\mathbf{D}$. Show that $d_i \neq 0$ $(i = 1, 2, \ldots, n-1)$ and that $d_n \neq 0$ if and only if $\mathbf{A}$ is nonsingular.

(b) Suppose that an $n \times n$ (symmetric) matrix $\mathbf{A}$ (where $n \geq 2$) has a unique U′DU decomposition, say $\mathbf{A} = \mathbf{U}'\mathbf{DU}$, and let $d_1, d_2, \ldots, d_n$ represent the first, second, ..., nth diagonal elements of $\mathbf{D}$. Show that $d_i \neq 0$ $(i = 1, 2, \ldots, n-1)$ and that $d_n \neq 0$ if and only if $\mathbf{A}$ is nonsingular.

Solution. Let us restrict attention to Part (a) — Part (b) can be proved in essentially the same way as Part (a).

Suppose — for purposes of establishing a contradiction — that, for some i $(1 \leq i \leq n-1)$, $d_i = 0$. Take $\mathbf{L}^*$ to be a unit lower triangular matrix and $\mathbf{U}^*$ a unit upper triangular matrix that are identical to $\mathbf{L}$ and $\mathbf{U}$, respectively, except that, for some j $(j > i)$ the ijth element of $\mathbf{U}^*$ differs from the ijth element of $\mathbf{U}$ and/or the jith element of $\mathbf{L}^*$ differs from the jith element of $\mathbf{L}$. Then, according to Theorem 14.5.5, $\mathbf{A} = \mathbf{L}^*\mathbf{D}^*\mathbf{U}^*$ is an LDU decomposition of $\mathbf{A}$. Since this decomposition differs from the supposedly unique LDU decomposition $\mathbf{A} = \mathbf{LDU}$, we have arrived at the desired contradiction. We conclude that $d_i \neq 0$ $(i = 1, \ldots, n-1)$. And, since (in light of Lemma 14.3.1) $\mathbf{A}$ is nonsingular if and only if all n diagonal elements of $\mathbf{D}$ are nonzero, we further conclude that $\mathbf{A}$ is nonsingular if and only if $d_n \neq 0$.

EXERCISE 17. Suppose that an $n \times n$ (symmetric) matrix $\mathbf{A}$ has a unique U′DU decomposition, say $\mathbf{A} = \mathbf{U}'\mathbf{DU}$. Use the result of Part (b) of Exercise 16 to show that $\mathbf{A}$ has no LDU decompositions other than $\mathbf{A} = \mathbf{U}'\mathbf{DU}$.

Solution. Let us restrict attention to the case where $n \geq 2$ — if $n = 1$, then it is clear that $\mathbf{A}$ has no LDU decompositions other than $\mathbf{A} = \mathbf{U}'\mathbf{D}\mathbf{U}$.

The result of Part (b) of Exercise 16 implies that the first $n-1$ diagonal elements of $\mathbf{D}$ are nonzero. We conclude, on the basis of Theorem 14.5.5, that $\mathbf{A}$ has no LDU decompositions other than $\mathbf{A} = \mathbf{U}'\mathbf{D}\mathbf{U}$.

EXERCISE 18. Show that if a nonsingular matrix has an LDU decomposition, then that decomposition is unique.

Solution. Any 1×1 matrix (nonsingular or not) has a unique LDU decomposition, as discussed in Section 14.5b. Consider now a nonsingular matrix $\mathbf{A}$ of order $n \geq 2$ that has an LDU decomposition, say $\mathbf{A} = \mathbf{LDU}$. Let $\mathbf{A}_{11}, \mathbf{L}_{11}, \mathbf{U}_{11}$, and $\mathbf{D}_1$ represent the $(n-1)$th-order leading principal submatrices of $\mathbf{A}$, $\mathbf{L}$, $\mathbf{U}$, and $\mathbf{D}$, respectively. Then, according to Theorem 14.5.3, $\mathbf{A}_{11} = \mathbf{L}_{11}\mathbf{D}_1\mathbf{U}_{11}$ is an LDU decomposition of $\mathbf{A}_{11}$. Since $\mathbf{A}$ is nonsingular, $\mathbf{D}$ is nonsingular, implying that $\mathbf{D}_1$ is nonsingular and hence that $\mathbf{A}_{11}$ is nonsingular. We conclude, on the basis of Corollary 14.5.6, that $\mathbf{A} = \mathbf{LDU}$ is the unique LDU decomposition of $\mathbf{A}$.

EXERCISE 19. Let $\mathbf{A}$ represent an $n \times n$ matrix (where $n \geq 2$). By for instance using the results of Exercises 16, 17, and 18, show that if $\mathbf{A}$ has a unique LDU decomposition or (in the special case where $\mathbf{A}$ is symmetric) a unique U′DU decomposition, then the leading principal submatrices (of $\mathbf{A}$) of orders $1, 2, \ldots, n-1$ are nonsingular and have unique LDU decompositions.

Solution. In light of the result of Exercise 17, it suffices to restrict attention to the case where $\mathbf{A}$ has a unique LDU decomposition, say $\mathbf{A} = \mathbf{LDU}$.

For $i = 1, 2, \ldots, n-1$, let $\mathbf{A}_i, \mathbf{L}_i, \mathbf{U}_i$, and $\mathbf{D}_i$ represent the ith-order leading principal submatrices of $\mathbf{A}$, $\mathbf{L}$, $\mathbf{U}$, and $\mathbf{D}$, respectively. Then, according to Theorem 14.5.3, an LDU decomposition of $\mathbf{A}_i$ is $\mathbf{A}_i = \mathbf{L}_i\mathbf{D}_i\mathbf{U}_i$, and, according to the result of Exercise 16, $\mathbf{D}_i$ is nonsingular. Thus, $\mathbf{A}_i$ is nonsingular and, in light of the result of Exercise 18, has a unique LDU decomposition.

EXERCISE 20. (a) Let $\mathbf{A} = \{a_{ij}\}$ represent an $m \times n$ nonnull matrix of rank r. Show that there exist an $m \times m$ permutation matrix $\mathbf{P}$ and an $n \times n$ permutation matrix $\mathbf{Q}$ such that

$$\mathbf{PAQ} = \begin{pmatrix} \mathbf{B}_{11} & \mathbf{B}_{12} \\ \mathbf{B}_{21} & \mathbf{B}_{22} \end{pmatrix},$$

where $\mathbf{B}_{11}$ is an $r \times r$ nonsingular matrix whose leading principal submatrices (of orders $1, 2, \ldots, r-1$) are nonsingular.

(b) Let $\mathbf{B} = \begin{pmatrix} \mathbf{B}_{11} & \mathbf{B}_{12} \\ \mathbf{B}_{21} & \mathbf{B}_{22} \end{pmatrix}$ represent any $m \times n$ nonnull matrix of rank r such that $\mathbf{B}_{11}$ is an $r \times r$ nonsingular matrix whose leading principal submatrices (of orders $1, 2, \ldots, r-1$) are nonsingular. Show that there exists a unique decomposition of

B of the form

$$\mathbf{B} = \begin{pmatrix} \mathbf{L}_1 \\ \mathbf{L}_2 \end{pmatrix} \mathbf{D} (\mathbf{U}_1, \ \mathbf{U}_2),$$

where $\mathbf{L}_1$ is an $r \times r$ unit lower triangular matrix, $\mathbf{U}_1$ is an $r \times r$ unit upper triangular matrix, and $\mathbf{D}$ is an $r \times r$ diagonal matrix. Show further that this decomposition is such that $\mathbf{B}_{11} = \mathbf{L}_1\mathbf{D}\mathbf{U}_1$ is the unique LDU decomposition of $\mathbf{B}_{11}$, $\mathbf{D}$ is nonsingular, $\mathbf{L}_2 = \mathbf{B}_{21}\mathbf{U}_1^{-1}\mathbf{D}^{-1}$, and $\mathbf{U}_2 = \mathbf{D}^{-1}\mathbf{L}_1^{-1}\mathbf{B}_{12}$.

Solution. (a) The matrix $\mathbf{A}$ contains r linearly independent rows, say rows i_1, $i_2, \ldots, i_r$. For $k = 1, \ldots, r$, denote by $\mathbf{A}_k$ the $k \times n$ matrix whose rows are respectively rows $i_1, i_2, \ldots, i_k$ of $\mathbf{A}$.

There exists a subset $j_1, j_2, \ldots, j_r$ of the first n positive integers such that, for $k = 1, \ldots, r$, the matrix, say $\mathbf{A}_k^*$, whose columns are respectively columns $j_1, j_2, \ldots, j_k$ of $\mathbf{A}_k$, is nonsingular. As evidence of this, let us outline a recursive scheme for constructing such a subset.

Row i_1 of $\mathbf{A}$ is nonnull, so that j_1 can be chosen in such a way that $a_{i_1 j_1} \neq 0$ and hence in such a way that $\mathbf{A}_1^* = (a_{i_1 j_1})$ is nonsingular. Suppose now that $j_1, j_2, \ldots, j_{k-1}$ have been chosen in such a way that $\mathbf{A}_1^*, \mathbf{A}_2^*, \ldots, \mathbf{A}_{k-1}^*$ are nonsingular. Since $\mathbf{A}_{k-1}^*$ is nonsingular, columns $j_1, j_2, \ldots, j_{k-1}$ of $\mathbf{A}_k$ are linearly independent, and, since rank $(\mathbf{A}_k) = k$, $\mathbf{A}_k$ has a column that is not expressible as a linear combination of columns $j_1, j_2, \ldots, j_{k-1}$. Thus, it follows from Corollary 3.2.3 that j_k can be chosen in such a way that $\mathbf{A}_k^*$ is nonsingular.

Take $\mathbf{P}$ to be any $m \times m$ permutation matrix whose first r rows are respectively rows $i_1, i_2, \ldots, i_r$ of $\mathbf{I}_m$, and take $\mathbf{Q}$ to be any $n \times n$ permutation matrix whose first r columns are respectively columns $j_1, j_2, \ldots, j_r$ of $\mathbf{I}_n$. Then,

$$\mathbf{PAQ} = \begin{pmatrix} \mathbf{B}_{11} & \mathbf{B}_{12} \\ \mathbf{B}_{21} & \mathbf{B}_{22} \end{pmatrix},$$

where $\mathbf{B}_{11} = \mathbf{A}_r^*$ is a nonsingular matrix whose leading principal submatrices (of orders $1, 2, \ldots, r-1$) are respectively the nonsingular matrices $\mathbf{A}_1^*, \mathbf{A}_2^*, \ldots, \mathbf{A}_{r-1}^*$.

(b) Clearly, showing that $\mathbf{B}$ has a unique decomposition of the form specified in the exercise is equivalent to showing that there exist a unique unit lower triangular matrix $\mathbf{L}_1$, a unique unit upper triangular matrix $\mathbf{U}_1$, a unique diagonal matrix $\mathbf{D}$, and unique matrices $\mathbf{L}_2$ and $\mathbf{U}_2$ such that

$$\begin{aligned} \mathbf{B}_{11} &= \mathbf{L}_1\mathbf{D}\mathbf{U}_1, \\ \mathbf{B}_{12} &= \mathbf{L}_1\mathbf{D}\mathbf{U}_2, \\ \mathbf{B}_{21} &= \mathbf{L}_2\mathbf{D}\mathbf{U}_1, \qquad \text{and} \\ \mathbf{B}_{22} &= \mathbf{L}_2\mathbf{D}\mathbf{U}_2. \end{aligned}$$

It follows from Corollary 14.5.7 that there exists a unique unit lower triangular matrix $\mathbf{L}_1$, a unique unit upper triangular matrix $\mathbf{U}_1$, and a unique diagonal matrix $\mathbf{D}$ such that $\mathbf{B}_{11} = \mathbf{L}_1\mathbf{D}\mathbf{U}_1$ — by definition, $\mathbf{B}_{11} = \mathbf{L}_1\mathbf{D}\mathbf{U}_1$ is the unique LDU decomposition of $\mathbf{B}_{11}$. Moreover, $\mathbf{D}$ is nonsingular (since $\mathbf{B}_{11} = \mathbf{L}_1\mathbf{D}\mathbf{U}_1$ is nonsingular). Thus, there exist unique matrices $\mathbf{L}_2$ and $\mathbf{U}_2$ such that $\mathbf{B}_{21} = \mathbf{L}_2\mathbf{D}\mathbf{U}_1$

and $\mathbf{B}_{12} = \mathbf{L}_1\mathbf{D}\mathbf{U}_2$, namely, $\mathbf{L}_2 = \mathbf{B}_{21}\mathbf{U}_1^{-1}\mathbf{D}^{-1}$ and $\mathbf{U}_2 = \mathbf{D}^{-1}\mathbf{L}_1^{-1}\mathbf{B}_{12}$. Finally, it follows from Lemma 9.2.2 that

$$\begin{aligned}\mathbf{B}_{22} = \mathbf{B}_{21}\mathbf{B}_{11}^{-1}\mathbf{B}_{12} &= \mathbf{L}_2\mathbf{D}\mathbf{U}_1(\mathbf{L}_1\mathbf{D}\mathbf{U}_1)^{-1}\mathbf{L}_1\mathbf{D}\mathbf{U}_2\\ &= \mathbf{L}_2\mathbf{D}\mathbf{U}_1\mathbf{U}_1^{-1}\mathbf{D}^{-1}\mathbf{L}_1^{-1}\mathbf{L}_1\mathbf{D}\mathbf{U}_2 = \mathbf{L}_2\mathbf{D}\mathbf{U}_2\,.\end{aligned}$$

EXERCISE 21. Show, by example, that there exist $n \times n$ (nonsymmetric) positive semidefinite matrices that do not have LDU decompositions.

Solution. Let

$$\mathbf{A} = \begin{pmatrix} 0 & -\mathbf{1}'_{n-1} \\ \mathbf{1}_{n-1} & \mathbf{I}_{n-1} \end{pmatrix}.$$

Consider the quadratic form $\mathbf{x}'\mathbf{A}\mathbf{x}$ in $\mathbf{x}$. Partitioning $\mathbf{x}$ as $\mathbf{x} = \begin{pmatrix} x_1 \\ \mathbf{x}_2 \end{pmatrix}$, where $\mathbf{x}_2$ is of dimensions $(n-1) \times 1$, we find that

$$\mathbf{x}'\mathbf{A}\mathbf{x} = -x_1(\mathbf{1}'\mathbf{x}_2) + x_1(\mathbf{x}_2'\mathbf{1}) + \mathbf{x}_2'\mathbf{I}\mathbf{x}_2 = \mathbf{x}_2'\mathbf{x}_2.$$

Thus, $\mathbf{x}'\mathbf{A}\mathbf{x} \geq 0$ for all $\mathbf{x}$, with equality holding when, for example, $x_1 = 1$ and $\mathbf{x}_2 = \mathbf{0}$, so that $\mathbf{x}'\mathbf{A}\mathbf{x}$ is a positive semidefinite quadratic form and hence $\mathbf{A}$ is a positive semidefinite matrix. Moreover, since the leading principal submatrix of $\mathbf{A}$ of order two is $\begin{pmatrix} 0 & -1 \\ 1 & 1 \end{pmatrix}$ and since $-1 \notin \mathcal{C}(0)$, it follows from Part (2) of Theorem 14.5.4 that $\mathbf{A}$ does not have an LDU decomposition.

EXERCISE 22. Let $\mathbf{A}$ represent an $n \times n$ nonnegative definite (possibly nonsymmetric) matrix that has an LDU decomposition, say $\mathbf{A} = \mathbf{LDU}$. Show that the diagonal elements of the diagonal matrix $\mathbf{D}$ are nonnegative.

Solution. Consider the matrix $\mathbf{B} = \mathbf{DU}(\mathbf{L}^{-1})'$. Since (in light of Corollary 8.5.9) $(\mathbf{L}^{-1})'$ — like $\mathbf{U}$ — is unit upper triangular, it follows from Lemma 1.3.1 that the diagonal elements of $\mathbf{B}$ are the same as the diagonal elements, say $d_1, \ldots, d_n$, of $\mathbf{D}$.

Moreover,

$$\mathbf{B} = \mathbf{L}^{-1}(\mathbf{LDU})(\mathbf{L}^{-1})' = \mathbf{L}^{-1}\mathbf{A}(\mathbf{L}^{-1})',$$

implying (in light of Theorem 14.2.9) that $\mathbf{B}$ is nonnegative definite. We conclude — on the basis of Corollary 14.2.13 — that $d_1, \ldots, d_n$ are nonnegative.

EXERCISE 23. Let $\mathbf{A}$ represent an $m \times k$ matrix of full column rank. And, let $\mathbf{A} = \mathbf{QR}$ represent the QR decomposition of $\mathbf{A}$; that is, let $\mathbf{Q}$ represent the unique $m \times k$ matrix whose columns are orthonormal with respect to the usual inner product and let $\mathbf{R}$ represent the unique $k \times k$ upper triangular matrix with positive diagonal elements such that $\mathbf{A} = \mathbf{QR}$. Show that $\mathbf{A}'\mathbf{A} = \mathbf{R}'\mathbf{R}$ (so that $\mathbf{A}'\mathbf{A} = \mathbf{R}'\mathbf{R}$ is the Cholesky decomposition of $\mathbf{A}'\mathbf{A}$).

Solution. Since the inner product with respect to which the columns of $\mathbf{Q}$ are orthonormal is the usual inner product, $\mathbf{Q}'\mathbf{Q} = \mathbf{I}_k$, and consequently

$$\mathbf{A}'\mathbf{A} = \mathbf{R}'\mathbf{Q}'\mathbf{Q}\mathbf{R} = \mathbf{R}'\mathbf{R}.$$

EXERCISE 24. Let $\mathbf{A}$ represent an $m \times k$ matrix of rank r (where r is possibly less than k). Consider the decomposition $\mathbf{A} = \mathbf{Q}\mathbf{R}_1$, where $\mathbf{Q}$ is an $m \times r$ matrix with orthonormal columns and $\mathbf{R}_1$ is an $r \times k$ submatrix whose rows are the r nonnull rows of a $k \times k$ upper triangular matrix $\mathbf{R}$ having r positive diagonal elements and $n - r$ null rows. (Such a decomposition can be obtained by using the results of Exercise 6.4 — refer to Exercise 6.5.) Generalize the result of Exercise 23 by showing that if the inner product with respect to which the columns of $\mathbf{Q}$ are orthonormal is the usual inner product, then $\mathbf{A}'\mathbf{A} = \mathbf{R}'\mathbf{R}$ (so that $\mathbf{A}'\mathbf{A} = \mathbf{R}'\mathbf{R}$ is the Cholesky decomposition of $\mathbf{A}'\mathbf{A}$).

Solution. Suppose that the inner product with respect to which the columns of $\mathbf{Q}$ are orthonormal is the usual inner product. Then, $\mathbf{Q}'\mathbf{Q} = \mathbf{I}_r$. Thus, recalling result (2.2.9), we find that

$$\mathbf{A}'\mathbf{A} = \mathbf{R}_1'\mathbf{Q}'\mathbf{Q}\mathbf{R}_1 = \mathbf{R}_1'\mathbf{R}_1 = \mathbf{R}'\mathbf{R}.$$

EXERCISE 25. Let $\mathbf{A} = \{a_{ij}\}$ represent an $n \times n$ matrix that has an LDU decomposition, say $\mathbf{A} = \mathbf{LDU}$. And, define $\mathbf{G} = \mathbf{U}^{-1}\mathbf{D}^{-}\mathbf{L}^{-1}$ (which is a generalized inverse of $\mathbf{A}$).

(a) Show that

$$\mathbf{G} = \mathbf{D}^{-}\mathbf{L}^{-1} + (\mathbf{I} - \mathbf{U})\mathbf{G} = \mathbf{U}^{-1}\mathbf{D}^{-} + \mathbf{G}(\mathbf{I} - \mathbf{L}) .$$

(b) For $i = 1, \ldots, n$, let d_i represent the ith diagonal element of the diagonal matrix $\mathbf{D}$; and, for $i, j = 1, \ldots, n$, let ℓ_{ij}, u_{ij}, and g_{ij} represent the ijth elements of $\mathbf{L}$, $\mathbf{U}$, and $\mathbf{G}$, respectively. Take $\mathbf{D}^{-} = \text{diag}(d_1^*, \ldots, d_n^*)$, where $d_i^* = 1/d_i$, if $d_i \neq 0$, and d_i^* is an arbitrary scalar, if $d_i = 0$. Show that

$$g_{ii} = d_i^* - \sum_{k=i+1}^{n} u_{ik} g_{ki} = d_i^* - \sum_{k=i+1}^{n} g_{ik} \ell_{ki} \tag{E.1}$$

and that

$$g_{ij} = \begin{cases} -\displaystyle\sum_{k=j+1}^{n} g_{ik} \ell_{kj}, & \text{for } j < i, \quad \text{(E.2a)} \\ -\displaystyle\sum_{k=i+1}^{n} u_{ik} g_{kj}, & \text{for } j > i \quad \text{(E.2b)} \end{cases}$$

(where the degenerate sums $\sum_{k=n+1}^{n} g_{ik}\ell_{ki}$ and $\sum_{k=n+1}^{n} u_{ik}g_{ki}$ are to be interpreted as 0).

(c) Devise a recursive procedure that uses the formulas from Part (b) to generate a generalized inverse of **A**.

Solution. (a) Clearly,

$$\mathbf{D}^{-}\mathbf{L}^{-1} + (\mathbf{I}-\mathbf{U})\mathbf{G} = \mathbf{D}^{-}\mathbf{L}^{-1} + \mathbf{G} - \mathbf{U}\mathbf{G} = \mathbf{D}^{-}\mathbf{L}^{-1} + \mathbf{G} - \mathbf{U}\mathbf{U}^{-1}\mathbf{D}^{-}\mathbf{L}^{-1} = \mathbf{G}.$$

Similarly,

$$\mathbf{U}^{-1}\mathbf{D}^{-} + \mathbf{G}(\mathbf{I}-\mathbf{L}) = \mathbf{U}^{-1}\mathbf{D}^{-} + \mathbf{G} - \mathbf{G}\mathbf{L} = \mathbf{U}^{-1}\mathbf{D}^{-} + \mathbf{G} - \mathbf{U}^{-1}\mathbf{D}^{-}\mathbf{L}^{-1}\mathbf{L} = \mathbf{G}.$$

(b) Since **L** and **U** are unit triangular, their diagonal elements equal 1 (and the diagonal elements of **I** − **L** and **I** − **U** equal 0). Thus, it follows from Part (a) that

$$g_{ij} = \begin{cases} d_i^* - \sum_{k=i+1}^{n} u_{ik} g_{ki}, & \text{if } j = i, \\ -\sum_{k=i+1}^{n} u_{ik} g_{kj}, & \text{if } j > i, \end{cases}$$

and similarly that

$$g_{ij} = \begin{cases} d_i^* - \sum_{k=i+1}^{n} g_{ik} \ell_{ki}, & \text{if } j = i, \\ -\sum_{k=i+1}^{n} g_{ik} \ell_{kj}, & \text{if } j < i. \end{cases}$$

(c) The formulas from Part (b) can be used to generate a generalized inverse of **A** in n steps. During the first step, the nth diagonal element g_{nn} is generated from the formula $g_{nn} = d_n^*$, and then $g_{n-1,n}, \ldots, g_{1n}$ and $g_{n,n-1}, \ldots, g_{n1}$ (the off-diagonal elements of the nth column and row of **G**) are generated recursively using formulas (E.2b) and (E.2a), respectively. During the $(n-s+1)$th step $(n-1 \le s \le 2)$, the sth diagonal element g_{ss} is generated from $g_{s+1,s}, \ldots, g_{ns}$ or alternatively from $g_{s,s+1}, \ldots, g_{sn}$ using result (E.1), and then $g_{s-1,s}, \ldots, g_{1s}$ and $g_{s,s-1}, \ldots, g_{s1}$ are generated recursively using formulas (E.2b) and (E.2a), respectively. During the nth (and final) step, the first diagonal element g_{11} is generated from the last $n-1$ elements of the first column or row using result (E.1).

EXERCISE 26. Verify that a principal submatrix of a skew-symmetric matrix is skew-symmetric.

Solution. Let $\mathbf{B} = \{b_{ij}\}$ represent the $r \times r$ principal submatrix of an $n \times n$ skew-symmetric matrix $\mathbf{A} = \{a_{ij}\}$ obtained by striking out all of its rows and columns except the k_1th, k_2th, ..., k_rth rows and columns (where $k_1 < k_2 < \ldots, k_r$). Then, for $i, j = 1, \ldots, r$,

$$b_{ji} = a_{k_j k_i} = -a_{k_i k_j} = -b_{ij}.$$

Since b_{ji} is the ijth element of $\mathbf{B}'$ and $-b_{ij}$ the ijth element of $-\mathbf{B}$, we conclude that $\mathbf{B}' = -\mathbf{B}$.

EXERCISE 27. (a) Show that the sum of skew-symmetric matrices is skew-symmetric.

(b) Show that the sum $\mathbf{A}_1 + \mathbf{A}_2 + \cdots + \mathbf{A}_k$ of $n \times n$ nonnegative definite matrices $\mathbf{A}_1, \mathbf{A}_2, \ldots, \mathbf{A}_k$ is skew-symmetric if and only if $\mathbf{A}_1, \mathbf{A}_2, \ldots, \mathbf{A}_k$ are skew-symmetric.

(c) Show that the sum $\mathbf{A}_1 + \mathbf{A}_2 + \cdots + \mathbf{A}_k$ of $n \times n$ symmetric nonnegative definite matrices $\mathbf{A}_1, \mathbf{A}_2, \ldots, \mathbf{A}_k$ is a null matrix if and only if $\mathbf{A}_1, \mathbf{A}_2, \ldots, \mathbf{A}_k$ are null matrices.

Solution. (a) Let $\mathbf{A}_1, \mathbf{A}_2, \ldots, \mathbf{A}_k$ represent $n \times n$ skew-symmetric matrices. Then,

$$\Big(\sum_i \mathbf{A}_i\Big)' = \sum_i \mathbf{A}_i' = \sum_i (-\mathbf{A}_i) = -\sum_i \mathbf{A}_i,$$

so that $\sum_i \mathbf{A}_i$ is skew-symmetric.

(b) If the nonnegative definite matrices $\mathbf{A}_1, \mathbf{A}_2, \ldots, \mathbf{A}_k$ are skew-symmetric, then it follows from Part (a) that their sum $\sum_i \mathbf{A}_i$ is skew-symmetric.

Conversely, suppose that $\sum_i \mathbf{A}_i$ is skew-symmetric. Let d_{ij} represent the jth diagonal element of $\mathbf{A}_i$ $(i = 1, \ldots, k;\ j = 1, \ldots, n)$. Since (according to the definition of skew-symmetry) the diagonal elements of $\sum_i \mathbf{A}_i$ equal zero, we have that

$$d_{1j} + d_{2j} + \cdots + d_{kj} = 0$$

$(j = 1, \ldots, n)$. Moreover, it follows from Corollary 14.2.13 that

$$d_{1j} \geq 0, \quad d_{2j} \geq 0, \quad \ldots, \quad d_{kj} \geq 0.$$

leading to the conclusion that $d_{1j}, d_{2j}, \ldots, d_{kj}$ equal zero $(j = 1, \ldots, n)$. Thus, it follows from Lemma 14.6.4 that $\mathbf{A}_1, \mathbf{A}_2, \ldots, \mathbf{A}_k$ are skew-symmetric.

(c) Since (according to Lemma 14.6.1) the only $n \times n$ symmetric matrix that is skew-symmetric is the $n \times n$ null matrix, Part (c) is a special case of Part (b).

EXERCISE 28. (a) Let $\mathbf{A}_1, \mathbf{A}_2, \ldots, \mathbf{A}_k$ represent $n \times n$ nonnegative definite matrices. Show that $\operatorname{tr}(\sum_{i=1}^k \mathbf{A}_i) \geq 0$, with equality holding if and only if $\sum_{i=1}^k \mathbf{A}_i$ is skew-symmetric or equivalently if and only if $\mathbf{A}_1, \mathbf{A}_2, \ldots, \mathbf{A}_k$ are skew-symmetric. [*Note*. That $\sum_{i=1}^k \mathbf{A}_i$ being skew-symmetric is equivalent to $\mathbf{A}_1, \mathbf{A}_2, \ldots, \mathbf{A}_k$ being skew-symmetric is the result of Part (b) of Exercise 27.]

(b) Let $\mathbf{A}_1, \mathbf{A}_2, \ldots, \mathbf{A}_k$ represent $n \times n$ symmetric nonnegative definite matrices. Show that $\operatorname{tr}(\sum_{i=1}^k \mathbf{A}_i) \geq 0$, with equality holding if and only if $\sum_{i=1}^k \mathbf{A}_i = \mathbf{0}$ or equivalently if and only if $\mathbf{A}_1, \mathbf{A}_2, \ldots, \mathbf{A}_k$ are null matrices.

Solution. (a) According to Corollary 14.2.5, $\sum_{i=1}^k \mathbf{A}_i$ is nonnegative definite. Thus, it follows from Theorem 14.7.2 that $\operatorname{tr}(\sum_{i=1}^k \mathbf{A}_i) \geq 0$, with equality holding if and only if $\sum_{i=1}^{k} \mathbf{A}_i$ is skew-symmetric or equivalently [in light of the result of Part (b) of Exercise 27] if and only if $\mathbf{A}_1, \mathbf{A}_2, \ldots, \mathbf{A}_k$ are skew-symmetric.

(b) Part (b) follows from Part (a) upon observing (on the basis of Lemma 14.6.1) that a symmetric matric is skew-symmetric if and only if it is null.

EXERCISE 29. Show, via an example, that (for $n > 1$) there exist $n \times n$ (non-symmetric) positive definite matrices $\mathbf{A}$ and $\mathbf{B}$ such that $\text{tr}(\mathbf{AB}) < 0$.

Solution. Take $\mathbf{A} = \{a_{ij}\}$ to be an $n \times n$ matrix such that

$$\begin{aligned} a_{ij} &= 1, \quad \text{for } j = i, \\ &= 2, \quad \text{for } j = i+1, \ldots, n, \\ &= -2, \quad \text{for } j = 1, \ldots, i-1 \end{aligned}$$

($i = 1, \ldots, n$), and take $\mathbf{B} = \mathbf{A}$. That is, take

$$\mathbf{B} = \mathbf{A} = \begin{pmatrix} 1 & 2 & 2 & \ldots & 2 \\ -2 & 1 & 2 & \ldots & 2 \\ -2 & -2 & 1 & & 2 \\ \vdots & \vdots & & \ddots & \\ -2 & -2 & & & 1 \end{pmatrix}.$$

Then, $(1/2)(\mathbf{A}+\mathbf{A}') = \mathbf{I}_n$, which is a positive definite matrix, implying (in light of Corollary 14.2.7) that $\mathbf{A}$ is positive definite. Moreover, all n diagonal elements of $\mathbf{AB}$ equal $1 - 4(n-1)$, which (for $n > 1$) is a negative number. Thus, $\text{tr}(\mathbf{AB}) < 0$.

EXERCISE 30. (a) Show, via an example, that (for $n > 1$) there exist $n \times n$ symmetric positive definite matrices $\mathbf{A}$ and $\mathbf{B}$ such that the product $\mathbf{AB}$ has one or more negative diagonal elements (and hence such that $\mathbf{AB}$ is not nonnegative definite).

(b) Show, however, that the product of two $n \times n$ symmetric positive definite matrices cannot be nonpositive definite.

Solution. (a) Take $\mathbf{A} = \text{diag}(\mathbf{A}_{11}, \mathbf{I}_{n-2})$ and $\mathbf{B} = \text{diag}(\mathbf{B}_{11}, \mathbf{I}_{n-2})$, where

$$\mathbf{A}_{11} = \begin{pmatrix} 2 & -1 \\ -1 & 2 \end{pmatrix} \quad \text{and} \quad \mathbf{B}_{11} = \begin{pmatrix} 12 & 3 \\ 3 & 1 \end{pmatrix},$$

and consider the quadratic forms $\mathbf{x}'\mathbf{Ax}$ and $\mathbf{x}'\mathbf{Bx}$ in the n-dimensional vector $\mathbf{x} = (x_1, x_2, \ldots, x_n)'$. We find that

$$\begin{aligned} \mathbf{x}'\mathbf{Ax} &= 2[x_1 - (1/2)\, x_2]^2 + (3/2)\, x_2^2 + x_3^2 + x_4^2 + \cdots + x_n^2, \\ \mathbf{x}'\mathbf{Bx} &= 12[x_1 + (1/4)\, x_2]^2 + (1/4)\, x_2^2 + x_3^2 + x_4^2 + \cdots + x_n^2. \end{aligned}$$

Clearly, $\mathbf{x}'\mathbf{Ax} \geq 0$ with equality holding only if $\mathbf{x} = \mathbf{0}$, and similarly $\mathbf{x}'\mathbf{Bx} \geq 0$ with equality holding only if $\mathbf{x} = \mathbf{0}$. Thus, the quadratic forms $\mathbf{x}'\mathbf{Ax}$ and $\mathbf{x}'\mathbf{Bx}$ are positive definite, and hence, by definition, the matrices **A** and **B** of the quadratic forms are positive definite.

Consider now the product **AB**. We find that $\mathbf{AB} = \text{diag}(\mathbf{A}_{11}\mathbf{B}_{11}, \mathbf{I}_{n-2})$ and that $\mathbf{A}_{11}\mathbf{B}_{11} = \begin{pmatrix} 21 & 5 \\ -6 & -1 \end{pmatrix}$, thereby revealing that the second diagonal element of **AB** equals the negative number -1.

(b) Let **A** and **B** represent $n \times n$ symmetric positive definite matrices. Suppose, for purposes of establishing a contradiction, that **AB** is nonpositive definite. Then, by definition, $-\mathbf{AB}$ is nonnegative definite, implying (in light of Theorem 14.7.2) that $\text{tr}(-\mathbf{AB}) \geq 0$ and hence that

$$\text{tr}(\mathbf{AB}) = -\text{tr}(-\mathbf{AB}) \leq 0.$$

However, according to Theorem 14.7.4, $\text{tr}(\mathbf{AB}) > 0$. Thus, we have arrived at the sought-after contradiction. We conclude that **AB** cannot be nonpositive definite.

EXERCISE 31. Let $\mathbf{A} = \{a_{ij}\}$ and $\mathbf{B} = \{b_{ij}\}$ represent $n \times n$ matrices, and take **C** to be the $n \times n$ matrix whose ijth element $c_{ij} = a_{ij}b_{ij}$ is the product of the ijth elements of **A** and **B** — **C** is the so-called Hadamard product of **A** and **B**. Show that if **A** is nonnegative definite and **B** is symmetric nonnegative definite, then **C** is nonnegative definite. Show further that if **A** is positive definite and **B** is symmetric positive definite, then **C** is positive definite. [*Hint.* Taking $\mathbf{x} = (x_1, \ldots, x_n)'$ to be an arbitrary $n \times 1$ vector and $\mathbf{F} = (\mathbf{f}_1, \ldots, \mathbf{f}_n)$ to be a matrix such that $\mathbf{B} = \mathbf{F}'\mathbf{F}$, begin by showing that $\mathbf{x}'\mathbf{Cx} = \text{tr}(\mathbf{AH})$, where $\mathbf{H} = \mathbf{G}'\mathbf{G}$ with $\mathbf{G} = (x_1\mathbf{f}_1, \ldots, x_n\mathbf{f}_n)$.]

Solution. Suppose that **B** is symmetric nonnegative definite. Then, according to Corollary 14.3.8, there exists a matrix $\mathbf{F} = (\mathbf{f}_1, \ldots, \mathbf{f}_n)$ such that $\mathbf{B} = \mathbf{F}'\mathbf{F}$.

Let $\mathbf{x} = (x_1, \ldots, x_n)'$ represent an arbitrary n-dimensional column vector, let $\mathbf{G} = (x_1\mathbf{f}_1, \ldots, x_n\mathbf{f}_n)$, and let $\mathbf{H} = \mathbf{G}'\mathbf{G}$. Then, the ijth element of **H** is

$$h_{ij} = (x_i\mathbf{f}_i)'x_j\mathbf{f}_j = x_ix_j\mathbf{f}_i'\mathbf{f}_j = x_ix_jb_{ij}.$$

Thus,

$$\begin{aligned}\mathbf{x}'\mathbf{Cx} = \sum_{i,\,j} c_{ij}x_ix_j = \sum_{i,\,j} a_{ij}b_{ij}x_ix_j &= \sum_{i,\,j} a_{ij}h_{ij} \\ &= \sum_{i,\,j} a_{ij}h_{ji} = \text{tr}(\mathbf{AH}).\end{aligned}$$

Clearly, the matrix **H** is symmetric nonnegative definite, implying (in light of Theorem 14.7.6) that if **A** is nonnegative definite, then $\text{tr}(\mathbf{AH}) \geq 0$ and consequently $\mathbf{x}'\mathbf{Cx} \geq 0$.

Consider now the special case where **B** is symmetric positive definite. In this special case, $\text{rank}(\mathbf{F}) = \text{rank}(\mathbf{B}) = n$, implying that the columns of **F** are linearly independent and hence nonnull. Thus, unless $\mathbf{x} = \mathbf{0}$, **G** is nonnull and hence **H** is nonnull. It follows (in light of Theorem 14.7.4) that if $\mathbf{A}$ is positive definite, then, unless $\mathbf{x} = \mathbf{0}$, $\text{tr}(\mathbf{AH}) > 0$ and consequently $\mathbf{x}'\mathbf{C}\mathbf{x} > 0$.

We conclude that if $\mathbf{A}$ is nonnegative definite and $\mathbf{B}$ is symmetric nonnegative definite, then $\mathbf{C}$ is nonnegative definite and that if $\mathbf{A}$ is positive definite and $\mathbf{B}$ is symmetric positive definite, then $\mathbf{C}$ is positive definite.

EXERCISE 32. Let $\mathbf{A}_1, \mathbf{A}_2, \ldots, \mathbf{A}_k$ and $\mathbf{B}_1, \mathbf{B}_2, \ldots, \mathbf{B}_k$ represent $n \times n$ symmetric nonnegative definite matrices. Show that $\text{tr}(\sum_{i=1}^k \mathbf{A}_i\mathbf{B}_i) \geq 0$, with equality holding if and only if, for $i = 1, 2, \ldots, k$, $\mathbf{A}_i\mathbf{B}_i = \mathbf{0}$, thereby generalizing and the results of Part (b) of Exercise 28.

Solution. According to Corollary 14.7.7, $\text{tr}(\mathbf{A}_i\mathbf{B}_i) \geq 0$ $(i = 1, 2, \ldots, k)$. Thus,

$$\text{tr}\left(\sum_{i=1}^k \mathbf{A}_i\mathbf{B}_i\right) = \sum_{i=1}^k \text{tr}(\mathbf{A}_i\mathbf{B}_i) \geq 0,$$

with equality holding if and only if, for $i = 1, 2, \ldots, k$, $\text{tr}(\mathbf{A}_i\mathbf{B}_i) = 0$ or equivalently (in light of Corollary 14.7.7) if and only if, for $i = 1, 2, \ldots, k$, $\mathbf{A}_i\mathbf{B}_i = \mathbf{0}$.

EXERCISE 33. Let $\mathbf{A}$ represent a symmetric nonnegative definite matrix that has been partitioned as

$$\mathbf{A} = \begin{pmatrix} \mathbf{T} & \mathbf{U} \\ \mathbf{V} & \mathbf{W} \end{pmatrix}$$

where $\mathbf{T}$ (and hence $\mathbf{W}$) is square. Show that $\mathbf{VT}^-\mathbf{U}$ and $\mathbf{UW}^-\mathbf{V}$ are symmetric and nonnegative definite.

Solution. According to Lemma 14.8.1, there exist matrices $\mathbf{R}$ and $\mathbf{S}$ such that

$$\mathbf{T} = \mathbf{R}'\mathbf{R}, \quad \mathbf{U} = \mathbf{R}'\mathbf{S}, \quad \mathbf{V} = \mathbf{S}'\mathbf{R}, \quad \mathbf{W} = \mathbf{S}'\mathbf{S}.$$

Thus, making use of Parts (6) and (3) of Theorem 12.3.4, we find that

$$\mathbf{VT}^-\mathbf{U} = \mathbf{S}'\mathbf{R}(\mathbf{R}'\mathbf{R})^-\mathbf{R}'\mathbf{S} = \mathbf{S}'\mathbf{P}_\mathbf{R}\mathbf{S} = \mathbf{S}'\mathbf{P}_\mathbf{R}\mathbf{P}_\mathbf{R}\mathbf{S} = \mathbf{S}'\mathbf{P}_\mathbf{R}'\mathbf{P}_\mathbf{R}\mathbf{S} = (\mathbf{P}_\mathbf{R}\mathbf{S})'\mathbf{P}_\mathbf{R}\mathbf{S}$$

and similarly that

$$\mathbf{UW}^-\mathbf{V} = \mathbf{R}'\mathbf{S}(\mathbf{S}'\mathbf{S})^-\mathbf{S}'\mathbf{R} = \mathbf{R}'\mathbf{P}_\mathbf{S}\mathbf{R} = \mathbf{R}'\mathbf{P}_\mathbf{S}\mathbf{P}_\mathbf{S}\mathbf{R} = \mathbf{R}'\mathbf{P}_\mathbf{S}'\mathbf{P}_\mathbf{S}\mathbf{R} = (\mathbf{P}_\mathbf{S}\mathbf{R})'\mathbf{P}_\mathbf{S}\mathbf{R}.$$

We conclude that $\mathbf{VT}^-\mathbf{U}$ and $\mathbf{UW}^-\mathbf{V}$ are symmetric and (in light of Corollary 14.2.14) nonnegative definite.

EXERCISE 34. Show, via an example, that there exists an $(m + n) \times (m + n)$ (nonsymmetric) positive semidefinite matrix $\mathbf{A}$ of the form $\mathbf{A} = \begin{pmatrix} \mathbf{T} & \mathbf{U} \\ \mathbf{V} & \mathbf{W} \end{pmatrix}$, where $\mathbf{T}$ is of dimensions $m \times m$, $\mathbf{W}$ of dimensions $n \times n$, $\mathbf{U}$ of dimensions $m \times n$, and $\mathbf{V}$ of dimensions $n \times m$, for which $\mathcal{C}(\mathbf{U}) \not\subset \mathcal{C}(\mathbf{T})$ and/or $\mathcal{R}(\mathbf{V}) \not\subset \mathcal{R}(\mathbf{T})$, the expression $\text{rank}(\mathbf{T}) + \text{rank}(\mathbf{W} - \mathbf{VT}^-\mathbf{U})$ does not necessarily equal rank(**A**), and the formula

$$\begin{pmatrix} \mathbf{T}^- + \mathbf{T}^-\mathbf{UQ}^-\mathbf{VT}^- & -\mathbf{T}^-\mathbf{UQ}^- \\ -\mathbf{Q}^-\mathbf{VT}^- & \mathbf{Q}^- \end{pmatrix}, \tag{$*$}$$

where $\mathbf{Q} = \mathbf{W} - \mathbf{VT}^-\mathbf{U}$, does not necessarily give a generalized inverse of **A**.

Solution. Consider the matrix $\mathbf{A} = \begin{pmatrix} \mathbf{T} & \mathbf{U} \\ \mathbf{V} & \mathbf{W} \end{pmatrix}$, where $\mathbf{T} = \mathbf{0}$, $\mathbf{U} = \mathbf{J}_{mn}$, $\mathbf{V} = -\mathbf{J}_{nm}$, and $\mathbf{W} = \mathbf{I}_n$. Clearly, $\mathcal{C}(\mathbf{U}) \not\subset \mathcal{C}(\mathbf{T})$ and $\mathcal{R}(\mathbf{V}) \not\subset \mathcal{R}(\mathbf{T})$. Moreover, $(1/2)(\mathbf{A} + \mathbf{A}') = \begin{pmatrix} \mathbf{0} & \mathbf{0} \\ \mathbf{0} & \mathbf{I}_n \end{pmatrix}$ (which is a positive semidefinite matrix), so that (according to Corollary 14.2.7) **A** is positive semidefinite.

Now, take $\mathbf{T}^- = \mathbf{0}$, in which case the Schur complement of **T** relative to $\mathbf{T}^-$ is $\mathbf{W} - \mathbf{VT}^-\mathbf{U} = \mathbf{I}_n$. Using Theorem 9.6.1, we find that

$$\begin{aligned} \text{rank}(\mathbf{A}) &= n + \text{rank}(\mathbf{T} - \mathbf{UW}^{-1}\mathbf{V}) \\ &= n + \text{rank}(\mathbf{J}_{mn}\mathbf{J}_{nm}) \\ &= n + \text{rank}(n\mathbf{J}_{mm}) = n + 1. \end{aligned}$$

However,

$$\text{rank}(\mathbf{T}) + \text{rank}(\mathbf{W} - \mathbf{VT}^-\mathbf{U}) = \text{rank}(\mathbf{0}) + \text{rank}(\mathbf{I}_n) = n.$$

Thus,

$$\text{rank}(\mathbf{A}) \neq \text{rank}(\mathbf{T}) + \text{rank}(\mathbf{W} - \mathbf{VT}^-\mathbf{U}).$$

Further, the matrix obtained by applying formula ($*$) [or equivalently formula (9.6.2)] is $\begin{pmatrix} \mathbf{0} & \mathbf{0} \\ \mathbf{0} & \mathbf{I}_n \end{pmatrix}$. Since

$$\mathbf{A}\begin{pmatrix} \mathbf{0} & \mathbf{0} \\ \mathbf{0} & \mathbf{I}_n \end{pmatrix}\mathbf{A} = \begin{pmatrix} -n\mathbf{J}_{mm} & \mathbf{J}_{mn} \\ -\mathbf{J}_{nm} & \mathbf{I}_n \end{pmatrix} \neq \mathbf{A},$$

the matrix $\begin{pmatrix} \mathbf{0} & \mathbf{0} \\ \mathbf{0} & \mathbf{I}_n \end{pmatrix}$ is not a generalized inverse of **A**.

EXERCISE 35. Show, via an example, that there exists an $(m + n) \times (m + n)$ symmetric partitioned matrix **A** of the form $\mathbf{A} = \begin{pmatrix} \mathbf{T} & \mathbf{U} \\ \mathbf{U}' & \mathbf{W} \end{pmatrix}$, where **T** is of dimensions $m \times m$, **U** of dimensions $m \times n$, and **W** of dimensions $n \times n$, such that **T** is nonnegative definite and (depending on the choice of $\mathbf{T}^-$) the Schur complement $\mathbf{W} - \mathbf{U}'\mathbf{T}^-\mathbf{U}$ of **T** relative to $\mathbf{T}^-$ is nonnegative definite, but **A** is not nonnegative definite.

Solution. Consider the symmetric matrix $\mathbf{A} = \begin{pmatrix} \mathbf{T} & \mathbf{U} \\ \mathbf{U}' & \mathbf{W} \end{pmatrix}$, where $\mathbf{T} = \mathbf{0}$, $\mathbf{U} = \mathbf{J}_{mn}$, and $\mathbf{W} = \mathbf{I}_n$. And, take $\mathbf{T}^- = \mathbf{0}$, in which case the Schur complement of **T** relative to $\mathbf{T}^-$ is $\mathbf{W} - \mathbf{U}'\mathbf{T}^-\mathbf{U} = \mathbf{I}_n$.

Then, clearly, $\mathbf{T}$ is nonnegative definite, and the Schur complement of $\mathbf{T}$ relative to $\mathbf{T}^-$ is nonnegative definite. However, $\mathbf{A}$ is not nonnegative definite, as is evident from Corollary 14.8.2.

EXERCISE 36. An $n \times n$ matrix $\mathbf{A} = \{a_{ij}\}$ is said to be *diagonally dominant* if, for $i = 1, 2, \ldots, n$, $|a_{ii}| > \sum_{j=1\,(j \neq i)}^{n} |a_{ij}|$. (In the degenerate special case where $n = 1$, $\mathbf{A}$ is said to be diagonally dominant if it is nonnull).

(a) Show that a principal submatrix of a diagonally dominant matrix is diagonally dominant.

(b) Let $\mathbf{A} = \{a_{ij}\}$ represent an $n \times n$ diagonally dominant matrix, partition $\mathbf{A}$ as $\mathbf{A} = \begin{pmatrix} \mathbf{A}_{11} & \mathbf{a} \\ \mathbf{b}' & a_{nn} \end{pmatrix}$ [so that $\mathbf{A}_{11}$ is of dimensions $(n-1) \times (n-1)$], and let $\mathbf{C} = \mathbf{A}_{11} - (1/a_{nn})\mathbf{a}\mathbf{b}'$ represent the Schur complement of a_{nn}. Show that $\mathbf{C}$ is diagonally dominant.

(c) Show that a diagonally dominant matrix is nonsingular.

(d) Show that a diagonally dominant matrix has a unique LDU decomposition.

(e) Let $\mathbf{A} = \{a_{ij}\}$ represent an $n \times n$ symmetric matrix. Show that if $\mathbf{A}$ is diagonally dominant and if the diagonal elements $a_{11}, a_{22}, \ldots, a_{nn}$ of $\mathbf{A}$ are all positive, then $\mathbf{A}$ is positive definite.

Solution. (a) Let $\mathbf{A} = \{a_{ij}\}$ represent an $n \times n$ diagonally dominant matrix, and let $\mathbf{B} = \{b_{k\ell}\}$ represent the $m \times m$ principal submatrix obtained by striking out all of the rows and columns of $\mathbf{A}$ except the i_1th, i_2th, $\ldots$, i_mth rows and columns (where $i_1 < i_2 < \cdots < i_m$). Then, for $k = 1, 2, \ldots, m$,

$$|b_{kk}| = |a_{i_k i_k}| > \sum_{j=1\,(j \neq i_k)}^{n} |a_{i_k j}| \geq \sum_{\ell=1\,(\ell \neq k)}^{m} |a_{i_k i_\ell}| = \sum_{\ell=1\,(\ell \neq k)}^{m} |b_{k\ell}|.$$

Thus, $\mathbf{B}$ is diagonally dominant.

(b) For $i, j = 1, 2, \ldots, n-1$, let c_{ij} represent the ijth element of $\mathbf{C}$. By definition,

$$c_{ij} = a_{ij} - a_{in}a_{nj}/a_{nn}.$$

Then, for $i = 1, 2, \ldots, n-1$,

$$\begin{aligned} \sum_{j=1\,(j \neq i)}^{n-1} |c_{ij}| &\leq \sum_{j=1\,(j \neq i)}^{n-1} \big(|a_{ij}| + |a_{in}a_{nj}/a_{nn}|\big) \\ &= \sum_{j=1\,(j \neq i)}^{n} |a_{ij}| - |a_{in}| + \sum_{j=1\,(j \neq i)}^{n-1} |a_{in}a_{nj}/a_{nn}| \\ &< |a_{ii}| - |a_{in}| + \sum_{j=1\,(j \neq i)}^{n-1} |a_{in}a_{nj}/a_{nn}| \end{aligned}$$

$$\leq |c_{ii}| - |a_{in}| + \sum_{j=1}^{n-1} |a_{in}a_{nj}/a_{nn}|$$

$$\begin{aligned}(\text{since } |a_{ii}| &= |a_{ii} - a_{in}a_{ni}/a_{nn} + a_{in}a_{ni}/a_{nn}| \\ &\leq |a_{ii} - a_{in}a_{ni}/a_{nn}| + |a_{in}a_{ni}/a_{nn}| \\ &= |c_{ii}| + |a_{in}a_{ni}/a_{nn}|)\end{aligned}$$

$$= |c_{ii}| - |a_{in}| + |a_{in}| \sum_{j=1}^{n-1} |a_{nj}/a_{nn}|$$

$$\leq |c_{ii}| - |a_{in}| + |a_{in}|$$

$$\begin{aligned}(\text{since } \textstyle\sum_{j=1}^{n-1} |a_{nj}/a_{nn}| &= \textstyle\sum_{j=1}^{n-1} |a_{nj}|/|a_{nn}| \\ &< |a_{nn}|/|a_{nn}| = 1)\end{aligned}$$

$$= |c_{ii}|.$$

(c) The proof is by mathematical induction. Clearly, any 1×1 diagonally dominant matrix is nonsingular. Suppose now that any $(n-1) \times (n-1)$ diagonally dominant matrix is nonsingular, and let $\mathbf{A} = \{a_{ij}\}$ represent an arbitrary $n \times n$ diagonally dominant matrix. It suffices to show that $\mathbf{A}$ is nonsingular.

Partition $\mathbf{A}$ as $\mathbf{A} = \begin{pmatrix} \mathbf{A}_{11} & \mathbf{a} \\ \mathbf{b}' & a_{nn} \end{pmatrix}$ [so that $\mathbf{A}_{11}$ is of dimensions $(n-1) \times (n-1)$]. Since $\mathbf{A}$ is diagonally dominant, $a_{nn} \neq 0$. Let $\mathbf{C} = \mathbf{A}_{11} - (1/a_{nn})\mathbf{ab}'$. It follows from Part (b) that $\mathbf{C}$ is diagonally dominant, so that by supposition $\mathbf{C}$ [which is of dimensions $(n-1) \times (n-1)$] is nonsingular. Based on Theorem 8.5.11, we conclude that $\mathbf{A}$ is nonsingular.

(d) It follows from Part (a) that every principal submatrix of a diagonally dominant matrix is diagonally dominant and hence—in light of Part (c)—nonsingular. We conclude—on the basis of Corollary 14.5.7—that a diagonally dominant matrix has a unique LDU decomposition.

(e) The proof is by mathematical induction. Clearly, any 1×1 diagonally dominant matrix with a positive (diagonal) element is positive definite. Suppose now that any $(n-1) \times (n-1)$ symmetric diagonally dominant matrix with positive diagonal elements is positive definite. Let $\mathbf{A} = \{a_{ij}\}$ represent an $n \times n$ symmetric diagonally dominant matrix with positive diagonal elements. It suffices to show that $\mathbf{A}$ is positive definite.

Partition $\mathbf{A}$ as $\mathbf{A} = \begin{pmatrix} \mathbf{A}_{11} & \mathbf{a} \\ \mathbf{a}' & a_{nn} \end{pmatrix}$ [so that $\mathbf{A}_{11}$ is of dimensions $(n-1) \times (n-1)$], and let $\mathbf{C} = \mathbf{A}_{11} - (1/a_{nn})\mathbf{aa}'$ represent the Schur complement of a_{nn}. It follows from Part (b) that $\mathbf{C}$ is diagonally dominant. Moreover, the ith diagonal element of $\mathbf{C}$ is

$$\begin{aligned} a_{ii} - a_{in}a_{ni}/a_{nn} &\geq a_{ii} - |a_{in}|\,|a_{ni}/a_{nn}| \\ &\geq a_{ii} - |a_{in}| \\ &> 0 \end{aligned}$$

($i = 1, 2, \ldots, n-1$). Thus, by supposition, $\mathbf{C}$ [which is symmetric and of dimensions $(n-1) \times (n-1)$] is positive definite. Based on Corollary 14.8.6, we conclude that $\mathbf{A}$ is positive definite.

EXERCISE 37. Let $\mathbf{A} = \{a_{ij}\}$ represent an $n \times n$ symmetric positive definite matrix. Show that $\det(\mathbf{A}) \leq \prod_{i=1}^n a_{ii}$, with equality holding if and only if $\mathbf{A}$ is diagonal.

Solution. That $\det(\mathbf{A}) = \prod_{i=1}^n a_{ii}$ if $\mathbf{A}$ is diagonal is an immediate consequence of Corollary 13.1.2. Thus, it suffices to show that if $\mathbf{A}$ is not diagonal, then $\det(\mathbf{A}) < \prod_{i=1}^n a_{ii}$. This is accomplished by mathematical induction.

Consider a 2×2 symmetric matrix

$$\mathbf{A} = \begin{pmatrix} a_{11} & a_{12} \\ a_{12} & a_{22} \end{pmatrix}$$

that is not diagonal. (Every 1×1 matrix is diagonal.) Even if $\mathbf{A}$ is not positive definite, we have that

$$\det(\mathbf{A}) = a_{11}a_{22} - a_{12}^2 < a_{11}a_{22}.$$

Suppose now that, for every $(n-1) \times (n-1)$ symmetric positive definite matrix that is not diagonal, the determinant of the matrix is less than the product of its diagonal elements, and consider the determinant of an $n \times n$ symmetric positive definite matrix $\mathbf{A} = \{a_{ij}\}$ that is not diagonal (where $n \geq 3$).

Partition $\mathbf{A}$ as

$$\mathbf{A} = \begin{pmatrix} \mathbf{A}_* & \mathbf{a} \\ \mathbf{a}' & a_{nn} \end{pmatrix}$$

[where $\mathbf{A}_*$ is of dimensions $(n-1) \times (n-1)$]. Then, in light of the discussion of Section 14.8a, it follows from Theorem 13.3.8 that

$$|\mathbf{A}| = |\mathbf{A}_*| \, (a_{nn} - \mathbf{a}'\mathbf{A}_*^{-1}\mathbf{a}). \tag{S.6}$$

And, it follows from Corollary 14.8.6 and Lemma 14.9.1 that $|\mathbf{A}_*| > 0$ and $a_{nn} - \mathbf{a}'\mathbf{A}_*^{-1}\,\mathbf{a} > 0$.

In the case where $\mathbf{A}_*$ is diagonal, we have (since $\mathbf{A}$ is not diagonal) that $\mathbf{a} \neq \mathbf{0}$, implying (since $\mathbf{A}_*^{-1}$ is positive definite) that $\mathbf{a}'\mathbf{A}_*^{-1}\,\mathbf{a} > 0$ and hence that $a_{nn} > a_{nn} - \mathbf{a}'\mathbf{A}_*^{-1}\,\mathbf{a}$, so that [in light of result (S.6)]

$$|\mathbf{A}| < a_{nn}|\mathbf{A}_*| = a_{nn}\prod_{i=1}^{n-1} a_{ii} = \prod_{i=1}^{n} a_{ii}.$$

In the alternative case where $\mathbf{A}_*$ is not diagonal, we have that $\mathbf{a}'\mathbf{A}_*^{-1}\,\mathbf{a} \geq 0$, implying that $a_{nn} \geq a_{nn} - \mathbf{a}'\mathbf{A}_*^{-1}\,\mathbf{a}$, and we have, by supposition, that $|\mathbf{A}_*| < \prod_{i=1}^{n-1} a_{ii}$, so that

$$|\mathbf{A}| < (a_{nn} - \mathbf{a}'\mathbf{A}_*^{-1}\,\mathbf{a})\prod_{i=1}^{n-1} a_{ii} \leq a_{nn}\prod_{i=1}^{n-1} a_{ii} = \prod_{i=1}^{n} a_{ii}.$$

Thus, in either case, $|\mathbf{A}| < \prod_{i=1}^{n} a_{ii}$.

EXERCISE 38. Let $\mathbf{A} = \begin{pmatrix} a & b \\ c & d \end{pmatrix}$, where a, b, c, and d are scalars.

(a) Show that $\mathbf{A}$ is positive definite if and only if $a > 0$, $d > 0$, and $|b + c| /2 < \sqrt{ad}$.

(b) Show that, in the special case where $\mathbf{A}$ is symmetric (i.e., where $c = b$), $\mathbf{A}$ is positive definite if and only if $a > 0$, $d > 0$, and $|b| < \sqrt{ad}$.

Solution. (a) Let

$$\mathbf{B} = (1/2)(\mathbf{A} + \mathbf{A}') = \begin{pmatrix} a & (b+c)/2 \\ (b+c)/2 & d \end{pmatrix}.$$

Observe that

$$\det(\mathbf{B}) = ad - [(b+c)/2]^2 \tag{S.7}$$

and (in light of Corollary 14.2.7) that $\mathbf{A}$ is positive definite if and only if $\mathbf{B}$ is positive definite.

Suppose that $\mathbf{A}$ is positive definite (and hence that $\mathbf{B}$ is positive definite). Then, it follows from Corollary 14.2.13 that $a > 0$ and $d > 0$, and [in light of equality (S.7)] it follows from Lemma 14.9.1 that ad $- [(b+c)/2]^2 > 0$, or equivalently that $[(b+c)/2]^2 <$ ad, and hence that $|b+c|/2 < \sqrt{ad}$.

Conversely, suppose that $a > 0$, $d > 0$, and $|b+c|/2 < \sqrt{ad}$, in which case $[(b+c)/2]^2 <$ ad or equivalently [in light of equality (S.7)] that $\det(\mathbf{B}) > 0$. Then, it follows from Theorem 14.9.5 that $\mathbf{B}$ is positive definite and hence that $\mathbf{A}$ is positive definite.

(b) Part (a) follows from Part (b) upon observing that, in the special case where $c = b$, the condition $|b+c|/2 < \sqrt{ad}$ simplifies to the condition $|b| < \sqrt{ad}$.

EXERCISE 39. By, for example, making use of the result of Exercise 38, show that if an $n \times n$ matrix $\mathbf{A} = \{a_{ij}\}$ is symmetric positive definite, then, for $j \neq i = 1, \ldots, n$,

$$|a_{ij}| < \sqrt{a_{ii}a_{jj}} \leq \max(a_{ii},\ a_{jj}) .$$

Solution. Suppose that $\mathbf{A}$ is symmetric positive definite. Clearly, the 2×2 matrix $\begin{pmatrix} a_{ii} & a_{ij} \\ a_{ji} & a_{jj} \end{pmatrix}$ is a principal submatrix of $\mathbf{A}$ and hence (in light of Corollary 14.2.12) is symmetric positive definite. Thus, it follows from Part (b) of Exercise 38 that $a_{ii} > 0$, $a_{jj} > 0$, and $|a_{ij}| < \sqrt{a_{ii}a_{jj}}$.

Moreover, if $a_{ii} \geq a_{jj}$, then

$$\sqrt{a_{ii}a_{jj}} \leq \sqrt{a_{ii}^2} = a_{ii} = \max(a_{ii}, a_{jj});$$

and similarly if $a_{ii} < a_{jj}$, then

$$\sqrt{a_{ii}a_{jj}} \leq \sqrt{a_{jj}^2} = a_{jj} = \max(a_{ii}, a_{jj}).$$

EXERCISE 40. Show, by example, that it is possible for the determinants of both leading principal submatrices of a 2×2 symmetric matrix to be nonnegative without the matrix being nonnegative definite and that, for $n \geq 3$, it is possible for the determinants of all n leading principal submatrices of an $n \times n$ symmetric matrix to be nonnegative *and* for the matrix to be nonsingular without the matrix being nonnegative definite.

Solution. Consider the 2×2 symmetric matrix $\begin{pmatrix} 0 & 0 \\ 0 & -1 \end{pmatrix}$. The determinants of both of its leading principal submatrices are zero (and hence nonnegative), but it is obviously not nonnegative definite.

Next, consider the 3×3 symmetric matrix

$$\mathbf{A}_* = \begin{pmatrix} 0 & 0 & 1 \\ 0 & -1 & 0 \\ 1 & 0 & 1 \end{pmatrix}.$$

By, for example, expanding $|\mathbf{A}_*|$ in terms of the cofactors of the three elements of the first row of $\mathbf{A}_*$, we find that $|\mathbf{A}_*| = 1$. Thus, the determinants of the leading principal submatrices of $\mathbf{A}_*$ (of orders 1, 2, and 3) are 0, 0, and 1, respectively, all of which are nonnegative; and $\mathbf{A}_*$ is nonsingular. However, $\mathbf{A}_*$ is not nonnegative definite (since, e.g., one of its diagonal elements is negative).

Finally, for $n \geq 4$, consider the $n \times n$ symmetric matrix

$$\mathbf{A} = \begin{pmatrix} \mathbf{A}_* & \mathbf{0} \\ \mathbf{0} & \mathbf{I}_{n-3} \end{pmatrix}.$$

Clearly, the leading principal submatrices of $\mathbf{A}$ of orders 1, 2, and 3 are the same as those of $\mathbf{A}_*$, so that their determinants are 0, 0, and 1, respectively. Moreover, it follows from results (13.3.5) and (13.1.9) that the determinants of all of the leading principal submatrices of $\mathbf{A}$ of order 4 or more equal $|\mathbf{A}_*|$ and hence equal 1. Thus, the determinants of all n leading principal submatrices of $\mathbf{A}$ are nonnegative, and $\mathbf{A}$ is nonsingular. However, $\mathbf{A}$ is not nonnegative definite (since, e.g., one of its diagonal elements is negative).

EXERCISE 41. Let $\mathcal{V}$ represent a subspace of $\mathcal{R}^{n\times 1}$ of dimension r (where $r \geq 1$). Take $\mathbf{B} = (\mathbf{b}_1, \mathbf{b}_2, \ldots, \mathbf{b}_r)$ to be any $n \times r$ matrix whose columns $\mathbf{b}_1, \mathbf{b}_2, \ldots, \mathbf{b}_r$ form a basis for $\mathcal{V}$, and let $\mathbf{L}$ represent any left inverse of $\mathbf{B}$. Let g represent a function that assigns the value $\mathbf{x} * \mathbf{y}$ to an arbitrary pair of vectors $\mathbf{x}$ and $\mathbf{y}$ in $\mathcal{V}$.

(a) Let f represent an arbitrary inner product for $\mathcal{R}^{r\times 1}$, and denote by $\mathbf{s} \bullet \mathbf{t}$ the value assigned by f to an arbitrary pair of r-dimensional vectors $\mathbf{s}$ and $\mathbf{t}$. Show that g is an inner product (for $\mathcal{V}$) if and only if there exists an f such that (for all $\mathbf{x}$ and $\mathbf{y}$ in $\mathcal{V}$)

$$\mathbf{x} * \mathbf{y} = (\mathbf{L}\mathbf{x}) \bullet (\mathbf{L}\mathbf{y}).$$

(b) Show that g is an inner product (for $\mathcal{V}$) if and only if there exists an $r \times r$

symmetric positive definite matrix $\mathbf{W}$ such that (for all $\mathbf{x}$ and $\mathbf{y}$ in $\mathcal{V}$)

$$\mathbf{x} * \mathbf{y} = \mathbf{x}'\mathbf{L}'\mathbf{W}\mathbf{L}\mathbf{y}.$$

(c) Show that g is an inner product (for $\mathcal{V}$) if and only if there exists an $n \times n$ symmetric positive definite matrix $\mathbf{W}$ such that (for all $\mathbf{x}$ and $\mathbf{y}$ in $\mathcal{V}$)

$$\mathbf{x} * \mathbf{y} = \mathbf{x}'\mathbf{W}\mathbf{y}.$$

Solution. (a) Suppose that, for some f, $\mathbf{x} * \mathbf{y} = (\mathbf{L}\mathbf{x}) \bullet (\mathbf{L}\mathbf{y})$ (for all $\mathbf{x}$ and $\mathbf{y}$ in $\mathcal{V}$). Then,

(1) $\mathbf{x} * \mathbf{y} = (\mathbf{L}\mathbf{x}) \bullet (\mathbf{L}\mathbf{y}) = (\mathbf{L}\mathbf{y}) \bullet (\mathbf{L}\mathbf{x}) = \mathbf{y} * \mathbf{x}$;

(2) $\mathbf{x} * \mathbf{x} = (\mathbf{L}\mathbf{x}) \bullet (\mathbf{L}\mathbf{x}) \geq 0$, with equality holding if and only if $\mathbf{L}\mathbf{x} = \mathbf{0}$ or equivalently (since $\mathbf{x} = \mathbf{B}\mathbf{k}$ for some vector $\mathbf{k}$, so that $\mathbf{L}\mathbf{x} = \mathbf{0} \Longrightarrow \mathbf{L}\mathbf{B}\mathbf{k} = \mathbf{0} \Longrightarrow \mathbf{I}\mathbf{k} = \mathbf{0} \Longrightarrow \mathbf{k} = \mathbf{0} \Longrightarrow \mathbf{B}\mathbf{k} = \mathbf{0} \Longrightarrow \mathbf{x} = \mathbf{0}$) if and only if $\mathbf{x} = \mathbf{0}$;

(3) $(k\mathbf{x}) * \mathbf{y} = (k\mathbf{L}\mathbf{x}) \bullet (\mathbf{L}\mathbf{y}) = k[(\mathbf{L}\mathbf{x}) \bullet (\mathbf{L}\mathbf{y})] = k(\mathbf{x} * \mathbf{y})$;

(4) $(\mathbf{x}+\mathbf{y}) * \mathbf{z} = (\mathbf{L}\mathbf{x}+\mathbf{L}\mathbf{y}) \bullet (\mathbf{L}\mathbf{z}) = [(\mathbf{L}\mathbf{x}) \bullet (\mathbf{L}\mathbf{z})] + [(\mathbf{L}\mathbf{y}) \bullet (\mathbf{L}\mathbf{z})] = (\mathbf{x} * \mathbf{z}) + (\mathbf{y} * \mathbf{z})$

(where $\mathbf{x}$, $\mathbf{y}$, and $\mathbf{z}$ represent arbitrary vectors in $\mathcal{V}$ and k represents an arbitrary scalar). Thus, g is an inner product.

Conversely, suppose that g is an inner product, and consider the function $\tilde{f}$ that assigns to an arbitrary pair of vectors $\mathbf{s}$ and $\mathbf{t}$ in $\mathcal{R}^{r \times 1}$ the value

$$\mathbf{s} \star \mathbf{t} = (\mathbf{B}\mathbf{s}) * (\mathbf{B}\mathbf{t}).$$

We find that

(1) $\mathbf{s} \star \mathbf{t} = (\mathbf{B}\mathbf{s}) * (\mathbf{B}\mathbf{t}) = (\mathbf{B}\mathbf{t}) * (\mathbf{B}\mathbf{s}) = \mathbf{t} \star \mathbf{s}$;

(2) $\mathbf{s} \star \mathbf{s} = (\mathbf{B}\mathbf{s}) * (\mathbf{B}\mathbf{s}) \geq 0$, with equality holding if and only if $\mathbf{B}\mathbf{s} = \mathbf{0}$ or equivalently (since the columns of $\mathbf{B}$ are linearly independent) if and only if $\mathbf{s} = \mathbf{0}$;

(3) $(k\mathbf{s}) \star \mathbf{t} = (k\mathbf{B}\mathbf{s}) * (\mathbf{B}\mathbf{t}) = k[(\mathbf{B}\mathbf{s}) * (\mathbf{B}\mathbf{t})] = k(\mathbf{s} \star \mathbf{t})$;

(4) $(\mathbf{s}+\mathbf{t}) \star \mathbf{u} = (\mathbf{B}\mathbf{s}+\mathbf{B}\mathbf{t}) * (\mathbf{B}\mathbf{u}) = [(\mathbf{B}\mathbf{s}) * (\mathbf{B}\mathbf{u})] + [(\mathbf{B}\mathbf{t}) * (\mathbf{B}\mathbf{u})] = (\mathbf{s} \star \mathbf{u}) + (\mathbf{t} \star \mathbf{u})$

(where $\mathbf{s}$, $\mathbf{t}$, and $\mathbf{u}$ represent arbitrary vectors in $\mathcal{R}^{r \times 1}$ and k represents an arbitrary scalar). Thus, $\tilde{f}$ is an inner product (for $\mathcal{R}^{r \times 1}$).

Now, set $f = \tilde{f}$. Then, letting $\mathbf{x}$ and $\mathbf{y}$ represent arbitrary vectors in $\mathcal{V}$ and defining $\mathbf{s}$ and $\mathbf{t}$ to be the unique vectors that satisfy $\mathbf{B}\mathbf{s} = \mathbf{x}$ and $\mathbf{B}\mathbf{t} = \mathbf{y}$ (so that $\mathbf{s} = \mathbf{I}\mathbf{s} = \mathbf{L}\mathbf{B}\mathbf{s} = \mathbf{L}\mathbf{x}$ and similarly $\mathbf{t} = \mathbf{L}\mathbf{y}$), we find that

$$\mathbf{x} * \mathbf{y} = (\mathbf{B}\mathbf{s}) * (\mathbf{B}\mathbf{t}) = \mathbf{s} \star \mathbf{t} = \mathbf{s} \bullet \mathbf{t} = (\mathbf{L}\mathbf{x}) \bullet (\mathbf{L}\mathbf{y}).$$

(b) Let f represent an arbitrary inner product for $\mathcal{R}^{r \times 1}$, and denote by $\mathbf{s} \bullet \mathbf{t}$ the value assigned by f to an arbitrary pair of r-dimensional vectors $\mathbf{s}$ and $\mathbf{t}$. According to Part (a), g is an inner product (for $\mathcal{V}$) if and only if there exists an f such that (for all $\mathbf{x}$ and $\mathbf{y}$ in $\mathcal{V}$) $\mathbf{x} * \mathbf{y} = (\mathbf{L}\mathbf{x}) \bullet (\mathbf{L}\mathbf{y})$. Moreover, according to the discussion of

Section 14.10a, every inner product for $\mathcal{R}^{r\times 1}$ is expressible as a bilinear form, and a bilinear form (in r-dimensional vectors) qualifies as an inner product for $\mathcal{R}^{r\times 1}$ if and only if the matrix of the bilinear form is symmetric and positive definite. Thus, g is an inner product (for $\mathcal{V}$) if and only if there exists an $r \times r$ symmetric positive definite matrix $\mathbf{W}$ such that (for all $\mathbf{x}$ and $\mathbf{y}$ in $\mathcal{V}$) $\mathbf{x} * \mathbf{y} = (\mathbf{Lx})'\mathbf{WLy}$.

(c) Suppose that there exists an $n \times n$ symmetric positive definite matrix $\mathbf{W}$ such that (for all $\mathbf{x}$ and $\mathbf{y}$ in $\mathcal{V}$) $\mathbf{x} * \mathbf{y} = \mathbf{x}'\mathbf{Wy}$. According to the discussion of Section 14.10a, the function that assigns the value $\mathbf{x}'\mathbf{Wy}$ to an arbitrary pair of vectors $\mathbf{x}$ and $\mathbf{y}$ in $\mathcal{R}^{n\times 1}$ is an inner product for $\mathcal{R}^{n\times 1}$. Thus, it follows from the discussion of Section 6.1b that g is an inner product (for $\mathcal{V}$).

Conversely, suppose that g is an inner product. According to Theorem 4.3.12, there exist $n-r$ n-dimensional column vectors $\mathbf{b}_{r+1}, \ldots, \mathbf{b}_n$ such that $\mathbf{b}_1, \ldots, \mathbf{b}_r, \mathbf{b}_{r+1}, \ldots, \mathbf{b}_n$ form a basis for $\mathcal{R}^{n\times 1}$. Let $\mathbf{C} = (\mathbf{b}_{r+1}, \mathbf{b}_{r+2}, \ldots, \mathbf{b}_n)$, and define

$$\mathbf{F} = (\mathbf{B},\ \mathbf{C}).$$

Partition $\mathbf{F}^{-1}$ as

$$\mathbf{F}^{-1} = \begin{pmatrix} \mathbf{L}_* \\ \mathbf{M} \end{pmatrix},$$

where $\mathbf{L}_*$ is of dimensions $r \times n$. (The matrix $\mathbf{F}$ is invertible since its columns are linearly independent.) Note that (by definition)

$$\mathbf{L}_*\mathbf{B} = \mathbf{I}_r\,,$$

so that $\mathbf{L}_*$ is a left inverse of $\mathbf{B}$, and

$$\mathbf{MB} = \mathbf{0}.$$

According to Part (b), there exists an $r \times r$ symmetric positive definite matrix $\mathbf{W}_*$ such that (for all $\mathbf{x}$ and $\mathbf{y}$ in $\mathcal{V}$)

$$\mathbf{x} * \mathbf{y} = \mathbf{x}'\mathbf{L}_*'\mathbf{W}_*\mathbf{L}_*\mathbf{y}.$$

Moreover, letting $\mathbf{x}$ and $\mathbf{y}$ represent arbitrary vectors in $\mathcal{V}$ and defining $\mathbf{s}$ and $\mathbf{t}$ to be the unique vectors that satisfy $\mathbf{Bs} = \mathbf{x}$ and $\mathbf{Bt} = \mathbf{y}$, we find that

$$\begin{aligned}
\mathbf{x}'\mathbf{L}_*'\mathbf{W}_*\mathbf{L}_*\mathbf{y} &= \mathbf{s}'(\mathbf{L}_*\mathbf{B})'\mathbf{W}_*\mathbf{L}_*\mathbf{Bt} \\
&= \mathbf{s}'\begin{pmatrix} \mathbf{L}_*\mathbf{B} \\ \mathbf{0} \end{pmatrix}'\begin{pmatrix} \mathbf{W}_* & \mathbf{0} \\ \mathbf{0} & \mathbf{I}_{n-r} \end{pmatrix}\begin{pmatrix} \mathbf{L}_*\mathbf{B} \\ \mathbf{0} \end{pmatrix}\mathbf{t} \\
&= \mathbf{s}'\left[\begin{pmatrix} \mathbf{L}_* \\ \mathbf{M} \end{pmatrix}\mathbf{B}\right]'\begin{pmatrix} \mathbf{W}_* & \mathbf{0} \\ \mathbf{0} & \mathbf{I}_{n-r} \end{pmatrix}\begin{pmatrix} \mathbf{L}_* \\ \mathbf{M} \end{pmatrix}\mathbf{Bt} \\
&= \mathbf{x}'\mathbf{Wy},
\end{aligned}$$

where $\mathbf{W} = (\mathbf{F}^{-1})'\mathrm{diag}(\mathbf{W}_*, \mathbf{I}_{n-r})\mathbf{F}^{-1}$. Thus, $\mathbf{x} * \mathbf{y} = \mathbf{x}'\mathbf{Wy}$. Furthermore, $\mathbf{W}$ is symmetric, and it follows from Lemma 14.8.3 and Corollary 14.2.10 that $\mathbf{W}$ is positive definite.

EXERCISE 42. Let $\mathcal{V}$ represent a linear space of $m \times n$ matrices, and let $\mathbf{A} \bullet \mathbf{B}$ represent the value assigned by a quasi-inner product to any pair of matrices $\mathbf{A}$ and $\mathbf{B}$ (in $\mathcal{V}$). Show that the set

$$\mathcal{U} = \{\mathbf{A} \in \mathcal{V} : \mathbf{A} \bullet \mathbf{A} = 0\},$$

which comprises every matrix in $\mathcal{V}$ with a zero quasi norm, is a linear space.

Solution. Let $\mathbf{A}$ and $\mathbf{B}$ represent arbitrary matrices in $\mathcal{U}$, and let k represent an arbitrary scalar.

Since (by definition) $\|\mathbf{A}\| = 0$, it follows from the discussion in Section 14.10c that $\mathbf{A} \bullet \mathbf{B} = 0$. Thus,

$$(\mathbf{A} + \mathbf{B}) \bullet (\mathbf{A} + \mathbf{B}) = (\mathbf{A} \bullet \mathbf{A}) + 2(\mathbf{A} \bullet \mathbf{B}) + (\mathbf{B} \bullet \mathbf{B}) = 0 + 0 + 0 = 0,$$

implying that $(\mathbf{A} + \mathbf{B}) \in \mathcal{U}$. Moreover,

$$(k\mathbf{A}) \bullet (k\mathbf{A}) = k^2(\mathbf{A} \bullet \mathbf{A}) = k^2(0) = 0,$$

so that $k\mathbf{A} \in \mathcal{U}$.

We conclude that $\mathcal{U}$ is a linear space.

EXERCISE 43. Let $\mathbf{W}$ represent an $m \times m$ symmetric positive definite matrix and $\mathbf{V}$ an $n \times n$ symmetric positive definite matrix.

(a) Show that the function that assigns the value $\text{tr}(\mathbf{A}'\mathbf{W}\mathbf{B}\mathbf{V})$ to an arbitrary pair of $m \times n$ matrices $\mathbf{A}$ and $\mathbf{B}$ qualifies as an inner product for the linear space $\mathcal{R}^{m \times n}$.

(b) Show that the function that assigns the value $\text{tr}(\mathbf{A}'\mathbf{W}\mathbf{B})$ to an arbitrary pair of $m \times n$ matrices $\mathbf{A}$ and $\mathbf{B}$ qualifies as an inner product for $\mathcal{R}^{m \times n}$.

(c) Show that the function that assigns the value $\text{tr}(\mathbf{A}'\mathbf{W}\mathbf{B}\mathbf{W})$ to an arbitrary pair of $m \times m$ matrices $\mathbf{A}$ and $\mathbf{B}$ qualifies as an inner product for $\mathcal{R}^{m \times m}$.

Solution. (a) Let us show that the function that assigns the value $\text{tr}(\mathbf{A}'\mathbf{W}\mathbf{B}\mathbf{V})$ to an arbitrary pair of $m \times n$ matrices $\mathbf{A}$ and $\mathbf{B}$ has the four basic properties (described in Section 6.1b) of an inner product. For this purpose, let $\mathbf{A}$, $\mathbf{B}$, and $\mathbf{C}$ represent arbitrary $m \times n$ matrices, and let k represent an arbitrary scalar.

(1) Using results (5.1.5) and (5.2.3), we find that

$$\text{tr}(\mathbf{A}'\mathbf{W}\mathbf{B}\mathbf{V}) = \text{tr}[(\mathbf{A}'\mathbf{W}\mathbf{B}\mathbf{V})'] = \text{tr}(\mathbf{V}\mathbf{B}'\mathbf{W}\mathbf{A}) = \text{tr}(\mathbf{B}'\mathbf{W}\mathbf{A}\mathbf{V}).$$

(2) According to Corollary 14.3.13, $\mathbf{W} = \mathbf{Q}'\mathbf{Q}$ for some $m \times m$ nonsingular matrix $\mathbf{Q}$, and $\mathbf{V} = \mathbf{P}'\mathbf{P}$ for some $n \times n$ nonsingular matrix $\mathbf{P}$. Thus, using results (5.2.3) and (5.2.5) along with Lemma 5.3.1, we find that

$$\begin{aligned}\text{tr}(\mathbf{A}'\mathbf{W}\mathbf{A}\mathbf{V}) = \text{tr}(\mathbf{A}'\mathbf{Q}'\mathbf{Q}\mathbf{A}\mathbf{P}'\mathbf{P}) &= \text{tr}(\mathbf{P}\mathbf{A}'\mathbf{Q}'\mathbf{Q}\mathbf{A}\mathbf{P}') \\ &= \text{tr}[(\mathbf{Q}\mathbf{A}\mathbf{P}')'\mathbf{Q}\mathbf{A}\mathbf{P}'] \geq 0,\end{aligned}$$

with equality holding if and only if $\mathbf{Q}\mathbf{A}\mathbf{P}' = \mathbf{0}$ or, equivalently, if and only if $\mathbf{A} = \mathbf{0}$.

(3) Clearly, $\mathrm{tr}[(k\mathbf{A})'\mathbf{WBV})] = k\,\mathrm{tr}(\mathbf{A}'\mathbf{WBV})$.

(4) Clearly, $\mathrm{tr}[(\mathbf{A}+\mathbf{B})'\mathbf{WCV}] = \mathrm{tr}(\mathbf{A}'\mathbf{WCV}) + \mathrm{tr}(\mathbf{B}'\mathbf{WCV})$.

(b) and (c) The functions described in Parts (b) and (c) are special cases of the function described in Part (a) — those where $\mathbf{V} = \mathbf{I}$ and $\mathbf{V} = \mathbf{W}$, respectively.

EXERCISE 44. Let $\mathbf{A}$ represent a $q \times p$ matrix, $\mathbf{B}$ a $p \times n$ matrix, and $\mathbf{C}$ an $m \times q$ matrix. Show that (a) $\mathbf{CAB}(\mathbf{CAB})^-\mathbf{C} = \mathbf{C}$ if and only if $\mathrm{rank}(\mathbf{CAB}) = \mathrm{rank}(\mathbf{C})$, and (b) $\mathbf{B}(\mathbf{CAB})^-\mathbf{CAB} = \mathbf{B}$ if and only if $\mathrm{rank}(\mathbf{CAB}) = \mathrm{rank}(\mathbf{B})$.

Solution. (a) Suppose that $\mathrm{rank}(\mathbf{CAB}) = \mathrm{rank}(\mathbf{C})$. Then, it follows from Corollary 4.4.7 that $\mathcal{C}(\mathbf{CAB}) = \mathcal{C}(\mathbf{C})$ and hence that $\mathbf{C} = \mathbf{CABR}$ for some matrix $\mathbf{R}$. Thus,

$$\mathbf{CAB}(\mathbf{CAB})^-\mathbf{C} = \mathbf{CAB}(\mathbf{CAB})^-\mathbf{CABR} = \mathbf{CABR} = \mathbf{C}.$$

Conversely, suppose that $\mathbf{CAB}(\mathbf{CAB})^-\mathbf{C} = \mathbf{C}$. Then,

$$\mathrm{rank}(\mathbf{CAB}) \geq \mathrm{rank}[\mathbf{CAB}(\mathbf{CAB})^-\mathbf{C}] = \mathrm{rank}(\mathbf{C}).$$

Since clearly $\mathrm{rank}(\mathbf{CAB}) \leq \mathrm{rank}(\mathbf{C})$, we have that $\mathrm{rank}(\mathbf{CAB}) = \mathrm{rank}(\mathbf{C})$.

(b) Similarly, suppose that $\mathrm{rank}(\mathbf{CAB}) = \mathrm{rank}(\mathbf{B})$. Then, it follows from Corollary 4.4.7 that $\mathcal{R}(\mathbf{CAB}) = \mathcal{R}(\mathbf{B})$ and hence that $\mathbf{B} = \mathbf{LCAB}$ for some matrix $\mathbf{L}$. Thus,

$$\mathbf{B}(\mathbf{CAB})^-\mathbf{CAB} = \mathbf{LCAB}(\mathbf{CAB})^-\mathbf{CAB} = \mathbf{LCAB} = \mathbf{B}.$$

Conversely, suppose that $\mathbf{B}(\mathbf{CAB})^-\mathbf{CAB} = \mathbf{B}$. Then,

$$\mathrm{rank}(\mathbf{CAB}) \geq \mathrm{rank}[\mathbf{B}(\mathbf{CAB})^-\mathbf{CAB}] = \mathrm{rank}(\mathbf{B}).$$

Since clearly $\mathrm{rank}(\mathbf{CAB}) \leq \mathrm{rank}(\mathbf{B})$, we have that $\mathrm{rank}(\mathbf{CAB}) = \mathrm{rank}(\mathbf{B})$.

EXERCISE 45. Let $\mathcal{U}$ represent a subspace of $\mathcal{R}^{n\times 1}$, let $\mathbf{X}$ represent an $n \times p$ matrix whose columns span $\mathcal{U}$, and let $\mathbf{W}$ and $\mathbf{V}$ represent $n \times n$ symmetric positive definite matrices. Show that each of the following two conditions is necessary and sufficient for the projection $\mathbf{P}_{\mathbf{X},\mathbf{W}}\mathbf{y}$ of $\mathbf{y}$ on $\mathcal{U}$ with respect to $\mathbf{W}$ to be the same (for every $\mathbf{y}$ in $\mathcal{R}^n$) as the projection $\mathbf{P}_{\mathbf{X},\mathbf{V}}\mathbf{y}$ of $\mathbf{y}$ on $\mathcal{U}$ with respect to $\mathbf{V}$:

(a) $$\mathbf{V} = \mathbf{P}'_{\mathbf{X},\mathbf{W}}\mathbf{V}\mathbf{P}_{\mathbf{X},\mathbf{W}} + (\mathbf{I} - \mathbf{P}_{\mathbf{X},\mathbf{W}})'\mathbf{V}(\mathbf{I} - \mathbf{P}_{\mathbf{X},\mathbf{W}});$$

(b) there exist a scalar c, a $p \times p$ matrix $\mathbf{K}$, and an $n \times n$ matrix $\mathbf{H}$ such that

$$\mathbf{V} = c\mathbf{W} + \mathbf{WXKX}'\mathbf{W} + (\mathbf{I} - \mathbf{P}_{\mathbf{X},\mathbf{W}})'\mathbf{H}(\mathbf{I} - \mathbf{P}_{\mathbf{X},\mathbf{W}}).$$

Solution. (a) In light of Theorem 14.12.18, it suffices to show that this condition is equivalent to the condition $(\mathbf{I} - \mathbf{P}_{\mathbf{X},\mathbf{W}})'\mathbf{V}\mathbf{P}_{\mathbf{X},\mathbf{W}} = \mathbf{0}$.

Suppose that $\mathbf{V} = \mathbf{P}'_{\mathbf{X},\mathbf{W}}\mathbf{V}\mathbf{P}_{\mathbf{X},\mathbf{W}} + (\mathbf{I} - \mathbf{P}_{\mathbf{X},\mathbf{W}})'\mathbf{V}(\mathbf{I} - \mathbf{P}_{\mathbf{X},\mathbf{W}})$. Then,

$$\begin{aligned}&(\mathbf{I} - \mathbf{P}_{\mathbf{X},\mathbf{W}})'\mathbf{V}\mathbf{P}_{\mathbf{X},\mathbf{W}}\\ &\quad= [\mathbf{P}_{\mathbf{X},\mathbf{W}}(\mathbf{I} - \mathbf{P}_{\mathbf{X},\mathbf{W}})]'\mathbf{V}\mathbf{P}^2_{\mathbf{X},\mathbf{W}} + [(\mathbf{I} - \mathbf{P}_{\mathbf{X},\mathbf{W}})']^2\mathbf{V}(\mathbf{I} - \mathbf{P}_{\mathbf{X},\mathbf{W}})\mathbf{P}_{\mathbf{X},\mathbf{W}}.\end{aligned}$$

Moreover, since [according to Part (6) of Theorem 14.12.11] $\mathbf{P_{X,W}}$ is idempotent,

$$\mathbf{P_{X,W}(I - P_{X,W}) = P_{X,W} - P^2_{X,W} = P_{X,W} - P_{X,W} = 0},$$

and similarly $\mathbf{(I - P_{X,W})P_{X,W} = 0}$. Thus, $\mathbf{(I - P_{X,W})'VP_{X,W} = 0}$.

Conversely, suppose that $\mathbf{(I - P_{X,W})'VP_{X,W} = 0}$. Then,

$$\begin{aligned}\mathbf{V} &= \mathbf{[P_{X,W} + (I - P_{X,W})]'V[P_{X,W} + (I - P_{X,W})]} \\ &= \mathbf{P'_{X,W}VP_{X,W} + (I - P_{X,W})'V(I - P_{X,W})} \\ &\qquad\qquad + \mathbf{(I - P_{X,W})'VP_{X,W} + [(I - P_{X,W})'VP_{X,W}]'} \\ &= \mathbf{P'_{X,W}VP_{X,W} + (I - P_{X,W})'V(I - P_{X,W})}.\end{aligned}$$

(b) Suppose that $\mathbf{P_{X,W}y}$ is the same for every $\mathbf{y}$ in $\mathcal{R}^n$ as $\mathbf{P_{X,V}y}$. Then, Condition (a) of the exercise is satisfied, so that

$$\begin{aligned}\mathbf{V} &= \mathbf{P'_{X,W}VP_{X,W} + (I - P_{X,W})'V(I - P_{X,W})} \\ &= c\mathbf{W} + \mathbf{WXKX'W + (I - P_{X,W})'H(I - P_{X,W})}\end{aligned}$$

for $c = 0$, $\mathbf{K} = [(\mathbf{X'WX})^{-}]'\mathbf{X'VX}(\mathbf{X'WX})^{-}$, and $\mathbf{H = V}$.

Conversely, suppose that

$$\mathbf{V} = c\mathbf{W} + \mathbf{WXKX'W + (I - P_{X,W})'H(I - P_{X,W})}$$

for some scalar c and some matrices $\mathbf{K}$ and $\mathbf{H}$. Then, since [according to Part (1) of Theorem 14.12.11] $\mathbf{X - P_{X,W}X = 0}$, we have that

$$\begin{aligned}\mathbf{VX} &= c\mathbf{WX} + \mathbf{WXKX'WX + (I - P_{X,W})'H(X - P_{X,W}X)} \\ &= c\mathbf{WX} + \mathbf{WXKX'WX} \\ &= \mathbf{WX}(c\mathbf{I} + \mathbf{KX'WX}) \\ &= \mathbf{WXQ}\end{aligned}$$

for $\mathbf{Q} = c\mathbf{I} + \mathbf{KX'WX}$. Thus, it follows from Theorem 14.12.18 that $\mathbf{P_{X,W}y}$ is the same for every $\mathbf{y}$ in $\mathcal{R}^n$ as $\mathbf{P_{X,V}y}$.

EXERCISE 46. Let $\mathbf{y}$ represent an n-dimensional column vector, let $\mathcal{U}$ represent a subspace of $\mathcal{R}^{n\times 1}$, and let $\mathbf{X}$ represent an $n \times p$ matrix whose columns span $\mathcal{U}$. "Recall" that, for any $m \times n$ matrix $\mathbf{L}$, for any subspace $\mathcal{W}$ of $\mathcal{R}^{n\times 1}$, and for $\mathcal{V} = \{\mathbf{v} \in \mathcal{R}^m : \mathbf{v} = \mathbf{Lx} \text{ for some } \mathbf{x} \in \mathcal{W}\}$,

$$\mathbf{x} \perp_{\mathbf{H}} \mathcal{W} \iff \mathbf{Lx} \perp_{\mathbf{I}} \mathcal{V}, \tag{*}$$

where $\mathbf{H = L'L}$ (and where $\mathbf{x}$ represents an arbitrary $n \times 1$ vector), and

$$\mathcal{C}(\mathbf{LZ}) = \mathcal{V}, \tag{**}$$

where $\mathbf{Z}$ is any $n \times q$ matrix whose columns span $\mathcal{W}$. Use results (*) and (**) to show that, for any $n \times n$ symmetric nonnegative definite matrix $\mathbf{W}$, $\mathbf{y} \perp_{\mathbf{W}} \mathcal{U}$ if and only if $\mathbf{X'Wy = 0}$.

Solution. According to Corollary 14.3.8, $\mathbf{W} = \mathbf{L}'\mathbf{L}$ for some matrix $\mathbf{L}$. Denote by m the number of rows in $\mathbf{L}$, and let

$$\mathcal{V} = \{\mathbf{v} \in \mathcal{R}^m : \mathbf{v} = \mathbf{L}\mathbf{x} \text{ for some } \mathbf{x} \in \mathcal{U}\}.$$

Then, it follows from result (∗) [or equivalently from Part (3) of Lemma 14.12.2] that $\mathbf{y} \perp_{\mathbf{W}} \mathcal{U}$ if and only if $\mathbf{L}\mathbf{y} \perp_{\mathbf{I}} \mathcal{V}$ and hence, in light of result (∗∗) [or equivalently in light of Part (5) of Lemma 14.12.2], if and only if $(\mathbf{L}\mathbf{X})'\mathbf{L}\mathbf{y} = \mathbf{0}$. Since $(\mathbf{L}\mathbf{X})'\mathbf{L}\mathbf{y} = \mathbf{X}'\mathbf{W}\mathbf{y}$, we conclude that $\mathbf{y} \perp_{\mathbf{W}} \mathcal{U}$ if and only if $\mathbf{X}'\mathbf{W}\mathbf{y} = \mathbf{0}$.

EXERCISE 47. Let $\mathbf{W}$ represent an $n \times n$ symmetric nonnegative definite matrix.

(a) Show that, for any $n \times p$ matrix $\mathbf{X}$ and any $n \times q$ matrix $\mathbf{U}$ such that $\mathcal{C}(\mathbf{U}) \subset \mathcal{C}(\mathbf{X})$,

(1) $\mathbf{W}\mathbf{P}_{\mathbf{X},\mathbf{W}}\mathbf{U} = \mathbf{W}\mathbf{U}$, and $\mathbf{U}'\mathbf{W}\mathbf{P}_{\mathbf{X},\mathbf{W}} = \mathbf{U}'\mathbf{W}$;

(2) $\mathbf{P}_{\mathbf{U},\mathbf{W}}\mathbf{P}_{\mathbf{X},\mathbf{W}} = \mathbf{P}_{\mathbf{U},\mathbf{W}}$, and $\mathbf{P}'_{\mathbf{X},\mathbf{W}}\mathbf{W}\mathbf{P}_{\mathbf{U},\mathbf{W}} = \mathbf{W}\mathbf{P}_{\mathbf{X},\mathbf{W}}\mathbf{P}_{\mathbf{U},\mathbf{W}} = \mathbf{W}\mathbf{P}_{\mathbf{U},\mathbf{W}}$.

(b) Show that, for any $n \times p$ matrix $\mathbf{X}$ and any $n \times q$ matrix $\mathbf{U}$ such that $\mathcal{C}(\mathbf{U}) = \mathcal{C}(\mathbf{X})$,

$$\mathbf{W}\mathbf{P}_{\mathbf{U},\mathbf{W}} = \mathbf{W}\mathbf{P}_{\mathbf{X},\mathbf{W}}.$$

Solution. (a) (1) According to Lemma 4.2.2, there exists a matrix $\mathbf{F}$ such that $\mathbf{U} = \mathbf{X}\mathbf{F}$. Thus, making use of Parts (1) and (5) of Theorem 14.12.25, we find that

$$\mathbf{W}\mathbf{P}_{\mathbf{X},\mathbf{W}}\mathbf{U} = \mathbf{W}\mathbf{P}_{\mathbf{X},\mathbf{W}}\mathbf{X}\mathbf{F} = \mathbf{W}\mathbf{X}\mathbf{F} = \mathbf{W}\mathbf{U}$$

and

$$\mathbf{U}'\mathbf{W}\mathbf{P}_{\mathbf{X},\mathbf{W}} = \mathbf{F}'\mathbf{X}'\mathbf{W}\mathbf{P}_{\mathbf{X},\mathbf{W}} = \mathbf{F}'\mathbf{X}'\mathbf{W} = \mathbf{U}'\mathbf{W}.$$

(2) Making use of Part (1), we find that

$$\mathbf{P}_{\mathbf{U},\mathbf{W}}\mathbf{P}_{\mathbf{X},\mathbf{W}} = \mathbf{U}(\mathbf{U}'\mathbf{W}\mathbf{U})^{-}\mathbf{U}'\mathbf{W}\mathbf{P}_{\mathbf{X},\mathbf{W}} = \mathbf{U}(\mathbf{U}'\mathbf{W}\mathbf{U})^{-}\mathbf{U}'\mathbf{W} = \mathbf{P}_{\mathbf{U},\mathbf{W}}$$

and similarly that

$$\mathbf{W}\mathbf{P}_{\mathbf{X},\mathbf{W}}\mathbf{P}_{\mathbf{U},\mathbf{W}} = \mathbf{W}\mathbf{P}_{\mathbf{X},\mathbf{W}}\mathbf{U}(\mathbf{U}'\mathbf{W}\mathbf{U})^{-}\mathbf{U}'\mathbf{W} = \mathbf{W}\mathbf{U}(\mathbf{U}'\mathbf{W}\mathbf{U})^{-}\mathbf{U}'\mathbf{W} = \mathbf{W}\mathbf{P}_{\mathbf{U},\mathbf{W}}.$$

Further, making use of Part (3) of Theorem 14.12.25, we find that

$$\mathbf{P}'_{\mathbf{X},\mathbf{W}}\mathbf{W}\mathbf{P}_{\mathbf{U},\mathbf{W}} = (\mathbf{W}\mathbf{P}_{\mathbf{X},\mathbf{W}})'\mathbf{P}_{\mathbf{U},\mathbf{W}} = \mathbf{W}\mathbf{P}_{\mathbf{X},\mathbf{W}}\mathbf{P}_{\mathbf{U},\mathbf{W}}.$$

(b) Making use of Part (a), we find that

$$\mathbf{W}\mathbf{P}_{\mathbf{U},\mathbf{W}} = \mathbf{W}(\mathbf{P}_{\mathbf{X},\mathbf{W}}\mathbf{P}_{\mathbf{U},\mathbf{W}}) = \mathbf{W}\mathbf{P}_{\mathbf{X},\mathbf{W}}.$$

EXERCISE 48. Let $\mathcal{U}$ represent a subspace of $\mathcal{R}^{n\times 1}$, let $\mathbf{A}$ represent an $n \times n$ matrix and $\mathbf{W}$ an $n \times n$ symmetric nonnegative definite matrix, and let $\mathbf{X}$ represent any $n \times p$ matrix whose columns span $\mathcal{U}$.

(a) Show that $\mathbf{A}$ is a projection matrix for $\mathcal{U}$ with respect to $\mathbf{W}$ if and only if

$$\mathbf{A} = \mathbf{P}_{\mathbf{X},\mathbf{W}} + (\mathbf{I} - \mathbf{P}_{\mathbf{X},\mathbf{W}})\mathbf{X}\mathbf{K}$$

for some $p \times n$ matrix $\mathbf{K}$.

(b) Show that if $\mathbf{A}$ is a projection matrix for $\mathcal{U}$ with respect to $\mathbf{W}$, then $\mathbf{W}\mathbf{A} = \mathbf{W}\mathbf{P}_{\mathbf{X},\mathbf{W}}$.

Solution. (a) Suppose that

$$\mathbf{A} = \mathbf{P}_{\mathbf{X},\mathbf{W}} + (\mathbf{I} - \mathbf{P}_{\mathbf{X},\mathbf{W}})\mathbf{X}\mathbf{K}$$

for some matrix $\mathbf{K}$. Then, for any n-dimensional column vector $\mathbf{y}$,

$$\mathbf{A}\mathbf{y} = \mathbf{P}_{\mathbf{X},\mathbf{W}}\mathbf{y} + (\mathbf{I} - \mathbf{P}_{\mathbf{X},\mathbf{W}})\mathbf{X}(\mathbf{K}\mathbf{y}).$$

Thus, it follows from Corollary 14.12.27 that $\mathbf{A}$ is a projection matrix for $\mathcal{U}$ with respect to $\mathbf{W}$.

Conversely, suppose that $\mathbf{A}$ is a projection matrix for $\mathcal{U}$ with respect to $\mathbf{W}$. Then, $\mathbf{A}\mathbf{y} \in \mathcal{U}$ for every $\mathbf{y}$ in $\mathcal{R}^n$, so that $\mathcal{C}(\mathbf{A}) \subset \mathcal{U} = \mathcal{C}(\mathbf{X})$ and hence $\mathbf{A} = \mathbf{X}\mathbf{F}$ for some matrix $\mathbf{F}$. Moreover, it follows from Parts (2) and (3) of Theorem 14.12.26 that $\mathbf{W}\mathbf{A}\mathbf{y} = \mathbf{W}\mathbf{X}(\mathbf{X}'\mathbf{W}\mathbf{X})^-\mathbf{X}'\mathbf{W}\mathbf{y}$ for every $\mathbf{y}$ in $\mathcal{R}^n$ [since one solution to linear system (12.4) is $\mathbf{X}'\mathbf{W}\mathbf{X})^-\mathbf{X}'\mathbf{W}\mathbf{y}$], implying that

$$\mathbf{W}\mathbf{A} = \mathbf{W}\mathbf{X}(\mathbf{X}'\mathbf{W}\mathbf{X})^-\mathbf{X}'\mathbf{W}$$

and hence that

$$\mathbf{W}\mathbf{X}\mathbf{F} = \mathbf{W}\mathbf{X}(\mathbf{X}'\mathbf{W}\mathbf{X})^-\mathbf{X}'\mathbf{W}. \tag{S.8}$$

Since [according to Part (1) of Theorem 14.12.25] $(\mathbf{X}'\mathbf{W}\mathbf{X})^-\mathbf{X}'$ is a generalized inverse of $\mathbf{W}\mathbf{X}$, we conclude, on the basis of Theorem 11.2.4 and Part (5) of Theorem 14.12.25, that there exists a matrix $\mathbf{K}$ such that

$$\mathbf{F} = (\mathbf{X}'\mathbf{W}\mathbf{X})^-\mathbf{X}'\mathbf{W} + [\mathbf{I} - (\mathbf{X}'\mathbf{W}\mathbf{X})^-\mathbf{X}'\mathbf{W}\mathbf{X}]\mathbf{K}$$

and hence such that

$$\mathbf{A} = \mathbf{X}(\mathbf{X}'\mathbf{W}\mathbf{X})^-\mathbf{X}'\mathbf{W} + \mathbf{X}[\mathbf{I} - (\mathbf{X}'\mathbf{W}\mathbf{X})^-\mathbf{X}'\mathbf{W}\mathbf{X}]\mathbf{K} = \mathbf{P}_{\mathbf{X},\mathbf{W}} + (\mathbf{I} - \mathbf{P}_{\mathbf{X},\mathbf{W}})\mathbf{X}\mathbf{K}.$$

(b) Suppose that $\mathbf{A}$ is a projection matrix for $\mathcal{U}$ with respect to $\mathbf{W}$. Then, it follows from Part (a) that

$$\mathbf{A} = \mathbf{P}_{\mathbf{X},\mathbf{W}} + (\mathbf{I} - \mathbf{P}_{\mathbf{X},\mathbf{W}})\mathbf{X}\mathbf{K}$$

for some matrix $\mathbf{K}$. Thus, making use of Part (1) of Theorem 14.12.25, we find that

$$\mathbf{W}\mathbf{A} = \mathbf{W}\mathbf{P}_{\mathbf{X},\mathbf{W}} + (\mathbf{W}\mathbf{X} - \mathbf{W}\mathbf{P}_{\mathbf{X},\mathbf{W}}\mathbf{X})\mathbf{K} = \mathbf{W}\mathbf{P}_{\mathbf{X},\mathbf{W}}.$$

EXERCISE 49. Let $\mathbf{A}$ represent an $n \times n$ matrix and $\mathbf{W}$ an $n \times n$ symmetric nonnegative definite matrix.

(a) Show (by, e.g., using the results of Exercise 48) that if $\mathbf{A}'\mathbf{W}\mathbf{A} = \mathbf{W}\mathbf{A}$ [or, equivalently, if $(\mathbf{I} - \mathbf{A})'\mathbf{W}\mathbf{A} = \mathbf{0}$], then $\mathbf{A}$ is a projection matrix with respect to $\mathbf{W}$, and in particular $\mathbf{A}$ is a projection matrix for $\mathcal{C}(\mathbf{A})$ with respect to $\mathbf{W}$, and, conversely, show that if $\mathbf{A}$ is a projection matrix with respect to $\mathbf{W}$, then $\mathbf{A}'\mathbf{W}\mathbf{A} = \mathbf{W}\mathbf{A}$.

(b) Show that if $\mathbf{A}$ is a projection matrix with respect to $\mathbf{W}$, then in particular $\mathbf{A}$ is a projection matrix for $\mathcal{C}(\mathbf{A})$ with respect to $\mathbf{W}$.

(c) Show that $\mathbf{A}$ is a projection matrix with respect to $\mathbf{W}$ if and only if $\mathbf{W}\mathbf{A}$ is symmetric and $\mathbf{W}\mathbf{A}^2 = \mathbf{W}\mathbf{A}$.

Solution. (a) Suppose that $\mathbf{A}'\mathbf{W}\mathbf{A} = \mathbf{W}\mathbf{A}$ and hence that

$$\mathbf{A}'\mathbf{W} = (\mathbf{W}\mathbf{A})' = (\mathbf{A}'\mathbf{W}\mathbf{A})' = \mathbf{A}'\mathbf{W}\mathbf{A}.$$

Then,

$$\begin{aligned}
\mathbf{A} &= \mathbf{P}_{\mathbf{A},\mathbf{W}} + \mathbf{A} - \mathbf{A}(\mathbf{A}'\mathbf{W}\mathbf{A})^{-}\mathbf{A}'\mathbf{W} \\
&= \mathbf{P}_{\mathbf{A},\mathbf{W}} + \mathbf{A} - \mathbf{A}(\mathbf{A}'\mathbf{W}\mathbf{A})^{-}\mathbf{A}'\mathbf{W}\mathbf{A} \\
&= \mathbf{P}_{\mathbf{A},\mathbf{W}} + (\mathbf{I} - \mathbf{P}_{\mathbf{A},\mathbf{W}})\mathbf{A}\mathbf{I}_n,
\end{aligned}$$

and it follows from Part (a) of Exercise 48 that $\mathbf{A}$ is a projection matrix for $\mathcal{C}(\mathbf{A})$ with respect to $\mathbf{W}$.

Conversely, suppose that $\mathbf{A}$ is a projection matrix with respect to $\mathbf{W}$. Let $\mathcal{U}$ represent any subspace of $\mathcal{R}^{n\times 1}$ for which $\mathbf{A}$ is a projection matrix with respect to $\mathbf{W}$, and let $\mathbf{X}$ represent any $n \times p$ matrix whose columns span $\mathcal{U}$. Then, according to Part (b) of Exercise 48,

$$\mathbf{W}\mathbf{A} = \mathbf{W}\mathbf{P}_{\mathbf{X},\mathbf{W}},$$

and, making use of Part (6$'$) of Theorem 14.12.25, we find that

$$\begin{aligned}
\mathbf{A}'\mathbf{W}\mathbf{A} = \mathbf{A}'\mathbf{W}\mathbf{P}_{\mathbf{X},\mathbf{W}} = (\mathbf{W}\mathbf{A})'\mathbf{P}_{\mathbf{X},\mathbf{W}} &= (\mathbf{W}\mathbf{P}_{\mathbf{X},\mathbf{W}})'\mathbf{P}_{\mathbf{X},\mathbf{W}} \\
&= \mathbf{P}_{\mathbf{X},\mathbf{W}}'\mathbf{W}\mathbf{P}_{\mathbf{X},\mathbf{W}} = \mathbf{W}\mathbf{P}_{\mathbf{X},\mathbf{W}} = \mathbf{W}\mathbf{A}.
\end{aligned}$$

(b) Suppose that $\mathbf{A}$ is a projection matrix with respect to $\mathbf{W}$. Then, it follows from Part (a) that $\mathbf{A}'\mathbf{W}\mathbf{A} = \mathbf{W}\mathbf{A}$, and we conclude [on the basis of Part (a)] that $\mathbf{A}$ is a projection matrix for $\mathcal{C}(\mathbf{A})$ with respect to $\mathbf{W}$.

(c) In light of Part (a), it suffices to show that $\mathbf{A}'\mathbf{W}\mathbf{A} = \mathbf{W}\mathbf{A}$ if and only if $\mathbf{W}\mathbf{A}$ is symmetric and $\mathbf{W}\mathbf{A}^2 = \mathbf{W}\mathbf{A}$.

If $\mathbf{W}\mathbf{A}$ is symmetric and $\mathbf{W}\mathbf{A}^2 = \mathbf{W}\mathbf{A}$, then

$$\mathbf{A}'\mathbf{W}\mathbf{A} = (\mathbf{W}\mathbf{A})'\mathbf{A} = \mathbf{W}\mathbf{A}\mathbf{A} = \mathbf{W}\mathbf{A}^2 = \mathbf{W}\mathbf{A}.$$

Conversely, if $\mathbf{A}'\mathbf{W}\mathbf{A} = \mathbf{W}\mathbf{A}$, then

$$(\mathbf{W}\mathbf{A})' = (\mathbf{A}'\mathbf{W}\mathbf{A})' = \mathbf{A}'\mathbf{W}\mathbf{A} = \mathbf{W}\mathbf{A}$$

(i.e., $\mathbf{WA}$ is symmetric), and

$$\mathbf{WA}^2 = \mathbf{WAA} = (\mathbf{WA})'\mathbf{A} = \mathbf{A}'\mathbf{WA} = \mathbf{WA}.$$

EXERCISE 50. Let $\mathcal{U}$ represent a subspace of $\mathcal{R}^{n\times 1}$, let $\mathbf{X}$ represent an $n \times p$ matrix whose columns span $\mathcal{U}$, and let $\mathbf{W}$ and $\mathbf{V}$ represent $n \times n$ symmetric nonnegative definite matrices. Show (by, e.g., making use of the result of Exercise 46) that each of the following two conditions is necessary and sufficient for every projection of $\mathbf{y}$ on $\mathcal{U}$ with respect to $\mathbf{W}$ to be a projection (for every $\mathbf{y}$ in $\mathcal{R}^n$) of $\mathbf{y}$ on $\mathcal{U}$ with respect to $\mathbf{V}$:

(a) $\mathbf{X}'\mathbf{V}\mathbf{P}_{\mathbf{X},\mathbf{W}} = \mathbf{X}'\mathbf{V}$, or, equivalently, $\mathbf{X}'\mathbf{V}(\mathbf{I} - \mathbf{P}_{\mathbf{X},\mathbf{W}}) = \mathbf{0}$;

(b) there exists a $p \times p$ matrix $\mathbf{Q}$ such that $\mathbf{VX} = \mathbf{WXQ}$, or, equivalently, $\mathcal{C}(\mathbf{VX}) \subset \mathcal{C}(\mathbf{WX})$.

Solution. (a) It follows from the result of Exercise 46 that a vector $\mathbf{z}$ (in $\mathcal{U}$) is a projection of a vector $\mathbf{y}$ (in $\mathcal{R}^n$) on $\mathcal{U}$ with respect to $\mathbf{V}$ if and only if

$$\mathbf{X}'\mathbf{V}(\mathbf{y} - \mathbf{z}) = \mathbf{0}.$$

Further, it follows from Corollary 14.12.27 that every projection of $\mathbf{y}$ on $\mathcal{U}$ with respect to $\mathbf{W}$ is a projection (for every $\mathbf{y}$ in $\mathcal{R}^n$) of $\mathbf{y}$ on $\mathcal{U}$ with respect to $\mathbf{V}$ if and only if, for every $\mathbf{y}$ and every vector-valued function $\mathbf{k}(\mathbf{y})$,

$$\mathbf{X}'\mathbf{V}[\mathbf{y} - \mathbf{P}_{\mathbf{X},\mathbf{W}}\mathbf{y} - (\mathbf{I} - \mathbf{P}_{\mathbf{X},\mathbf{W}})\mathbf{X}\mathbf{k}(\mathbf{y})] = \mathbf{0},$$

or, equivalently, if and only if, for every $\mathbf{y}$ and every vector-valued function $\mathbf{k}(\mathbf{y})$,

$$\mathbf{X}'\mathbf{V}(\mathbf{I} - \mathbf{P}_{\mathbf{X},\mathbf{W}})[\mathbf{y} - \mathbf{X}\mathbf{k}(\mathbf{y})] = \mathbf{0}. \tag{S.9}$$

Thus, it suffices to show that condition (S.9) is satisfied for every $\mathbf{y}$ and every vector-valued function $\mathbf{k}(\mathbf{y})$ if and only if $\mathbf{X}'\mathbf{V}(\mathbf{I} - \mathbf{P}_{\mathbf{X},\mathbf{W}}) = \mathbf{0}$.

If $\mathbf{X}'\mathbf{V}(\mathbf{I} - \mathbf{P}_{\mathbf{X},\mathbf{W}}) = \mathbf{0}$, then condition (S.9) is obviously satisfied. Conversely, suppose that condition (S.9) is satisfied for every $\mathbf{y}$ and every vector-valued function $\mathbf{k}(\mathbf{y})$. Then, since one choice for $\mathbf{k}(\mathbf{y})$ is $\mathbf{k}(\mathbf{y}) \equiv \mathbf{0}$,

$$\mathbf{X}'\mathbf{V}(\mathbf{I} - \mathbf{P}_{\mathbf{X},\mathbf{W}})\mathbf{y} = \mathbf{0}$$

for every $\mathbf{y}$, implying that $\mathbf{X}'\mathbf{V}(\mathbf{I} - \mathbf{P}_{\mathbf{X},\mathbf{W}}) = \mathbf{0}$.

(b) It suffices to show that Condition (b) is equivalent to Condition (a) or, equivalently [since $\mathbf{VX} = (\mathbf{X}'\mathbf{V})'$ and $\mathbf{P}_{\mathbf{X},\mathbf{W}}'\mathbf{VX} = (\mathbf{X}'\mathbf{V}\mathbf{P}_{\mathbf{X},\mathbf{W}})'$], to the condition

$$\mathbf{VX} = \mathbf{P}_{\mathbf{X},\mathbf{W}}'\mathbf{VX}. \tag{S.10}$$

If condition (S.10) is satisfied, then

$$\mathbf{VX} = \mathbf{WX}[(\mathbf{X}'\mathbf{WX})^-]'\mathbf{X}'\mathbf{VX} = \mathbf{WXQ}$$

for $\mathbf{Q} = [(\mathbf{X}'\mathbf{W}\mathbf{X})^-]'\mathbf{X}'\mathbf{V}\mathbf{X}$. Conversely, if $\mathbf{V}\mathbf{X} = \mathbf{W}\mathbf{X}\mathbf{Q}$ for some matrix $\mathbf{Q}$, then, making use of Part (4) of Theorem 14.12.25, we find that

$$\mathbf{P}'_{\mathbf{X},\mathbf{W}}\mathbf{V}\mathbf{X} = \mathbf{P}'_{\mathbf{X},\mathbf{W}}\mathbf{W}\mathbf{X}\mathbf{Q} = \mathbf{W}\mathbf{X}\mathbf{Q} = \mathbf{V}\mathbf{X},$$

that is, condition (S.10) is satisfied.

EXERCISE 51. Let $\mathbf{X}$ represent an $n \times p$ matrix and $\mathbf{W}$ an $n \times n$ symmetric nonnegative definite matrix. As in the special case where $\mathbf{W}$ is positive definite, let

$$\mathcal{C}^{\perp}_{\mathbf{W}}(\mathbf{X}) = \{\mathbf{y} \in \mathcal{R}^{n\times 1} : \mathbf{y} \perp_{\mathbf{W}} \mathcal{C}(\mathbf{X})\}.$$

(a) By, for example, making use of the result of Exercise 46, show that

$$\mathcal{C}^{\perp}_{\mathbf{W}}(\mathbf{X}) = \mathcal{N}(\mathbf{X}'\mathbf{W}) = \mathcal{C}(\mathbf{I} - \mathbf{P}_{\mathbf{X},\mathbf{W}}).$$

(b) Show that

$$\dim[\mathcal{C}^{\perp}_{\mathbf{W}}(\mathbf{X})] = n - \operatorname{rank}(\mathbf{W}\mathbf{X}) \geq n - \operatorname{rank}(\mathbf{X}) = n - \dim[\mathcal{C}(\mathbf{X})].$$

(c) By, for example, making use of the result of Exercise 46, show that, for any solution $\mathbf{b}^*$ to the linear system $\mathbf{X}'\mathbf{W}\mathbf{X}\mathbf{b} = \mathbf{X}'\mathbf{W}\mathbf{y}$ (in $\mathbf{b}$), the vector $\mathbf{y} - \mathbf{X}\mathbf{b}^*$ is a projection of $\mathbf{y}$ on $\mathcal{C}^{\perp}_{\mathbf{W}}(\mathbf{X})$ with respect to $\mathbf{W}$.

Solution. (a) It follows from the result of Exercise 46 that an n-dimensional column vector $\mathbf{y}$ and $\mathcal{C}(\mathbf{X})$ are orthogonal with respect to $\mathbf{W}$ if and only if $\mathbf{X}'\mathbf{W}\mathbf{y} = \mathbf{0}$. Thus, $\mathcal{C}^{\perp}_{\mathbf{W}}(\mathbf{X}) = \mathcal{N}(\mathbf{X}'\mathbf{W})$. Moreover, since [according to Part (5) of Theorem 14.12.25] $\mathbf{X}(\mathbf{X}'\mathbf{W}\mathbf{X})^-$ is a generalized inverse of $\mathbf{X}'\mathbf{W}$, we have (in light of Corollary 11.2.2) that $\mathcal{N}(\mathbf{X}'\mathbf{W}) = \mathcal{C}(\mathbf{I} - \mathbf{P}_{\mathbf{X},\mathbf{W}})$.

(b) Making use of Part (a), together with Part (10) of Theorem 14.12.25 and Corollary 4.4.5, we find that

$$\begin{aligned}\dim[\mathcal{C}^{\perp}_{\mathbf{W}}(\mathbf{X})] &= \dim[\mathcal{C}(\mathbf{I} - \mathbf{P}_{\mathbf{X},\mathbf{W}})] \\ &= \operatorname{rank}(\mathbf{I} - \mathbf{P}_{\mathbf{X},\mathbf{W}}) \\ &= n - \operatorname{rank}(\mathbf{W}\mathbf{X}) \geq n - \operatorname{rank}(\mathbf{X}) = n - \dim[\mathcal{C}(\mathbf{X})].\end{aligned}$$

(c) Let $\mathbf{z} = \mathbf{X}\mathbf{b}^*$. According to Theorem 14.12.26, $\mathbf{z}$ is a projection of $\mathbf{y}$ on $\mathcal{C}(\mathbf{X})$ with respect to $\mathbf{W}$. Thus, $(\mathbf{y} - \mathbf{z}) \perp_{\mathbf{W}} \mathcal{C}(\mathbf{X})$, and, consequently, $(\mathbf{y} - \mathbf{z}) \in \mathcal{C}^{\perp}_{\mathbf{W}}(\mathbf{X})$. It remains to show that $[\mathbf{y} - (\mathbf{y} - \mathbf{z})] \perp_{\mathbf{W}} \mathcal{C}^{\perp}_{\mathbf{W}}(\mathbf{X})$ or, equivalently, that $\mathbf{z} \perp_{\mathbf{W}} \mathcal{C}^{\perp}_{\mathbf{W}}(\mathbf{X})$.

Making use of Part (4) of Theorem 14.12.25, we find that

$$(\mathbf{I} - \mathbf{P}_{\mathbf{X},\mathbf{W}})'\mathbf{W}\mathbf{z} = (\mathbf{I} - \mathbf{P}_{\mathbf{X},\mathbf{W}})'\mathbf{W}\mathbf{X}\mathbf{b}^* = \mathbf{0},$$

implying (in light of the result of Exercise 46) that

$$\mathbf{z} \perp_{\mathbf{W}} \mathcal{C}(\mathbf{I} - \mathbf{P}_{\mathbf{X},\mathbf{W}})$$

and hence [in light of the result of Part (a)] that

$$\mathbf{z} \perp \mathcal{C}^{\perp}_{\mathbf{W}}(\mathbf{X}).$$

15
Matrix Differentiation

EXERCISE 1. Using the result of Part (c) of Exercise 6.2, verify that every neighborhood of a point $\mathbf{x}$ in $\mathcal{R}^{m\times 1}$ is an open set.

Solution. Take the norm for $\mathcal{R}^{m\times 1}$ to be the usual norm, let N represent the neighborhood of $\mathbf{x}$ of radius r, and let $\mathbf{y}$ represent an arbitrary point in N. Further, take M to be the neighborhood of $\mathbf{y}$ of radius

$$s = r - \| \mathbf{y} - \mathbf{x} \|,$$

and let $\mathbf{z}$ represent an arbitrary point in M. Then, using the result of Part (c) of Exercise 6.2, we find that

$$\begin{aligned} \| \mathbf{z} - \mathbf{x} \| &\le \| \mathbf{z} - \mathbf{y} \| + \| \mathbf{y} - \mathbf{x} \| \\ &< s + \| \mathbf{y} - \mathbf{x} \| \\ &= r - \| \mathbf{y} - \mathbf{x} \| + \| \mathbf{y} - \mathbf{x} \| \\ &= r, \end{aligned}$$

implying that $\mathbf{z} \in N$. It follows that $M \subset N$ and hence that $\mathbf{y}$ is an interior point of N. We conclude that N is an open set.

EXERCISE 2. Let f represent a function, defined on a set S, of a vector $\mathbf{x} = (x_1, \ldots, x_m)'$ of m variables, suppose that the set S contains at least some interior points, and let $\mathbf{c}$ represent an arbitrary one of those points. Verify that if f is k times continuously differentiable at $\mathbf{c}$, then f is k times continuously differentiable at every point in some neighborhood of $\mathbf{c}$.

Solution. Suppose that f is k times continuously differentiable at $\mathbf{c}$. Then, there exists a neighborhood N of $\mathbf{c}$ such that all of the first- through kth-order partial derivatives of f exist and are continuous at every point in N.

Let $\mathbf{x}^*$ represent an arbitrary point in N. Since any neighborhood is an open set, $\mathbf{x}^*$ is an interior point of N. Thus, there exists a neighborhood N^* of $\mathbf{x}^*$, all of whose points belong to N. It follows that the first- through kth-order partial derivatives of f exist and are continuous at every point in N^* and hence that f is continuously differentiable at $\mathbf{x}^*$.

EXERCISE 3. Let $\mathbf{X} = \{x_{ij}\}$ represent an $m \times n$ matrix of mn variables, and let $\mathbf{x}$ represent an mn-dimensional column vector obtained by rearranging the elements of $\mathbf{X}$ (in the form of a column vector). Further, let S represent a set of $\mathbf{X}$-values, and let S^* represent the corresponding set of $\mathbf{x}$-values (i.e., the set obtained by rearranging the elements of each $m \times n$ matrix in S in the form of a column vector). Verify that an mn-dimensional column vector is an interior point of S^* if and only if it is a rearrangement of an $m \times n$ matrix that is an interior point of S.

Solution. Let $\mathbf{C} = \{c_{ij}\}$ represent a value of $\mathbf{X}$, and let $\mathbf{c}$ represent the corresponding value of $\mathbf{x}$. Then (when the inner products for $\mathcal{R}^{mn\times 1}$ and $\mathcal{R}^{m\times n}$ are taken to be the usual inner products)

$$\begin{aligned}\|\mathbf{x}-\mathbf{c}\| = [(\mathbf{x}-\mathbf{c})'(\mathbf{x}-\mathbf{c})]^{1/2} &= \left[\sum_{i,j}(x_{ij}-c_{ij})^2\right]^{1/2} \\ &= \{\mathrm{tr}[(\mathbf{X}-\mathbf{C})'(\mathbf{X}-\mathbf{C})]\}^{1/2} \\ &= \|\mathbf{X}-\mathbf{C}\|.\end{aligned}$$

Thus, a set of $\mathbf{X}$-values is a neighborhood of $\mathbf{C}$ of radius r if and only if the corresponding set of $\mathbf{x}$-values is a neighborhood of $\mathbf{c}$ of radius r. It follows that there exists a neighborhood of $\mathbf{c}$, all of whose points belong to S^*, if and only if there exists a neighborhood of $\mathbf{C}$, all of whose points belong to S. We conclude that $\mathbf{c}$ is an interior point of S^* if and only if $\mathbf{C}$ is an interior point of S.

EXERCISE 4. Let f represent a function whose domain is a set S in $\mathcal{R}^{m\times 1}$ (that contains at least some interior points). Show that the Hessian matrix $\mathbf{H}f$ of f is the gradient matrix of the gradient vector $(\mathbf{D}f)'$ of f.

Solution. The gradient vector of f is $(\mathbf{D}f)' = (D_1 f, \ldots, D_m f)'$. The gradient matrix of this vector is the $m \times m$ matrix whose ijth element is the ith (first-order) partial derivative $D_{ij}^2 f$ of $D_j f$, which by definition is the Hessian matrix of f.

EXERCISE 5. Let g represent a function, defined on a set S, of a vector $\mathbf{x} = (x_1, \ldots, x_m)'$ of m variables, let $S^* = \{\mathbf{x} \in S : g(\mathbf{x}) \neq 0\}$, and let $\mathbf{c}$ represent any interior point of S^* at which g is continuously differentiable. Show that (for any positive integer k) g^{-k} is continuously differentiable at $\mathbf{c}$ and

$$\frac{\partial g^{-k}}{\partial x_j} = -kg^{-k-1}\frac{\partial g}{\partial x_j}.$$

Do so based on the result that

$$\frac{\partial(1/g)}{\partial x_j} = -(1/g)^2\frac{\partial g}{\partial x_j}, \tag{$*$}$$

and, letting f (like g) represent a function (defined on S) of $\mathbf{x}$, the results that if f is continuously differentiable at $\mathbf{c}$, then the ratio f/g is also continuously differentiable at $\mathbf{c}$, that, for any positive integer k, f^k is continuously differentiable at any point at which f is continuously differentiable, and that

$$\frac{\partial f^k}{\partial x_j} = kf^{k-1}\frac{\partial f}{\partial x_j}. \tag{$**$}$$

Solution. In light of the results cited in the statement of the exercise (or equivalently in light of Lemma 15.2.2 and the ensuing discussion), we have that $1/g$ is continuously differentiable at $\mathbf{c}$ and that, as a consequence, $(1/g)^k$ or equivalently g^{-k} is continuously differentiable at $\mathbf{c}$. Moreover, using results ($**$) and ($*$) [or, equivalently, results (2.16) and (2.8)], we find that

$$\begin{aligned}\frac{\partial(g^{-k})}{\partial x_j} = \frac{\partial[(1/g)^k]}{\partial x_j} = k(1/g)^{k-1}\frac{\partial(1/g)}{\partial x_j} &= k(1/g)^{k-1}(-1)(1/g^2)\frac{\partial g}{\partial x_j}\\ &= -k(1/g)^{k+1}\frac{\partial g}{\partial x_j}\\ &= -kg^{-k-1}\frac{\partial g}{\partial x_j}.\end{aligned}$$

EXERCISE 6. Let $\mathbf{F}$ represent a $p \times p$ matrix of functions, defined on a set S, of a vector $\mathbf{x} = (x_1, \ldots, x_m)'$ of m variables. Let $\mathbf{c}$ represent any interior point of S at which $\mathbf{F}$ is continuously differentiable. Show that if $\mathbf{F}$ is idempotent at all points in some neighborhood of $\mathbf{c}$, then (at $\mathbf{x} = \mathbf{c}$)

$$\mathbf{F}\frac{\partial \mathbf{F}}{\partial x_j}\mathbf{F} = \mathbf{0}.$$

Solution. Suppose that $\mathbf{F}$ is idempotent at all points in some neighborhood of $\mathbf{c}$. Then, differentiating both sides of the equality $\mathbf{F} = \mathbf{FF}$ [with the help of result (4.3)], we find that (at $\mathbf{x} = \mathbf{c}$)

$$\frac{\partial \mathbf{F}}{\partial x_j} = \mathbf{F}\frac{\partial \mathbf{F}}{\partial x_j} + \frac{\partial \mathbf{F}}{\partial x_j}\mathbf{F}. \tag{S.1}$$

Premultiplying both sides of equality (S.1) by **F** gives

$$\mathbf{F}\frac{\partial \mathbf{F}}{\partial x_j} = \mathbf{F}^2\frac{\partial \mathbf{F}}{\partial x_j} + \mathbf{F}\frac{\partial \mathbf{F}}{\partial x_j}\mathbf{F} = \mathbf{F}\frac{\partial \mathbf{F}}{\partial x_j} + \mathbf{F}\frac{\partial \mathbf{F}}{\partial x_j}\mathbf{F}$$

or equivalently

$$\mathbf{F}\frac{\partial \mathbf{F}}{\partial x_j}\mathbf{F} = \mathbf{0}.$$

EXERCISE 7. Let g represent a function, defined on a set S, of a vector $\mathbf{x} = (x_1, \ldots, x_m)'$ of m variables, and let $\mathbf{f}$ represent a $p \times 1$ vector of functions (defined on S) of $\mathbf{x}$. Let $\mathbf{c}$ represent any interior point (of S) at which g and $\mathbf{f}$ are continuously differentiable. Show that $g\mathbf{f}$ is continuously differentiable at $\mathbf{c}$ and that (at $\mathbf{x} = \mathbf{c}$)

$$\frac{\partial(g\mathbf{f})}{\partial \mathbf{x}'} = \mathbf{f}\frac{\partial g}{\partial \mathbf{x}'} + g\frac{\partial \mathbf{f}}{\partial \mathbf{x}'}.$$

Solution. It follows from result (4.9) (and the discussion thereof) that $g\mathbf{f}$ is continuously differentiable at $\mathbf{c}$ and that (at $\mathbf{x} = \mathbf{c}$)

$$\frac{\partial(g\mathbf{f})}{\partial x_j} = \frac{\partial g}{\partial x_j}\mathbf{f} + g\frac{\partial \mathbf{f}}{\partial x_j}$$

($j = 1, \ldots, m$). Moreover, since $\partial(g\mathbf{f})/\partial x_j$, $(\partial g/\partial x_j)\mathbf{f}$, and $g(\partial \mathbf{f}/\partial \mathbf{x}_j)$ are the jth columns of $\partial(g\mathbf{f})/\partial \mathbf{x}'$, $\mathbf{f}(\partial g/\partial \mathbf{x}')$, and $g(\partial \mathbf{f}/\partial \mathbf{x}')$, respectively, we have that

$$\frac{\partial(g\mathbf{f})}{\partial \mathbf{x}'} = \mathbf{f}\frac{\partial g}{\partial \mathbf{x}'} + g\frac{\partial \mathbf{f}}{\partial \mathbf{x}'}.$$

EXERCISE 8. (a) Let $\mathbf{X} = \{x_{ij}\}$ represent an $m \times n$ matrix of mn "independent" variables, and suppose that $\mathbf{X}$ is free to range over all of $\mathcal{R}^{m \times n}$.

(1) Show that, for any $p \times m$ and $n \times p$ matrices of constants $\mathbf{A}$ and $\mathbf{B}$,

$$\frac{\partial\, \mathrm{tr}(\mathbf{AXB})}{\partial \mathbf{X}} = \mathbf{A}'\mathbf{B}'.$$

[*Hint*. Observe that $\mathrm{tr}(\mathbf{AXB}) = \mathrm{tr}(\mathbf{BAX})$.]

(2) Show that, for any m- and n-dimensional column vectors $\mathbf{a}$ and $\mathbf{b}$,

$$\frac{\partial(\mathbf{a}'\mathbf{X}\mathbf{b})}{\partial \mathbf{X}} = \mathbf{a}\mathbf{b}'.$$

[*Hint*. Observe that $\mathbf{a}'\mathbf{X}\mathbf{b} = \mathrm{tr}(\mathbf{a}'\mathbf{X}\mathbf{b})$.]

(b) Suppose now that $\mathbf{X}$ is a symmetric (but otherwise unrestricted) matrix (of dimensions $m \times m$).

(1) Show that, for any $p \times m$ and $m \times p$ matrices of constants $\mathbf{A}$ and $\mathbf{B}$,

$$\frac{\partial \operatorname{tr}(\mathbf{AXB})}{\partial \mathbf{X}} = \mathbf{C} + \mathbf{C}' - \operatorname{diag}(c_{11}, c_{22}, \ldots, c_{mm}),$$

where $\mathbf{C} = \{c_{ij}\} = \mathbf{BA}$.

(2) Show that, for any m-dimensional column vectors $\mathbf{a} = \{a_i\}$ and $\mathbf{b} = \{b_i\}$,

$$\frac{\partial (\mathbf{a}'\mathbf{X}\mathbf{b})}{\partial \mathbf{X}} = \mathbf{ab}' + \mathbf{ba}' - \operatorname{diag}(a_1 b_1, a_2 b_2, \ldots, a_m b_m).$$

Solution. (a) (1) Using result (6.5), we find that

$$\frac{\partial \operatorname{tr}(\mathbf{AXB})}{\partial \mathbf{X}} = \frac{\partial \operatorname{tr}(\mathbf{BAX})}{\partial \mathbf{X}} = (\mathbf{BA})' = \mathbf{A}'\mathbf{B}'.$$

(2) Observing that $\mathbf{a}'\mathbf{X}\mathbf{b} = \operatorname{tr}(\mathbf{a}'\mathbf{X}\mathbf{b})$ and applying Part (1) (with $\mathbf{A} = \mathbf{a}'$ and $\mathbf{B} = \mathbf{b}$), we find that

$$\frac{\partial (\mathbf{a}'\mathbf{X}\mathbf{b})}{\partial \mathbf{X}} = \mathbf{ab}'.$$

(b) (1) Using result (6.7), we find that

$$\frac{\partial \operatorname{tr}(\mathbf{AXB})}{\partial \mathbf{X}} = \frac{\partial \operatorname{tr}(\mathbf{BAX})}{\partial \mathbf{X}} = \mathbf{C} + \mathbf{C}' - \operatorname{diag}(c_{11}, c_{22}, \ldots, c_{mm}).$$

(2) Observing that $\mathbf{a}'\mathbf{X}\mathbf{b} = \operatorname{tr}(\mathbf{a}'\mathbf{X}\mathbf{b})$ and applying Part (1) (with $\mathbf{A} = \mathbf{a}'$ and $\mathbf{B} = \mathbf{b}$), we find that

$$\frac{\partial (\mathbf{a}'\mathbf{X}\mathbf{b})}{\partial \mathbf{X}} = \mathbf{ba}' + \mathbf{ab}' - \operatorname{diag}(a_1 b_1, a_2 b_2, \ldots, a_m b_m).$$

EXERCISE 9. (a) Let $\mathbf{X} = \{x_{st}\}$ represent an $m \times n$ matrix of "independent" variables, and suppose that $\mathbf{X}$ is free to range over all of $\mathcal{R}^{m \times n}$. Show that, for any $n \times m$ matrix of constants $\mathbf{A}$,

$$\frac{\partial \operatorname{tr}[(\mathbf{AX})^2]}{\partial \mathbf{X}} = 2(\mathbf{AXA})'.$$

(b) Let $\mathbf{X} = \{x_{st}\}$ represent an $m \times m$ symmetric (but otherwise unrestricted) matrix of variables. Show that, for any $m \times m$ matrix of constants $\mathbf{A}$,

$$\frac{\partial \operatorname{tr}[(\mathbf{AX})^2]}{\partial \mathbf{X}} = 2[\mathbf{B} + \mathbf{B}' - \operatorname{diag}(b_{11}, b_{22}, \ldots, b_{mm})],$$

where $\mathbf{B} = \{b_{st}\} = \mathbf{AXA}$.

Solution. Let $\mathbf{u}_j$ represent the jth column of an identity matrix (of unspecified dimensions).

(a) According to results (4.7) and (5.3),

$$\frac{\partial(\mathbf{AXA})}{\partial x_{ij}} = \mathbf{A}\frac{\partial \mathbf{X}}{\partial x_{ij}}\mathbf{A} = \mathbf{A}\mathbf{u}_i\mathbf{u}_j'\mathbf{A}.$$

Thus, making use of results (6.3), (5.3), and (5.2.3), we find that

$$\begin{aligned}\frac{\partial\,\mathrm{tr}[(\mathbf{AX})^2]}{\partial x_{ij}} &= \frac{\partial\,\mathrm{tr}[(\mathbf{AXA})\mathbf{X}]}{\partial x_{ij}}\\ &= \mathrm{tr}\left(\mathbf{AXA}\frac{\partial \mathbf{X}}{\partial x_{ij}}\right) + \mathrm{tr}\left[\mathbf{X}\frac{\partial(\mathbf{AXA})}{\partial x_{ij}}\right]\\ &= \mathrm{tr}(\mathbf{AXA}\mathbf{u}_i\mathbf{u}_j') + \mathrm{tr}(\mathbf{XA}\mathbf{u}_i\mathbf{u}_j'\mathbf{A})\\ &= 2\,\mathrm{tr}(\mathbf{u}_j'\mathbf{AXA}\mathbf{u}_i)\\ &= 2\,\mathbf{u}_j'\mathbf{AXA}\mathbf{u}_i.\end{aligned}$$

Since $\mathbf{u}_j'\mathbf{AXA}\mathbf{u}_i$ is the jith element of $\mathbf{AXA}$ or, equivalently, the ijth element of $(\mathbf{AXA})'$, we find that

$$\frac{\partial\,\mathrm{tr}[(\mathbf{AX})^2]}{\partial \mathbf{X}} = 2(\mathbf{AXA})'.$$

(b) For purposes of differentiating $\mathrm{tr}[(\mathbf{AX})^2]$, $\mathrm{tr}[(\mathbf{AX})^2]$ is interpreted as a function of an $m(m+1)/2$-dimensional column vector $\mathbf{x}$ whose elements are x_{ij} $(j \le i = 1, \dots, m)$.

According to results (4.7), (5.6), and (5.7),

$$\frac{\partial(\mathbf{AXA})}{\partial x_{ij}} = \mathbf{A}\frac{\partial \mathbf{X}}{\partial x_{ij}}\mathbf{A} = \begin{cases}\mathbf{A}\mathbf{u}_i\mathbf{u}_i'\mathbf{A}, & \text{if } j = i,\\ \mathbf{A}(\mathbf{u}_i\mathbf{u}_j' + \mathbf{u}_j\mathbf{u}_i')\mathbf{A}, & \text{if } j < i,\end{cases}$$

and, according to result (6.3),

$$\begin{aligned}\frac{\partial\,\mathrm{tr}[(\mathbf{AX})^2]}{\partial x_{ij}} &= \frac{\partial\,\mathrm{tr}[(\mathbf{AXA})\mathbf{X}]}{\partial x_{ij}}\\ &= \mathrm{tr}\left(\mathbf{AXA}\frac{\partial \mathbf{X}}{\partial x_{ij}}\right) + \mathrm{tr}\left[\mathbf{X}\frac{\partial(\mathbf{AXA})}{\partial x_{ij}}\right].\end{aligned}$$

Thus, making use of results (5.6), (5.7), and (5.2.3), we find that

$$\begin{aligned}\frac{\partial\,\mathrm{tr}[(\mathbf{AX})^2]}{\partial x_{ii}} &= \mathrm{tr}(\mathbf{AXA}\mathbf{u}_i\mathbf{u}_i') + \mathrm{tr}(\mathbf{XA}\mathbf{u}_i\mathbf{u}_i'\mathbf{A})\\ &= 2\,\mathrm{tr}(\mathbf{u}_i'\mathbf{AXA}\mathbf{u}_i)\\ &= 2\,\mathbf{u}_i'\mathbf{B}\mathbf{u}_i\end{aligned}$$

and that (for $j < i$)

$$\frac{\partial\,\mathrm{tr}[(\mathbf{AX})^2]}{\partial x_{ij}} = \mathrm{tr}[\mathbf{AXA}(\mathbf{u}_i\mathbf{u}_j' + \mathbf{u}_j\mathbf{u}_i')] + \mathrm{tr}[\mathbf{XA}(\mathbf{u}_i\mathbf{u}_j' + \mathbf{u}_j\mathbf{u}_i')\mathbf{A}]$$

$$
\begin{aligned}
&= 2\,\mathrm{tr}[\mathbf{AXA}(\mathbf{u}_i\mathbf{u}_j' + \mathbf{u}_j\mathbf{u}_i')] \\
&= 2\,\mathrm{tr}(\mathbf{AXAu}_i\mathbf{u}_j') + 2\,\mathrm{tr}(\mathbf{AXAu}_j\mathbf{u}_i') \\
&= 2\,\mathrm{tr}(\mathbf{u}_j'\mathbf{AXAu}_i) + 2\,\mathrm{tr}(\mathbf{u}_i'\mathbf{AXAu}_j) \\
&= 2\,\mathbf{u}_j'\mathbf{Bu}_i + 2\,\mathbf{u}_i'\mathbf{Bu}_j.
\end{aligned}
$$

Since (for $i, j = 1, \ldots, m$) $\mathbf{u}_j'\mathbf{Bu}_i$ is the jith element of $\mathbf{B}$ or, equivalently, the ijth element of $\mathbf{B}'$ and since $\mathbf{u}_i'\mathbf{Bu}_j$ is the ijth element of $\mathbf{B}$, we conclude that

$$\frac{\partial\,\mathrm{tr}[(\mathbf{AX})^2]}{\partial\mathbf{X}} = 2\,[\mathbf{B} + \mathbf{B}' - \mathrm{diag}(b_{11}, b_{22}, \ldots, b_{mm})].$$

EXERCISE 10. Let $\mathbf{X} = \{x_{ij}\}$ represent an $m \times n$ matrix of "independent" variables, and suppose that $\mathbf{X}$ is free to range over all of $\mathcal{R}^{m\times n}$. Show that, for $k = 2, 3, \ldots,$

$$\frac{\partial\,\mathrm{tr}(\mathbf{X}^k)}{\partial\mathbf{X}} = k(\mathbf{X}')^{k-1}.$$

Solution. Let $\mathbf{u}_j$ represent the jth column of $\mathbf{I}_m$. Then, making use of results (6.1), (4.8), (5.2.3), and (5.3), we find that

$$
\begin{aligned}
\frac{\partial\,\mathrm{tr}(\mathbf{X}^k)}{\partial x_{ij}} &= \mathrm{tr}\left(\frac{\partial\mathbf{X}^k}{\partial x_{ij}}\right) \\
&= \mathrm{tr}\left(\mathbf{X}^{k-1}\frac{\partial\mathbf{X}}{\partial x_{ij}} + \mathbf{X}^{k-2}\frac{\partial\mathbf{X}}{\partial x_{ij}}\mathbf{X} + \cdots + \frac{\partial\mathbf{X}}{\partial x_{ij}}\mathbf{X}^{k-1}\right) \\
&= \mathrm{tr}\left(\mathbf{X}^{k-1}\frac{\partial\mathbf{X}}{\partial x_{ij}}\right) + \mathrm{tr}\left(\mathbf{X}^{k-2}\frac{\partial\mathbf{X}}{\partial x_{ij}}\mathbf{X}\right) + \cdots + \mathrm{tr}\left(\frac{\partial\mathbf{X}}{\partial x_{ij}}\mathbf{X}^{k-1}\right) \\
&= \mathrm{tr}\left(\mathbf{X}^{k-1}\frac{\partial\mathbf{X}}{\partial x_{ij}}\right) + \mathrm{tr}\left(\mathbf{X}^{k-1}\frac{\partial\mathbf{X}}{\partial x_{ij}}\right) + \cdots + \mathrm{tr}\left(\mathbf{X}^{k-1}\frac{\partial\mathbf{X}}{\partial x_{ij}}\right) \\
&= k\,\mathrm{tr}\left(\mathbf{X}^{k-1}\frac{\partial\mathbf{X}}{\partial x_{ij}}\right) \\
&= k\,\mathrm{tr}(\mathbf{X}^{k-1}\mathbf{u}_i\mathbf{u}_j') \\
&= k\,\mathrm{tr}(\mathbf{u}_j'\mathbf{X}^{k-1}\mathbf{u}_i) \\
&= k(\mathbf{u}_j'\mathbf{X}^{k-1}\mathbf{u}_i).
\end{aligned}
$$

Moreover, $\mathbf{u}_j'\mathbf{X}^{k-1}\mathbf{u}_i$ equals the jith element of $\mathbf{X}^{k-1}$ or equivalently the ijth element of $(\mathbf{X}^{k-1})'$. Since $(\mathbf{X}^{k-1})' = (\mathbf{X}')^{k-1}$, we conclude that $\partial\,\mathrm{tr}(\mathbf{X}^k)/\partial x_{ij}$, which is the ijth element of $\partial\,\mathrm{tr}(\mathbf{X}^k)/\partial\mathbf{X}$, equals the ijth element of $k(\mathbf{X}')^{k-1}$ and hence that

$$\frac{\partial\,\mathrm{tr}(\mathbf{X}^k)}{\partial\mathbf{X}} = k(\mathbf{X}')^{k-1}.$$

EXERCISE 11. Let $\mathbf{X} = \{x_{st}\}$ represent an $m \times n$ matrix of "independent" variables, and suppose that $\mathbf{X}$ is free to range over all of $\mathcal{R}^{m\times n}$.

(a) Show that, for any $m \times m$ matrix of constants $\mathbf{A}$,

$$\frac{\partial\,\mathrm{tr}(\mathbf{X}'\mathbf{A}\mathbf{X})}{\partial \mathbf{X}} = (\mathbf{A}+\mathbf{A}')\mathbf{X}.$$

(b) Show that, for any $n \times n$ matrix of constants $\mathbf{A}$,

$$\frac{\partial\,\mathrm{tr}(\mathbf{X}\mathbf{A}\mathbf{X}')}{\partial \mathbf{X}} = \mathbf{X}(\mathbf{A}+\mathbf{A}').$$

(c) Show (in the special case where $n = m$) that, for any $m \times m$ matrix of constants $\mathbf{A}$,

$$\frac{\partial\,\mathrm{tr}(\mathbf{X}\mathbf{A}\mathbf{X})}{\partial \mathbf{X}} = (\mathbf{A}\mathbf{X})' + (\mathbf{X}\mathbf{A})'.$$

(d) Use the results of Parts (a)-(c) to devise simple formulas for $\partial\,\mathrm{tr}(\mathbf{X}'\mathbf{X})/\partial\mathbf{X}$, $\partial\,\mathrm{tr}(\mathbf{X}\mathbf{X}')/\partial\mathbf{X}$, and (in the special case where $n = m$) $\partial\,\mathrm{tr}(\mathbf{X}^2)/\partial\mathbf{X}$.

Solution. Let $\mathbf{u}_j$ represent the jth column of an identity matrix (of unspecified dimensions).

(a) Since $\mathrm{tr}(\mathbf{X}'\mathbf{A}\mathbf{X}) = \mathrm{tr}(\mathbf{A}\mathbf{X}\mathbf{X}')$, it follows from result (6.2) that

$$\frac{\partial\,\mathrm{tr}(\mathbf{X}'\mathbf{A}\mathbf{X})}{\partial x_{ij}} = \mathrm{tr}\left[\mathbf{A}\frac{\partial(\mathbf{X}\mathbf{X}')}{\partial x_{ij}}\right].$$

Thus, making use of results (4.3), (4.10), (5.3), and (5.2.3), we find that

$$\begin{aligned}\frac{\partial\,\mathrm{tr}(\mathbf{X}'\mathbf{A}\mathbf{X})}{\partial x_{ij}} &= \mathrm{tr}\left[\mathbf{A}\mathbf{X}\left(\frac{\partial\mathbf{X}}{\partial x_{ij}}\right)'\right] + \mathrm{tr}\left(\mathbf{A}\frac{\partial\mathbf{X}}{\partial x_{ij}}\mathbf{X}'\right)\\ &= \mathrm{tr}(\mathbf{A}\mathbf{X}\mathbf{u}_j\mathbf{u}_i') + \mathrm{tr}(\mathbf{A}\mathbf{u}_i\mathbf{u}_j'\mathbf{X}')\\ &= \mathrm{tr}(\mathbf{u}_i'\mathbf{A}\mathbf{X}\mathbf{u}_j) + \mathrm{tr}(\mathbf{u}_j'\mathbf{X}'\mathbf{A}\mathbf{u}_i)\\ &= \mathbf{u}_i'\mathbf{A}\mathbf{X}\mathbf{u}_j + \mathbf{u}_j'\mathbf{X}'\mathbf{A}\mathbf{u}_i.\end{aligned}$$

Since $\mathbf{u}_i'\mathbf{A}\mathbf{X}\mathbf{u}_j$ is the ijth element of $\mathbf{A}\mathbf{X}$ and since $\mathbf{u}_j'\mathbf{X}'\mathbf{A}\mathbf{u}_i$ is the jith element of $\mathbf{X}'\mathbf{A}$ or equivalently the ijth element of $(\mathbf{X}'\mathbf{A})'$, we conclude that

$$\frac{\partial\,\mathrm{tr}(\mathbf{X}'\mathbf{A}\mathbf{X})}{\partial \mathbf{X}} = \mathbf{A}\mathbf{X} + (\mathbf{X}'\mathbf{A})' = (\mathbf{A}+\mathbf{A}')\mathbf{X}.$$

(b) Since $\mathrm{tr}(\mathbf{X}\mathbf{A}\mathbf{X}') = \mathrm{tr}(\mathbf{A}\mathbf{X}'\mathbf{X})$, it follows from result (6.2) that

$$\frac{\partial\,\mathrm{tr}(\mathbf{X}\mathbf{A}\mathbf{X}')}{\partial x_{ij}} = \mathrm{tr}\left[\mathbf{A}\frac{\partial(\mathbf{X}'\mathbf{X})}{\partial x_{ij}}\right].$$

Thus, making use of results (4.3), (5.3), and (5.2.3), we find that

$$\begin{aligned}\frac{\partial\,\mathrm{tr}(\mathbf{XAX}')}{\partial x_{ij}} &= \mathrm{tr}\left(\mathbf{AX}'\frac{\partial\mathbf{X}}{\partial x_{ij}}\right) + \mathrm{tr}\left[\mathbf{A}\left(\frac{\partial\mathbf{X}}{\partial x_{ij}}\right)'\mathbf{X}\right]\\ &= \mathrm{tr}(\mathbf{AX}'\mathbf{u}_i\mathbf{u}_j') + \mathrm{tr}(\mathbf{Au}_j\mathbf{u}_i'\mathbf{X})\\ &= \mathbf{u}_j'\mathbf{AX}'\mathbf{u}_i + \mathbf{u}_i'\mathbf{XAu}_j.\end{aligned}$$

Since $\mathbf{u}_j'\mathbf{AX}'\mathbf{u}_i$ is the jith element of $\mathbf{AX}'$ or equivalently the ijth element of $(\mathbf{AX}')'$ and since $\mathbf{u}_i'\mathbf{XAu}_j$ is the ijth element of $\mathbf{XA}$, we conclude that

$$\frac{\partial\,\mathrm{tr}(\mathbf{XAX}')}{\partial\mathbf{X}} = (\mathbf{AX}')' + \mathbf{XA} = \mathbf{X}(\mathbf{A}+\mathbf{A}').$$

(c) Since $\mathrm{tr}(\mathbf{XAX}) = \mathrm{tr}(\mathbf{AXX})$, it follows from result (6.2) that

$$\frac{\partial\,\mathrm{tr}(\mathbf{XAX})}{\partial x_{ij}} = \mathrm{tr}\left[\mathbf{A}\frac{\partial(\mathbf{XX})}{\partial x_{ij}}\right].$$

Thus, making use of results (4.3), (5.3), and (5.2.3), we find that

$$\begin{aligned}\frac{\partial\,\mathrm{tr}(\mathbf{XAX})}{\partial x_{ij}} &= \mathrm{tr}\left(\mathbf{AX}\frac{\partial\mathbf{X}}{\partial x_{ij}}\right) + \mathrm{tr}\left(\mathbf{A}\frac{\partial\mathbf{X}}{\partial x_{ij}}\mathbf{X}\right)\\ &= \mathrm{tr}(\mathbf{AXu}_i\mathbf{u}_j') + \mathrm{tr}(\mathbf{Au}_i\mathbf{u}_j'\mathbf{X})\\ &= \mathbf{u}_j'\mathbf{AXu}_i + \mathbf{u}_j'\mathbf{XAu}_i\,.\end{aligned}$$

Since $\mathbf{u}_j'\mathbf{AXu}_i$ is the jith element of $\mathbf{AX}$ or equivalently the ijth element of $(\mathbf{AX})'$ and since $\mathbf{u}_j'\mathbf{XAu}_i$ is the jith element of $\mathbf{XA}$ or equivalently the ijth element of $(\mathbf{XA})'$, we conclude that

$$\frac{\partial\,\mathrm{tr}(\mathbf{XAX})}{\partial\mathbf{X}} = (\mathbf{AX})' + (\mathbf{XA})'.$$

(d) Upon setting $\mathbf{A} = \mathbf{I}$ in the formulas from Parts (a)-(c), we find that

$$\frac{\partial\,\mathrm{tr}(\mathbf{X}'\mathbf{X})}{\partial\mathbf{X}} = \frac{\partial\,\mathrm{tr}(\mathbf{XX}')}{\partial\mathbf{X}} = 2\mathbf{X}$$

and (in the special case where $n = m$)

$$\frac{\partial\,\mathrm{tr}(\mathbf{X}^2)}{\partial\mathbf{X}} = 2\mathbf{X}'.$$

EXERCISE 12. Let $\mathbf{X} = \{x_{ij}\}$ represent an $m \times m$ matrix. Let f represent a function of $\mathbf{X}$ defined on a set S comprising some or all $m \times m$ symmetric matrices. Suppose that, for purposes of differentiation, f is to be interpreted as a function of the $[m(m+1)/2]$-dimensional column vector $\mathbf{x}$ whose elements are x_{ij} ($j \le i = 1, \ldots, m$). Suppose further that there exists a function g, whose domain is a set T of not-necessarily-symmetric matrices that contains S as a proper subset, such that $g(\mathbf{X}) = f(\mathbf{X})$ for $\mathbf{X} \in S$, so that g is a function of $\mathbf{X}$ and f is the function obtained by restricting the domain of g to S. Define $S^* = \{\mathbf{x} : \mathbf{X} \in S\}$. Let $\mathbf{c}$ represent an interior point of S^*, and let $\mathbf{C}$ represent the corresponding value of $\mathbf{X}$. Show that if $\mathbf{C}$ is an interior point of T and if g is continuously differentiable at $\mathbf{C}$, then f is continuously differentiable at $\mathbf{c}$ and that (at $\mathbf{x} = \mathbf{c}$)

$$\frac{\partial f}{\partial \mathbf{X}} = \frac{\partial g}{\partial \mathbf{X}} + \left(\frac{\partial g}{\partial \mathbf{X}}\right)' - \operatorname{diag}\left(\frac{\partial g}{\partial x_{11}}, \frac{\partial g}{\partial x_{22}}, \ldots, \frac{\partial g}{\partial x_{mm}}\right).$$

Solution. Let $\mathbf{H}$ represent the $m \times m$ matrix of functions defined, on S^*, by $\mathbf{H}(\mathbf{x}) = \mathbf{X}$. Then, $\mathbf{H}$ is continuously differentiable at $\mathbf{c}$. Thus, it follows from the results of Section 15.7 that if $\mathbf{C}$ is an interior point of T and if g is continuously differentiable at $\mathbf{C}$, then f is continuously differentiable at $\mathbf{c}$ and (at $\mathbf{x} = \mathbf{c}$)

$$\frac{\partial f}{\partial x_{ij}} = \operatorname{tr}\left[\left(\frac{\partial g}{\partial \mathbf{X}}\right)' \frac{\partial \mathbf{H}}{\partial x_{ij}}\right]$$

($j \le i = 1, \ldots, m$).

Moreover, in light of results (5.6), (5.7), and (5.2.3), we have that

$$\operatorname{tr}\left[\left(\frac{\partial g}{\partial \mathbf{X}}\right)' \frac{\partial \mathbf{H}}{\partial x_{ii}}\right] = \operatorname{tr}\left[\left(\frac{\partial g}{\partial \mathbf{X}}\right)' \mathbf{u}_i \mathbf{u}_i'\right] = \mathbf{u}_i' \left(\frac{\partial g}{\partial \mathbf{X}}\right)' \mathbf{u}_i$$

and that (for $j < i$)

$$\begin{aligned}\operatorname{tr}\left[\left(\frac{\partial g}{\partial \mathbf{X}}\right)' \frac{\partial \mathbf{H}}{\partial x_{ij}}\right] &= \operatorname{tr}\left[\left(\frac{\partial g}{\partial \mathbf{X}}\right)' \mathbf{u}_i \mathbf{u}_j'\right] + \operatorname{tr}\left[\left(\frac{\partial g}{\partial \mathbf{X}}\right)' \mathbf{u}_j \mathbf{u}_i'\right] \\ &= \mathbf{u}_j' \left(\frac{\partial g}{\partial \mathbf{X}}\right)' \mathbf{u}_i + \mathbf{u}_i' \left(\frac{\partial g}{\partial \mathbf{X}}\right)' \mathbf{u}_j.\end{aligned}$$

Since $\mathbf{u}_i'(\partial g/\partial \mathbf{X})'\mathbf{u}_i$ is the ith diagonal element of $(\partial g/\partial \mathbf{X})'$ (or equivalently the ith diagonal element of $\partial g/\partial \mathbf{X}$) and since $\mathbf{u}_j'(\partial g/\partial \mathbf{X})'\mathbf{u}_i$ and $\mathbf{u}_i'(\partial g/\partial \mathbf{X})'\mathbf{u}_j$ are the ijth elements of $\partial g/\partial \mathbf{X}$ and $(\partial g/\partial \mathbf{X})'$, respectively, it follows that (at $\mathbf{x} = \mathbf{c}$)

$$\frac{\partial f}{\partial \mathbf{X}} = \frac{\partial g}{\partial \mathbf{X}} + \left(\frac{\partial g}{\partial \mathbf{X}}\right)' - \operatorname{diag}\left(\frac{\partial g}{\partial x_{11}}, \frac{\partial g}{\partial x_{22}}, \ldots, \frac{\partial g}{\partial x_{mm}}\right).$$

EXERCISE 13. Let $\mathbf{h} = \{h_i\}$ represent an $n \times 1$ vector of functions, defined on a set S, of a vector $\mathbf{x} = (x_1, \ldots, x_m)'$ of m variables. Let g represent a function, defined on a set T, of a vector $\mathbf{y} = (y_1, \ldots, y_n)'$ of n variables. Suppose that $\mathbf{h}(\mathbf{x}) \in T$ for every $\mathbf{x}$ in S, and take f to be the composite function defined (on

S) by $f(\mathbf{x}) = g[\mathbf{h}(\mathbf{x})]$. Show that if $\mathbf{h}$ is twice continuously differentiable at an interior point $\mathbf{c}$ of S and if [assuming that $\mathbf{h}(\mathbf{c})$ is an interior point of T] g is twice continuously differentiable at $\mathbf{h}(\mathbf{c})$, then f is twice continuously differentiable at $\mathbf{c}$ and

$$\mathbf{H}f(\mathbf{c}) = [\mathbf{Dh}(\mathbf{c})]'\mathbf{H}g[\mathbf{h}(\mathbf{c})]\mathbf{Dh}(\mathbf{c}) + \sum_{i=1}^{n} D_i g[\mathbf{h}(\mathbf{c})]\mathbf{H}h_i(\mathbf{c}).$$

Solution. Suppose that $\mathbf{h}$ is twice continuously differentiable at $\mathbf{c}$ or equivalently that $h_1, \ldots, h_n$ are twice continuously differentiable at $\mathbf{c}$. Suppose further that g is twice continuously differentiable at $\mathbf{h}(\mathbf{c})$.

Then, g is continuously differentiable at $\mathbf{h}(\mathbf{c})$ and hence is continuously differentiable at every point in some neighborhood N_g of $\mathbf{h}(\mathbf{c})$. Moreover, $h_1, \ldots, h_n$ are continuously differentiable at $\mathbf{c}$ and (in light of Lemma 15.1.1) continuous at $\mathbf{c}$. Consequently, there exists a neighborhood N_h of $\mathbf{c}$ such that (i) $h_1, \ldots, h_n$ are continuously differentiable at every point in N_h and (ii) $\mathbf{h}(\mathbf{x}) \in N_g$ for every $\mathbf{x}$ in N_h.

Thus, it follows from Theorem 15.7.1 that f is continuously differentiable at every $\mathbf{x}$ in N_h and that (for $\mathbf{x} \in N_h$)

$$D_j f(\mathbf{x}) = \sum_{i=1}^{n} u_i(\mathbf{x}) D_j h_i(\mathbf{x}),$$

where $u_i(\mathbf{x}) = D_i g[\mathbf{h}(\mathbf{x})]$. Since $D_i g$ is continuously differentiable at $h(\mathbf{c})$, we have (as a further consequence of Theorem 15.7.1) that

$$D_s u_i(\mathbf{c}) = \sum_{k=1}^{n} D^2_{ki} g[\mathbf{h}(\mathbf{c})] D_s h_k(\mathbf{c}).$$

Since $D_j h_i$ is continuously differentiable at $\mathbf{c}$, we conclude that $D_j f$ (like f) is continuously differentiable at $\mathbf{c}$ and hence that f is twice continuously differentiable at $\mathbf{c}$. Moreover,

$$\begin{aligned}
D^2_{sj} f(\mathbf{c}) &= D_s D_j f(\mathbf{c}) \\
&= \sum_{i=1}^{n} [u_i(\mathbf{c}) D^2_{sj} h_i(\mathbf{c}) + D_s u_i(\mathbf{c}) D_j h_i(\mathbf{c})] \\
&= \sum_{i=1}^{n} D_i g[\mathbf{h}(\mathbf{c})] D^2_{sj} h_i(\mathbf{c}) + \sum_{k=1}^{n}\sum_{i=1}^{n} D^2_{ki} g[\mathbf{h}(\mathbf{c})] D_s h_k(\mathbf{c}) D_j h_i(\mathbf{c}) \\
&= \sum_{i=1}^{n} D_i g[\mathbf{h}(\mathbf{c})] D^2_{sj} h_i(\mathbf{c}) + [D_s \mathbf{h}(\mathbf{c})]'\mathbf{H}g[\mathbf{h}(\mathbf{c})] D_j \mathbf{h}(\mathbf{c}).
\end{aligned}$$

To complete the argument, observe that $D^2_{sj} h_i(\mathbf{c})$ is the sjth element of $\mathbf{H}h_i(\mathbf{c})$ and that $[D_s\mathbf{h}(\mathbf{c})]'\mathbf{H}g[\mathbf{h}(\mathbf{c})]D_j\mathbf{h}(\mathbf{c})$ is the sjth element of $[\mathbf{Dh}(\mathbf{c})]'\mathbf{H}g[\mathbf{h}(\mathbf{c})]\mathbf{Dh}(\mathbf{c})$

and hence that

$$\mathbf{H}f(\mathbf{c}) = [\mathbf{Dh}(\mathbf{c})]'\mathbf{H}g[\mathbf{h}(\mathbf{c})]\mathbf{Dh}(\mathbf{c}) + \sum_{i=1}^{n} D_i g[\mathbf{h}(\mathbf{c})]\mathbf{H}h_i(\mathbf{c}).$$

EXERCISE 14. Let $\mathbf{X} = \{x_{ij}\}$ represent an $m \times m$ matrix of m^2 "independent" variables (where $m \geq 2$), and suppose that the range of $\mathbf{X}$ comprises all of $\mathcal{R}^{m\times m}$. Show that (for any positive integer k) the function f defined (on $\mathcal{R}^{m\times m}$) by $f(\mathbf{X}) = |\mathbf{X}|^k$ is continuously differentiable at every $\mathbf{X}$ and that

$$\frac{\partial |\mathbf{X}|^k}{\partial \mathbf{X}} = k|\mathbf{X}|^{k-1}[\text{adj}(\mathbf{X})]'.$$

Solution. For purposes of differentiation, rearrange the elements of $\mathbf{X}$ in the form of an m^2-dimensional column vector $\mathbf{x}$, and reinterpret f as a function of $\mathbf{x}$ (in which case the domain of f comprises all of $\mathcal{R}^{m^2}$). Let h represent a function of $\mathbf{x}$ defined (on $\mathcal{R}^{m^2}$) by $h(\mathbf{x}) = \det(\mathbf{X})$, let g represent a function of a variable y defined (for all y) by $g(y) = y^k$, and express f as the composite of g and h, so that $f(\mathbf{x}) = g[h(\mathbf{x})]$.

The function g is continuously differentiable at every y, and

$$\frac{\partial g}{\partial y} = ky^{k-1}.$$

And, the function h is continuously differentiable at every $\mathbf{x}$, and

$$\frac{\partial h}{\partial x_{ij}} = \xi_{ij},$$

where ξ_{ij} is the cofactor of the ijth element x_{ij} of $\mathbf{X}$. Thus, it follows from the chain rule that f is continuously differentiable at every $\mathbf{x}$ (or equivalently at every $\mathbf{X}$) and that

$$\frac{\partial f}{\partial x_{ij}} = k|\mathbf{X}|^{k-1}\xi_{ij},$$

or equivalently that

$$\frac{\partial f}{\partial \mathbf{X}} = k|\mathbf{X}|^{k-1}[\text{adj}(\mathbf{X})]'.$$

EXERCISE 15. Let $\mathbf{F} = \{f_{is}\}$ represent a $p \times p$ matrix of functions, defined on a set S, of a vector $\mathbf{x} = (x_1, \ldots, x_m)'$ of m variables. Let $\mathbf{c}$ represent any interior point (of S) at which $\mathbf{F}$ is continuously differentiable. Use the results of Exercise 13.10 to show that (a) if $\text{rank}[\mathbf{F}(\mathbf{c})] = p - 1$, then (at $\mathbf{x} = \mathbf{c}$)

$$\frac{\partial \det(\mathbf{F})}{\partial x_j} = k\mathbf{y}'\frac{\partial \mathbf{F}}{\partial x_j}\mathbf{z},$$

where $\mathbf{z} = \{z_s\}$ and $\mathbf{y} = \{y_i\}$ are any nonnull p-dimensional vectors such that $\mathbf{F}(\mathbf{c})\mathbf{z} = \mathbf{0}$ and $[\mathbf{F}(\mathbf{c})]'\mathbf{y} = \mathbf{0}$ and where [letting ϕ_{is} represent the cofactor of $f_{is}(\mathbf{c})$] k is a scalar that is expressible as $k = \phi_{is}/(y_i z_s)$ for any i and s such that $y_i \neq 0$ and $z_s \neq 0$; and (b) if $\text{rank}[\mathbf{F}(\mathbf{c})] \leq p-2$, then (at $\mathbf{x} = \mathbf{c}$)

$$\frac{\partial \det(\mathbf{F})}{\partial x_j} = 0 .$$

Solution. Recall that $\det(\mathbf{F})$ is continuously differentiable at $\mathbf{c}$ and that (at $\mathbf{x} = \mathbf{c}$)

$$\frac{\partial \det(\mathbf{F})}{\partial x_j} = \text{tr}\left[\text{adj}(\mathbf{F})\frac{\partial \mathbf{F}}{\partial x_j}\right].$$

(a) Suppose that $\text{rank}[\mathbf{F}(\mathbf{c})] = p-1$. Then, according to the result of Part (a) of Exercise 13.10,

$$\text{adj}[\mathbf{F}(\mathbf{c})] = k\mathbf{z}\mathbf{y}',$$

so that (at $\mathbf{x} = \mathbf{c}$)

$$\frac{\partial \det(\mathbf{F})}{\partial x_j} = \text{tr}\left(k\mathbf{z}\mathbf{y}'\frac{\partial \mathbf{F}}{\partial x_j}\right) = k\,\text{tr}\left(\mathbf{y}'\frac{\partial \mathbf{F}}{\partial x_j}\mathbf{z}\right) = k\mathbf{y}'\frac{\partial \mathbf{F}}{\partial x_j}\mathbf{z}.$$

(b) Suppose that $\text{rank}[\mathbf{F}(\mathbf{c})] \leq p-2$. Then, it follows from the result of Part (b) of Exercise 13.10 that (at $\mathbf{x} = \mathbf{c}$)

$$\frac{\partial \det(\mathbf{F})}{\partial x_j} = \text{tr}\left(\mathbf{0}\frac{\partial \mathbf{F}}{\partial x_j}\right) = 0 .$$

EXERCISE 16. Let $\mathbf{X} = \{x_{st}\}$ represent an $m \times n$ matrix of "independent" variables, let $\mathbf{A}$ represent an $m \times m$ matrix of constants, and suppose that the range of $\mathbf{X}$ is a set S comprising some or all $\mathbf{X}$-values for which $\det(\mathbf{X}'\mathbf{A}\mathbf{X}) > 0$. Show that $\log\det(\mathbf{X}'\mathbf{A}\mathbf{X})$ is continuously differentiable at any interior point $\mathbf{C}$ of S and that (at $\mathbf{X} = \mathbf{C}$)

$$\frac{\partial \log\det(\mathbf{X}'\mathbf{A}\mathbf{X})}{\partial \mathbf{X}} = \mathbf{A}\mathbf{X}(\mathbf{X}'\mathbf{A}\mathbf{X})^{-1} + [(\mathbf{X}'\mathbf{A}\mathbf{X})^{-1}\mathbf{X}'\mathbf{A}]'.$$

Solution. For purposes of differentiating a function of $\mathbf{X}$, rearrange the elements of $\mathbf{X}$ in the form of an mn-dimensional column vector $\mathbf{x}$ and reinterpret the function as a function of $\mathbf{x}$, in which case the domain of the function is the set S^* obtained by rearranging the elements of each $m \times n$ matrix in S in the form of a column vector.

Let $\mathbf{c}$ represent the value of $\mathbf{x}$ corresponding to the interior point $\mathbf{C}$ of S (and note that $\mathbf{c}$ is an interior point of S^*). Since $\mathbf{X}$ is continuously differentiable at $\mathbf{c}$, $\mathbf{X}'\mathbf{A}\mathbf{X}$ is continuously differentiable at $\mathbf{c}$, and hence $\log\det(\mathbf{X}'\mathbf{A}\mathbf{X})$ is continuously differentiable at $\mathbf{c}$ (or equivalently at $\mathbf{C}$).

Moreover, making use of results (8.6), (4.6), (4.10), (5.3), and (5.2.3) and letting $\mathbf{u}_j$ represent the jth column of $\mathbf{I}_m$ or $\mathbf{I}_n$, we find that (at $\mathbf{x} = \mathbf{c}$)

$$\begin{aligned}
\frac{\partial \log \det(\mathbf{X}'\mathbf{A}\mathbf{X})}{\partial x_{ij}} &= \mathrm{tr}\left[(\mathbf{X}'\mathbf{A}\mathbf{X})^{-1}\frac{\partial(\mathbf{X}'\mathbf{A}\mathbf{X})}{\partial x_{ij}}\right] \\
&= \mathrm{tr}\left\{(\mathbf{X}'\mathbf{A}\mathbf{X})^{-1}\left[\mathbf{X}'\mathbf{A}\frac{\partial \mathbf{X}}{\partial x_{ij}} + \left(\frac{\partial \mathbf{X}}{\partial x_{ij}}\right)'\mathbf{A}\mathbf{X}\right]\right\} \\
&= \mathrm{tr}[(\mathbf{X}'\mathbf{A}\mathbf{X})^{-1}\mathbf{X}'\mathbf{A}\mathbf{u}_i\mathbf{u}_j'] + \mathrm{tr}[(\mathbf{X}'\mathbf{A}\mathbf{X})^{-1}\mathbf{u}_j\mathbf{u}_i'\mathbf{A}\mathbf{X}] \\
&= \mathbf{u}_j'(\mathbf{X}'\mathbf{A}\mathbf{X})^{-1}\mathbf{X}'\mathbf{A}\mathbf{u}_i + \mathbf{u}_i'\mathbf{A}\mathbf{X}(\mathbf{X}'\mathbf{A}\mathbf{X})^{-1}\mathbf{u}_j.
\end{aligned}$$

Upon observing that $\mathbf{u}_i'\mathbf{A}\mathbf{X}(\mathbf{X}'\mathbf{A}\mathbf{X})^{-1}\mathbf{u}_j$ and $\mathbf{u}_j'(\mathbf{X}'\mathbf{A}\mathbf{X})^{-1}\mathbf{X}'\mathbf{A}\mathbf{u}_i$ are the ijth elements of $\mathbf{A}\mathbf{X}(\mathbf{X}'\mathbf{A}\mathbf{X})^{-1}$ and $[(\mathbf{X}'\mathbf{A}\mathbf{X})^{-1}\mathbf{X}'\mathbf{A}]'$, respectively, we conclude that (at $\mathbf{x} = \mathbf{c}$)

$$\frac{\partial \log \det(\mathbf{X}'\mathbf{A}\mathbf{X})}{\partial \mathbf{X}} = \mathbf{A}\mathbf{X}(\mathbf{X}'\mathbf{A}\mathbf{X})^{-1} + [(\mathbf{X}'\mathbf{A}\mathbf{X})^{-1}\mathbf{X}'\mathbf{A}]'.$$

EXERCISE 17. (a) Let $\mathbf{X}$ represent an $m \times n$ matrix of "independent" variables, let $\mathbf{A}$ and $\mathbf{B}$ represent $q \times m$ and $n \times q$ matrices of constants, and suppose that the range of $\mathbf{X}$ is a set S comprising some or all $\mathbf{X}$-values for which $\det(\mathbf{A}\mathbf{X}\mathbf{B}) > 0$. Show that $\log \det(\mathbf{A}\mathbf{X}\mathbf{B})$ is continuously differentiable at any interior point $\mathbf{C}$ of S and that (at $\mathbf{X} = \mathbf{C}$)

$$\frac{\partial \log \det(\mathbf{A}\mathbf{X}\mathbf{B})}{\partial \mathbf{X}} = [\mathbf{B}(\mathbf{A}\mathbf{X}\mathbf{B})^{-1}\mathbf{A}]'.$$

(b) Suppose now that $\mathbf{X}$ is an $m \times m$ symmetric matrix; that $\mathbf{A}$ and $\mathbf{B}$ are $q \times m$ and $m \times q$ matrices of constants; that, for purposes of differentiating any function of $\mathbf{X}$, the function is to be interpreted as a function of the column vector $\mathbf{x}$ whose elements are x_{ij} ($j \le i = 1, \ldots, m$); and that the range of $\mathbf{x}$ is a set S comprising some or all $\mathbf{x}$-values for which $\det(\mathbf{A}\mathbf{X}\mathbf{B}) > 0$. Show that $\log \det(\mathbf{A}\mathbf{X}\mathbf{B})$ is continuously differentiable at any interior point $\mathbf{c}$ (of S) and that (at $\mathbf{x} = \mathbf{c}$)

$$\frac{\partial \log \det(\mathbf{A}\mathbf{X}\mathbf{B})}{\partial \mathbf{X}} = \mathbf{K} + \mathbf{K}' - \mathrm{diag}(k_{11}, k_{22}, \ldots, k_{qq}),$$

where $\mathbf{K} = \{k_{ij}\} = \mathbf{B}(\mathbf{A}\mathbf{X}\mathbf{B})^{-1}\mathbf{A}$.

Solution. (a) For purposes of differentiating a function of $\mathbf{X}$, rearrange the elements of $\mathbf{X}$ in the form of an mn-dimensional column vector $\mathbf{x}$ and reinterpret the function as a function of $\mathbf{x}$, in which case the domain of the function is the set S^* obtained by rearranging the elements of each $m \times n$ matrix in S in the form of a column vector.

Let $\mathbf{c}$ represent the value of $\mathbf{x}$ corresponding to the interior point $\mathbf{C}$ of S (and note that $\mathbf{c}$ is an interior point of S^*). Then, $\mathbf{X}$ is continuously differentiable at $\mathbf{c}$, implying that $\mathbf{A}\mathbf{X}\mathbf{B}$ is continuously differentiable at $\mathbf{c}$ and hence that $\log \det(\mathbf{A}\mathbf{X}\mathbf{B})$ is continuously differentiable at $\mathbf{c}$ (or equivalently at $\mathbf{C}$).

Moreover, in light of results (8.6), (4.7), (5.3), and (5.2.3), we have that (at $\mathbf{x} = \mathbf{c}$)

$$\frac{\partial \log \det(\mathbf{AXB})}{\partial x_{ij}} = \operatorname{tr}\left[(\mathbf{AXB})^{-1}\frac{\partial(\mathbf{AXB})}{\partial x_{ij}}\right] = \operatorname{tr}\left[(\mathbf{AXB})^{-1}\mathbf{A}\frac{\partial \mathbf{X}}{\partial x_{ij}}\mathbf{B}\right]$$
$$= \operatorname{tr}\left[(\mathbf{AXB})^{-1}\mathbf{A}\mathbf{u}_i\mathbf{u}_j'\mathbf{B}\right]$$
$$= \mathbf{u}_j'\mathbf{B}(\mathbf{AXB})^{-1}\mathbf{A}\mathbf{u}_i$$

and hence {since $\mathbf{u}_j'\mathbf{B}(\mathbf{AXB})^{-1}\mathbf{A}\mathbf{u}_i$ is the ijth element of $[\mathbf{B}(\mathbf{AXB})^{-1}\mathbf{A}]'$} that (at $\mathbf{x} = \mathbf{c}$)

$$\frac{\partial \log \det(\mathbf{AXB})}{\partial \mathbf{X}} = [\mathbf{B}(\mathbf{AXB})^{-1}\mathbf{A}]'.$$

(b) By employing essentially the same reasoning as in Part (a), it can be established that $\log \det(\mathbf{AXB})$ is continuously differentiable at the interior point $\mathbf{c}$ and that (at $\mathbf{x} = \mathbf{c}$)

$$\frac{\partial \log \det(\mathbf{AXB})}{\partial x_{ij}} = \operatorname{tr}\left[(\mathbf{AXB})^{-1}\mathbf{A}\frac{\partial \mathbf{X}}{\partial x_{ij}}\mathbf{B}\right].$$

Moreover, in light of results (5.6), (5.7), and (5.2.3), we have that

$$\operatorname{tr}\left[(\mathbf{AXB})^{-1}\mathbf{A}\frac{\partial \mathbf{X}}{\partial x_{ii}}\mathbf{B}\right] = \operatorname{tr}[(\mathbf{AXB})^{-1}\mathbf{A}\mathbf{u}_i\mathbf{u}_i'\mathbf{B}] = \mathbf{u}_i'\mathbf{K}\mathbf{u}_i$$

and that (for $j < i$)

$$\operatorname{tr}\left[(\mathbf{AXB})^{-1}\mathbf{A}\frac{\partial \mathbf{X}}{\partial x_{ij}}\mathbf{B}\right] = \operatorname{tr}[(\mathbf{AXB})^{-1}\mathbf{A}\mathbf{u}_i\mathbf{u}_j'\mathbf{B}] + \operatorname{tr}[(\mathbf{AXB})^{-1}\mathbf{A}\mathbf{u}_j\mathbf{u}_i'\mathbf{B}]$$
$$= \mathbf{u}_j'\mathbf{K}\mathbf{u}_i + \mathbf{u}_i'\mathbf{K}\mathbf{u}_j.$$

Since $\mathbf{u}_i'\mathbf{K}\mathbf{u}_i$ is the ith diagonal element of $\mathbf{K}$ and since $\mathbf{u}_i'\mathbf{K}\mathbf{u}_j$ and $\mathbf{u}_j'\mathbf{K}\mathbf{u}_i$ are the ijth elements of $\mathbf{K}$ and $\mathbf{K}'$, respectively, it follows that (at $\mathbf{x} = \mathbf{c}$)

$$\frac{\partial \log \det(\mathbf{AXB})}{\partial \mathbf{X}} = \mathbf{K} + \mathbf{K}' - \operatorname{diag}(k_{11}, k_{22}, \ldots, k_{qq}).$$

EXERCISE 18. Let $\mathbf{F} = \{f_{is}\}$ represent a $p \times p$ matrix of functions, defined on a set S, of a vector $\mathbf{x} = (x_1, \ldots, x_m)'$ of m variables, and let $\mathbf{A}$ and $\mathbf{B}$ represent $q \times p$ and $p \times q$ matrices of constants. Suppose that S is the set of all $\mathbf{x}$-values for which $\mathbf{F}(\mathbf{x})$ is nonsingular and $\det[\mathbf{AF}^{-1}(\mathbf{x})\mathbf{B}] > 0$ or is a subset of that set. Show that if $\mathbf{F}$ is continuously differentiable at an interior point $\mathbf{c}$ of S, then $\log \det(\mathbf{AF}^{-1}\mathbf{B})$ is continuously differentiable at $\mathbf{c}$ and (at $\mathbf{x} = \mathbf{c}$)

$$\frac{\partial \log \det(\mathbf{AF}^{-1}\mathbf{B})}{\partial x_j} = -\operatorname{tr}\left[\mathbf{F}^{-1}\mathbf{B}(\mathbf{AF}^{-1}\mathbf{B})^{-1}\mathbf{AF}^{-1}\frac{\partial \mathbf{F}}{\partial x_j}\right].$$

Solution. Suppose that $\mathbf{F}$ is continuously differentiable at $\mathbf{c}$. Then, in light of the results of Section 15.8, $\mathbf{AF}^{-1}\mathbf{B}$ is continuously differentiable at $\mathbf{c}$ and hence $\log\det(\mathbf{AF}^{-1}\mathbf{B})$ is continuously differentiable at $\mathbf{c}$. Moreover, making use of results (8.6), (8.18), and (5.2.3), we find that (at $\mathbf{x} = \mathbf{c}$)

$$\begin{aligned}\frac{\partial \log\det(\mathbf{AF}^{-1}\mathbf{B})}{\partial x_j} &= \operatorname{tr}\left[(\mathbf{AF}^{-1}\mathbf{B})^{-1}\frac{\partial(\mathbf{AF}^{-1}\mathbf{B})}{\partial x_j}\right] \\ &= \operatorname{tr}\left[(\mathbf{AF}^{-1}\mathbf{B})^{-1}(-\mathbf{AF}^{-1}\frac{\partial \mathbf{F}}{\partial x_j}\mathbf{F}^{-1}\mathbf{B})\right] \\ &= -\operatorname{tr}\left[\mathbf{F}^{-1}\mathbf{B}(\mathbf{AF}^{-1}\mathbf{B})^{-1}\mathbf{AF}^{-1}\frac{\partial \mathbf{F}}{\partial x_j}\right].\end{aligned}$$

EXERCISE 19. Let $\mathbf{A}$ and $\mathbf{B}$ represent $q \times m$ and $m \times q$ matrices of constants.

(a) Let $\mathbf{X}$ represent an $m \times m$ matrix of m^2 "independent" variables, and suppose that the range of $\mathbf{X}$ is a set S comprising some or all $\mathbf{X}$-values for which $\mathbf{X}$ is nonsingular and $\det(\mathbf{AX}^{-1}\mathbf{B}) > 0$. Use the result of Exercise 18 to show that $\log\det(\mathbf{AX}^{-1}\mathbf{B})$ is continuously differentiable at any interior point $\mathbf{C}$ of S and that (at $\mathbf{X} = \mathbf{C}$)

$$\frac{\partial \log\det(\mathbf{AX}^{-1}\mathbf{B})}{\partial \mathbf{X}} = -[\mathbf{X}^{-1}\mathbf{B}(\mathbf{AX}^{-1}\mathbf{B})^{-1}\mathbf{AX}^{-1}]'.$$

(b) Suppose now that $\mathbf{X}$ is an $m \times m$ symmetric matrix; that, for purposes of differentiating any function of $\mathbf{X}$, the function is to be interpreted as a function of the column vector $\mathbf{x}$ whose elements are x_{ij} ($j \le i = 1, \ldots, m$); and that the range of $\mathbf{x}$ is a set S comprising some or all $\mathbf{x}$-values for which $\mathbf{X}$ is nonsingular and $\det(\mathbf{AX}^{-1}\mathbf{B}) > 0$. Use the result of Exercise 18 to show that $\log\det(\mathbf{AX}^{-1}\mathbf{B})$ is continuously differentiable at any interior point $\mathbf{c}$ of S and that (at $\mathbf{x} = \mathbf{c}$)

$$\frac{\partial \log\det(\mathbf{AX}^{-1}\mathbf{B})}{\partial \mathbf{X}} = -\mathbf{K} - \mathbf{K}' + \operatorname{diag}(k_{11}, k_{22}, \ldots, k_{qq}),$$

where $\mathbf{K} = \{k_{ij}\} = \mathbf{X}^{-1}\mathbf{B}(\mathbf{AX}^{-1}\mathbf{B})^{-1}\mathbf{AX}^{-1}$.

Solution. (a) For purposes of differentiating a function of $\mathbf{X}$, rearrange the elements of $\mathbf{X}$ in the form of an m^2-dimensional column vector $\mathbf{x}$ and reinterpret the function as a function of $\mathbf{x}$, in which case the domain of the function is the set S^* obtained by rearranging the elements of each $m \times m$ matrix in S in the form of a column vector.

Let $\mathbf{c}$ represent the value of $\mathbf{x}$ corresponding to the interior point $\mathbf{C}$ of S (and note that $\mathbf{c}$ is an interior point of S^*). Since $\mathbf{X}$ is continuously differentiable at $\mathbf{c}$, it follows from the result of Exercise 18 that $\log\det(\mathbf{AX}^{-1}\mathbf{B})$ is continuously differentiable at $\mathbf{c}$ and that (at $\mathbf{x} = \mathbf{c}$)

$$\frac{\partial \log\det(\mathbf{AX}^{-1}\mathbf{B})}{\partial x_{ij}} = -\operatorname{tr}\left[\mathbf{X}^{-1}\mathbf{B}(\mathbf{AX}^{-1}\mathbf{B})^{-1}\mathbf{AX}^{-1}\frac{\partial \mathbf{X}}{\partial x_{ij}}\right].$$

Moreover, in light of results (5.3) and (5.2.3), we have that

$$\operatorname{tr}\left[\mathbf{X}^{-1}\mathbf{B}(\mathbf{A}\mathbf{X}^{-1}\mathbf{B})^{-1}\mathbf{A}\mathbf{X}^{-1}\frac{\partial \mathbf{X}}{\partial x_{ij}}\right] = \operatorname{tr}[\mathbf{X}^{-1}\mathbf{B}(\mathbf{A}\mathbf{X}^{-1}\mathbf{B})^{-1}\mathbf{A}\mathbf{X}^{-1}\mathbf{u}_i\mathbf{u}_j']$$
$$= \mathbf{u}_j'\mathbf{X}^{-1}\mathbf{B}(\mathbf{A}\mathbf{X}^{-1}\mathbf{B})^{-1}\mathbf{A}\mathbf{X}^{-1}\mathbf{u}_i\,.$$

And, upon observing that $\mathbf{u}_j'\mathbf{X}^{-1}\mathbf{B}(\mathbf{A}\mathbf{X}^{-1}\mathbf{B})^{-1}\mathbf{A}\mathbf{X}^{-1}\mathbf{u}_i$ is the ijth element of $[\mathbf{X}^{-1}\mathbf{B}(\mathbf{A}\mathbf{X}^{-1}\mathbf{B})^{-1}\mathbf{A}\mathbf{X}^{-1}]'$, we conclude that

$$\frac{\partial \log\det(\mathbf{A}\mathbf{X}^{-1}\mathbf{B})}{\partial \mathbf{X}} = -[\mathbf{X}^{-1}\mathbf{B}(\mathbf{A}\mathbf{X}^{-1}\mathbf{B})^{-1}\mathbf{A}\mathbf{X}^{-1}]'.$$

(b) By employing essentially the same reasoning as in Part (a), it can be established that $\log\det(\mathbf{A}\mathbf{X}^{-1}\mathbf{B})$ is continuously differentiable at the interior point $\mathbf{c}$ and that (at $\mathbf{x} = \mathbf{c}$)

$$\frac{\partial \log\det(\mathbf{A}\mathbf{X}^{-1}\mathbf{B})}{\partial x_{ij}} = -\operatorname{tr}\left(\mathbf{K}\frac{\partial \mathbf{X}}{\partial x_{ij}}\right).$$

Moreover, in light of results (5.6), (5.7), and (5.2.3), we have that

$$\operatorname{tr}\left(\mathbf{K}\frac{\partial \mathbf{X}}{\partial x_{ii}}\right) = \operatorname{tr}(\mathbf{K}\mathbf{u}_i\mathbf{u}_i') = \mathbf{u}_i'\mathbf{K}\mathbf{u}_i$$

and that (for $j < i$)

$$\operatorname{tr}\left(\mathbf{K}\frac{\partial \mathbf{X}}{\partial x_{ij}}\right) = \operatorname{tr}(\mathbf{K}\mathbf{u}_i\mathbf{u}_j') + \operatorname{tr}(\mathbf{K}\mathbf{u}_j\mathbf{u}_i') = \mathbf{u}_j'\mathbf{K}\mathbf{u}_i + \mathbf{u}_i'\mathbf{K}\mathbf{u}_j.$$

Since $\mathbf{u}_i'\mathbf{K}\mathbf{u}_i$ is the ith diagonal element of $\mathbf{K}$ and since $\mathbf{u}_i'\mathbf{K}\mathbf{u}_j$ and $\mathbf{u}_j'\mathbf{K}\mathbf{u}_i$ are the ijth elements of $\mathbf{K}$ and $\mathbf{K}'$, respectively, it follows that (at $\mathbf{x} = \mathbf{c}$)

$$\frac{\partial \log\det(\mathbf{A}\mathbf{X}^{-1}\mathbf{B})}{\partial \mathbf{X}} = -\mathbf{K} - \mathbf{K}' + \operatorname{diag}(k_{11}, k_{22}, \ldots, k_{qq}).$$

EXERCISE 20. Let $\mathbf{F} = \{f_{is}\}$ represent a $p \times p$ matrix of functions, defined on a set S, of a vector $\mathbf{x} = (x_1, \ldots, x_m)'$ of m variables. Let $\mathbf{c}$ represent any interior point (of S) at which $\mathbf{F}$ is continuously differentiable. By, for instance, using the result of Part (b) of Exercise 13.10, show that if $\operatorname{rank}[\mathbf{F}(\mathbf{c})] \le p - 3$, then

$$\frac{\partial \operatorname{adj}(\mathbf{F})}{\partial x_j} = \mathbf{0}\,.$$

Solution. Let ϕ_{si} represent the cofactor of f_{si} and hence the isth element of $\operatorname{adj}(\mathbf{F})$, and let $\mathbf{F}_{si}$ represent the $(p-1) \times (p-1)$ submatrix of $\mathbf{F}$ obtained by striking out the sth row and the ith column (of $\mathbf{F}$). Then, as discussed in Section 15.8, ϕ_{si} is continuously differentiable at $\mathbf{c}$ and (at $\mathbf{x} = \mathbf{c}$)

$$\frac{\partial \phi_{si}}{\partial x_j} = (-1)^{s+i} \operatorname{tr}\left[\operatorname{adj}(\mathbf{F}_{si}) \frac{\partial \mathbf{F}_{si}}{\partial x_j}\right].$$

Now, suppose that $\operatorname{rank}[\mathbf{F}(\mathbf{c})] \leq p-3$. Then, $\operatorname{rank}[\mathbf{F}_{si}(\mathbf{c})] \leq p-3$ [since otherwise $\mathbf{F}_{si}(\mathbf{c})$, and hence $\mathbf{F}(\mathbf{c})$, would contain an $r \times r$ nonsingular submatrix, where $r > p-3$, in which case the rank of $\mathbf{F}(\mathbf{c})$ would exceed $p-3$]. Thus, it follows from Part (b) of Exercise 13.10 that $\operatorname{adj}[\mathbf{F}_{si}(\mathbf{c})] = \mathbf{0}$, leading to the conclusion that (at $\mathbf{x} = \mathbf{c}$) $\partial \phi_{si}/\partial x_j = 0$ and hence that (at $\mathbf{x} = \mathbf{c}$)

$$\frac{\partial \operatorname{adj}(\mathbf{F})}{\partial x_j} = \mathbf{0}.$$

EXERCISE 21. (a) Let $\mathbf{X}$ represent an $m \times m$ matrix of m^2 "independent" variables, and suppose that the range of $\mathbf{X}$ is a set S comprising some or all $\mathbf{X}$-values for which $\mathbf{X}$ is nonsingular. Show that (when the elements of $\mathbf{X}^{-1}$ are regarded as functions of $\mathbf{X}$) $\mathbf{X}^{-1}$ is continuously differentiable at any interior point $\mathbf{C}$ of S and that (at $\mathbf{X} = \mathbf{C}$)

$$\frac{\partial \mathbf{X}^{-1}}{\partial x_{ij}} = -\mathbf{y}_i \mathbf{z}'_j,$$

where $\mathbf{y}_i$ represents the ith column and $\mathbf{z}'_j$ the jth row of $\mathbf{X}^{-1}$.

(b) Suppose now that $\mathbf{X}$ is an $m \times m$ symmetric matrix; that, for purposes of differentiating a function of $\mathbf{X}$, the function is to be interpreted as a function of the column vector $\mathbf{x}$ whose elements are x_{ij} ($j \leq i = 1, \ldots, m$); and that the range of $\mathbf{x}$ is a set S comprising some or all $\mathbf{x}$-values for which $\mathbf{X}$ is nonsingular. Show that $\mathbf{X}^{-1}$ is continuously differentiable at any interior point $\mathbf{c}$ of S and that (at $\mathbf{x} = \mathbf{c}$)

$$\frac{\partial \mathbf{X}^{-1}}{\partial x_{ij}} = \begin{cases} -\mathbf{y}_i \mathbf{y}'_i, & \text{if } j = i, \\ -\mathbf{y}_i \mathbf{y}'_j - \mathbf{y}_j \mathbf{y}'_i, & \text{if } j < i \end{cases}$$

(where $\mathbf{y}_i$ represents the ith column of $\mathbf{X}^{-1}$).

Solution. Denote by $\mathbf{u}_j$ the jth column of $\mathbf{I}_m$.

(a) For purposes of differentiating a function of $\mathbf{X}$, rearrange the elements of $\mathbf{X}$ in the form of an m^2-dimensional column vector $\mathbf{x}$ and regard the function as a function of $\mathbf{x}$, in which case the domain of the function is the set obtained by rearranging the elements of each $m \times m$ matrix in S in the form of a column vector.

Let $\mathbf{c}$ represent the value of $\mathbf{x}$ corresponding to the interior point $\mathbf{C}$ of S. Then, $\mathbf{X}$ is continuously differentiable at $\mathbf{c}$, implying that $\mathbf{X}^{-1}$ is continuously differentiable at $\mathbf{c}$ and [in light of results (8.15) and (5.3)] that (at $\mathbf{x} = \mathbf{c}$)

$$\frac{\partial \mathbf{X}^{-1}}{\partial x_{ij}} = -\mathbf{X}^{-1} \frac{\partial \mathbf{X}}{\partial x_{ij}} \mathbf{X}^{-1} = -\mathbf{X}^{-1} \mathbf{u}_i \mathbf{u}'_j \mathbf{X}^{-1} = -\mathbf{y}_i \mathbf{z}'_j.$$

(b) The matrix $\mathbf{X}$ is continuously differentiable at the interior point $\mathbf{c}$, implying that $\mathbf{X}^{-1}$ is continuously differentiable at $\mathbf{c}$ and [in light of results (8.15), (5.6), and (5.7)] that (at $\mathbf{x} = \mathbf{c}$)

$$\frac{\partial \mathbf{X}^{-1}}{\partial x_{ii}} = -\mathbf{X}^{-1}\frac{\partial \mathbf{X}}{\partial x_{ii}}\mathbf{X}^{-1} = -\mathbf{X}^{-1}\mathbf{u}_i\mathbf{u}_i'\mathbf{X}^{-1} = \mathbf{y}_i\mathbf{y}_i'$$

and similarly (for $j < i$)

$$\frac{\partial \mathbf{X}^{-1}}{\partial x_{ij}} = -\mathbf{X}^{-1}(\mathbf{u}_i\mathbf{u}_j' + \mathbf{u}_j\mathbf{u}_i')\mathbf{X}^{-1} = -\mathbf{y}_i\mathbf{y}_j' - \mathbf{y}_j\mathbf{y}_i' .$$

EXERCISE 22. Let $\mathbf{X}$ represent an $m \times m$ matrix of m^2 "independent" variables. Suppose that the range of $\mathbf{X}$ is a set S comprising some or all $\mathbf{X}$-values for which $\mathbf{X}$ is nonsingular, and let $\mathbf{C}$ represent an interior point of S. Denote the ijth element of $\mathbf{X}^{-1}$ by y_{ij}, the jth column of $\mathbf{X}^{-1}$ by $\mathbf{y}_j$, and the ith row of $\mathbf{X}^{-1}$ by $\mathbf{z}_i'$.

(a) Show that $\mathbf{X}^{-1}$ is twice continuously differentiable at $\mathbf{C}$ and that (at $\mathbf{X} = \mathbf{C}$)

$$\frac{\partial^2 \mathbf{X}^{-1}}{\partial x_{ij}\partial x_{st}} = y_{js}\mathbf{y}_i\mathbf{z}_t' + y_{ti}\mathbf{y}_s\mathbf{z}_j' .$$

(b) Suppose that $\det(\mathbf{X}) > 0$ for every $\mathbf{X}$ in S. Show that $\log\det(\mathbf{X})$ is twice continuously differentiable at $\mathbf{C}$ and that (at $\mathbf{X} = \mathbf{C}$)

$$\frac{\partial^2 \log\det(\mathbf{X})}{\partial x_{ij}\partial x_{st}} = -y_{ti}y_{js} .$$

Solution. For purposes of differentiating a function of $\mathbf{X}$, rearrange the elements of $\mathbf{X}$ in the form of an m^2-dimensional column vector $\mathbf{x}$ and reinterpret the function as a function of $\mathbf{X}$, in which case the domain of the function is the set S^* obtained by rearranging the elements of each $m \times m$ matrix in S in the form of a column vector.

Let $\mathbf{c}$ represent the value of $\mathbf{x}$ corresponding to the interior point $\mathbf{C}$ of S (and note that $\mathbf{c}$ is an interior point of S^*). Denote by $\mathbf{u}_j$ the jth column of $\mathbf{I}_m$.

It follows from the results of Section 15.5 (together with Lemma 15.4.1) that $\mathbf{X}$ is twice continuously differentiable at $\mathbf{c}$ and that (at $\mathbf{x} = \mathbf{c}$) $\partial \mathbf{X}/\partial x_{ij} = \mathbf{u}_i\mathbf{u}_j'$ and $\partial^2\mathbf{X}/\partial x_{ij}\partial x_{st} = \mathbf{0}$.

(a) Based on the results of Section 15.9, we conclude that $\mathbf{X}^{-1}$ is twice continuously differentiable at $\mathbf{c}$ and that (at $\mathbf{x} = \mathbf{c}$)

$$\begin{aligned}\frac{\partial^2 \mathbf{X}^{-1}}{\partial x_{ij}\partial x_{st}} &= \mathbf{X}^{-1}\mathbf{u}_i\mathbf{u}_j'\mathbf{X}^{-1}\mathbf{u}_s\mathbf{u}_t'\mathbf{X}^{-1} + \mathbf{X}^{-1}\mathbf{u}_s\mathbf{u}_t'\mathbf{X}^{-1}\mathbf{u}_i\mathbf{u}_j'\mathbf{X}^{-1} \\ &= y_{js}\mathbf{y}_i\mathbf{z}_t' + y_{ti}\mathbf{y}_s\mathbf{z}_j' .\end{aligned}$$

(b) Similarly, based on the results of Section 15.9 (along with Lemma 5.2.1), we conclude that $\log\det(\mathbf{X})$ is twice continuously differentiable at $\mathbf{c}$ and that (at $\mathbf{x} = \mathbf{c}$)

$$\frac{\partial^2 \log\det(\mathbf{X})}{\partial x_{ij}\partial x_{st}} = -\mathrm{tr}(\mathbf{X}^{-1}\mathbf{u}_i\mathbf{u}_j'\mathbf{X}^{-1}\mathbf{u}_s\mathbf{u}_t') = -\mathbf{u}_t'\mathbf{X}^{-1}\mathbf{u}_i\mathbf{u}_j'\mathbf{X}^{-1}\mathbf{u}_s$$
$$= -y_{ti}y_{js}\,.$$

EXERCISE 23. Let $\mathbf{F} = \{f_{is}\}$ represent a $p \times p$ matrix of functions, defined on a set S, of a vector $\mathbf{x} = (x_1, \ldots, x_m)'$ of m variables. For any nonempty set $T = \{t_1, \ldots, t_s\}$, whose members are integers between 1 and m, inclusive, define $\mathbf{D}(T) = \partial^s\mathbf{F}/\partial x_{t_1}\cdots\partial x_{t_s}$. Let k represent a positive integer and, for $i = 1, \ldots, k$, let j_i represent an arbitrary integer between 1 and m, inclusive.

(a) Suppose that $\mathbf{F}$ is nonsingular for every $\mathbf{x}$ in S, and denote by $\mathbf{c}$ any interior point (of S) at which $\mathbf{F}$ is k times continuously differentiable. Show that $\mathbf{F}^{-1}$ is k times continuously differentiable at $\mathbf{c}$ and that (at $\mathbf{x} = \mathbf{c}$)

$$\frac{\partial^k\mathbf{F}^{-1}}{\partial x_{j_1}\cdots\partial x_{j_k}} = \sum_{r=1}^{k}\sum_{T_1,\ldots,T_r}(-1)^r\mathbf{F}^{-1}\mathbf{D}(T_1)\mathbf{F}^{-1}\mathbf{D}(T_2)\cdots\mathbf{F}^{-1}\mathbf{D}(T_r)\mathbf{F}^{-1}, \quad \text{(E.1)}$$

where $T_1, \ldots, T_r$ are r nonempty mutually exclusive and exhaustive subsets of $\{j_1, \ldots, j_k\}$ (and where the second summation is over all possible choices for $T_1, \ldots, T_r$).

(b) Suppose that $\det(\mathbf{F}) > 0$ for every $\mathbf{x}$ in S, and denote by $\mathbf{c}$ any interior point (of S) at which $\mathbf{F}$ is k times continuously differentiable. Show that $\log\det(\mathbf{F})$ is k times continuously differentiable at $\mathbf{c}$ and that (at $\mathbf{x} = \mathbf{c}$)

$$\frac{\partial^k \log\det(\mathbf{F})}{\partial x_{j_1}\cdots\partial x_{j_k}}$$
$$= \sum_{r=1}^{k}\sum_{T_1,\ldots,T_r}(-1)^{r+1}\mathrm{tr}[\mathbf{F}^{-1}\mathbf{D}(T_1)\mathbf{F}^{-1}\mathbf{D}(T_2)\cdots\mathbf{F}^{-1}\mathbf{D}(T_r)], \quad \text{(E.2)}$$

where $T_1, \ldots, T_r$ are r nonempty mutually exclusive and exhaustive subsets of $\{j_1, \ldots, j_k\}$ with $j_k \in T_r$ (and where the second summation is over all possible choices for $T_1, \ldots, T_r$).

Solution. (a) The proof is by mathematical induction. For $k = 1$ and $k = 2$, it follows from the results of Sections 15.8 and 15.9 that $\mathbf{F}^{-1}$ is k times continuously differentiable and formula (E.1) valid at any interior point at which $\mathbf{F}$ is k times continuously differentiable.

Suppose now that, for an arbitrary value of k, $\mathbf{F}^{-1}$ is k times continuously differentiable and formula (E.1) valid at any interior point at which $\mathbf{F}$ is k times continuously differentiable. Denote by $\mathbf{c}^*$ an interior point at which $\mathbf{F}$ is $k+1$ times continuously differentiable. Then, it suffices to show that $\mathbf{F}^{-1}$ is $k+1$ times continuously differentiable at $\mathbf{c}^*$ and that (at $\mathbf{x} = \mathbf{c}^*$)

$$\frac{\partial^{k+1}\mathbf{F}^{-1}}{\partial x_{j_1} \cdots \partial x_{j_{k+1}}} = \sum_{r=1}^{k+1} \sum_{T_1^*, \ldots, T_r^*} (-1)^r \mathbf{F}^{-1}\mathbf{D}(T_1^*)\mathbf{F}^{-1}\mathbf{D}(T_2^*) \cdots \mathbf{F}^{-1}\mathbf{D}(T_r^*)\mathbf{F}^{-1}, \tag{S.2}$$

where j_{k+1} is an integer between 1 and m, inclusive, and where $T_1^*, \ldots, T_r^*$ are r nonempty mutually exclusive and exhaustive subsets of $\{j_1, \ldots, j_{k+1}\}$.

The matrix $\mathbf{F}$ is k times continuously differentiable at $\mathbf{c}^*$ and hence at every point in some neighborhood N of $\mathbf{c}^*$. By supposition, $\mathbf{F}^{-1}$ is k times continuously differentiable and formula (E.1) valid at every point in N. Moreover, all partial derivatives of $\mathbf{F}$ of order less than or equal to k are continuously differentiable at $\mathbf{c}^*$. Thus, it follows from results (4.8) and (8.15) that $\partial^k\mathbf{F}^{-1}/\partial x_{j_1} \cdots \partial x_{j_k}$ is continuously differentiable at $\mathbf{c}^*$ and that (at $\mathbf{x} = \mathbf{c}^*$)

$$\begin{aligned}
\frac{\partial^{k+1}\mathbf{F}^{-1}}{\partial x_{j_1} \cdots \partial x_{j_{k+1}}} &= \sum_{r=1}^{k} \sum_{T_1, \ldots, T_r} (-1)^r \\
&\quad \times \Big[-\mathbf{F}^{-1}\frac{\partial \mathbf{F}}{\partial x_{j_{k+1}}}\mathbf{F}^{-1}\mathbf{D}(T_1)\mathbf{F}^{-1}\mathbf{D}(T_2) \cdots \mathbf{F}^{-1}\mathbf{D}(T_r)\mathbf{F}^{-1} \\
&\quad -\mathbf{F}^{-1}\mathbf{D}(T_1)\mathbf{F}^{-1}\frac{\partial \mathbf{F}}{\partial x_{j_{k+1}}}\mathbf{F}^{-1}\mathbf{D}(T_2) \cdots \mathbf{F}^{-1}\mathbf{D}(T_r)\mathbf{F}^{-1} \\
&\quad \vdots \\
&\quad -\mathbf{F}^{-1}\mathbf{D}(T_1)\mathbf{F}^{-1}\mathbf{D}(T_2) \cdots \mathbf{F}^{-1}\mathbf{D}(T_r)\mathbf{F}^{-1}\frac{\partial \mathbf{F}}{\partial x_{j_{k+1}}}\mathbf{F}^{-1} \\
&\quad +\mathbf{F}^{-1}\mathbf{D}(T_1 \cup \{j_{k+1}\})\mathbf{F}^{-1}\mathbf{D}(T_2) \cdots \mathbf{F}^{-1}\mathbf{D}(T_r)\mathbf{F}^{-1} \\
&\quad +\mathbf{F}^{-1}\mathbf{D}(T_1)\mathbf{F}^{-1}\mathbf{D}(T_2 \cup \{j_{k+1}\}) \cdots \mathbf{F}^{-1}\mathbf{D}(T_r)\mathbf{F}^{-1} \\
&\quad \vdots \\
&\quad +\mathbf{F}^{-1}\mathbf{D}(T_1)\mathbf{F}^{-1}\mathbf{D}(T_2) \cdots \mathbf{F}^{-1}\mathbf{D}(T_r \cup \{j_{k+1}\})\mathbf{F}^{-1} \Big].
\end{aligned} \tag{S.3}$$

The terms of sum (S.3) can be put into one-to-one correspondence with the terms of sum (S.2) (in such a way that the corresponding terms are identical), so that formula (S.2) is valid and the mathematical induction argument is complete.

(b) The proof is by mathematical induction. For $k = 1$ and $k = 2$, it follows from the results of Sections 15.8 and 15.9 that $\log\det(\mathbf{F})$ is k times continuously differentiable and formula (E.2) valid at any interior point at which $\mathbf{F}$ is k times continuously differentiable.

Suppose now that, for an arbitrary value of k, $\log\det(\mathbf{F})$ is k times continuously differentiable and formula (E.2) valid at any interior point at which $\mathbf{F}$ is k times continuously differentiable. Denote by $\mathbf{c}^*$ an interior point at which $\mathbf{F}$ is $k+1$ times continuously differentiable. Then, it suffices to show that $\log\det(\mathbf{F})$ is $k+1$ times continuously differentiable at $\mathbf{c}^*$ and that (at $\mathbf{x} = \mathbf{c}^*$)

$$\frac{\partial^k \log\det(\mathbf{F})}{\partial x_{j_1} \cdots \partial x_{j_{k+1}}} = \sum_{r=1}^{k+1} \sum_{T_1^*,\ldots,T_r^*} (-1)^{r+1} \operatorname{tr}[\mathbf{F}^{-1}\mathbf{D}(T_1^*)\mathbf{F}^{-1}\mathbf{D}(T_2^*) \cdots \mathbf{F}^{-1}\mathbf{D}(T_r^*)], \tag{S.4}$$

where $T_1^*,\ldots,T_r^*$ are r nonempty mutually exclusive and exhaustive subsets of $\{j_1,\ldots,j_{k+1}\}$ with $j_{k+1} \in T_r^*$.

The matrix $\mathbf{F}$ is k times continuously differentiable at $\mathbf{c}^*$ and hence at every point in some neighborhood N of $\mathbf{c}^*$. By supposition, $\log\det(\mathbf{F})$ is k times continuously differentiable and formula (E.2) valid at every point in N. Moreover, all partial derivatives of $\mathbf{F}$ of order less than or equal to k are continuously differentiable at $\mathbf{c}^*$, and $\mathbf{F}^{-1}$ is continuously differentiable at $\mathbf{c}^*$. Thus, it follows from results (4.8) and (8.15) that $\partial^k \log\det(\mathbf{F})/\partial x_{j_1} \ldots \partial x_{j_k}$ is continuously differentiable at $\mathbf{c}^*$ and that (at $\mathbf{x} = \mathbf{c}^*$)

$$\begin{aligned}
\frac{\partial^{k+1} \log\det(\mathbf{F})}{\partial x_{j_1} \cdots \partial x_{j_{k+1}}}
&= \sum_{r=1}^{k} \sum_{T_1,\ldots,T_r} (-1)^{r+1} \\
&\quad \times \operatorname{tr}\Big[-\mathbf{F}^{-1}\frac{\partial \mathbf{F}}{\partial x_{j_{k+1}}}\mathbf{F}^{-1}\mathbf{D}(T_1)\mathbf{F}^{-1}\mathbf{D}(T_2)\cdots\mathbf{F}^{-1}\mathbf{D}(T_r) \\
&\quad -\mathbf{F}^{-1}\mathbf{D}(T_1)\mathbf{F}^{-1}\frac{\partial \mathbf{F}}{\partial x_{j_{k+1}}}\mathbf{F}^{-1}\mathbf{D}(T_2)\cdots\mathbf{F}^{-1}\mathbf{D}(T_r) \\
&\quad \vdots \\
&\quad -\mathbf{F}^{-1}\mathbf{D}(T_1)\mathbf{F}^{-1}\mathbf{D}(T_2)\cdots\mathbf{F}^{-1}\frac{\partial \mathbf{F}}{\partial x_{j_{k+1}}}\mathbf{F}^{-1}\mathbf{D}(T_r) \\
&\quad +\mathbf{F}^{-1}\mathbf{D}(T_1 \cup \{j_{k+1}\})\mathbf{F}^{-1}\mathbf{D}(T_2)\cdots\mathbf{F}^{-1}\mathbf{D}(T_r) \\
&\quad +\mathbf{F}^{-1}\mathbf{D}(T_1)\mathbf{F}^{-1}\mathbf{D}(T_2 \cup \{j_{k+1}\})\cdots\mathbf{F}^{-1}\mathbf{D}(T_r) \\
&\quad \vdots \\
&\quad +\mathbf{F}^{-1}\mathbf{D}(T_1)\mathbf{F}^{-1}\mathbf{D}(T_2)\cdots\mathbf{F}^{-1}\mathbf{D}(T_r \cup \{j_{k+1}\})\Big].
\end{aligned} \tag{S.5}$$

The terms of sum (S.5) can be put into one-to-one correspondence with the terms of sum (S.4) (in such a way that the corresponding terms are identical), so that formula (S.4) is valid and the mathematical induction argument is complete.

EXERCISE 24. Let $\mathbf{X} = \{x_{ij}\}$ represent an $m \times m$ symmetric matrix, and let $\mathbf{x}$ represent the $m(m+1)/2$-dimensional column vector whose elements are x_{ij} ($j \leq i = 1, \dots, m$). Define S to be the set of all $\mathbf{x}$-values for which $\mathbf{X}$ is nonsingular and S^* to be the set of all $\mathbf{x}$-values for which $\mathbf{X}$ is positive definite. Show that S and S^* are both open sets.

Solution. Let $\mathbf{c}$ represent an arbitrary point in S, and $\mathbf{c}^*$ an arbitrary point in S^*. It suffices to show that $\mathbf{c}$ and $\mathbf{c}^*$ are interior points (of S and S^*, respectively). Denote by $\mathbf{C}$ and $\mathbf{C}^*$ the values of $\mathbf{X}$ at $\mathbf{x} = \mathbf{c}$ and $\mathbf{x} = \mathbf{c}^*$, respectively.

According to Lemma 15.10.2, there exists a neighborhood N of $\mathbf{c}$ such that $\mathbf{X}$ is nonsingular for $\mathbf{x} \in N$. And, it follows from the very definition of S that $N \subset S$. Thus, $\mathbf{c}$ is an interior point of S.

Now, let $\mathbf{X}_k$ and $\mathbf{C}_k^*$ represent the kth-order leading principal submatrices of $\mathbf{X}$ and $\mathbf{C}^*$, respectively. Then, $\det(\mathbf{X}_k)$ is a continuous function of $\mathbf{x}$ (at all points in $\mathcal{R}^{m(m+1)/2}$) and hence

$$\lim_{\mathbf{x} \to \mathbf{c}^*} \det(\mathbf{X}_k) = \det(\mathbf{C}_k^*).$$

Since (according to Theorem 14.9.5) $\det(\mathbf{C}_k^*) > 0$, there exists a neighborhood N^* of $\mathbf{c}^*$ such that $|\det(\mathbf{X}_k) - \det(\mathbf{C}_k^*)| < \det(\mathbf{C}_k^*)$ for $\mathbf{x} \in N^*$ and hence {since $-[\det(\mathbf{X}_k) - \det(\mathbf{C}_k^*)] \leq |\det(\mathbf{X}_k) - \det(\mathbf{C}_k^*)|$} such that $-\det(\mathbf{X}_k) + \det(\mathbf{C}_k^*) < \det(\mathbf{C}_k^*)$ for $\mathbf{x} \in N^*$. Thus, $-\det(\mathbf{X}_k) < 0$ for $\mathbf{x} \in N^*$ or, equivalently, $\det(\mathbf{X}_k) > 0$ for $\mathbf{x} \in N^*$ ($k = 1, \dots, m$). Based on Theorem 14.9.5, we conclude that $\mathbf{X}$ is positive definite for $\mathbf{x} \in N^*$, or equivalently that $N^* \subset S^*$, and hence that $\mathbf{c}^*$ is an interior point of S^*.

EXERCISE 25. Let $\mathbf{X}$ represent an $n \times p$ matrix of constants, and let $\mathbf{W}$ represent an $n \times n$ symmetric positive definite matrix whose elements are functions, defined on a set S, of a vector $\mathbf{z} = (z_1, \dots, z_m)'$ of m variables. Further, let $\mathbf{c}$ represent any interior point (of S) at which $\mathbf{W}$ is twice continuously differentiable. Show that $\mathbf{W} - \mathbf{W}\mathbf{P}_{\mathbf{X},\mathbf{W}}$ is twice continuously differentiable at $\mathbf{c}$ and that (at $\mathbf{z} = \mathbf{c}$)

$$\begin{aligned}\frac{\partial^2(\mathbf{W} - \mathbf{W}\mathbf{P}_{\mathbf{X},\mathbf{W}})}{\partial z_i \partial z_j} &\\ &= (\mathbf{I} - \mathbf{P}'_{\mathbf{X},\mathbf{W}}) \frac{\partial^2 \mathbf{W}}{\partial z_i \partial z_j} (\mathbf{I} - \mathbf{P}_{\mathbf{X},\mathbf{W}}) \\ &\quad - (\mathbf{I} - \mathbf{P}'_{\mathbf{X},\mathbf{W}}) \frac{\partial \mathbf{W}}{\partial z_i} \mathbf{X}(\mathbf{X}'\mathbf{W}\mathbf{X})^- \mathbf{X}' \frac{\partial \mathbf{W}}{\partial z_j} (\mathbf{I} - \mathbf{P}_{\mathbf{X},\mathbf{W}}) \\ &\quad - [(\mathbf{I} - \mathbf{P}'_{\mathbf{X},\mathbf{W}}) \frac{\partial \mathbf{W}}{\partial z_i} \mathbf{X}(\mathbf{X}'\mathbf{W}\mathbf{X})^- \mathbf{X}' \frac{\partial \mathbf{W}}{\partial z_j} (\mathbf{I} - \mathbf{P}_{\mathbf{X},\mathbf{W}})]'.\end{aligned}$$

Solution. Since $\mathbf{W}$ is twice continuously differentiable at $\mathbf{c}$, it is continuously differentiable at $\mathbf{c}$ and hence continuously differentiable at every point in some neighborhood N of $\mathbf{c}$. Then, it follows from Theorem 15.11.1 that $\mathbf{W} - \mathbf{W}\mathbf{P}_{\mathbf{X},\mathbf{W}}$ is continuously differentiable at every point in N and that (for $\mathbf{z} \in N$)

$$\frac{\partial(\mathbf{W} - \mathbf{W}\mathbf{P}_{\mathbf{X},\mathbf{W}})}{\partial z_j} = (\mathbf{I} - \mathbf{P}'_{\mathbf{X},\mathbf{W}})\frac{\partial \mathbf{W}}{\partial z_j}(\mathbf{I} - \mathbf{P}_{\mathbf{X},\mathbf{W}}).$$

Further, $\mathbf{P}_{\mathbf{X},\mathbf{W}}$ and $\partial\mathbf{W}/\partial z_j$ are continuously differentiable at $\mathbf{c}$.

Thus, $\partial(\mathbf{W} - \mathbf{W}\mathbf{P}_{\mathbf{X},\mathbf{W}})/\partial z_j$ is continuously differentiable at $\mathbf{c}$, and hence $\mathbf{W} - \mathbf{W}\mathbf{P}_{\mathbf{X},\mathbf{W}}$ is twice continuously differentiable at $\mathbf{c}$. Moreover, making use of results (4.6) and (11.1) [along with Part (3′) of Theorem 14.12.11], we find that (at $\mathbf{z} = \mathbf{c}$)

$$\begin{aligned}
\frac{\partial^2(\mathbf{W} - \mathbf{W}\mathbf{P}_{\mathbf{X},\mathbf{W}})}{\partial z_i \partial z_j} &= \frac{\partial[\partial(\mathbf{W} - \mathbf{W}\mathbf{P}_{\mathbf{X},\mathbf{W}})/\partial z_j]}{\partial z_i} \\
&= -(\mathbf{I} - \mathbf{P}'_{\mathbf{X},\mathbf{W}})\frac{\partial \mathbf{W}}{\partial z_j}\frac{\partial \mathbf{P}_{\mathbf{X},\mathbf{W}}}{\partial z_i} \\
&\quad + (\mathbf{I} - \mathbf{P}'_{\mathbf{X},\mathbf{W}})\frac{\partial^2 \mathbf{W}}{\partial z_i \partial z_j}(\mathbf{I} - \mathbf{P}_{\mathbf{X},\mathbf{W}}) \\
&\quad - \left(\frac{\partial \mathbf{P}_{\mathbf{X},\mathbf{W}}}{\partial z_i}\right)'\frac{\partial \mathbf{W}}{\partial z_j}(\mathbf{I} - \mathbf{P}_{\mathbf{X},\mathbf{W}}) \\
&= -\left[\left(\frac{\partial \mathbf{P}_{\mathbf{X},\mathbf{W}}}{\partial z_i}\right)'\frac{\partial \mathbf{W}}{\partial z_j}(\mathbf{I} - \mathbf{P}_{\mathbf{X},\mathbf{W}})\right]' \\
&\quad + (\mathbf{I} - \mathbf{P}_{\mathbf{X},\mathbf{W}})'\frac{\partial^2 \mathbf{W}}{\partial z_i \partial z_j}(\mathbf{I} - \mathbf{P}_{\mathbf{X},\mathbf{W}}) \\
&\quad - \left(\frac{\partial \mathbf{P}_{\mathbf{X},\mathbf{W}}}{\partial z_i}\right)'\frac{\partial \mathbf{W}}{\partial z_j}(\mathbf{I} - \mathbf{P}_{\mathbf{X},\mathbf{W}}) \\
&= -[(\mathbf{I} - \mathbf{P}'_{\mathbf{X},\mathbf{W}})\frac{\partial \mathbf{W}}{\partial z_i}\mathbf{X}(\mathbf{X}'\mathbf{W}\mathbf{X})^-\mathbf{X}'\frac{\partial \mathbf{W}}{\partial z_j}(\mathbf{I} - \mathbf{P}_{\mathbf{X},\mathbf{W}})]' \\
&\quad + (\mathbf{I} - \mathbf{P}'_{\mathbf{X},\mathbf{W}})\frac{\partial^2 \mathbf{W}}{\partial z_i \partial z_j}(\mathbf{I} - \mathbf{P}_{\mathbf{X},\mathbf{W}}) \\
&\quad - (\mathbf{I} - \mathbf{P}'_{\mathbf{X},\mathbf{W}})\frac{\partial \mathbf{W}}{\partial z_i}\mathbf{X}(\mathbf{X}'\mathbf{W}\mathbf{X})^-\mathbf{X}'\frac{\partial \mathbf{W}}{\partial z_j}(\mathbf{I} - \mathbf{P}_{\mathbf{X},\mathbf{W}}).
\end{aligned}$$

EXERCISE 26. Let $\mathbf{X}$ represent an $n \times p$ matrix and $\mathbf{W}$ an $n \times n$ symmetric positive definite matrix, and suppose that the elements of $\mathbf{X}$ and $\mathbf{W}$ are functions, defined on a set S, of a vector $\mathbf{z} = (z_1, \ldots, z_m)'$ of m variables. And, let $\mathbf{c}$ represent any interior point (of S) at which $\mathbf{W}$ and $\mathbf{X}$ are continuously differentiable, and suppose that $\mathbf{X}$ has constant rank on some neighborhood of $\mathbf{c}$. Further, let $\mathbf{B}$ represent any $p \times n$ matrix such that $\mathbf{X}'\mathbf{W}\mathbf{X}\mathbf{B} = \mathbf{X}'\mathbf{W}$. Then, at $\mathbf{z} = \mathbf{c}$,

$$\begin{aligned}
\frac{\partial(\mathbf{W}\mathbf{P}_{\mathbf{X},\mathbf{W}})}{\partial z_j} &= \frac{\partial \mathbf{W}}{\partial z_j} - (\mathbf{I} - \mathbf{P}'_{\mathbf{X},\mathbf{W}})\frac{\partial \mathbf{W}}{\partial z_j}(\mathbf{I} - \mathbf{P}_{\mathbf{X},\mathbf{W}}) \\
&\quad + \mathbf{W}(\mathbf{I} - \mathbf{P}_{\mathbf{X},\mathbf{W}})\frac{\partial \mathbf{X}}{\partial z_j}\mathbf{B} + [\mathbf{W}(\mathbf{I} - \mathbf{P}_{\mathbf{X},\mathbf{W}})\frac{\partial \mathbf{X}}{\partial z_j}\mathbf{B}]'. \qquad (*)
\end{aligned}$$

Derive result $(*)$ by using the result

$$\mathbf{P}'_{\mathbf{X},\mathbf{W}}\mathbf{W}\mathbf{P}_{\mathbf{X},\mathbf{W}} = \mathbf{W}\mathbf{P}_{\mathbf{X},\mathbf{W}} \qquad (**)$$

to obtain the representation

$$\frac{\partial(\mathbf{W}\mathbf{P}_{\mathbf{X},\mathbf{W}})}{\partial z_j} = \mathbf{P}'_{\mathbf{X},\mathbf{W}}\mathbf{W}\frac{\partial \mathbf{P}_{\mathbf{X},\mathbf{W}}}{\partial z_j} + \mathbf{P}'_{\mathbf{X},\mathbf{W}}\frac{\partial \mathbf{W}}{\partial z_j}\mathbf{P}_{\mathbf{X},\mathbf{W}} + \left(\frac{\partial \mathbf{P}_{\mathbf{X},\mathbf{W}}}{\partial z_j}\right)'\mathbf{W}\mathbf{P}_{\mathbf{X},\mathbf{W}},$$

and by then making use of the result

$$\left(\frac{\partial \mathbf{P}_{\mathbf{X},\mathbf{W}}}{\partial z_j}\right)'\mathbf{W}\mathbf{P}_{\mathbf{X},\mathbf{W}} = (\mathbf{I} - \mathbf{P}'_{\mathbf{X},\mathbf{W}})\frac{\partial \mathbf{W}}{\partial z_j}\mathbf{P}_{\mathbf{X},\mathbf{W}} + \mathbf{W}(\mathbf{I} - \mathbf{P}_{\mathbf{X},\mathbf{W}})\frac{\partial \mathbf{X}}{\partial z_j}\mathbf{B}. \qquad (\star)$$

Solution. According to result $(**)$ [or, equivalently, according to Part (6′) of Theorem 14.12.11], $\mathbf{W}\mathbf{P}_{\mathbf{X},\mathbf{W}} = \mathbf{P}'_{\mathbf{X},\mathbf{W}}\mathbf{W}\mathbf{P}_{\mathbf{X},\mathbf{W}}$. Thus, it follows from results (4.6) and 4.10) that

$$\frac{\partial(\mathbf{W}\mathbf{P}_{\mathbf{X},\mathbf{W}})}{\partial z_j} = \mathbf{P}'_{\mathbf{X},\mathbf{W}}\mathbf{W}\frac{\partial \mathbf{P}_{\mathbf{X},\mathbf{W}}}{\partial z_j} + \mathbf{P}'_{\mathbf{X},\mathbf{W}}\frac{\partial \mathbf{W}}{\partial z_j}\mathbf{P}_{\mathbf{X},\mathbf{W}} + \left(\frac{\partial \mathbf{P}_{\mathbf{X},\mathbf{W}}}{\partial z_j}\right)'\mathbf{W}\mathbf{P}_{\mathbf{X},\mathbf{W}}.$$

Substituting from result $(\star)$ [or equivalently from result (11.16)], we find that (for any $p \times n$ matrix $\mathbf{B}$ such that $\mathbf{X}'\mathbf{W}\mathbf{X}\mathbf{B} = \mathbf{X}'\mathbf{W}$)

$$\begin{aligned}
\frac{\partial(\mathbf{W}\mathbf{P}_{\mathbf{X},\mathbf{W}})}{\partial z_j} &= [(\mathbf{I} - \mathbf{P}'_{\mathbf{X},\mathbf{W}})\frac{\partial \mathbf{W}}{\partial z_j}\mathbf{P}_{\mathbf{X},\mathbf{W}} + \mathbf{W}(\mathbf{I} - \mathbf{P}_{\mathbf{X},\mathbf{W}})\frac{\partial \mathbf{X}}{\partial z_j}\mathbf{B}]' \\
&\qquad + \mathbf{P}'_{\mathbf{X},\mathbf{W}}\frac{\partial \mathbf{W}}{\partial z_j}\mathbf{P}_{\mathbf{X},\mathbf{W}} + (\mathbf{I} - \mathbf{P}'_{\mathbf{X},\mathbf{W}})\frac{\partial \mathbf{W}}{\partial z_j}\mathbf{P}_{\mathbf{X},\mathbf{W}} \\
&\qquad + \mathbf{W}(\mathbf{I} - \mathbf{P}_{\mathbf{X},\mathbf{W}})\frac{\partial \mathbf{X}}{\partial z_j}\mathbf{B} \\
&= \mathbf{P}'_{\mathbf{X},\mathbf{W}}\frac{\partial \mathbf{W}}{\partial z_j}(\mathbf{I} - \mathbf{P}_{\mathbf{X},\mathbf{W}}) + [\mathbf{W}(\mathbf{I} - \mathbf{P}_{\mathbf{X},\mathbf{W}})\frac{\partial \mathbf{X}}{\partial z_j}\mathbf{B}]' \\
&\qquad + \frac{\partial \mathbf{W}}{\partial z_j}\mathbf{P}_{\mathbf{X},\mathbf{W}} + \mathbf{W}(\mathbf{I} - \mathbf{P}_{\mathbf{X},\mathbf{W}})\frac{\partial \mathbf{X}}{\partial z_j}\mathbf{B}.
\end{aligned}$$

And, upon reexpressing $(\partial \mathbf{W}/\partial z_j)\mathbf{P}_{\mathbf{X},\mathbf{W}}$ as

$$\frac{\partial \mathbf{W}}{\partial z_j}\mathbf{P}_{\mathbf{X},\mathbf{W}} = \frac{\partial \mathbf{W}}{\partial z_j} - \frac{\partial \mathbf{W}}{\partial z_j}(\mathbf{I} - \mathbf{P}_{\mathbf{X},\mathbf{W}}),$$

it is clear that

$$\begin{aligned}
\frac{\partial(\mathbf{W}\mathbf{P}_{\mathbf{X},\mathbf{W}})}{\partial z_j} &= \frac{\partial \mathbf{W}}{\partial z_j} - (\mathbf{I} - \mathbf{P}'_{\mathbf{X},\mathbf{W}})\frac{\partial \mathbf{W}}{\partial z_j}(\mathbf{I} - \mathbf{P}_{\mathbf{X},\mathbf{W}}) \\
&\qquad + \mathbf{W}(\mathbf{I} - \mathbf{P}_{\mathbf{X},\mathbf{W}})\frac{\partial \mathbf{X}}{\partial z_j}\mathbf{B} + [\mathbf{W}(\mathbf{I} - \mathbf{P}_{\mathbf{X},\mathbf{W}})\frac{\partial \mathbf{X}}{\partial z_j}\mathbf{B}]'.
\end{aligned}$$

16

Kronecker Products and the Vec and Vech Operators

EXERCISE 1. (a) Verify that, for any $m \times n$ matrices $\mathbf{A}$ and $\mathbf{B}$ and any $p \times q$ matrices $\mathbf{C}$ and $\mathbf{D}$,

$$(\mathbf{A} + \mathbf{B}) \otimes (\mathbf{C} + \mathbf{D}) = (\mathbf{A} \otimes \mathbf{C}) + (\mathbf{A} \otimes \mathbf{D}) + (\mathbf{B} \otimes \mathbf{C}) + (\mathbf{B} \otimes \mathbf{D}).$$

(b) Verify that, for any $m \times n$ matrices $\mathbf{A}_1, \mathbf{A}_2, \ldots, \mathbf{A}_r$ and $p \times q$ matrices $\mathbf{B}_1, \mathbf{B}_2, \ldots, \mathbf{B}_s$,

$$\left(\sum_{i=1}^{r} \mathbf{A}_i\right) \otimes \left(\sum_{j=1}^{s} \mathbf{B}_j\right) = \sum_{i=1}^{r} \sum_{j=1}^{s} (\mathbf{A}_i \otimes \mathbf{B}_j).$$

Solution. (a) It follows from results (1.11) and (1.12) that

$$\begin{aligned}(\mathbf{A} + \mathbf{B}) \otimes (\mathbf{C} + \mathbf{D}) &= [\mathbf{A} \otimes (\mathbf{C} + \mathbf{D})] + [\mathbf{B} \otimes (\mathbf{C} + \mathbf{D})] \\ &= (\mathbf{A} \otimes \mathbf{C}) + (\mathbf{A} \otimes \mathbf{D}) + (\mathbf{B} \otimes \mathbf{C}) + (\mathbf{B} \otimes \mathbf{D}).\end{aligned}$$

(b) Let us begin by showing that, for any $m \times n$ matrix $\mathbf{A}$,

$$\mathbf{A} \otimes \left(\sum_{j=1}^{s} \mathbf{B}_j\right) = \sum_{j=1}^{s} (\mathbf{A} \otimes \mathbf{B}_j). \tag{S.1}$$

The proof is by mathematical induction. Result (S.1) is valid for $s = 2$, as is evident from result (1.12). Suppose now that result (S.1) is valid for $s = s^*$. Then, making use of result (1.12), we find that

$$\mathbf{A} \otimes \left(\sum_{j=1}^{s^*+1} \mathbf{B}_j\right) = \mathbf{A} \otimes \left(\sum_{j=1}^{s^*} \mathbf{B}_j + \mathbf{B}_{s^*+1}\right)$$

$$= \left[\mathbf{A} \otimes \left(\sum_{j=1}^{s^*} \mathbf{B}_j\right)\right] + (\mathbf{A} \otimes \mathbf{B}_{s^*+1})$$

$$= \sum_{j=1}^{s^*+1} (\mathbf{A} \otimes \mathbf{B}_j),$$

which indicates that result (S.1) is valid for $s = s^* + 1$, thereby completing the induction argument. Moreover, it can be shown in analogous fashion that, for any $p \times q$ matrix $\mathbf{B}$,

$$\left(\sum_{i=1}^{r} \mathbf{A}_i\right) \otimes \mathbf{B} = \sum_{i=1}^{r} (\mathbf{A}_i \otimes \mathbf{B}). \tag{S.2}$$

Now, making use of results (S.2) and (S.1), we find that

$$\left(\sum_{i=1}^{r} \mathbf{A}_i\right) \otimes \left(\sum_{j=1}^{s} \mathbf{B}_j\right) = \sum_{i=1}^{r} \left[\mathbf{A}_i \otimes \left(\sum_{j=1}^{s} \mathbf{B}_j\right)\right] = \sum_{i=1}^{r} \sum_{j=1}^{s} (\mathbf{A}_i \otimes \mathbf{B}_j).$$

EXERCISE 2. Show that, for any $m \times n$ matrix $\mathbf{A}$ and $p \times q$ matrix $\mathbf{B}$,

$$\mathbf{A} \otimes \mathbf{B} = (\mathbf{A} \otimes \mathbf{I}_p)\, \mathrm{diag}(\mathbf{B}, \mathbf{B}, \ldots, \mathbf{B}).$$

Solution. Making use of results (1.20) and (1.7), we find that

$$\mathbf{A} \otimes \mathbf{B} = (\mathbf{A} \otimes \mathbf{I}_p)(\mathbf{I}_n \otimes \mathbf{B}) = (\mathbf{A} \otimes \mathbf{I}_p)\, \mathrm{diag}(\mathbf{B}, \mathbf{B}, \ldots, \mathbf{B}).$$

EXERCISE 3. Show that, for any $m \times 1$ vector $\mathbf{a}$ and any $p \times 1$ vector $\mathbf{b}$, (1) $\mathbf{a} \otimes \mathbf{b} = (\mathbf{a} \otimes \mathbf{I}_p)\mathbf{b}$ and (2) $\mathbf{a}' \otimes \mathbf{b}' = \mathbf{b}'(\mathbf{a}' \otimes \mathbf{I}_p)$.

Solution. Making use of results (1.20) and (1.1), we find (1) that

$$\mathbf{a} \otimes \mathbf{b} = (\mathbf{a} \otimes \mathbf{I}_p)(1 \otimes \mathbf{b}) = (\mathbf{a} \otimes \mathbf{I}_p)\mathbf{b}$$

and similarly (2) that

$$\mathbf{a}' \otimes \mathbf{b}' = (1 \otimes \mathbf{b}')(\mathbf{a}' \otimes \mathbf{I}_p) = \mathbf{b}'(\mathbf{a}' \otimes \mathbf{I}_p).$$

EXERCISE 4. Let $\mathbf{A}$ and $\mathbf{B}$ represent square matrices.

(a) Show that if $\mathbf{A}$ and $\mathbf{B}$ are orthogonal, then $\mathbf{A} \otimes \mathbf{B}$ is orthogonal.

(b) Show that if $\mathbf{A}$ and $\mathbf{B}$ are idempotent, then $\mathbf{A} \otimes \mathbf{B}$ is idempotent.

Solution. Note that (since $\mathbf{A}$ and $\mathbf{B}$ are square) $\mathbf{A} \otimes \mathbf{B}$ is square.

(a) If $\mathbf{A}$ and $\mathbf{B}$ are orthogonal, then we have [in light of results (1.15), (1.19), and (1.8)] that

$$(\mathbf{A} \otimes \mathbf{B})'(\mathbf{A} \otimes \mathbf{B}) = (\mathbf{A}' \otimes \mathbf{B}')(\mathbf{A} \otimes \mathbf{B}) = (\mathbf{A}'\mathbf{A}) \otimes (\mathbf{B}'\mathbf{B}) = \mathbf{I} \otimes \mathbf{I} = \mathbf{I}$$

and hence that $\mathbf{A} \otimes \mathbf{B}$ is orthogonal.

(b) If $\mathbf{A}$ and $\mathbf{B}$ are idempotent, then we have [in light of result (1.19)] that

$$(\mathbf{A} \otimes \mathbf{B})(\mathbf{A} \otimes \mathbf{B}) = (\mathbf{AA}) \otimes (\mathbf{BB}) = \mathbf{A} \otimes \mathbf{B}$$

and hence that $\mathbf{A} \otimes \mathbf{B}$ is idempotent.

EXERCISE 5. Letting m, n, p, and q represent arbitrary positive integers, show (a) that, for any $p \times q$ matrix $\mathbf{B}$ (having $p > 1$ or $q > 1$), there exists an $m \times n$ matrix $\mathbf{A}$ such that $\mathbf{A} \otimes \mathbf{B}$ has generalized inverses that are not expressible in the form $\mathbf{A}^- \otimes \mathbf{B}^-$ and (b) that, for any $m \times n$ matrix $\mathbf{A}$ (having $m > 1$ or $n > 1$), there exists a $p \times q$ matrix $\mathbf{B}$ such that $\mathbf{A} \otimes \mathbf{B}$ has generalized inverses that are not expressible in the form $\mathbf{A}^- \otimes \mathbf{B}^-$.

Solution. (a) Take $\mathbf{A} = \mathbf{0}$. Then, $\mathbf{A} \otimes \mathbf{B} = \mathbf{0}$, so that any $nq \times mp$ matrix is a generalized inverse of $\mathbf{A} \otimes \mathbf{B}$. Since every one of the mn ($q \times p$ dimensional) blocks of the Kronecker product $\mathbf{A}^- \otimes \mathbf{B}^-$ is a scalar multiple of the same $q \times p$ matrix (namely, $\mathbf{B}^-$), $\mathbf{A} \otimes \mathbf{B}$ has generalized inverses that are not expressible in the form $\mathbf{A}^- \otimes \mathbf{B}^-$. Consider, for example, an $nq \times mp$ partitioned matrix comprising mn ($q \times p$ dimensional) blocks, including one block that has a single nonzero entry and a second block that also has a single nonzero entry but in a different location than the first. Clearly, this matrix is a generalized inverse of $\mathbf{A} \otimes \mathbf{B}$ that is not expressible in the form $\mathbf{A}^- \otimes \mathbf{B}^-$.

(b) Take $\mathbf{B} = \mathbf{0}$. Then, $\mathbf{A} \otimes \mathbf{B} = \mathbf{0}$, so that any $nq \times mp$ matrix is a generalized inverse of $\mathbf{A} \otimes \mathbf{B}$. Now, letting c_{ts} represent the tsth element of $\mathbf{B}^-$ and observing that (for $t = 1, \ldots, q$ and $s = 1, \ldots, p$) the $n \times m$ submatrix of $\mathbf{A}^- \otimes \mathbf{B}^-$ obtained by striking out all of the rows and columns except the tth, $(q + t)$th, $\ldots$, $[(n-1)q + t]$th rows and sth, $(p + s)$th, $\ldots$, $[(m-1)p + s]$th columns equals $c_{ts}\mathbf{A}^-$, it follows that $\mathbf{A} \otimes \mathbf{B}$ has generalized inverses that are not expressible in the form $\mathbf{A}^- \otimes \mathbf{B}^-$. Consider, for example, an $nq \times mp$ matrix for which the $n \times m$ submatrix obtained by striking out (for some t and s) all of the rows and columns except the tth, $(q + t)$th, $\ldots$, $[(n-1)q + t]$th rows and sth, $(p + s)$th, $\ldots$, $[(m-1)p + s]$th columns has a single nonzero entry and for which the $n \times m$ submatrix obtained by striking out (for some t' and s' with $t' \neq t$ or $s' \neq s$) all of the rows and columns except the t'th, $(q + t')$th, $\ldots$, $[(n-1)q + t']$th rows and s'th, $(p + s')$th, $\ldots$, $[(m-1)p + s']$th columns also has a single nonzero entry but in a different location than the first submatrix. Clearly, this matrix is a generalized inverse of $\mathbf{A} \otimes \mathbf{B}$ that is not expressible in the form $\mathbf{A}^- \otimes \mathbf{B}^-$.

EXERCISE 6. Let $\mathbf{X} = \mathbf{A} \otimes \mathbf{B}$, where $\mathbf{A}$ is an $m \times n$ matrix and $\mathbf{B}$ a $p \times q$ matrix. Show that $\mathbf{P_X} = \mathbf{P_A} \otimes \mathbf{P_B}$.

Solution. According to result (1.15), $\mathbf{X}' = \mathbf{A}' \otimes \mathbf{B}'$. Thus, making use of result (1.19), we find that

$$\mathbf{X}'\mathbf{X} = (\mathbf{A}' \otimes \mathbf{B}')(\mathbf{A} \otimes \mathbf{B}) = (\mathbf{A}'\mathbf{A}) \otimes (\mathbf{B}'\mathbf{B}),$$

so that $(\mathbf{A}'\mathbf{A})^- \otimes (\mathbf{B}'\mathbf{B})^-$ is a generalized inverse of $\mathbf{X}'\mathbf{X}$. And, again making use of result (1.19), it follows that

$$\begin{aligned}\mathbf{P_X} &= \mathbf{X}(\mathbf{X}'\mathbf{X})^-\mathbf{X}' \\ &= (\mathbf{A}\otimes\mathbf{B})[(\mathbf{A}'\mathbf{A})^- \otimes (\mathbf{B}'\mathbf{B})^-](\mathbf{A}'\otimes\mathbf{B}') \\ &= [\mathbf{A}(\mathbf{A}'\mathbf{A})^-\mathbf{A}'] \otimes [\mathbf{B}(\mathbf{B}'\mathbf{B})^-\mathbf{B}'] \\ &= \mathbf{P_A} \otimes \mathbf{P_B}.\end{aligned}$$

EXERCISE 7. Show that the Kronecker product $\mathbf{A}\otimes\mathbf{B}$ of an $m \times m$ matrix $\mathbf{A}$ and an $n \times n$ matrix $\mathbf{B}$ is (a) symmetric nonnegative definite if $\mathbf{A}$ and $\mathbf{B}$ are both symmetric nonnegative definite or both symmetric nonpositive definite and (b) symmetric positive definite if $\mathbf{A}$ and $\mathbf{B}$ are both symmetric positive definite or both symmetric negative definite.

Solution. (a) Suppose that $\mathbf{A}$ and $\mathbf{B}$ are both symmetric nonnegative definite. Then, according to Corollary 14.3.8, there exist matrices $\mathbf{P}$ and $\mathbf{Q}$ such that $\mathbf{A} = \mathbf{P}'\mathbf{P}$ and $\mathbf{B} = \mathbf{Q}'\mathbf{Q}$. Thus, making use of results (1.19) and (1.15), we find that

$$\mathbf{A}\otimes\mathbf{B} = (\mathbf{P}'\otimes\mathbf{Q}')(\mathbf{P}\otimes\mathbf{Q}) = (\mathbf{P}\otimes\mathbf{Q})'(\mathbf{P}\otimes\mathbf{Q}).$$

We conclude (in light of Corollary 14.3.8 or 14.2.14) that $\mathbf{A}\otimes\mathbf{B}$ is symmetric nonnegative definite.

Alternatively, if $\mathbf{A}$ and $\mathbf{B}$ are both symmetric nonpositive definite, then $-\mathbf{A}$ and $-\mathbf{B}$ are symmetric and (by definition) nonnegative definite, and the proof [of Part (a)] is complete upon observing [in light of result (1.10)] that $\mathbf{A}\otimes\mathbf{B} = (-\mathbf{A})\otimes(-\mathbf{B})$.

(b) Suppose now that $\mathbf{A}$ and $\mathbf{B}$ are both symmetric positive definite. Then, according to Corollary 14.3.13, there exist nonsingular matrices $\mathbf{P}$ and $\mathbf{Q}$ such that $\mathbf{A} = \mathbf{P}'\mathbf{P}$ and $\mathbf{B} = \mathbf{Q}'\mathbf{Q}$. Further, $\mathbf{A}\otimes\mathbf{B} = (\mathbf{P}\otimes\mathbf{Q})'(\mathbf{P}\otimes\mathbf{Q})$, and $\mathbf{P}\otimes\mathbf{Q}$ is nonsingular. We conclude (in light of Corollary 14.3.13 or 14.2.14) that $\mathbf{A}\otimes\mathbf{B}$ is symmetric positive definite.

Alternatively, if $\mathbf{A}$ and $\mathbf{B}$ are both symmetric negative definite, then $-\mathbf{A}$ and $-\mathbf{B}$ are symmetric and (by definition) positive definite, and the proof is complete upon observing that $\mathbf{A}\otimes\mathbf{B} = (-\mathbf{A})\otimes(-\mathbf{B})$.

EXERCISE 8. Let $\mathbf{A}$ and $\mathbf{B}$ represent $m \times m$ symmetric matrices and $\mathbf{C}$ and $\mathbf{D}$ $n \times n$ symmetric matrices. Using the result of Exercise 7 (or otherwise), show that if $\mathbf{A}-\mathbf{B}$, $\mathbf{C}-\mathbf{D}$, $\mathbf{B}$, and $\mathbf{C}$ are nonnegative definite, then $(\mathbf{A}\otimes\mathbf{C}) - (\mathbf{B}\otimes\mathbf{D})$ is symmetric nonnegative definite.

Solution. Using properties (1.10) - (1.12), we find that

$$\begin{aligned}(\mathbf{A}\otimes\mathbf{C}) - (\mathbf{B}\otimes\mathbf{D}) &= \{[(\mathbf{A}-\mathbf{B})+\mathbf{B}]\otimes\mathbf{C}\} - \{\mathbf{B}\otimes[\mathbf{C}-(\mathbf{C}-\mathbf{D})]\} \\ &= [(\mathbf{A}-\mathbf{B})\otimes\mathbf{C}] + (\mathbf{B}\otimes\mathbf{C}) \\ &\qquad -\{(\mathbf{B}\otimes\mathbf{C}) - [\mathbf{B}\otimes(\mathbf{C}-\mathbf{D})]\} \\ &= [(\mathbf{A}-\mathbf{B})\otimes\mathbf{C}] + [\mathbf{B}\otimes(\mathbf{C}-\mathbf{D})]. \qquad \text{(S.3)}\end{aligned}$$

Now, suppose that $\mathbf{A}-\mathbf{B}$, $\mathbf{C}-\mathbf{D}$, $\mathbf{B}$, and $\mathbf{C}$ are nonnegative definite. Then, it follows from the result of Part (a) of Exercise 7 that $(\mathbf{A}-\mathbf{B})\otimes\mathbf{C}$ and $\mathbf{B}\otimes(\mathbf{C}-\mathbf{D})$ are both symmetric nonnegative definite and hence (in light of Lemma 14.2.4) that their sum $[(\mathbf{A}-\mathbf{B})\otimes\mathbf{C}]+[\mathbf{B}\otimes(\mathbf{C}-\mathbf{D})]$ is symmetric nonnegative definite. And, based on equality (S.3), we conclude that $(\mathbf{A}\otimes\mathbf{C})-(\mathbf{B}\otimes\mathbf{D})$ is symmetric nonnegative definite.

EXERCISE 9. Let $\mathbf{A}$ represent an $m\times n$ matrix and $\mathbf{B}$ a $p\times q$ matrix. Show that, in the case of the usual norm,

$$\|\mathbf{A}\otimes\mathbf{B}\|=\|\mathbf{A}\|\;\|\mathbf{B}\|\,.$$

Solution. Making use of results (1.15), (1.19), and (1.25), we find that

$$\begin{aligned}\|\mathbf{A}\otimes\mathbf{B}\| &= \{\mathrm{tr}[(\mathbf{A}\otimes\mathbf{B})'(\mathbf{A}\otimes\mathbf{B})]\}^{\frac{1}{2}}\\ &= \{\mathrm{tr}[(\mathbf{A}'\otimes\mathbf{B}')(\mathbf{A}\otimes\mathbf{B})]\}^{\frac{1}{2}}\\ &= \{\mathrm{tr}[(\mathbf{A}'\mathbf{A})\otimes(\mathbf{B}'\mathbf{B})]\}^{\frac{1}{2}}\\ &= \{\mathrm{tr}(\mathbf{A}'\mathbf{A})\}^{\frac{1}{2}}\{\mathrm{tr}(\mathbf{B}'\mathbf{B})\}^{\frac{1}{2}}\\ &= \|\mathbf{A}\|\;\|\mathbf{B}\|\,.\end{aligned}$$

EXERCISE 10. Verify that, for an $m\times n$ partitioned matrix

$$\mathbf{A}=\begin{pmatrix}\mathbf{A}_{11} & \mathbf{A}_{12} & \dots & \mathbf{A}_{1c}\\ \mathbf{A}_{21} & \mathbf{A}_{22} & \dots & \mathbf{A}_{2c}\\ \vdots & \vdots & & \vdots\\ \mathbf{A}_{r1} & \mathbf{A}_{r2} & \dots & \mathbf{A}_{rc}\end{pmatrix}$$

and a $p\times q$ matrix $\mathbf{B}$,

$$\mathbf{A}\otimes\mathbf{B}=\begin{pmatrix}\mathbf{A}_{11}\otimes\mathbf{B} & \mathbf{A}_{12}\otimes\mathbf{B} & \dots & \mathbf{A}_{1c}\otimes\mathbf{B}\\ \mathbf{A}_{21}\otimes\mathbf{B} & \mathbf{A}_{22}\otimes\mathbf{B} & \dots & \mathbf{A}_{2c}\otimes\mathbf{B}\\ \vdots & \vdots & & \vdots\\ \mathbf{A}_{r1}\otimes\mathbf{B} & \mathbf{A}_{r2}\otimes\mathbf{B} & \dots & \mathbf{A}_{rc}\otimes\mathbf{B}\end{pmatrix},$$

that is, $\mathbf{A}\otimes\mathbf{B}$ equals the $mp\times nq$ matrix obtained by replacing each block $\mathbf{A}_{ij}$ of $\mathbf{A}$ with the Kronecker product of $\mathbf{A}_{ij}$ and $\mathbf{B}$.

Solution. For $i=1,\dots,r$, let m_i represent the number of rows in $\mathbf{A}_{i1}$, $\mathbf{A}_{i2}$, $\dots$, $\mathbf{A}_{ic}$; and, for $j=1,\dots,c$, let n_j represent the number of columns in $\mathbf{A}_{1j}$, $\mathbf{A}_{2j}$, $\dots$, $\mathbf{A}_{rj}$. Define $\mathbf{F}=\mathbf{A}\otimes\mathbf{B}$; and partition $\mathbf{F}$ as

$$\mathbf{F}=\begin{pmatrix}\mathbf{F}_{11} & \mathbf{F}_{12} & \dots & \mathbf{F}_{1c}\\ \mathbf{F}_{21} & \mathbf{F}_{22} & \dots & \mathbf{F}_{2c}\\ \vdots & \vdots & & \vdots\\ \mathbf{F}_{r1} & \mathbf{F}_{r2} & \dots & \mathbf{F}_{rc}\end{pmatrix};$$

that is, partition $\mathbf{F}$ into r rows and c columns of blocks, the ijth of which is of dimensions $m_i p \times n_j q$ and is denoted by $\mathbf{F}_{ij}$. Then (for $i = 1, \ldots, r$ and $j = 1, \ldots, c$), $\mathbf{F}_{ij}$ equals a partitioned matrix comprising m_i rows and n_j columns of $p \times q$ dimensional blocks, the uvth of which is

$$a_{m_1+\cdots+m_{i-1}+u,\, n_1+\cdots+n_{j-1}+v}\mathbf{B}.$$

(When $i = 1$ or $j = 1$, interpret the degenerate sum $m_1 + \cdots + m_{i-1}$ or $n_1 + \cdots + n_{j-1}$ as zero.) It follows that $\mathbf{F}_{ij} = \mathbf{A}_{ij} \otimes \mathbf{B}$.

EXERCISE 11. Show that (a) if $\mathbf{T}$ and $\mathbf{U}$ are both upper triangular matrices, then $\mathbf{T} \otimes \mathbf{U}$ is an upper triangular matrix and (b) if $\mathbf{T}$ and $\mathbf{L}$ are both lower triangular matrices, then $\mathbf{T} \otimes \mathbf{L}$ is a lower triangular matrix.

Solution. (a) Suppose that $\mathbf{T} = \{t_{ij}\}$ is an upper triangular matrix of order m and $\mathbf{U} = \{u_{ij}\}$ an upper triangular matrix of order n. Then, $\mathbf{T} \otimes \mathbf{U}$ is a square matrix of order mn, and the element that appears in the $[n(i-1)+r]$th row and $[n(j-1)+s]$th column of $\mathbf{T} \otimes \mathbf{U}$ is $t_{ij}u_{rs}$.

Clearly, $t_{ij}u_{rs} \neq 0$ only if $j \geq i$ and $s \geq r$. Thus, the element that appears in the $[n(i-1)+r]$th row and $[n(j-1)+s]$th column of $\mathbf{T} \otimes \mathbf{U}$ is nonzero only if $n(j-1)+s \geq n(i-1)+r$. It follows that $\mathbf{T} \otimes \mathbf{U}$ is an upper triangular matrix.

(b) Suppose that $\mathbf{T}$ and $\mathbf{L}$ are both lower triangular matrices. Then, $\mathbf{T}'$ and $\mathbf{L}'$ are both upper triangular, and consequently it follows from Part (a) that $\mathbf{T}' \otimes \mathbf{L}'$ is upper triangular. Since [in light of result (1.15)] $\mathbf{T} \otimes \mathbf{L} = (\mathbf{T}' \otimes \mathbf{L}')'$, we conclude that $\mathbf{T} \otimes \mathbf{L}$ is lower triangular.

EXERCISE 12. Let $\mathbf{A}$ represent an $m \times m$ matrix and $\mathbf{B}$ an $n \times n$ matrix. Suppose that $\mathbf{A}$ and $\mathbf{B}$ have LDU decompositions, say $\mathbf{A} = \mathbf{L}_1\mathbf{D}_1\mathbf{U}_1$ and $\mathbf{B} = \mathbf{L}_2\mathbf{D}_2\mathbf{U}_2$. Using the results of Exercise 11, show that $\mathbf{A} \otimes \mathbf{B}$ has the LDU decomposition $\mathbf{A} \otimes \mathbf{B} = \mathbf{LDU}$, where $\mathbf{L} = \mathbf{L}_1 \otimes \mathbf{L}_2$, $\mathbf{D} = \mathbf{D}_1 \otimes \mathbf{D}_2$, and $\mathbf{U} = \mathbf{U}_1 \otimes \mathbf{U}_2$.

Solution. That $\mathbf{A} \otimes \mathbf{B} = \mathbf{LDU}$ is an immediate consequence of result (1.19). Moreover, $\mathbf{D}$ is (by definition) the Kronecker product of two diagonal matrices (namely, $\mathbf{D}_1$ and $\mathbf{D}_2$) and hence is diagonal. And, $\mathbf{U}$ is (by definition) the Kronecker product of two upper triangular matrices (namely, $\mathbf{U}_1$ and $\mathbf{U}_2$) and hence [as a consequence of Part (a) of Exercise 11] is upper triangular. Similarly, $\mathbf{L}$ is (by definition) the Kronecker product of two lower triangular matrices (namely, $\mathbf{L}_1$ and $\mathbf{L}_2$) and hence [as a consequence of Part (b) of Exercise 11] is lower triangular.

It remains to show that the diagonal elements of $\mathbf{L}$ and $\mathbf{U}$ equal one. In this regard, observe that the $[(n-1)i+r]$th diagonal element of $\mathbf{L}$ is the product of the ith diagonal element of $\mathbf{L}_1$ and the rth diagonal element of $\mathbf{L}_2$ and that the $[(n-1)i+r]$th diagonal element of $\mathbf{U}$ is the product of the ith diagonal element of $\mathbf{U}_1$ and the rth diagonal element of $\mathbf{U}_2$. Since $\mathbf{L}_1, \mathbf{L}_2, \mathbf{U}_1$, and $\mathbf{U}_2$ are unit triangular, their diagonal elements equal one. Thus, the diagonal elements of $\mathbf{L}$ and $\mathbf{U}$ equal one.

EXERCISE 13. Let $\mathbf{A}_1, \mathbf{A}_2, \ldots, \mathbf{A}_k$ represent k matrices (of the same dimensions). Show that $\mathbf{A}_1, \mathbf{A}_2, \ldots, \mathbf{A}_k$ are linearly independent if and only if $\text{vec}(\mathbf{A}_1)$, $\text{vec}(\mathbf{A}_2), \ldots, \text{vec}(\mathbf{A}_k)$ are linearly independent.

Solution. It suffices to show that $\mathbf{A}_1, \mathbf{A}_2, \ldots, \mathbf{A}_k$ are linearly dependent if and only if $\text{vec}(\mathbf{A}_1)$, $\text{vec}(\mathbf{A}_2), \ldots, \text{vec}(\mathbf{A}_k)$ are linearly dependent.

Suppose that $\mathbf{A}_1, \mathbf{A}_2, \ldots, \mathbf{A}_k$ are linearly dependent. Then, there exist scalars $c_1, c_2, \ldots, c_k$, not all zero, such that $\sum_{i=1}^k c_i \mathbf{A}_i = \mathbf{0}$. Since [in light of result (2.6)]

$$\sum_{i=1}^k c_i \,\text{vec}(\mathbf{A}_i) = \text{vec}\Big(\sum_{i=1}^k c_i \mathbf{A}_i\Big) = \text{vec}(\mathbf{0}) = \mathbf{0},$$

we conclude that $\text{vec}(\mathbf{A}_1), \text{vec}(\mathbf{A}_2), \ldots, \text{vec}(\mathbf{A}_k)$ are linearly dependent.

Conversely, suppose that $\text{vec}(\mathbf{A}_1), \text{vec}(\mathbf{A}_2), \ldots, \text{vec}(\mathbf{A}_k)$ are linearly dependent. Then, there exist scalars $c_1, c_2, \ldots, c_k$, not all zero, such that

$$\sum_{i=1}^k c_i \,\text{vec}(\mathbf{A}_i) = \mathbf{0},$$

or equivalently [in light of result (2.6)] such that $\text{vec}(\sum_{i=1}^k c_i \mathbf{A}_i) = \mathbf{0}$, and hence such that $\sum_{i=1}^k c_i \mathbf{A}_i = \mathbf{0}$. We conclude that $\mathbf{A}_1, \mathbf{A}_2, \ldots, \mathbf{A}_k$ are linearly dependent.

EXERCISE 14. Let m represent a positive integer, let $\mathbf{e}_i$ represent the ith column of $\mathbf{I}_m$ $(i = 1, \ldots, m)$, and (for $i,\ j = 1, \ldots, m$) let $\mathbf{U}_{ij} = \mathbf{e}_i \mathbf{e}_j'$ (in which case $\mathbf{U}_{ij}$ is an $m \times m$ matrix whose ijth element is 1 and whose remaining $m^2 - 1$ elements are 0).

(a) Show that

$$\text{vec}(\mathbf{I}_m) = \sum_{i=1}^m \mathbf{e}_i \otimes \mathbf{e}_i \,.$$

(b) Show that (for $i, j, r, s = 1, \ldots, m$)

$$\text{vec}(\mathbf{U}_{ri})[\text{vec}(\mathbf{U}_{sj})]' = \mathbf{U}_{ij} \otimes \mathbf{U}_{rs} \,.$$

(c) Show that

$$\sum_{i=1}^m \sum_{j=1}^m \mathbf{U}_{ij} \otimes \mathbf{U}_{ij} = \text{vec}(\mathbf{I}_m)[\text{vec}(\mathbf{I}_m)]'.$$

Solution. (a) Making use of results (2.4.4), (2.6), and (2.3), we find that

$$\text{vec}(\mathbf{I}_m) = \text{vec}\Big(\sum_i \mathbf{e}_i \mathbf{e}_i'\Big) = \sum_i \text{vec}(\mathbf{e}_i \mathbf{e}_i') = \sum_i \mathbf{e}_i \otimes \mathbf{e}_i \,.$$

(b) Making use of results (2.3), (1.15), and (1.19), we find that

$$\text{vec}(\mathbf{U}_{ri})[\text{vec}(\mathbf{U}_{sj})]' = \text{vec}(\mathbf{e}_r \mathbf{e}_i')[\text{vec}(\mathbf{e}_s \mathbf{e}_j')]'$$

$$= (\mathbf{e}_i \otimes \mathbf{e}_r)(\mathbf{e}_j \otimes \mathbf{e}_s)'$$
$$= (\mathbf{e}_i \otimes \mathbf{e}_r)(\mathbf{e}_j' \otimes \mathbf{e}_s')$$
$$= (\mathbf{e}_i\mathbf{e}_j') \otimes \mathbf{e}_r\mathbf{e}_s') = \mathbf{U}_{ij} \otimes \mathbf{U}_{rs}\,.$$

(c) Making use of Part (b) and results (2.6) and (2.4.4), we find that

$$\begin{aligned}\sum_{i,j} \mathbf{U}_{ij} \otimes \mathbf{U}_{ij} &= \sum_i \text{vec}(\mathbf{U}_{ii}) \sum_j [\text{vec}(\mathbf{U}_{jj})]' \\ &= \sum_i \text{vec}(\mathbf{U}_{ii})[\sum_j \text{vec}(\mathbf{U}_{jj})]' \\ &= \text{vec}(\sum_i \mathbf{U}_{ii})[\text{vec}(\sum_j \mathbf{U}_{jj})]' \\ &= \text{vec}(\mathbf{I}_m)[\text{vec}(\mathbf{I}_m)]'.\end{aligned}$$

EXERCISE 15. Let $\mathbf{A}$ represent an $n \times n$ matrix.

(a) Show that if $\mathbf{A}$ is orthogonal, then $(\text{vec}\ \mathbf{A})'\text{vec}\ \mathbf{A} = n$.

(b) Show that if $\mathbf{A}$ is idempotent, then $[\text{vec}(\mathbf{A}')]'\text{vec}\ \mathbf{A} = \text{rank}(\mathbf{A})$.

Solution. (a) If $\mathbf{A}$ is orthogonal, then, making use of result (2.14), we find that

$$(\text{vec}\ \mathbf{A})'\text{vec}\ \mathbf{A} = \text{tr}(\mathbf{A}'\mathbf{A}) = \text{tr}(\mathbf{I}_n) = n.$$

(b) If $\mathbf{A}$ is idempotent, then, making use of result (2.14) and Corollary 10.2.2, we find that

$$[\text{vec}(\mathbf{A}')]'\text{vec}\ \mathbf{A} = \text{tr}(\mathbf{A}\mathbf{A}) = \text{tr}(\mathbf{A}) = \text{rank}(\mathbf{A}).$$

EXERCISE 16. Show that for any $m \times n$ matrix $\mathbf{A}$, $p \times n$ matrix $\mathbf{X}$, $p \times p$ matrix $\mathbf{B}$, and $n \times m$ matrix $\mathbf{C}$,

$$\text{tr}(\mathbf{A}\mathbf{X}'\mathbf{B}\mathbf{X}\mathbf{C}) = (\text{vec}\ \mathbf{X})'[(\mathbf{A}'\mathbf{C}') \otimes \mathbf{B}]\text{vec}\ \mathbf{X} = (\text{vec}\ \mathbf{X})'[(\mathbf{C}\mathbf{A}) \otimes \mathbf{B}']\text{vec}\ \mathbf{X}\,.$$

Solution. Making use of results (5.2.3) and (2.15), we find that

$$\text{tr}(\mathbf{A}\mathbf{X}'\mathbf{B}\mathbf{X}\mathbf{C}) = \text{tr}(\mathbf{X}'\mathbf{B}\mathbf{X}\mathbf{C}\mathbf{A}) = \text{tr}[\mathbf{X}'\mathbf{B}\mathbf{X}(\mathbf{A}'\mathbf{C}')'] = (\text{vec}\ \mathbf{X})'[(\mathbf{A}'\mathbf{C}') \otimes \mathbf{B}]\text{vec}\ \mathbf{X}\,.$$

Further, as a consequence of Lemma 14.1.1 and result (1.15), we have that

$$\begin{aligned}(\text{vec}\ \mathbf{X})'[(\mathbf{A}'\mathbf{C}') \otimes \mathbf{B}]\text{vec}\ \mathbf{X} &= (\text{vec}\ \mathbf{X})'[(\mathbf{A}'\mathbf{C}') \otimes \mathbf{B}]'\text{vec}\ \mathbf{X} \\ &= (\text{vec}\ \mathbf{X})'[(\mathbf{A}'\mathbf{C}')' \otimes \mathbf{B}']\text{vec}\ \mathbf{X} \\ &= (\text{vec}\ \mathbf{X})'[(\mathbf{C}\mathbf{A}) \otimes \mathbf{B}']\text{vec}\ \mathbf{X}\,.\end{aligned}$$

EXERCISE 17. (a) Let $\mathcal{V}$ represent a linear space of $m \times n$ matrices, and let g represent a function that assigns the value $\mathbf{A} * \mathbf{B}$ to each pair of matrices $\mathbf{A}$ and $\mathbf{B}$ in $\mathcal{V}$. Take $\mathcal{U}$ to be the linear space of $mn \times 1$ vectors defined by

$$\mathcal{U} = \{\mathbf{x} \in \mathcal{R}^{mn\times 1} : \mathbf{x} = \text{vec}(\mathbf{A}) \text{ for some } \mathbf{A} \in \mathcal{V}\},$$

and let $\mathbf{x} \bullet \mathbf{y}$ represent the value assigned to each pair of vectors $\mathbf{x}$ and $\mathbf{y}$ in $\mathcal{U}$ by an arbitrary inner product f. Show that g is an inner product (for $\mathcal{V}$) if and only if there exists an f such that (for all $\mathbf{A}$ and $\mathbf{B}$ in $\mathcal{V}$)

$$\mathbf{A} * \mathbf{B} = \operatorname{vec}(\mathbf{A}) \bullet \operatorname{vec}(\mathbf{B}) .$$

(b) Let g represent a function that assigns the value $\mathbf{A} * \mathbf{B}$ to an arbitrary pair of matrices $\mathbf{A}$ and $\mathbf{B}$ in $\mathcal{R}^{m \times n}$. Show that g is an inner product (for $\mathcal{R}^{m \times n}$) if and only if there exists an $mn \times mn$ partitioned symmetric positive definite matrix

$$\mathbf{W} = \begin{pmatrix} \mathbf{W}_{11} & \mathbf{W}_{12} & \dots & \mathbf{W}_{1n} \\ \mathbf{W}_{21} & \mathbf{W}_{22} & \dots & \mathbf{W}_{2n} \\ \vdots & \vdots & \ddots & \vdots \\ \mathbf{W}_{n1} & \mathbf{W}_{n2} & \dots & \mathbf{W}_{nn} \end{pmatrix}$$

(where each submatrix is of dimensions $m \times m$) such that (for all $\mathbf{A}$ and $\mathbf{B}$ in $\mathcal{R}^{m \times n}$)

$$\mathbf{A} * \mathbf{B} = \sum_{i,j} \mathbf{a}_i' \mathbf{W}_{ij} \mathbf{b}_j ,$$

where $\mathbf{a}_1, \mathbf{a}_2, \dots, \mathbf{a}_n$ and $\mathbf{b}_1, \mathbf{b}_2, \dots, \mathbf{b}_n$ represent the first, second, ..., nth columns of $\mathbf{A}$ and $\mathbf{B}$, respectively.

(c) Let g represent a function that assigns the value $\mathbf{x}' * \mathbf{y}'$ to an arbitrary pair of (row) vectors in $\mathcal{R}^{1 \times n}$. Show that g is an inner product (for $\mathcal{R}^{1 \times n}$) if and only if there exists an $n \times n$ symmetric positive definite matrix $\mathbf{W}$ such that (for every pair of n-dimensional row vectors $\mathbf{x}'$ and $\mathbf{y}'$)

$$\mathbf{x}' * \mathbf{y}' = \mathbf{x}' \mathbf{W} \mathbf{y} .$$

Solution. (a) Suppose that, for some f,

$$\mathbf{A} * \mathbf{B} = \operatorname{vec}(\mathbf{A}) \bullet \operatorname{vec}(\mathbf{B})$$

(for all $\mathbf{A}$ and $\mathbf{B}$ in $\mathcal{V}$). Then,

(1) $\mathbf{A} * \mathbf{B} = \operatorname{vec}(\mathbf{A}) \bullet \operatorname{vec}(\mathbf{B}) = \operatorname{vec}(\mathbf{B}) \bullet \operatorname{vec}(\mathbf{A}) = \mathbf{B} * \mathbf{A}$;

(2) $\mathbf{A} * \mathbf{A} = \operatorname{vec}(\mathbf{A}) \bullet \operatorname{vec}(\mathbf{A}) \geq 0$, with equality holding if and only if $\operatorname{vec}(\mathbf{A}) = \mathbf{0}$ or equivalently if and only if $\mathbf{A} = \mathbf{0}$;

(3) $$(k\mathbf{A}) * \mathbf{B} = \operatorname{vec}(k\mathbf{A}) \bullet \operatorname{vec}(\mathbf{B}) = [k \operatorname{vec}(\mathbf{A})] \bullet \operatorname{vec}(\mathbf{B}) = k [\operatorname{vec}(\mathbf{A}) \bullet \operatorname{vec}(\mathbf{B})] = k(\mathbf{A} * \mathbf{B}) ;$$

(4) $$\begin{aligned}(\mathbf{A} + \mathbf{B}) * \mathbf{C} &= \operatorname{vec}(\mathbf{A} + \mathbf{B}) \bullet \operatorname{vec}(\mathbf{C}) \\ &= [\operatorname{vec}(\mathbf{A}) + \operatorname{vec}(\mathbf{B})] \bullet \operatorname{vec}(\mathbf{C}) \\ &= [\operatorname{vec}(\mathbf{A}) \bullet \operatorname{vec}(\mathbf{C})] + [\operatorname{vec}(\mathbf{B}) \bullet \operatorname{vec}(\mathbf{C})] \\ &= (\mathbf{A} * \mathbf{C}) + (\mathbf{B} * \mathbf{C})\end{aligned}$$

(where $\mathbf{A}$, $\mathbf{B}$, and $\mathbf{C}$ represent arbitrary matrices in $\mathcal{V}$ and k represents an arbitrary scalar). Thus, g is an inner product.

Conversely, suppose that g is an inner product, and consider the function $\tilde{f}$ that assigns to each pair of vectors $\mathbf{x}$ and $\mathbf{y}$ in $\mathcal{U}$ the value

$$\mathbf{x} \star \mathbf{y} = \mathbf{X} * \mathbf{Y},$$

where $\mathbf{X}$ and $\mathbf{Y}$ are the (unique) $m \times n$ matrices such that $\mathbf{x} = \text{vec}(\mathbf{X})$ and $\mathbf{y} = \text{vec}(\mathbf{Y})$. Then, letting $\mathbf{x}$, $\mathbf{y}$, and $\mathbf{z}$ represent arbitrary vectors in $\mathcal{U}$, taking $\mathbf{X}$, $\mathbf{Y}$, and $\mathbf{Z}$ to be $m \times n$ matrices such that $\mathbf{x} = \text{vec}(\mathbf{X})$, $\mathbf{y} = \text{vec}(\mathbf{Y})$, and $\mathbf{z} = \text{vec}(\mathbf{Z})$, and denoting by k an arbitrary scalar, we find that

(1) $\mathbf{x} \star \mathbf{y} = \mathbf{X} * \mathbf{Y} = \mathbf{Y} * \mathbf{X} = \mathbf{y} \star \mathbf{x}$;

(2) $\mathbf{x} \star \mathbf{x} = \mathbf{X} * \mathbf{X} \geq 0$, with equality holding if and only if $\mathbf{X} = \mathbf{0}$ or equivalently if and only if $\mathbf{x} = \mathbf{0}$;

(3) $(k\mathbf{x}) \star \mathbf{y} = (k\mathbf{X}) * \mathbf{Y} = k(\mathbf{X} * \mathbf{Y}) = k(\mathbf{x} \star \mathbf{y})$;

(4) $(\mathbf{x} + \mathbf{y}) \star \mathbf{z} = (\mathbf{X} + \mathbf{Y}) * \mathbf{Z} = (\mathbf{X} * \mathbf{Z}) + (\mathbf{Y} * \mathbf{Z}) = (\mathbf{x} \star \mathbf{z}) + (\mathbf{y} \star \mathbf{z})$.

Thus, $\tilde{f}$ is an inner product (for $\mathcal{U}$). Moreover, for $f = \tilde{f}$, we have that

$$\mathbf{A} * \mathbf{B} = \text{vec}(\mathbf{A}) \star \text{vec}(\mathbf{B}) = \text{vec}(\mathbf{A}) \bullet \text{vec}(\mathbf{B})$$

(for all $\mathbf{A}$ and $\mathbf{B}$ in $\mathcal{V}$).

(b) Let f represent an arbitrary inner product for $\mathcal{R}^{mn \times 1}$, and let $\mathbf{x} \bullet \mathbf{y}$ represent the value assigned by f to an arbitrary pair of mn-dimensional column vectors $\mathbf{x}$ and $\mathbf{y}$. According to Part (a), g is an inner product (for $\mathcal{R}^{m \times n}$) if and only if there exists an f such that (for all $\mathbf{A}$ and $\mathbf{B}$ in $\mathcal{V}$)

$$\mathbf{A} * \mathbf{B} = \text{vec}(\mathbf{A}) \bullet \text{vec}(\mathbf{B}).$$

Moreover, according to the discussion of Section 14.10a, every inner product for $\mathcal{R}^{mn \times 1}$ is expressible as a bilinear form, and a bilinear form (in $mn \times 1$ vectors) qualifies as an inner product for $\mathcal{R}^{mn \times 1}$ if and only if the matrix of the bilinear form is symmetric and positive definite. Thus, g is an inner product (for $\mathcal{R}^{m \times n}$) if and only if there exists an $mn \times mn$ symmetric positive definite matrix $\mathbf{W}$ such that (for all $\mathbf{A}$ and $\mathbf{B}$ in $\mathcal{V}$)

$$\mathbf{A} * \mathbf{B} = (\text{vec}\, \mathbf{A})' \mathbf{W} \,\text{vec}\, \mathbf{B}.$$

Further, partitioning $\mathbf{W}$ as

$$\mathbf{W} = \begin{pmatrix} \mathbf{W}_{11} & \mathbf{W}_{12} & \dots & \mathbf{W}_{1n} \\ \mathbf{W}_{21} & \mathbf{W}_{22} & \dots & \mathbf{W}_{2n} \\ \vdots & \vdots & \ddots & \vdots \\ \mathbf{W}_{n1} & \mathbf{W}_{n2} & \dots & \mathbf{W}_{nn} \end{pmatrix},$$

and denoting by $\mathbf{a}_1, \mathbf{a}_2, \ldots, \mathbf{a}_n$ and $\mathbf{b}_1, \mathbf{b}_2, \ldots, \mathbf{b}_n$ the first, second, ..., nth columns of $\mathbf{A}$ and $\mathbf{B}$, respectively, we find that

$$(\text{vec } \mathbf{A})' \mathbf{W} \text{ vec } \mathbf{B} = \sum_{i,j} \mathbf{a}_i' \mathbf{W}_{ij} \mathbf{b}_j .$$

(c) It follows from Part (b) that g is an inner product (for $\mathcal{R}^{1\times n}$) if and only if there exists an $n \times n$ symmetric positive definite matrix $\mathbf{W} = \{w_{ij}\}$ such that (for every pair of n-dimensional row vectors $\mathbf{x}' = \{x_i\}$ and $\mathbf{y}' = \{y_i\}$)

$$\mathbf{x}' * \mathbf{y}' = \sum_{i,j} x_i w_{ij} y_j$$

or equivalently such that (for every $\mathbf{x}'$ and $\mathbf{y}'$)

$$\mathbf{x}' * \mathbf{y}' = \mathbf{x}'\mathbf{W}\mathbf{y}.$$

EXERCISE 18. (a) Define (for $m \geq 2$) $\mathbf{P}$ to be the $mn \times mn$ permutation matrix such that, for every $m \times n$ matrix $\mathbf{A}$,

$$\begin{pmatrix} \text{vec } \mathbf{A}_* \\ \mathbf{r} \end{pmatrix} = \mathbf{P} \text{ vec } \mathbf{A},$$

where $\mathbf{A}_*$ is the $(m-1) \times n$ matrix whose rows are respectively the first, ..., $(m-1)$th rows of $\mathbf{A}$ and $\mathbf{r}'$ is the mth row of $\mathbf{A}$ [and hence where $\mathbf{A} = \begin{pmatrix} \mathbf{A}_* \\ \mathbf{r}' \end{pmatrix}$ and $\mathbf{A}' = (\mathbf{A}_*', \mathbf{r})$] .

(1) Show that $\mathbf{K}_{mn} = \begin{pmatrix} \mathbf{K}_{m-1,n} & \mathbf{0} \\ \mathbf{0} & \mathbf{I}_n \end{pmatrix} \mathbf{P}$.

(2) Show that $|\mathbf{P}| = (-1)^{(m-1)n(n-1)/2}$.

(3) Show that $|\mathbf{K}_{mn}| = (-1)^{(m-1)n(n-1)/2} |\mathbf{K}_{m-1,n}|$.

(b) Show that $|\mathbf{K}_{mn}| = (-1)^{m(m-1)n(n-1)/4}$.

(c) Show that $|\mathbf{K}_{mm}| = (-1)^{m(m-1)/2}$.

Solution. (a) (1) Since $\text{vec } (\mathbf{A}_*') = \mathbf{K}_{m-1,n} \text{vec } \mathbf{A}_*$, we have [in light of the defining relation (3.1)] that

$$\mathbf{K}_{mn} \text{vec } \mathbf{A} = \text{vec } (\mathbf{A}') = \begin{pmatrix} \text{vec } (\mathbf{A}_*') \\ \mathbf{r} \end{pmatrix} = \begin{pmatrix} \mathbf{K}_{m-1,n} & \mathbf{0} \\ \mathbf{0} & \mathbf{I}_n \end{pmatrix} \begin{pmatrix} \text{vec } \mathbf{A}_* \\ \mathbf{r} \end{pmatrix}$$
$$= \begin{pmatrix} \mathbf{K}_{m-1,n} & \mathbf{0} \\ \mathbf{0} & \mathbf{I}_n \end{pmatrix} \mathbf{P} \text{ vec } \mathbf{A}.$$

Thus,

$$\mathbf{K}_{mn}\mathbf{a} = \begin{pmatrix} \mathbf{K}_{m-1,n} & \mathbf{0} \\ \mathbf{0} & \mathbf{I}_n \end{pmatrix} \mathbf{P}\mathbf{a}$$

for every mn-dimensional column vector $\mathbf{a}$, implying (in light of Lemma 2.3.2) that

$$\mathbf{K}_{mn} = \begin{pmatrix} \mathbf{K}_{m-1,n} & \mathbf{0} \\ \mathbf{0} & \mathbf{I}_n \end{pmatrix} \mathbf{P}.$$

(2) The vector $\mathbf{r}$ is the $n \times 1$ vector whose first, second, ..., nth elements are respectively the mth, $(2m)$th, ..., (nm)th elements of vec $\mathbf{A}$, and vec $\mathbf{A}_*$ is the $(m-1)n$-dimensional subvector of vec $\mathbf{A}$ obtained by striking out those n elements. Accordingly, $\mathbf{P} = \begin{pmatrix} \mathbf{P}_1 \\ \mathbf{P}_2 \end{pmatrix}$, where $\mathbf{P}_2$ is the $n \times mn$ matrix whose first, second, ..., nth rows are respectively the mth, $(2m)$th, ..., (nm)th rows of $\mathbf{I}_{mn}$, and $\mathbf{P}_1$ is the $(m-1)n \times mn$ submatrix of $\mathbf{I}_{mn}$ obtained by striking out those n rows. Now, applying Lemma 13.1.3, we find that $|\mathbf{P}| = (-1)^{\phi}$, where

$$\phi = (m-1)1 + (m-1)2 + \cdots + (m-1)(n-1) = (m-1)n(n-1)/2.$$

(3) It follows from Parts (1) and (2) that

$$|\mathbf{K}_{mn}| = |\mathbf{P}| \begin{vmatrix} \mathbf{K}_{m-1,n} & \mathbf{0} \\ \mathbf{0} & \mathbf{I}_n \end{vmatrix} = (-1)^{(m-1)n(n-1)/2} |\mathbf{K}_{m-1,n}|.$$

(b) It follows from Part (a) that, for $i \geq 2$,

$$|\mathbf{K}_{in}| = (-1)^{(i-1)n(n-1)/2} |\mathbf{K}_{i-1,n}|.$$

By applying this equality $m-1$ times (with $i = m, m-1, \ldots, 2$, respectively), we find that, for $m \geq 2$,

$$\begin{aligned} |\mathbf{K}_{mn}| &= (-1)^{(m-1)n(n-1)/2} |\mathbf{K}_{m-1,n}| \\ &= (-1)^{(m-1)n(n-1)/2} (-1)^{(m-2)n(n-1)/2} |\mathbf{K}_{m-2,n}| \\ &= (-1)^{[(m-1)+(m-2)+\cdots+1]n(n-1)/2} |\mathbf{K}_{1n}| \\ &= (-1)^{[m(m-1)/2]n(n-1)/2} |\mathbf{K}_{1n}| \\ &= (-1)^{m(m-1)n(n-1)/4} |\mathbf{K}_{1n}|. \end{aligned}$$

Since $\mathbf{K}_{1n} = \mathbf{I}_n$ [and since $(-1)^0 = 1$], we conclude that (for $m \geq 1$)

$$|\mathbf{K}_{mn}| = (-1)^{m(m-1)n(n-1)/4}.$$

(c) It follows from Part (b) that

$$|\mathbf{K}_{mm}| = (-1)^{[m(m-1)/2]^2}.$$

Since the product of two odd numbers is odd and the product of two even numbers even, we conclude that

$$|\mathbf{K}_{mm}| = (-1)^{m(m-1)/2}.$$

EXERCISE 19. Show that, for any $m \times n$ matrix $\mathbf{A}$, $p \times 1$ vector $\mathbf{a}$, and $q \times 1$ vector $\mathbf{b}$,

(1) $\mathbf{b}' \otimes \mathbf{A} \otimes \mathbf{a} = \mathbf{K}_{mp}[(\mathbf{ab}') \otimes \mathbf{A}]$;

(2) $\mathbf{a} \otimes \mathbf{A} \otimes \mathbf{b}' = \mathbf{K}_{pm}[\mathbf{A} \otimes (\mathbf{ab}')]$.

Solution. Making use of Corollary 16.3.3 and of results (1.16) and (1.4), we find that

$$\begin{aligned}\mathbf{b}' \otimes \mathbf{A} \otimes \mathbf{a} = (\mathbf{b}' \otimes \mathbf{A}) \otimes \mathbf{a} &= \mathbf{K}_{mp}[\mathbf{a} \otimes (\mathbf{b}' \otimes \mathbf{A})]\\ &= \mathbf{K}_{mp}[(\mathbf{a} \otimes \mathbf{b}') \otimes \mathbf{A}] = \mathbf{K}_{mp}[(\mathbf{ab}') \otimes \mathbf{A}]\end{aligned}$$

and similarly that

$$\begin{aligned}\mathbf{a} \otimes \mathbf{A} \otimes \mathbf{b}' = \mathbf{a} \otimes (\mathbf{A} \otimes \mathbf{b}') &= \mathbf{K}_{pm}[(\mathbf{A} \otimes \mathbf{b}') \otimes \mathbf{a}]\\ &= \mathbf{K}_{pm}[\mathbf{A} \otimes (\mathbf{b}' \otimes \mathbf{a})] = \mathbf{K}_{pm}[\mathbf{A} \otimes (\mathbf{ab}')].\end{aligned}$$

EXERCISE 20. Let m and n represent positive integers, and let $\mathbf{e}_i$ represent the ith column of $\mathbf{I}_m$ ($i = 1, \ldots, m$) and $\mathbf{u}_j$ represent the jth column of $\mathbf{I}_n$ ($j = 1, \ldots, n$). Show that

$$\mathbf{K}_{mn} = \sum_{j=1}^{n} \mathbf{u}_j' \otimes \mathbf{I}_m \otimes \mathbf{u}_j = \sum_{i=1}^{m} \mathbf{e}_i \otimes \mathbf{I}_n \otimes \mathbf{e}_i' .$$

Solution. Starting with result (3.3) and using results (1.4) and (2.4.4), we find that

$$\begin{aligned}\mathbf{K}_{mn} &= \sum_{i,j} (\mathbf{e}_i\mathbf{u}_j') \otimes (\mathbf{u}_j\mathbf{e}_i')\\ &= \sum_{i,j} \mathbf{u}_j' \otimes \mathbf{e}_i \otimes \mathbf{e}_i' \otimes \mathbf{u}_j\\ &= \sum_{j} \mathbf{u}_j' \otimes (\sum_{i} \mathbf{e}_i \otimes \mathbf{e}_i') \otimes \mathbf{u}_j\\ &= \sum_{j} \mathbf{u}_j' \otimes (\sum_{i} \mathbf{e}_i\mathbf{e}_i') \otimes \mathbf{u}_j = \sum_{j} \mathbf{u}_j' \otimes \mathbf{I}_m \otimes \mathbf{u}_j .\end{aligned}$$

Similarly,

$$\begin{aligned}\mathbf{K}_{mn} &= \sum_{i,j} (\mathbf{e}_i\mathbf{u}_j') \otimes (\mathbf{u}_j\mathbf{e}_i')\\ &= \sum_{i,j} \mathbf{e}_i \otimes \mathbf{u}_j' \otimes \mathbf{u}_j \otimes \mathbf{e}_i'\\ &= \sum_{i} \mathbf{e}_i \otimes (\sum_{j} \mathbf{u}_j' \otimes \mathbf{u}_j) \otimes \mathbf{e}_i'\\ &= \sum_{i} \mathbf{e}_i \otimes (\sum_{j} \mathbf{u}_j\mathbf{u}_j') \otimes \mathbf{e}_i' = \sum_{i} \mathbf{e}_i \otimes \mathbf{I}_n \otimes \mathbf{e}_i' .\end{aligned}$$

EXERCISE 21. Let m, n, and p represent positive integers. Using the result of Exercise 20, show that

(a) $\mathbf{K}_{mp,n} = \mathbf{K}_{p,mn}\mathbf{K}_{m,np}$;

(b) $\mathbf{K}_{mp,n}\mathbf{K}_{np,m}\mathbf{K}_{mn,p} = \mathbf{I}$;

(c) $\mathbf{K}_{n,mp} = \mathbf{K}_{np,m}\mathbf{K}_{mn,p}$;

(d) $\mathbf{K}_{p,mn}\mathbf{K}_{m,np} = \mathbf{K}_{m,np}\mathbf{K}_{p,mn}$;

(e) $\mathbf{K}_{np,m}\mathbf{K}_{mn,p} = \mathbf{K}_{mn,p}\mathbf{K}_{np,m}$;

(f) $\mathbf{K}_{m,np}\mathbf{K}_{mp,n} = \mathbf{K}_{mp,n}\mathbf{K}_{m,np}$.

[*Hint.* Begin by letting $\mathbf{u}_j$ represent the jth column of $\mathbf{I}_n$ and showing that $\mathbf{K}_{mp,n} = \sum_j (\mathbf{u}'_j \otimes \mathbf{I}_p) \otimes (\mathbf{I}_m \otimes \mathbf{u}_j)$ and then making use of the result that, for any $m \times n$ matrix $\mathbf{A}$ and $p \times q$ matrix $\mathbf{B}$, $\mathbf{B} \otimes \mathbf{A} = \mathbf{K}_{pm}(\mathbf{A} \otimes \mathbf{B})\mathbf{K}_{nq}$.]

Solution. (a) Letting $\mathbf{u}_j$ represent the jth column of $\mathbf{I}_n$ and making use of the result of Exercise 20 and the result cited in the hint [or equivalently result (3.10)], we find [in light of results (1.8), (1.16), (1.4), and (2.4.4)] that

$$\begin{aligned}
\mathbf{K}_{mp,n} &= \sum_j \mathbf{u}'_j \otimes \mathbf{I}_{mp} \otimes \mathbf{u}_j \\
&= \sum_j \mathbf{u}'_j \otimes (\mathbf{I}_p \otimes \mathbf{I}_m) \otimes \mathbf{u}_j \\
&= \sum_j (\mathbf{u}'_j \otimes \mathbf{I}_p) \otimes (\mathbf{I}_m \otimes \mathbf{u}_j) \\
&= \sum_j \mathbf{K}_{p,mn}[(\mathbf{I}_m \otimes \mathbf{u}_j) \otimes (\mathbf{u}'_j \otimes \mathbf{I}_p)]\mathbf{K}_{m,np} \\
&= \sum_j \mathbf{K}_{p,mn}[\mathbf{I}_m \otimes (\mathbf{u}_j\mathbf{u}'_j) \otimes \mathbf{I}_p]\mathbf{K}_{m,np} \\
&= \mathbf{K}_{p,mn}[\mathbf{I}_m \otimes (\sum_j \mathbf{u}_j\mathbf{u}'_j) \otimes \mathbf{I}_p]\mathbf{K}_{m,np} \\
&= \mathbf{K}_{p,mn}(\mathbf{I}_m \otimes \mathbf{I}_n \otimes \mathbf{I}_p)\mathbf{K}_{m,np} \\
&= \mathbf{K}_{p,mn}\mathbf{I}_{mnp}\mathbf{K}_{m,np} = \mathbf{K}_{p,mn}\mathbf{K}_{m,np} .
\end{aligned}$$

(b) Making use of Part (a) and result (3.6), we find that

$$\begin{aligned}
\mathbf{K}_{mp,n}\mathbf{K}_{np,m}\mathbf{K}_{mn,p} &= \mathbf{K}_{p,mn}\mathbf{K}_{m,np}\mathbf{K}_{np,m}\mathbf{K}_{mn,p} \\
&= \mathbf{K}_{p,mn}\,\mathbf{I}\,\mathbf{K}_{mn,p} = \mathbf{K}_{p,mn}\mathbf{K}_{mn,p} = \mathbf{I} .
\end{aligned}$$

(c) Making use of Part (b) and result (3.6), we find that

$$\begin{aligned}
\mathbf{K}_{n,mp} = \mathbf{K}_{n,mp}\mathbf{I}_{mnp} &= \mathbf{K}_{n,mp}\mathbf{K}_{mp,n}\mathbf{K}_{np,m}\mathbf{K}_{mn,p} \\
&= \mathbf{I}\,\mathbf{K}_{np,m}\mathbf{K}_{mn,p} = \mathbf{K}_{np,m}\mathbf{K}_{mn,p} .
\end{aligned}$$

(d) Using Part (a) (twice), we find that

$$\mathbf{K}_{p,mn}\mathbf{K}_{m,np} = \mathbf{K}_{mp,n} = \mathbf{K}_{pm,n} = \mathbf{K}_{m,pn}\mathbf{K}_{p,nm} = \mathbf{K}_{m,np}\mathbf{K}_{p,mn} .$$

(e) Using Part (c) (twice), we find that

$$\mathbf{K}_{np,m}\mathbf{K}_{mn,p} = \mathbf{K}_{n,mp} = \mathbf{K}_{n,pm} = \mathbf{K}_{nm,p}\mathbf{K}_{pn,m} = \mathbf{K}_{mn,p}\mathbf{K}_{np,m}\,.$$

(f) Making use of Parts (c) and (a) and result (3.6), we find that

$$\begin{aligned}\mathbf{K}_{m,np}\mathbf{K}_{mp,n} &= (\mathbf{K}_{mp,n}\mathbf{K}_{nm,p})(\mathbf{K}_{p,mn}\mathbf{K}_{m,np}) \\ &= \mathbf{K}_{mp,n}\mathbf{K}_{mn,p}\mathbf{K}_{p,mn}\mathbf{K}_{m,np} = \mathbf{K}_{mp,n}\mathbf{K}_{m,np}\,.\end{aligned}$$

EXERCISE 22. Let $\mathbf{A}$ represent an $m \times n$ matrix, and define $\mathbf{B} = \mathbf{K}_{mn}(\mathbf{A}' \otimes \mathbf{A})$. Show (a) that $\mathbf{B}$ is symmetric, (b) that $\text{rank}(\mathbf{B}) = [\text{rank}(\mathbf{A})]^2$, (c) that $\mathbf{B}^2 = (\mathbf{AA}') \otimes (\mathbf{A}'\mathbf{A})$, and (d) that $\text{tr}(\mathbf{B}) = \text{tr}(\mathbf{A}'\mathbf{A})$.

Solution. (a) Making use of results (1.15), (3.6), and (3.9), we find that

$$\mathbf{B}' = (\mathbf{A}' \otimes \mathbf{A})'\mathbf{K}'_{mn} = (\mathbf{A} \otimes \mathbf{A}')\mathbf{K}_{nm} = \mathbf{K}_{mn}(\mathbf{A}' \otimes \mathbf{A}) = \mathbf{B}\,.$$

(b) Since $\mathbf{K}_{mn}$ is nonsingular, $\text{rank}(\mathbf{B}) = \text{rank}(\mathbf{A}' \otimes \mathbf{A})$. Moreover, it follows from result (1.26) that $\text{rank}(\mathbf{A}' \otimes \mathbf{A}) = \text{rank}(\mathbf{A}')\,\text{rank}(\mathbf{A})$. Since $\text{rank}(\mathbf{A}') = \text{rank}(\mathbf{A})$, we conclude that $\text{rank}(\mathbf{B}) = [\text{rank}(\mathbf{A})]^2$.

(c) Making use of results (3.10) and (1.19), we find that

$$\mathbf{B}^2 = \mathbf{K}_{mn}(\mathbf{A}' \otimes \mathbf{A})\mathbf{K}_{mn}(\mathbf{A}' \otimes \mathbf{A}) = (\mathbf{A} \otimes \mathbf{A}')(\mathbf{A}' \otimes \mathbf{A}) = (\mathbf{AA}') \otimes (\mathbf{A}'\mathbf{A}).$$

(d) That $\text{tr}(\mathbf{B}) = \text{tr}(\mathbf{A}'\mathbf{A})$ is an immediate consequence of the second equality in result (3.15).

EXERCISE 23. Show that, for any $m \times n$ matrix $\mathbf{A}$ and any $p \times q$ matrix $\mathbf{B}$,

$$\text{vec}(\mathbf{A} \otimes \mathbf{B}) = (\mathbf{I}_n \otimes \mathbf{G})\text{vec}\,\mathbf{A} = (\mathbf{H} \otimes \mathbf{I}_p)\text{vec}\,\mathbf{B}\,,$$

where $\mathbf{G} = (\mathbf{K}_{qm} \otimes \mathbf{I}_p)(\mathbf{I}_m \otimes \text{vec}\,\mathbf{B})$ and $\mathbf{H} = (\mathbf{I}_n \otimes \mathbf{K}_{qm})[\text{vec}(\mathbf{A}) \otimes \mathbf{I}_q]$.

Solution. Making use of results (1.20), (1.1), and (1.8), we find that

$$\begin{aligned}\text{vec}(\mathbf{A}) \otimes \text{vec}(\mathbf{B}) &= (\mathbf{I}_{mn} \otimes \text{vec}\,\mathbf{B})[\text{vec}(\mathbf{A}) \otimes 1] \\ &= (\mathbf{I}_{mn} \otimes \text{vec}\,\mathbf{B})\text{vec}\,\mathbf{A} = (\mathbf{I}_n \otimes \mathbf{I}_m \otimes \text{vec}\,\mathbf{B})\text{vec}\,\mathbf{A}\end{aligned}$$

and similarly that

$$\begin{aligned}\text{vec}(\mathbf{A}) \otimes \text{vec}(\mathbf{B}) &= [\text{vec}(\mathbf{A}) \otimes \mathbf{I}_{pq}](1 \otimes \text{vec}\,\mathbf{B}) \\ &= [\text{vec}(\mathbf{A}) \otimes \mathbf{I}_{pq}]\text{vec}\,\mathbf{B} = [\text{vec}(\mathbf{A}) \otimes \mathbf{I}_q \otimes \mathbf{I}_p]\text{vec}\,\mathbf{B}\,.\end{aligned}$$

Now, substituting these expressions [for $\text{vec}(\mathbf{A}) \otimes \text{vec}(\mathbf{B})$] into formula (3.16) and making use of result (1.19), we obtain

$$\begin{aligned}\text{vec}(\mathbf{A} \otimes \mathbf{B}) &= [\mathbf{I}_n \otimes (\mathbf{K}_{qm} \otimes \mathbf{I}_p)][\mathbf{I}_n \otimes (\mathbf{I}_m \otimes \text{vec}\,\mathbf{B})]\text{vec}\,\mathbf{A} \\ &= (\mathbf{I}_n \otimes \mathbf{G})\text{vec}\,\mathbf{A}\end{aligned}$$

and

$$\begin{aligned}\operatorname{vec}(\mathbf{A} \otimes \mathbf{B}) &= [(\mathbf{I}_n \otimes \mathbf{K}_{qm}) \otimes \mathbf{I}_p]\{[\operatorname{vec}(\mathbf{A}) \otimes \mathbf{I}_q] \otimes \mathbf{I}_p\}\operatorname{vec} \mathbf{B} \\ &= (\mathbf{H} \otimes \mathbf{I}_p)\operatorname{vec} \mathbf{B}.\end{aligned}$$

EXERCISE 24. Show that, for $\mathbf{H}_n = (\mathbf{G}_n'\mathbf{G}_n)^{-1}\mathbf{G}_n'$,

$$\mathbf{G}_n\mathbf{H}_n\mathbf{H}_n' = \mathbf{H}_n'.$$

Solution. For $\mathbf{H}_n = (\mathbf{G}_n'\mathbf{G}_n)^{-1}\mathbf{G}_n'$,

$$\mathbf{G}_n\mathbf{H}_n\mathbf{H}_n' = \mathbf{G}_n(\mathbf{G}_n'\mathbf{G}_n)^{-1}\mathbf{G}_n'\mathbf{G}_n(\mathbf{G}_n'\mathbf{G}_n)^{-1} = \mathbf{G}_n(\mathbf{G}_n'\mathbf{G}_n)^{-1} = \mathbf{H}_n'.$$

EXERCISE 25. There exists a unique matrix $\mathbf{L}_n$ such that

$$\operatorname{vech} \mathbf{A} = \mathbf{L}_n \operatorname{vec} \mathbf{A}$$

for *every* $n \times n$ matrix $\mathbf{A}$ (symmetric or not). [The matrix $\mathbf{L}_n$ is one choice for the matrix $\mathbf{H}_n$, i.e., for a left inverse of $\mathbf{G}_n$. It is referred to by Magnus and Neudecker (1980) as the *elimination matrix*—the effect of premultiplying the vec of an $n \times n$ matrix $\mathbf{A}$ by $\mathbf{L}_n$ is to eliminate (from vec $\mathbf{A}$) the "supradiagonal" elements of $\mathbf{A}$.]

(a) Write out the elements of $\mathbf{L}_1$, $\mathbf{L}_2$, and $\mathbf{L}_3$.

(b) For an arbitrary positive integer n, describe $\mathbf{L}_n$ in terms of its rows.

Solution. (a) $\mathbf{L}_1 = (1)$,

$$\mathbf{L}_2 = \begin{pmatrix} 1 & 0 & 0 & 0 \\ 0 & 1 & 0 & 0 \\ 0 & 0 & 0 & 1 \end{pmatrix}, \quad \text{and} \quad \mathbf{L}_3 = \begin{pmatrix} 1 & 0 & 0 & 0 & 0 & 0 & 0 & 0 & 0 \\ 0 & 1 & 0 & 0 & 0 & 0 & 0 & 0 & 0 \\ 0 & 0 & 1 & 0 & 0 & 0 & 0 & 0 & 0 \\ 0 & 0 & 0 & 0 & 1 & 0 & 0 & 0 & 0 \\ 0 & 0 & 0 & 0 & 0 & 1 & 0 & 0 & 0 \\ 0 & 0 & 0 & 0 & 0 & 0 & 0 & 0 & 1 \end{pmatrix}.$$

(b) For $i \geq j$, the $[(j-1)(n-j/2)+i]$th row of $\mathbf{L}_n$ is the $[(j-1)n+i]$th row of $\mathbf{I}_{n^2}$.

EXERCISE 26. Let $\mathbf{A}$ represent an $n \times n$ matrix and $\mathbf{b}$ an $n \times 1$ vector.

(a) Show that $(1/2)[(\mathbf{A} \otimes \mathbf{b}') + (\mathbf{b}' \otimes \mathbf{A})]\mathbf{G}_n = (\mathbf{A} \otimes \mathbf{b}')\mathbf{G}_n$.

(b) Show that, for $\mathbf{H}_n = (\mathbf{G}_n'\mathbf{G}_n)^{-1}\mathbf{G}_n'$,

(1) $(1/2)\mathbf{H}_n[(\mathbf{b} \otimes \mathbf{A}) + (\mathbf{A} \otimes \mathbf{b})] = \mathbf{H}_n(\mathbf{b} \otimes \mathbf{A})$;

(2) $(\mathbf{A} \otimes \mathbf{b}')\mathbf{G}_n\mathbf{H}_n = (1/2)[(\mathbf{A} \otimes \mathbf{b}') + (\mathbf{b}' \otimes \mathbf{A})]$;

(3) $\mathbf{G}_n\mathbf{H}_n(\mathbf{b} \otimes \mathbf{A}) = (1/2)[(\mathbf{b} \otimes \mathbf{A}) + (\mathbf{A} \otimes \mathbf{b})]$.

Solution. (a) Using results (3.13) and (4.16), we find that

$$(1/2)[(\mathbf{A}\otimes\mathbf{b}') + (\mathbf{b}'\otimes\mathbf{A})]\mathbf{G}_n = (\mathbf{A}\otimes\mathbf{b}')[(1/2)(\mathbf{I}_{n^2}+\mathbf{K}_{nn})]\mathbf{G}_n = (\mathbf{A}\otimes\mathbf{b}')\mathbf{G}_n.$$

(b) Using results (3.12), (4.17), (4.22), and (3.13), we find that, for $\mathbf{H}_n = (\mathbf{G}_n'\mathbf{G}_n)^{-1}\mathbf{G}_n'$,

(1) $(1/2)\mathbf{H}_n[(\mathbf{b}\otimes\mathbf{A})+(\mathbf{A}\otimes\mathbf{b})] = \mathbf{H}_n[(1/2)(\mathbf{I}_{n^2}+\mathbf{K}_{nn})](\mathbf{b}\otimes\mathbf{A}) = \mathbf{H}_n(\mathbf{b}\otimes\mathbf{A})$;

(2) $(\mathbf{A}\otimes\mathbf{b}')\mathbf{G}_n\mathbf{H}_n = (\mathbf{A}\otimes\mathbf{b}')[(1/2)(\mathbf{I}_{n^2}+\mathbf{K}_{nn})] = (1/2)[(\mathbf{A}\otimes\mathbf{b}')+(\mathbf{b}'\otimes\mathbf{A})]$;

(3) $\mathbf{G}_n\mathbf{H}_n(\mathbf{b}\otimes\mathbf{A}) = (1/2)(\mathbf{I}_{n^2}+\mathbf{K}_{nn})(\mathbf{b}\otimes\mathbf{A}) = (1/2)[(\mathbf{b}\otimes\mathbf{A})+(\mathbf{A}\otimes\mathbf{b})]$.

EXERCISE 27. Let $\mathbf{A} = \{a_{ij}\}$ represent an $n\times n$ (possibly nonsymmetric) matrix.

(a) Show that, for $\mathbf{H}_n = (\mathbf{G}_n'\mathbf{G}_n)^{-1}\mathbf{G}_n'$,

$$\mathbf{H}_n \operatorname{vec} \mathbf{A} = (1/2)\operatorname{vech}(\mathbf{A}+\mathbf{A}').$$

(b) Show that

$$\mathbf{G}_n'\mathbf{G}_n \operatorname{vech} \mathbf{A} = \operatorname{vech}[2\mathbf{A} - \operatorname{diag}(a_{11}, a_{22}, \ldots, a_{nn})].$$

(c) Show that

$$\mathbf{G}_n' \operatorname{vec} \mathbf{A} = \operatorname{vech}[\mathbf{A}+\mathbf{A}' - \operatorname{diag}(a_{11}, a_{22}, \ldots, a_{nn})].$$

Solution. (a) Since $(1/2)(\mathbf{A}+\mathbf{A}')$ is an $n\times n$ symmetric matrix, we have [in light of result (4.17)] that, for $\mathbf{H}_n = (\mathbf{G}_n'\mathbf{G}_n)^{-1}\mathbf{G}_n'$,

$$\begin{aligned}
\mathbf{H}_n \operatorname{vec} \mathbf{A} &= \mathbf{H}_n[(1/2)(\mathbf{I}_{n^2}+\mathbf{K}_{nn})]\operatorname{vec} \mathbf{A} \\
&= (1/2)\mathbf{H}_n(\operatorname{vec} \mathbf{A} + \mathbf{K}_{nn}\operatorname{vec} \mathbf{A}) \\
&= (1/2)\mathbf{H}_n(\operatorname{vec} \mathbf{A} + \operatorname{vec} \mathbf{A}') \\
&= (1/2)\mathbf{H}_n \operatorname{vec}(\mathbf{A}+\mathbf{A}') \\
&= (1/2)\mathbf{H}_n\mathbf{G}_n \operatorname{vech}(\mathbf{A}+\mathbf{A}') = (1/2)\operatorname{vech}(\mathbf{A}+\mathbf{A}').
\end{aligned}$$

(b) The matrix $\mathbf{G}_n'\mathbf{G}_n$ is diagonal. Further, the $[(i-1)(n-i/2)+i]$th diagonal element of $\mathbf{G}_n'\mathbf{G}_n$ equals 1, and the $[(i-1)(n-i/2)+i]$th elements of vech $\mathbf{A}$ and vech$[2\mathbf{A}-\operatorname{diag}(a_{11}, a_{22}, \ldots, a_{nn})]$ both equal a_{ii}, so that the $[(i-1)(n-i/2)+i]$th elements of $\mathbf{G}_n'\mathbf{G}_n$ vech $\mathbf{A}$ and vech $[2\mathbf{A} - \operatorname{diag}(a_{11}, a_{22}, \ldots, a_{nn})]$ both equal a_{ii}. And, for $i > j$, the $[(j-1)(n-j/2)+i]$th diagonal element of $\mathbf{G}_n'\mathbf{G}_n$ equals 2, the $[(j-1)(n-j/2)+i]$th element of vech $\mathbf{A}$ equals a_{ij}, and the $[(j-1)(n-j/2)+i]$th element of vech $[2\mathbf{A} - \operatorname{diag}(a_{11}, a_{22}, \ldots, a_{nn})]$ equals $2a_{ij}$, so that (for $i > j$) the $[(j-1)(n-j/2)+i]$th elements of $\mathbf{G}_n'\mathbf{G}_n$ vech $\mathbf{A}$ and vech$[2\mathbf{A} - \operatorname{diag}(a_{11}, a_{22}, \ldots, a_{nn})]$ both equal $2a_{ij}$. We conclude that

$$\mathbf{G}_n'\mathbf{G}_n \operatorname{vech} \mathbf{A} = \operatorname{vech}[2\mathbf{A} - \operatorname{diag}(a_{11}, a_{22}, \ldots, a_{nn})].$$

(c) Using the results of Parts (a) and (b), we find that

$$\begin{aligned}\mathbf{G}_n' \operatorname{vec} \mathbf{A} &= \mathbf{G}_n'\mathbf{G}_n[(\mathbf{G}_n'\mathbf{G}_n)^{-1}\mathbf{G}_n' \operatorname{vec} \mathbf{A}] \\ &= (1/2)\mathbf{G}_n'\mathbf{G}_n \operatorname{vech} (\mathbf{A} + \mathbf{A}') \\ &= (1/2)\operatorname{vech}[2(\mathbf{A} + \mathbf{A}') - \operatorname{diag} (2a_{11}, 2a_{22}, \ldots, 2a_{nn})] \\ &= \operatorname{vech}[\mathbf{A} + \mathbf{A}' - \operatorname{diag} (a_{11}, a_{22}, \ldots, a_{nn})].\end{aligned}$$

EXERCISE 28. Let $\mathbf{A}$ represent a square matrix of order n. Show that, for $\mathbf{H}_n = (\mathbf{G}_n'\mathbf{G}_n)^{-1}\mathbf{G}_n'$,

$$\mathbf{G}_n\mathbf{H}_n(\mathbf{A} \otimes \mathbf{A})\mathbf{H}_n' = (\mathbf{A} \otimes \mathbf{A})\mathbf{H}_n' .$$

Solution. Using result (4.26), we find that, for $\mathbf{H}_n = (\mathbf{G}_n'\mathbf{G}_n)^{-1}\mathbf{G}_n'$,

$$\begin{aligned}\mathbf{G}_n\mathbf{H}_n(\mathbf{A} \otimes \mathbf{A})\mathbf{H}_n' = \mathbf{G}_n\mathbf{H}_n(\mathbf{A} \otimes \mathbf{A})\mathbf{G}_n(\mathbf{G}_n'\mathbf{G}_n)^{-1} &= (\mathbf{A} \otimes \mathbf{A})\mathbf{G}_n(\mathbf{G}_n'\mathbf{G}_n)^{-1} \\ &= (\mathbf{A} \otimes \mathbf{A})\mathbf{H}_n' .\end{aligned}$$

EXERCISE 29. Show that if an $n \times n$ matrix $\mathbf{A} = \{a_{ij}\}$ is upper triangular, lower triangular, or diagonal, then $\mathbf{H}_n(\mathbf{A} \otimes \mathbf{A})\mathbf{G}_n$ is respectively upper triangular, lower triangular, or diagonal with diagonal elements $a_{ii}a_{jj}$ $(i = 1, \ldots, n;\ j = i, \ldots, n)$.

Solution. Let us use mathematical induction to show that for any $n \times n$ upper triangular matrix $\mathbf{A} = \{a_{ij}\}$, $\mathbf{H}_n(\mathbf{A} \otimes \mathbf{A})\mathbf{G}_n$ is upper triangular with diagonal elements $a_{ii}a_{jj}$ $(i = 1, \ldots, n;\ j = i, \ldots, n)$.

For every 1×1 upper triangular matrix $\mathbf{A} = (a_{11})$, $\mathbf{H}_1(\mathbf{A} \otimes \mathbf{A})\mathbf{G}_1$ is the 1×1 matrix (a_{11}^2), which is upper triangular with diagonal element $a_{ii}a_{jj}$ $(i = 1;\ j = 1)$. Suppose now that, for every $n \times n$ upper triangular matrix $\mathbf{A} = \{a_{ij}\}$, $\mathbf{H}_n(\mathbf{A} \otimes \mathbf{A})\mathbf{G}_n$ is upper triangular with diagonal elements $a_{ii}a_{jj}$ $(i = 1, \ldots, n;\ j = i, \ldots, n)$, and let $\mathbf{B} = \{b_{ij}\}$ represent an $(n+1) \times (n+1)$ upper triangular matrix. Then, to complete the induction argument, it suffices to show that $\mathbf{H}_{n+1}(\mathbf{B} \otimes \mathbf{B})\mathbf{G}_{n+1}$ is upper triangular with diagonal elements $b_{ii}b_{jj}$ $(i = 1, \ldots, n+1;\ j = i, \ldots, n+1)$.

For this purpose, partition $\mathbf{B}$ as

$$\mathbf{B} = \begin{pmatrix} c & \mathbf{b}' \\ \mathbf{a} & \mathbf{A} \end{pmatrix}$$

(where $\mathbf{A}$ is $n \times n$ with ijth element $b_{i+1,j+1}$). Then (since $\mathbf{B}$ is upper triangular), $\mathbf{a} = \mathbf{0}$, and it follows from result (4.29) that

$$\mathbf{H}_{n+1}(\mathbf{B} \otimes \mathbf{B})\mathbf{G}_{n+1} = \begin{pmatrix} c^2 & 2c\mathbf{b}' & (\mathbf{b}' \otimes \mathbf{b}')\mathbf{G}_n \\ \mathbf{0} & c\mathbf{A} & (\mathbf{b}' \otimes \mathbf{A})\mathbf{G}_n \\ \mathbf{0} & \mathbf{0} & \mathbf{H}_n(\mathbf{A} \otimes \mathbf{A})\mathbf{G}_n \end{pmatrix}.$$

Moreover, $\mathbf{A}$ is upper triangular, and hence (by supposition) $\mathbf{H}_n(\mathbf{A} \otimes \mathbf{A})\mathbf{G}_n$ is upper triangular with diagonal elements $b_{ii}b_{jj}(i = 2, \ldots, n+1;\ j = i, \ldots, n+1)$.

Thus, $\mathbf{H}_{n+1}(\mathbf{B} \otimes \mathbf{B})\mathbf{G}_{n+1}$ is upper triangular. And, its diagonal elements are $c^2 = b_{11}^2$, $cb_{jj} = b_{11}b_{jj} (j = 2, \ldots, n+1)$, and $b_{ii}b_{jj} (i = 2, \ldots, n+1;\ j = i, \ldots, n+1)$; that is, its diagonal elements are $b_{ii}b_{jj} (i = 1, \ldots, n+1;\ j = i, \ldots, n+1)$.

It can be established via an analogous argument that, for any $n \times n$ lower triangular matrix $\mathbf{A} = \{a_{ij}\}$, $\mathbf{H}_n(\mathbf{A} \otimes \mathbf{A})\mathbf{G}_n$ is lower triangular with diagonal elements $a_{ii}a_{jj}$ $(i = 1, \ldots, n;\ j = i, \ldots, n)$.

Finally, note that if an $n \times n$ matrix $\mathbf{A} = \{a_{ij}\}$ is diagonal, then $\mathbf{A}$ is both upper and lower triangular, in which case $\mathbf{H}_n(\mathbf{A} \otimes \mathbf{A})\mathbf{G}_n$ is both upper and lower triangular, and hence diagonal, with diagonal elements $a_{ii}a_{jj}$ $(i = 1, \ldots, n;\ j = i, \ldots, n)$.

EXERCISE 30. Let $\mathbf{A}_1, \ldots, \mathbf{A}_k$, and $\mathbf{B}$ represent $m \times n$ matrices, and let $\mathbf{b} = \text{vec}\,\mathbf{B}$.

(a) Show that the matrix equation $\sum_{i=1}^k x_i\mathbf{A}_i = \mathbf{B}$ (in unknowns $x_1, \ldots, x_k$) is equivalent to a linear system of the form $\mathbf{Ax} = \mathbf{b}$, where $\mathbf{x} = (x_1, \ldots, x_k)'$ is a vector of unknowns.

(b) Show that if $\mathbf{A}_1, \ldots, \mathbf{A}_k$, and $\mathbf{B}$ are symmetric, then the matrix equation $\sum_{i=1}^k x_i\mathbf{A}_i = \mathbf{B}$ (in unknowns $x_1, \ldots, x_k$) is equivalent to a linear system of the form $\mathbf{A}^*\mathbf{x} = \mathbf{b}^*$, where $\mathbf{b}^* = \text{vech}\,\mathbf{B}$ and $\mathbf{x} = (x_1, \ldots, x_k)'$ is a vector of unknowns.

Solution. Let $\mathbf{A} = (\text{vec}\,\mathbf{A}_1, \ldots, \text{vec}\,\mathbf{A}_k)$.

(a) Making use of result (2.6), we find that

$$\text{vec}(\sum_i x_i\mathbf{A}_i) = \sum_i x_i \,\text{vec}\,\mathbf{A}_i = \mathbf{Ax}.$$

Since clearly the (matrix) equation $\sum_i x_i\mathbf{A}_i = \mathbf{B}$ is equivalent to the (vector) equation $\text{vec}(\sum_i x_i\mathbf{A}_i) = \text{vec}\,\mathbf{B}$, we conclude that the equation $\sum_i x_i\mathbf{A}_i = \mathbf{B}$ is equivalent to the linear system $\mathbf{Ax} = \mathbf{b}$.

(b) Suppose that $\mathbf{A}_1, \ldots, \mathbf{A}_k$, and $\mathbf{B}$ are symmetric (in which case $m = n$). And, let $\mathbf{A}^* = (\text{vech}\,\mathbf{A}_1, \ldots, \text{vech}\,\mathbf{A}_k)$. Then, for any value of $\mathbf{x}$ such that $\mathbf{A}^*\mathbf{x} = \mathbf{b}^*$,

$$\mathbf{Ax} = \mathbf{G}_n\mathbf{A}^*\mathbf{x} = \mathbf{G}_n\mathbf{b}^* = \mathbf{b},$$

and conversely, for any value of $\mathbf{x}$ such that $\mathbf{Ax} = \mathbf{b}$,

$$\mathbf{A}^*\mathbf{x} = \mathbf{H}_n\mathbf{Ax} = \mathbf{H}_n\mathbf{b} = \mathbf{b}^*.$$

We conclude that the linear system $\mathbf{A}^*\mathbf{x} = \mathbf{b}^*$ is equivalent to the linear system $\mathbf{Ax} = \mathbf{b}$ and hence [in light of Part (a)] equivalent to the equation $\sum_i x_i\mathbf{A}_i = \mathbf{B}$.

EXERCISE 31. Let $\mathbf{F}$ represent a $p \times p$ matrix of functions, defined on a set S, of a vector $\mathbf{x} = (x_1, \ldots, x_m)'$ of m variables. Show that, for $k = 2, 3, \ldots,$

$$\frac{\partial\, \text{vec}(\mathbf{F}^k)}{\partial \mathbf{x}'} = \sum_{s=1}^{k} [(\mathbf{F}^{s-1})' \otimes \mathbf{F}^{k-s}] \frac{\partial\, \text{vec}\,\mathbf{F}}{\partial \mathbf{x}'}$$

(where $\mathbf{F}^0 = \mathbf{I}_p$).

Solution. Making use of results (6.1), (15.4.8), and (2.10), we find that

$$\begin{aligned}
\frac{\partial \operatorname{vec}(\mathbf{F}^k)}{\partial x_j} &= \operatorname{vec}\left[\frac{\partial(\mathbf{F}^k)}{\partial x_j}\right] \\
&= \operatorname{vec}\left(\mathbf{F}^{k-1}\frac{\partial \mathbf{F}}{\partial x_j} + \mathbf{F}^{k-2}\frac{\partial \mathbf{F}}{\partial x_j}\mathbf{F} + \cdots + \frac{\partial \mathbf{F}}{\partial x_j}\mathbf{F}^{k-1}\right) \\
&= \operatorname{vec}\left(\sum_{s=1}^{k} \mathbf{F}^{k-s}\frac{\partial \mathbf{F}}{\partial x_j}\mathbf{F}^{s-1}\right) \\
&= \sum_{s=1}^{k} \operatorname{vec}\left(\mathbf{F}^{k-s}\frac{\partial \mathbf{F}}{\partial x_j}\mathbf{F}^{s-1}\right) \\
&= \sum_{s=1}^{k} [(\mathbf{F}^{s-1})' \otimes \mathbf{F}^{k-s}] \operatorname{vec}\left(\frac{\partial \mathbf{F}}{\partial x_j}\right) \\
&= \sum_{s=1}^{k} [(\mathbf{F}^{s-1})' \otimes \mathbf{F}^{k-s}] \frac{\partial \operatorname{vec} \mathbf{F}}{\partial x_j},
\end{aligned}$$

implying that

$$\frac{\partial \operatorname{vec}(\mathbf{F}^k)}{\partial \mathbf{x}'} = \sum_{s=1}^{k} [(\mathbf{F}^{s-1})' \otimes \mathbf{F}^{k-s}] \frac{\partial \operatorname{vec} \mathbf{F}}{\partial \mathbf{x}'}.$$

EXERCISE 32. Let $\mathbf{F} = \{f_{is}\}$ and $\mathbf{G}$ represent $p \times q$ and $r \times s$ matrices of functions, defined on a set S, of a vector $\mathbf{x} = (x_1, \ldots, x_m)'$ of m variables.

(a) Show that (for $j = 1, \ldots, m$)

$$\frac{\partial(\mathbf{F} \otimes \mathbf{G})}{\partial x_j} = \left(\mathbf{F} \otimes \frac{\partial \mathbf{G}}{\partial x_j}\right) + \left(\frac{\partial \mathbf{F}}{\partial x_j} \otimes \mathbf{G}\right).$$

(b) Show that (for $j = 1, \ldots, m$)

$$\frac{\partial \operatorname{vec}(\mathbf{F} \otimes \mathbf{G})}{\partial x_j} = (\mathbf{I}_q \otimes \mathbf{K}_{sp} \otimes \mathbf{I}_r)\left[(\operatorname{vec} \mathbf{F}) \otimes \frac{\partial \operatorname{vec} \mathbf{G}}{\partial x_j} + \frac{\partial \operatorname{vec} \mathbf{F}}{\partial x_j} \otimes (\operatorname{vec} \mathbf{G})\right].$$

(c) Show that

$$\frac{\partial \operatorname{vec}(\mathbf{F} \otimes \mathbf{G})}{\partial \mathbf{x}'} = (\mathbf{I}_q \otimes \mathbf{K}_{sp} \otimes \mathbf{I}_r)\left[(\operatorname{vec} \mathbf{F}) \otimes \frac{\partial \operatorname{vec} \mathbf{G}}{\partial \mathbf{x}'} + \frac{\partial \operatorname{vec} \mathbf{F}}{\partial \mathbf{x}'} \otimes (\operatorname{vec} \mathbf{G})\right].$$

(d) Show that, in the special case where $\mathbf{x}' = [(\operatorname{vec} \mathbf{X})',\ (\operatorname{vec} \mathbf{Y})']$, $\mathbf{F}(\mathbf{x}) = \mathbf{X}$, and $\mathbf{G}(\mathbf{x}) = \mathbf{Y}$ for some $p \times q$ and $r \times s$ matrices $\mathbf{X}$ and $\mathbf{Y}$ of variables, the formula in Part (c) simplifies to

$$\frac{\partial \operatorname{vec}(\mathbf{X} \otimes \mathbf{Y})}{\partial \mathbf{x}'} = (\mathbf{I}_q \otimes \mathbf{K}_{sp} \otimes \mathbf{I}_r)[\mathbf{I}_{pq} \otimes (\operatorname{vec} \mathbf{Y}),\ \ (\operatorname{vec} \mathbf{X}) \otimes \mathbf{I}_{rs}].$$

Solution. (a) Partition each of the three matrices $\partial(\mathbf{F}\otimes\mathbf{G})/\partial x_j$, $\mathbf{F}\otimes(\partial\mathbf{G}/\partial x_j)$, and $(\partial\mathbf{F}/\partial x_j)\otimes\mathbf{G}$ into p rows and q columns of $r \times s$ dimensional blocks. Then, for $i = 1, \ldots, p$ and $s = 1, \ldots, q$, the isth blocks of $\partial(\mathbf{F}\otimes\mathbf{G})/\partial x_j$, $\mathbf{F}\otimes(\partial\mathbf{G}/\partial x_j)$, and $(\partial\mathbf{F}/\partial x_j)\otimes\mathbf{G}$ are respectively $\partial(f_{is}\mathbf{G})/\partial x_j$, $f_{is}(\partial\mathbf{G}/\partial x_j)$, and $(\partial f_{is}/\partial x_j)\mathbf{G}$, implying [in light of result (15.4.9)] that the isth block of $\partial(\mathbf{F}\otimes\mathbf{G})/\partial x_j$ equals the sum of the isth blocks of $\mathbf{F}\otimes(\partial\mathbf{G}/\partial x_j)$ and $(\partial\mathbf{F}/\partial x_j)\otimes\mathbf{G}$. We conclude that

$$\frac{\partial(\mathbf{F}\otimes\mathbf{G})}{\partial x_j} = \left(\mathbf{F}\otimes\frac{\partial\mathbf{G}}{\partial x_j}\right) + \left(\frac{\partial\mathbf{F}}{\partial x_j}\otimes\mathbf{G}\right).$$

(b) Making use of Part (a) and Theorem 16.3.5, we find that

$$\begin{aligned}
\frac{\partial\,\text{vec}\,(\mathbf{F}\otimes\mathbf{G})}{\partial x_j}
&= \text{vec}\left[\frac{\partial(\mathbf{F}\otimes\mathbf{G})}{\partial x_j}\right]\\
&= \text{vec}\left(\mathbf{F}\otimes\frac{\partial\mathbf{G}}{\partial x_j}\right) + \text{vec}\left(\frac{\partial\mathbf{F}}{\partial x_j}\otimes\mathbf{G}\right)\\
&= (\mathbf{I}_q\otimes\mathbf{K}_{sp}\otimes\mathbf{I}_r)\left[(\text{vec}\,\mathbf{F})\otimes\text{vec}\left(\frac{\partial\mathbf{G}}{\partial x_j}\right) + \text{vec}\left(\frac{\partial\mathbf{F}}{\partial x_j}\right)\otimes(\text{vec}\,\mathbf{G})\right]\\
&= (\mathbf{I}_q\otimes\mathbf{K}_{sp}\otimes\mathbf{I}_r)\left[(\text{vec}\,\mathbf{F})\otimes\frac{\partial\,\text{vec}\,\mathbf{G}}{\partial x_j} + \frac{\partial\,\text{vec}\,\mathbf{F}}{\partial x_j}\otimes(\text{vec}\,\mathbf{G})\right].
\end{aligned}$$

(c) In light of result (1.28), $(\text{vec}\,\mathbf{F})\otimes[\partial(\text{vec}\,\mathbf{G})/\partial x_j]$ is the jth column of $(\text{vec}\,\mathbf{F})\otimes[\partial(\text{vec}\,\mathbf{G})/\partial\mathbf{x}']$. And, in light of result (1.27), $[\partial(\text{vec}\,\mathbf{F})/\partial x_j]\otimes(\text{vec}\,\mathbf{G})$ is the jth column of $[\partial(\text{vec}\,\mathbf{F})/\partial\mathbf{x}']\otimes(\text{vec}\,\mathbf{G})$. Thus, it follows from Part (b) that

$$\frac{\partial\,\text{vec}\,(\mathbf{F}\otimes\mathbf{G})}{\partial\mathbf{x}'} = (\mathbf{I}_q\otimes\mathbf{K}_{sp}\otimes\mathbf{I}_r)\left[(\text{vec}\,\mathbf{F})\otimes\frac{\partial\,\text{vec}\,\mathbf{G}}{\partial\mathbf{x}'} + \frac{\partial\,\text{vec}\,\mathbf{F}}{\partial\mathbf{x}'}\otimes(\text{vec}\,\mathbf{G})\right].$$

(d) In this special case,

$$\frac{\partial\,\text{vec}\,\mathbf{G}}{\partial\mathbf{x}'} = (\mathbf{0},\ \mathbf{I}_{rs}), \qquad \frac{\partial\,\text{vec}\,\mathbf{F}}{\partial\mathbf{x}'} = (\mathbf{I}_{pq},\ \mathbf{0}),$$

implying [in light of results (1.28) and (1.27)] that

$$(\text{vec}\,\mathbf{F})\otimes\frac{\partial\,\text{vec}\,\mathbf{G}}{\partial\mathbf{x}'} = [\mathbf{0},\ (\text{vec}\,\mathbf{F})\otimes\mathbf{I}_{rs}]$$

and that

$$\frac{\partial\,\text{vec}\,\mathbf{F}}{\partial\mathbf{x}'}\otimes(\text{vec}\,\mathbf{G}) = [\mathbf{I}_{pq}\otimes(\text{vec}\,\mathbf{G}),\ \mathbf{0}],$$

so that it follows from Part (c) that

$$\frac{\partial\,\text{vec}\,(\mathbf{F}\otimes\mathbf{G})}{\partial\mathbf{x}'} = (\mathbf{I}_q\otimes\mathbf{K}_{sp}\otimes\mathbf{I}_r)[\mathbf{I}_{pq}\otimes(\text{vec}\,\mathbf{G}),\ (\text{vec}\,\mathbf{F})\otimes\mathbf{I}_{rs}].$$

[illegible] of the [illegible] blocks of [illegible] matrices [illegible] and [illegible] rows [illegible] columns of [illegible] dimension [illegible]. Then, for [illegible] blocks [illegible] and [illegible] [illegible] the [illegible] block [illegible] is equal to the sum of [illegible] with the [illegible] and [illegible]. We conclude that

[illegible]

[illegible] and Theorem [illegible] we find that

[illegible]

(b) In light of result [illegible] and [illegible] the [illegible] column of [illegible] is the [illegible] column of [illegible]. Thus, [illegible]

[illegible]

(b) [illegible]

[illegible]

[illegible] results [illegible] and [illegible] that

[illegible]

and that

[illegible]

so that it follows [illegible] that

[illegible]

17

Intersections and Sums of Subspaces

EXERCISE 1. Let $\mathcal{U}$ and $\mathcal{V}$ represent subspaces of $\mathcal{R}^{m\times n}$.

(a) Show that $\mathcal{U} \cup \mathcal{V} \subset \mathcal{U} + \mathcal{V}$.

(b) Show that $\mathcal{U} + \mathcal{V}$ is the smallest subspace (of $\mathcal{R}^{m\times n}$) that contains $\mathcal{U} \cup \mathcal{V}$, or, equivalently [in light of Part (a)], show that, for any subspace $\mathcal{W}$ such that $\mathcal{U} \cup \mathcal{V} \subset \mathcal{W}$, $\mathcal{U} + \mathcal{V} \subset \mathcal{W}$.

Solution. (a) Let $\mathbf{A}$ represent an arbitrary $m \times n$ matrix in $\mathcal{U} \cup \mathcal{V}$, so that (by definition) $\mathbf{A} \in \mathcal{U}$ or $\mathbf{A} \in \mathcal{V}$. Upon observing that $\mathbf{A} = \mathbf{A} + \mathbf{0} = \mathbf{0} + \mathbf{A}$ and that the $m \times n$ null matrix $\mathbf{0}$ is a member of $\mathcal{V}$ and also of $\mathcal{U}$, we conclude that $\mathbf{A} \in \mathcal{U} + \mathcal{V}$.

(b) Let $\mathbf{A}$ represent an arbitrary matrix in $\mathcal{U} + \mathcal{V}$. Then, $\mathbf{A} = \mathbf{U} + \mathbf{V}$ for some matrix $\mathbf{U} \in \mathcal{U}$ and some matrix $\mathbf{V} \in \mathcal{V}$. Moreover, both $\mathbf{U}$ and $\mathbf{V}$ are in $\mathcal{U} \cup \mathcal{V}$, and hence both are in $\mathcal{W}$ (where $\mathcal{W}$ is an arbitrary subspace such that $\mathcal{U} \cup \mathcal{V} \subset \mathcal{W}$). Since $\mathcal{W}$ is a linear space, it follows that $\mathbf{A}$ $(= \mathbf{U} + \mathbf{V})$ is in $\mathcal{W}$.

EXERCISE 2. Let $\mathbf{A} = \begin{pmatrix} 1 & 0 \\ 0 & 1 \\ 0 & 0 \end{pmatrix}$ and $\mathbf{B} = \begin{pmatrix} 0 & 2 \\ 1 & 1 \\ 2 & 3 \end{pmatrix}$. Find (a) a basis for $\mathcal{C}(\mathbf{A}) + \mathcal{C}(\mathbf{B})$, (b) a basis for $\mathcal{C}(\mathbf{A}) \cap \mathcal{C}(\mathbf{B})$, and (c) a vector in $\mathcal{C}(\mathbf{A}) + \mathcal{C}(\mathbf{B})$ that is not in $\mathcal{C}(\mathbf{A}) \cup \mathcal{C}(\mathbf{B})$.

Solution. (a) According to result (1.4), $\mathcal{C}(\mathbf{A}) + \mathcal{C}(\mathbf{B}) = \mathcal{C}(\mathbf{A}, \mathbf{B})$. Since the partitioned matrix $(\mathbf{A}, \mathbf{B})$ has only 3 rows, its rank cannot exceed 3. Further, the first 3 columns of $(\mathbf{A}, \mathbf{B})$ are linearly independent. We conclude that $\text{rank}(\mathbf{A}, \mathbf{B}) = 3$ and that the first 3 columns of $(\mathbf{A}, \mathbf{B})$, namely, $(1, 0, 0)'$, $(0, 1, 0)'$, and $(0, 1, 2)'$, form a basis for $\mathcal{C}(\mathbf{A}, \mathbf{B})$ and hence for $\mathcal{C}(\mathbf{A}) + \mathcal{C}(\mathbf{B})$.

(b) The column space $\mathcal{C}(\mathbf{A})$ of $\mathbf{A}$ comprises vectors of the form $(x_1, x_2, 0)'$ (where x_1 and x_2 are arbitrary scalars), and $\mathcal{C}(\mathbf{B})$ comprises vectors of the form $(2y_2, y_1+y_2, 2y_1+3y_2)'$ (where y_1 and y_2 are arbitrary scalars). Thus, $\mathcal{C}(\mathbf{A}) \cap \mathcal{C}(\mathbf{B})$ comprises those vectors that are expressible as $(2y_2, y_1+y_2, 2y_1+3y_2)'$ for some scalars y_1 and y_2 such that $2y_1 + 3y_2 = 0$ or equivalently (since $2y_1 + 3y_2 = 0 \Leftrightarrow y_2 = -2y_1/3$) of those vectors that are expressible as $(-4y_1/3, y_1/3, 0)'$ [$= y_1(-4/3, 1/3, 0)'$] for some scalar y_1. We conclude that $\mathcal{C}(\mathbf{A}) \cap \mathcal{C}(\mathbf{B})$ is of dimension one and that the set whose only member is $(-4, 1, 0)'$ (obtained by setting $y_1 = 3$) is a basis for $\mathcal{C}(\mathbf{A}) \cap \mathcal{C}(\mathbf{B})$.

(c) In light of the solution to Part (a), it suffices to find any 3-dimensional column vector that is not contained in $\mathcal{C}(\mathbf{A})$ or $\mathcal{C}(\mathbf{B})$. It follows from the solution to Part (b) that the vector $(2y_2, z, 2y_1 + 3y_2)'$, where y_1, y_2, and z are any scalars such that $2y_1+3y_2 \neq 0$ and $z \neq y_1+y_2$, is not contained in $\mathcal{C}(\mathbf{A})$ or $\mathcal{C}(\mathbf{B})$. For example, the vector $(0, 2, 2)'$ (obtained by taking $y_1 = 1$, $y_2 = 0$, and $z = 2$) is not contained in $\mathcal{C}(\mathbf{A})$ or $\mathcal{C}(\mathbf{B})$.

EXERCISE 3. Let $\mathcal{U}$, $\mathcal{W}$, and $\mathcal{X}$ represent subspaces of a linear space $\mathcal{V}$ of matrices, and let $\mathbf{Y}$ represent an arbitrary matrix in $\mathcal{V}$.

(a) Show (1) that if $\mathbf{Y} \perp \mathcal{W}$ and $\mathbf{Y} \perp \mathcal{X}$, then $\mathbf{Y} \perp (\mathcal{W} + \mathcal{X})$, and (2) that if $\mathcal{U} \perp \mathcal{W}$ and $\mathcal{U} \perp \mathcal{X}$, then $\mathcal{U} \perp (\mathcal{W} + \mathcal{X})$.

(b) Show (1) that $(\mathcal{U}+\mathcal{W})^{\perp} = \mathcal{U}^{\perp} \cap \mathcal{W}^{\perp}$ and (2) that $(\mathcal{U} \cap \mathcal{W})^{\perp} = \mathcal{U}^{\perp} + \mathcal{W}^{\perp}$.

Solution. (a) (1) Suppose that $\mathbf{Y} \perp \mathcal{W}$ and $\mathbf{Y} \perp \mathcal{X}$. Let $\mathbf{Z}$ represent an arbitrary matrix in $\mathcal{W} + \mathcal{X}$. Then, there exists a matrix $\mathbf{W}$ in $\mathcal{W}$ and a matrix $\mathbf{X}$ in $\mathcal{X}$ such that $\mathbf{Z} = \mathbf{W} + \mathbf{X}$. Moreover, $\mathbf{Y} \perp \mathbf{W}$ and $\mathbf{Y} \perp \mathbf{X}$, implying that $\mathbf{Y} \perp \mathbf{Z}$. We conclude that $\mathbf{Y} \perp (\mathcal{W} + \mathcal{X})$.

(2) Suppose that $\mathcal{U} \perp \mathcal{W}$ and $\mathcal{U} \perp \mathcal{X}$. Let $\mathbf{U}$ represent an arbitrary matrix in $\mathcal{U}$. Then, $\mathbf{U} \perp \mathcal{W}$ and $\mathbf{U} \perp \mathcal{X}$, implying [in light of Part (1)] that $\mathbf{U} \perp (\mathcal{W} + \mathcal{X})$. We conclude that $\mathcal{U} \perp (\mathcal{W} + \mathcal{X})$.

(b) (1) Observing that $\mathcal{U} \subset (\mathcal{U} + \mathcal{W})$ and $\mathcal{W} \subset (\mathcal{U} + \mathcal{W})$ and making use of Part (a)-(1), we find that

$$\begin{aligned} \mathbf{Y} \in (\mathcal{U}+\mathcal{W})^{\perp} &\Leftrightarrow \mathbf{Y} \perp (\mathcal{U}+\mathcal{W}) \\ &\Leftrightarrow \mathbf{Y} \perp \mathcal{U} \text{ and } \mathbf{Y} \perp \mathcal{W} \\ &\Leftrightarrow \mathbf{Y} \in \mathcal{U}^{\perp} \text{ and } \mathbf{Y} \in \mathcal{W}^{\perp} \\ &\Leftrightarrow \mathbf{Y} \in (\mathcal{U}^{\perp} \cap \mathcal{W}^{\perp}). \end{aligned}$$

We conclude that $(\mathcal{U} + \mathcal{W})^{\perp} = \mathcal{U}^{\perp} \cap \mathcal{W}^{\perp}$.

(2) Making use of Part (1) and Theorem 12.5.4, we find that

$$\mathcal{U}^{\perp} + \mathcal{W}^{\perp} = [(\mathcal{U}^{\perp} + \mathcal{W}^{\perp})^{\perp}]^{\perp} = [(\mathcal{U}^{\perp})^{\perp} \cap (\mathcal{W}^{\perp})^{\perp}]^{\perp} = (\mathcal{U} \cap \mathcal{W})^{\perp}.$$

EXERCISE 4. Let $\mathcal{U}$, $\mathcal{W}$, and $\mathcal{X}$ represent subspaces of a linear space $\mathcal{V}$ of matrices.

(a) Show that $(\mathcal{U} \cap \mathcal{W}) + (\mathcal{U} \cap \mathcal{X}) \subset \mathcal{U} \cap (\mathcal{W} + \mathcal{X})$.

(b) Show (via an example) that $\mathcal{U} \cap \mathcal{W} = \{\mathbf{0}\}$ and $\mathcal{U} \cap \mathcal{X} = \{\mathbf{0}\}$ does not necessarily imply that $\mathcal{U} \cap (\mathcal{W} + \mathcal{X}) = \{\mathbf{0}\}$.

(c) Show that if $\mathcal{W} \subset \mathcal{U}$, then (1) $\mathcal{U} + \mathcal{W} = \mathcal{U}$ and (2) $\mathcal{U} \cap (\mathcal{W} + \mathcal{X}) = \mathcal{W} + (\mathcal{U} \cap \mathcal{X})$.

Solution. (a) Let $\mathbf{Y}$ represent an arbitrary matrix in $(\mathcal{U} \cap \mathcal{W}) + (\mathcal{U} \cap \mathcal{X})$. Then, $\mathbf{Y} = \mathbf{W} + \mathbf{X}$ for some matrix $\mathbf{W}$ in $\mathcal{U} \cap \mathcal{W}$ and some matrix $\mathbf{X}$ in $\mathcal{U} \cap \mathcal{X}$. Since both $\mathbf{W}$ and $\mathbf{X}$ are in $\mathcal{U}$, $\mathbf{Y}$ is in $\mathcal{U}$, and since $\mathbf{W}$ is in $\mathcal{W}$ and $\mathbf{X}$ in $\mathcal{X}$, $\mathbf{Y}$ is in $\mathcal{W} + \mathcal{X}$. Thus, $\mathbf{Y}$ is in $\mathcal{U} \cap (\mathcal{W} + \mathcal{X})$. We conclude that $(\mathcal{U} \cap \mathcal{W}) + (\mathcal{U} \cap \mathcal{X}) \subset \mathcal{U} \cap (\mathcal{W} + \mathcal{X})$.

(b) Suppose that $\mathcal{V} = \mathcal{R}^{1\times 2}$ and that $\mathcal{U}$, $\mathcal{W}$, and $\mathcal{X}$ are the one-dimensional subspaces spanned by $(1, 1)$, $(1, 0)$, and $(0, 1)$, respectively. Then, clearly, $\mathcal{U} \cap \mathcal{W} = \{\mathbf{0}\}$ and $\mathcal{U} \cap \mathcal{X} = \{\mathbf{0}\}$. However, $\mathcal{W} + \mathcal{X} = \mathcal{R}^{1\times 2}$, and consequently $\mathcal{U} \cap (\mathcal{W} + \mathcal{X}) = \mathcal{U} \neq \{\mathbf{0}\}$.

(c) Suppose that $\mathcal{W} \subset \mathcal{U}$.

(1) Since clearly $\mathcal{U} \subset \mathcal{U} + \mathcal{W}$, it suffices to show that $\mathcal{U} + \mathcal{W} \subset \mathcal{U}$. Let $\mathbf{Y}$ represent an arbitrary matrix in $\mathcal{U} + \mathcal{W}$. Then, $\mathbf{Y} = \mathbf{U} + \mathbf{W}$ for some matrix $\mathbf{U}$ in $\mathcal{U}$ and some matrix $\mathbf{W}$ in $\mathcal{W}$. Moreover, $\mathbf{W}$ is in $\mathcal{U}$ (since $\mathcal{W} \subset \mathcal{U}$), and consequently $\mathbf{Y}$ is in $\mathcal{U}$. We conclude that $\mathcal{U} + \mathcal{W} \subset \mathcal{U}$.

(2) It follows from Part (a) (and the supposition that $\mathcal{W} \subset \mathcal{U}$) that $\mathcal{W} + (\mathcal{U} \cap \mathcal{X}) \subset \mathcal{U} \cap (\mathcal{W} + \mathcal{X})$. Thus, it suffices to show that $\mathcal{U} \cap (\mathcal{W} + \mathcal{X}) \subset \mathcal{W} + (\mathcal{U} \cap \mathcal{X})$.

Let $\mathbf{Y}$ represent an arbitrary matrix in $\mathcal{U} \cap (\mathcal{W} + \mathcal{X})$. Then, $\mathbf{Y} \in \mathcal{W} + \mathcal{X}$, so that $\mathbf{Y} = \mathbf{W} + \mathbf{X}$ for some matrix $\mathbf{W}$ in $\mathcal{W}$ and some matrix $\mathbf{X}$ in $\mathcal{X}$, and also $\mathbf{Y} \in \mathcal{U}$. Thus, $\mathbf{X} = \mathbf{Y} - \mathbf{W}$, and, since (in light of the supposition that $\mathcal{W} \subset \mathcal{U}$) $\mathbf{W}$ (like $\mathbf{Y}$) is in $\mathcal{U}$, $\mathbf{X}$ is in $\mathcal{U}$ (as well as in $\mathcal{X}$) and hence is in $\mathcal{U} \cap \mathcal{X}$. It follows that $\mathbf{Y}$ is in $\mathcal{W} + (\mathcal{U} \cap \mathcal{X})$. We conclude that $\mathcal{U} \cap (\mathcal{W} + \mathcal{X}) \subset \mathcal{W} + (\mathcal{U} \cap \mathcal{X})$.

EXERCISE 5. Let $\mathcal{U}_1, \mathcal{U}_2, \ldots, \mathcal{U}_k$ represent subspaces of $\mathcal{R}^{m\times n}$. Show that if, for $j = 1, 2, \ldots, k$, $\mathcal{U}_j$ is spanned by a (finite nonempty) set of $(m \times n)$ matrices $\mathbf{U}_1^{(j)}, \ldots, \mathbf{U}_{r_j}^{(j)}$, then

$$\mathcal{U}_1 + \mathcal{U}_2 + \cdots + \mathcal{U}_k = \text{sp}(\mathbf{U}_1^{(1)}, \ldots, \mathbf{U}_{r_1}^{(1)}, \mathbf{U}_1^{(2)}, \ldots, \mathbf{U}_{r_2}^{(2)}, \ldots, \mathbf{U}_1^{(k)}, \ldots, \mathbf{U}_{r_k}^{(k)}).$$

Solution. Suppose that, for $j = 1, 2, \ldots, k$, $\mathcal{U}_j$ is spanned by the set $\{\mathbf{U}_1^{(j)}, \ldots, \mathbf{U}_{r_j}^{(j)}\}$. The proof that $\mathcal{U}_1 + \mathcal{U}_2 + \cdots + \mathcal{U}_k$ is spanned by $\mathbf{U}_1^{(1)}, \ldots, \mathbf{U}_{r_1}^{(1)}, \mathbf{U}_1^{(2)}, \ldots, \mathbf{U}_{r_2}^{(2)}, \ldots, \mathbf{U}_1^{(k)}, \ldots, \mathbf{U}_{r_k}^{(k)}$ is by mathematical induction.

It follows from Lemma 17.1.1 that

$$\mathcal{U}_1 + \mathcal{U}_2 = \text{sp}(\mathbf{U}_1^{(1)}, \ldots, \mathbf{U}_{r_1}^{(1)}, \mathbf{U}_1^{(2)}, \ldots, \mathbf{U}_{r_2}^{(2)}).$$

Now, suppose that (for an arbitrary integer j between 2 and $k-1$, inclusive)

$$\mathcal{U}_1 + \mathcal{U}_2 + \cdots + \mathcal{U}_j = \text{sp}(\mathbf{U}_1^{(1)}, \ldots, \mathbf{U}_{r_1}^{(1)}, \mathbf{U}_1^{(2)}, \ldots, \mathbf{U}_{r_2}^{(2)}, \ldots, \mathbf{U}_1^{(j)}, \ldots, \mathbf{U}_{r_j}^{(j)}).$$

Then, the proof is complete upon observing (in light of Lemma 17.1.1) that

$$\begin{aligned}\mathcal{U}_1 &+ \mathcal{U}_2 + \cdots + \mathcal{U}_{j+1} \\ &= (\mathcal{U}_1 + \mathcal{U}_2 + \cdots + \mathcal{U}_j) + \mathcal{U}_{j+1} \\ &= \text{sp}(\mathbf{U}_1^{(1)}, \ldots, \mathbf{U}_{r_1}^{(1)}, \mathbf{U}_1^{(2)}, \ldots, \mathbf{U}_{r_2}^{(2)}, \ldots, \mathbf{U}_1^{(j+1)}, \ldots, \mathbf{U}_{r_{j+1}}^{(j+1)}).\end{aligned}$$

EXERCISE 6. Let $\mathcal{U}_1, \ldots, \mathcal{U}_k$ represent subspaces of $\mathcal{R}^{m \times n}$. The k subspaces $\mathcal{U}_1, \ldots, \mathcal{U}_k$ are said to be *independent* if, for matrices $\mathbf{U}_1 \in \mathcal{U}_1, \ldots, \mathbf{U}_k \in \mathcal{U}_k$, the only solution to the matrix equation

$$\mathbf{U}_1 + \cdots + \mathbf{U}_k = \mathbf{0} \tag{E.1}$$

is $\mathbf{U}_1 = \cdots = \mathbf{U}_k = \mathbf{0}$.

(a) Show that $\mathcal{U}_1, \ldots, \mathcal{U}_k$ are independent if and only if, for $i = 2, \ldots, k$, $\mathcal{U}_i$ and $\mathcal{U}_1 + \cdots + \mathcal{U}_{i-1}$ are essentially disjoint.

(b) Show that $\mathcal{U}_1, \ldots, \mathcal{U}_k$ are independent if and only if, for $i = 1, \ldots, k$, $\mathcal{U}_i$ and $\mathcal{U}_1 + \cdots + \mathcal{U}_{i-1} + \mathcal{U}_{i+1} + \cdots + \mathcal{U}_k$ are essentially disjoint.

(c) Use the results of Exercise 3 [along with Part (a) or (b)] to show that if $\mathcal{U}_1, \ldots, \mathcal{U}_k$ are (pairwise) orthogonal, then they are independent.

(d) Assuming that $\mathcal{U}_1, \mathcal{U}_2, \ldots, \mathcal{U}_k$ are of dimension one or more and letting $\{\mathbf{U}_1^{(j)}, \ldots, \mathbf{U}_{r_j}^{(j)}\}$ represent any linearly independent set of matrices in $\mathcal{U}_j$ ($j = 1, 2, \ldots, k$), show that if $\mathcal{U}_1, \mathcal{U}_2, \ldots, \mathcal{U}_k$ are independent, then the combined set $\{\mathbf{U}_1^{(1)}, \ldots, \mathbf{U}_{r_1}^{(1)}, \mathbf{U}_1^{(2)}, \ldots, \mathbf{U}_{r_2}^{(2)}, \ldots, \mathbf{U}_1^{(k)}, \ldots, \mathbf{U}_{r_k}^{(k)}\}$ is linearly independent.

(e) Assuming that $\mathcal{U}_1, \mathcal{U}_2, \ldots, \mathcal{U}_k$ are of dimension one or more, show that $\mathcal{U}_1, \mathcal{U}_2, \ldots, \mathcal{U}_k$ are independent if and only if, for every nonnull matrix $\mathbf{U}_1$ in $\mathcal{U}_1$, every nonnull matrix $\mathbf{U}_2$ in $\mathcal{U}_2, \ldots,$ and every nonnull matrix $\mathbf{U}_k$ in $\mathcal{U}_k$, $\mathbf{U}_1, \mathbf{U}_2, \ldots, \mathbf{U}_k$ are linearly independent.

(f) For $j = 1, \ldots, k$, let $p_j = \dim(\mathcal{U}_j)$, and let S_j represent a basis for $\mathcal{U}_j$ ($j = 1, \ldots, k$). Define S to be the set of $\sum_{j=1}^{k} p_j$ matrices obtained by combining all of the matrices in $S_1, \ldots, S_k$ into a single set. Use the result of Exercise 5 [along with Part (d)] to show that (1) if $\mathcal{U}_1, \ldots, \mathcal{U}_k$ are independent, then S is a basis for $\mathcal{U}_1 + \ldots + \mathcal{U}_k$; and (2) if $\mathcal{U}_1, \ldots, \mathcal{U}_k$ are not independent, then S contains a proper subset that is a basis for $\mathcal{U}_1 + \ldots + \mathcal{U}_k$.

(g) Show that (1) if $\mathcal{U}_1, \ldots, \mathcal{U}_k$ are independent, then

$$\dim(\mathcal{U}_1 + \cdots + \mathcal{U}_k) = \dim(\mathcal{U}_1) + \cdots + \dim(\mathcal{U}_k);$$

and (2) if $\mathcal{U}_1, \ldots, \mathcal{U}_k$ are not independent, then

$$\dim(\mathcal{U}_1 + \cdots + \mathcal{U}_k) < \dim(\mathcal{U}_1) + \cdots + \dim(\mathcal{U}_k).$$

Solution. (a) It suffices to show that $\mathcal{U}_1, \ldots, \mathcal{U}_k$ are not independent if and only if, for some i $(2 \le i \le k)$, $\mathcal{U}_i$ and $\mathcal{U}_1 + \cdots + \mathcal{U}_{i-1}$ are not essentially disjoint.

Suppose that $\mathcal{U}_1, \ldots, \mathcal{U}_k$ are not independent. Then, by definition, equation (E.1) has a solution, say $\mathbf{U}_1 = \mathbf{U}_1^*, \ldots, \mathbf{U}_k = \mathbf{U}_k^*$, other than $\mathbf{U}_1 = \cdots = \mathbf{U}_k = \mathbf{0}$. Let r represent the largest value of i for which $\mathbf{U}_i^*$ is nonnull. (Clearly, $r \ge 2$.) Since $\mathbf{U}_1^* + \cdots + \mathbf{U}_r^* = \mathbf{U}_1^* + \cdots + \mathbf{U}_k^* = \mathbf{0}$,

$$\mathbf{U}_r^* = -\mathbf{U}_1^* + \cdots + (-\mathbf{U}_{r-1}^*) \in \mathcal{U}_1 + \cdots + \mathcal{U}_{r-1}.$$

Thus, for $i = r$, $\mathcal{U}_i$ and $\mathcal{U}_1 + \cdots + \mathcal{U}_{i-1}$ are not essentially disjoint.

Conversely, suppose that for some i, say $i = s$, $\mathcal{U}_i$ and $\mathcal{U}_1 + \cdots + \mathcal{U}_{i-1}$ are not essentially disjoint. Then, there exists a nonnull matrix $\mathbf{U}_s$ such that $\mathbf{U}_s \in \mathcal{U}_s$ and $\mathbf{U}_s \in \mathcal{U}_1 + \cdots + \mathcal{U}_{s-1}$. Further, there exist matrices $\mathbf{U}_1 \in \mathcal{U}_1, \ldots, \mathbf{U}_{s-1} \in \mathcal{U}_{s-1}$ such that $\mathbf{U}_s = \mathbf{U}_1 + \cdots + \mathbf{U}_{s-1}$ or equivalently such that $\mathbf{U}_1 + \cdots + \mathbf{U}_{s-1} + (-\mathbf{U}_s) = \mathbf{0}$. Thus, equation (E.1) has a solution other than $\mathbf{U}_1 = \cdots = \mathbf{U}_k = \mathbf{0}$.

(b) It suffices to show that $\mathcal{U}_1, \ldots, \mathcal{U}_k$ are not independent if and only if, for some i $(1 \le i \le k)$, $\mathcal{U}_i$ and $\mathcal{U}_1 + \cdots + \mathcal{U}_{i-1} + \mathcal{U}_{i+1} + \cdots + \mathcal{U}_k$ are not essentially disjoint.

Suppose that $\mathcal{U}_1, \ldots, \mathcal{U}_k$ are not independent. Then, by definition, equation (E.1) has a solution, say $\mathbf{U}_1 = \mathbf{U}_1^*, \ldots, \mathbf{U}_k = \mathbf{U}_k^*$, other than $\mathbf{U}_1 = \cdots = \mathbf{U}_k = \mathbf{0}$. Let r represent an integer (between 1 and k, inclusive) such that $\mathbf{U}_r^* \ne \mathbf{0}$. Since $\mathbf{U}_r^* = -\sum_{i \ne r} \mathbf{U}_i^*$, $\mathbf{U}_r^*$ is in the subspace $\sum_{i \ne r} \mathcal{U}_i$, as well as the subspace $\mathcal{U}_r$. Thus, for $i = r$, $\mathcal{U}_i$ and $\mathcal{U}_1 + \cdots + \mathcal{U}_{i-1} + \mathcal{U}_{i+1} + \cdots + \mathcal{U}_k$ are not essentially disjoint.

Conversely, suppose that for some i, say $i = s$, $\mathcal{U}_i$ and $\mathcal{U}_1 + \cdots + \mathcal{U}_{i-1} + \mathcal{U}_{i+1} + \cdots + \mathcal{U}_k$ are not essentially disjoint. Then, there exists a nonnull matrix $\mathbf{U}_s$ such that $\mathbf{U}_s \in \mathcal{U}_s$ and $\mathbf{U}_s \in \sum_{i \ne s} \mathcal{U}_i$. Further, there exist matrices $\mathbf{U}_1 \in \mathcal{U}_1, \ldots, \mathbf{U}_{s-1} \in \mathcal{U}_{s-1}, \mathbf{U}_{s+1} \in \mathcal{U}_{s+1}, \ldots, \mathbf{U}_k \in \mathcal{U}_k$ such that $\mathbf{U}_s = \sum_{i \ne s} \mathbf{U}_i$ or equivalently such that $\mathbf{U}_1 + \cdots + \mathbf{U}_{s-1} + (-\mathbf{U}_s) + \mathbf{U}_{s+1} + \cdots + \mathbf{U}_k = \mathbf{0}$. Thus, equation (E.1) has a solution other than $\mathbf{U}_1 = \cdots = \mathbf{U}_k = \mathbf{0}$.

(c) Suppose that $\mathcal{U}_1, \ldots, \mathcal{U}_k$ are orthogonal. Then, applying the result of Part (a)-(2) of Exercise 3 ($i - 2$ times), we find that $\mathcal{U}_i$ and $\mathcal{U}_1 + \cdots + \mathcal{U}_{i-1}$ are orthogonal, implying (in light of Lemma 17.1.9) that $\mathcal{U}_i$ and $\mathcal{U}_1 + \cdots + \mathcal{U}_{i-1}$ are essentially disjoint $(i = 2, \ldots, k)$. Based on Part (a), we conclude that $\mathcal{U}_1, \ldots, \mathcal{U}_k$ are independent.

(d) Suppose that $\mathcal{U}_1, \mathcal{U}_2, \ldots, \mathcal{U}_k$ are independent. The proof that the set $\{\mathbf{U}_1^{(1)}, \ldots, \mathbf{U}_{r_1}^{(1)}, \mathbf{U}_1^{(2)}, \ldots, \mathbf{U}_{r_2}^{(2)}, \ldots, \mathbf{U}_1^{(k)}, \ldots, \mathbf{U}_{r_k}^{(k)}\}$ is linearly independent is by mathematical induction.

By definition, the set $\{\mathbf{U}_1^{(1)}, \ldots, \mathbf{U}_{r_1}^{(1)}\}$ is linearly independent. Now, suppose that (for an arbitrary integer j between 1 and $k-1$, inclusive) the set $\{\mathbf{U}_1^{(1)}, \ldots, \mathbf{U}_{r_1}^{(1)}, \mathbf{U}_1^{(2)}, \ldots, \mathbf{U}_{r_2}^{(2)}, \ldots, \mathbf{U}_1^{(j)}, \ldots, \mathbf{U}_{r_j}^{(j)}\}$ is linearly independent. Then, it suffices to show that the set $\{\mathbf{U}_1^{(1)}, \ldots, \mathbf{U}_{r_1}^{(1)}, \mathbf{U}_1^{(2)}, \ldots, \mathbf{U}_{r_2}^{(2)}, \ldots, \mathbf{U}_1^{(j+1)}, \ldots, \mathbf{U}_{r_{j+1}}^{(j+1)}\}$ is linearly independent.

According to Part (a), $\mathcal{U}_{j+1}$ and $\mathcal{U}_1 + \cdots + \mathcal{U}_j$ are essentially disjoint. And, clearly, $\mathbf{U}_1^{(1)}, \ldots, \mathbf{U}_{r_1}^{(1)}, \mathbf{U}_1^{(2)}, \ldots, \mathbf{U}_{r_2}^{(2)}, \ldots, \mathbf{U}_1^{(j)}, \ldots, \mathbf{U}_{r_j}^{(j)}$ are in the subspace $\mathcal{U}_1 + \mathcal{U}_2 + \cdots + \mathcal{U}_j$. Thus, it follows from Lemma 17.1.3 that the set $\{\mathbf{U}_1^{(1)}, \ldots, \mathbf{U}_{r_1}^{(1)}, \mathbf{U}_1^{(2)}, \ldots, \mathbf{U}_{r_2}^{(2)}, \ldots, \mathbf{U}_1^{(j+1)}, \ldots, \mathbf{U}_{r_{j+1}}^{(j+1)}\}$ is linearly independent.

(e) Suppose that $\mathcal{U}_1, \mathcal{U}_2, \ldots, \mathcal{U}_k$ are independent. Let $\mathbf{U}_1, \mathbf{U}_2, \ldots, \mathbf{U}_k$ represent nonnull matrices in $\mathcal{U}_1, \mathcal{U}_2, \ldots, \mathcal{U}_k$, respectively. Then, it follows from Part (d) that the set $\{\mathbf{U}_1, \mathbf{U}_2, \ldots, \mathbf{U}_k\}$ is linearly independent.

Conversely, suppose that, for every nonnull matrix $\mathbf{U}_1$ in $\mathcal{U}_1$, every nonnull matrix $\mathbf{U}_2$ in $\mathcal{U}_2$, ..., and every nonnull matrix $\mathbf{U}_k$ in $\mathcal{U}_k$, $\mathbf{U}_1, \mathbf{U}_2, \ldots, \mathbf{U}_k$ are linearly independent. If $\mathcal{U}_1, \mathcal{U}_2, \ldots, \mathcal{U}_k$ were not independent, then, for some nonempty subset $\{j_1, \ldots, j_r\}$ of the first k positive integers, there would exist nonnull matrices $\mathbf{U}_{j_1}, \ldots, \mathbf{U}_{j_r}$ in $\mathcal{U}_{j_1}, \ldots, \mathcal{U}_{j_r}$, respectively, such that

$$\mathbf{U}_{j_1} + \cdots + \mathbf{U}_{j_r} = \mathbf{0},$$

and the set $\{\mathbf{U}_1, \mathbf{U}_2, \ldots, \mathbf{U}_k\}$ (where, for $j \notin \{j_1, \ldots, j_r\}$, $\mathbf{U}_j$ is an arbitrary nonnull matrix in $\mathcal{U}_j$) would be linearly dependent, which would be contradictory. Thus, $\mathcal{U}_1, \mathcal{U}_2, \ldots, \mathcal{U}_k$ are independent.

(f) It is clear from the result of Exercise 5 that S spans $\mathcal{U}_1 + \cdots + \mathcal{U}_k$.

(1) Now, suppose that $\mathcal{U}_1, \ldots, \mathcal{U}_k$ are independent. Then, it is evident from Part (d) that S is a linearly independent set. Thus, S is a basis for $\mathcal{U}_1 + \cdots + \mathcal{U}_k$.

(2) Alternatively, suppose that $\mathcal{U}_1, \ldots, \mathcal{U}_k$ are not independent. Then, for some (nonempty) subset $\{j_r, \ldots, j_r\}$ of the first k positive integers, there exist nonnull matrices $\mathbf{U}_{j_1}, \ldots, \mathbf{U}_{j_r}$ in $\mathcal{U}_{j_1}, \ldots, \mathcal{U}_{j_r}$, respectively, such that

$$\mathbf{U}_{j_1} + \cdots + \mathbf{U}_{j_r} = \mathbf{0}.$$

Further, for $m = 1, \ldots, r$,

$$\mathbf{U}_{j_m} = \sum_{i=1}^{p_{j_m}} c_i^{(m)} \mathbf{U}_i^{(m)},$$

where $c_1^{(m)}, \ldots, c_{p_{j_m}}^{(m)}$ are scalars (not all of which can be zero) and $\mathbf{U}_1^{(m)}, \ldots, \mathbf{U}_{p_{j_m}}^{(m)}$ are the matrices in S_{j_m}. Thus,

$$\sum_{m=1}^{r} \sum_{i=1}^{p_{j_m}} c_i^{(m)} \mathbf{U}_i^{(m)} = \mathbf{0},$$

implying that S is a linearly dependent set. We conclude that S itself is not a basis and consequently (in light of Theorem 4.3.11) that S contains a proper subset that is a basis for $\mathcal{U}_1 + \cdots + \mathcal{U}_k$.

(g) Part (g) is an immediate consequence of Part (f).

EXERCISE 7. Let $\mathbf{A}_1, \ldots, \mathbf{A}_k$ represent matrices having the same number of rows, and let $\mathbf{B}_1, \ldots, \mathbf{B}_k$ represent matrices having the same number of columns.

Adopting the terminology of Exercise 6, use Part (g) of that exercise to show (a) that if $\mathcal{C}(\mathbf{A}_1), \ldots, \mathcal{C}(\mathbf{A}_k)$ are independent, then

$$\operatorname{rank}(\mathbf{A}_1, \ldots, \mathbf{A}_k) = \operatorname{rank}(\mathbf{A}_1) + \cdots + \operatorname{rank}(\mathbf{A}_k),$$

and if $\mathcal{C}(\mathbf{A}_1), \ldots, \mathcal{C}(\mathbf{A}_k)$ are not independent, then

$$\operatorname{rank}(\mathbf{A}_1, \ldots, \mathbf{A}_k) < \operatorname{rank}(\mathbf{A}_1) + \cdots + \operatorname{rank}(\mathbf{A}_k)$$

and (b) that if $\mathcal{R}(\mathbf{B}_1), \ldots, \mathcal{R}(\mathbf{B}_k)$ are independent, then

$$\operatorname{rank}\begin{pmatrix}\mathbf{B}_1\\ \vdots \\ \mathbf{B}_k\end{pmatrix} = \operatorname{rank}(\mathbf{B}_1) + \cdots + \operatorname{rank}(\mathbf{B}_k),$$

and if $\mathcal{R}(\mathbf{B}_1), \ldots, \mathcal{R}(\mathbf{B}_k)$ are not independent, then

$$\operatorname{rank}\begin{pmatrix}\mathbf{B}_1\\ \vdots \\ \mathbf{B}_k\end{pmatrix} < \operatorname{rank}(\mathbf{B}_1) + \cdots + \operatorname{rank}(\mathbf{B}_k).$$

Solution. (a) Clearly,

$$\operatorname{rank}(\mathbf{A}_1) + \cdots + \operatorname{rank}(\mathbf{A}_k) = \dim[\mathcal{C}(\mathbf{A}_1)] + \cdots + \dim[\mathcal{C}(\mathbf{A}_k)].$$

And, in light of equality (1.6),

$$\operatorname{rank}(\mathbf{A}_1, \ldots, \mathbf{A}_k) = \dim[\mathcal{C}(\mathbf{A}_1 \ldots, \mathbf{A}_k)] = \dim[\mathcal{C}(\mathbf{A}_1) + \cdots + \mathcal{C}(\mathbf{A}_k)].$$

Thus, it follows from Part (g) of Exercise 6 that if $\mathcal{C}(\mathbf{A}_1), \ldots, \mathcal{C}(\mathbf{A}_k)$ are independent, then

$$\operatorname{rank}(\mathbf{A}_1, \ldots, \mathbf{A}_k) = \operatorname{rank}(\mathbf{A}_1) + \cdots + \operatorname{rank}(\mathbf{A}_k),$$

and if $\mathcal{C}(\mathbf{A}_1), \ldots, \mathcal{C}(\mathbf{A}_k)$ are not independent, then

$$\operatorname{rank}(\mathbf{A}_1, \ldots, \mathbf{A}_k) < \operatorname{rank}(\mathbf{A}_1) + \cdots + \operatorname{rank}(\mathbf{A}_k).$$

(b) The proof of Part (b) is analogous to that of Part (a).

EXERCISE 8. Letting $\mathbf{A}$ represent an $m \times n$ matrix and $\mathbf{B}$ an $m \times p$ matrix, show, by for instance using the result of Part (c)-(2) of Exercise 4 in combination with the result

$$\mathcal{C}(\mathbf{A},\ \mathbf{B}) = \mathcal{C}[\mathbf{A},\ (\mathbf{I} - \mathbf{A}\mathbf{A}^-)\mathbf{B}] = \mathcal{C}(\mathbf{A}) \oplus \mathcal{C}[(\mathbf{I} - \mathbf{A}\mathbf{A}^-)\mathbf{B}], \qquad (*)$$

that

(a) $\mathcal{C}[(\mathbf{I} - \mathbf{A}\mathbf{A}^-)\mathbf{B}] = \mathcal{C}(\mathbf{I} - \mathbf{A}\mathbf{A}^-) \cap \mathcal{C}(\mathbf{A}, \mathbf{B})$ and

(b) $\mathcal{C}[(\mathbf{I}-\mathbf{P_A})\mathbf{B}] = \mathcal{N}(\mathbf{A}') \cap \mathcal{C}(\mathbf{A}, \mathbf{B})$.

Solution. (a) According to result (∗) [or, equivalently, the first part of Corollary 17.2.9],

$$\mathcal{C}(\mathbf{A}, \mathbf{B}) = \mathcal{C}(\mathbf{A}) + \mathcal{C}[(\mathbf{I}-\mathbf{AA}^-)\mathbf{B}].$$

Thus, observing that $\mathcal{C}[(\mathbf{I}-\mathbf{AA}^-)\mathbf{B}] \subset \mathcal{C}(\mathbf{I}-\mathbf{AA}^-)$ and making use of the result of Part (c)-(2) of Exercise 4 (and also of Lemma 17.2.7), we find that

$$\begin{aligned}
\mathcal{C}(\mathbf{I}-\mathbf{AA}^-) \cap \mathcal{C}(\mathbf{A}, \mathbf{B}) &= \mathcal{C}(\mathbf{I}-\mathbf{AA}^-) \cap \{\mathcal{C}[(\mathbf{I}-\mathbf{AA}^-)\mathbf{B}] + \mathcal{C}(\mathbf{A})\} \\
&= \mathcal{C}[(\mathbf{I}-\mathbf{AA}^-)\mathbf{B}] + [\mathcal{C}(\mathbf{I}-\mathbf{AA}^-) \cap \mathcal{C}(\mathbf{A})] \\
&= \mathcal{C}[(\mathbf{I}-\mathbf{AA}^-)\mathbf{B}] + \{\mathbf{0}\} \\
&= \mathcal{C}[(\mathbf{I}-\mathbf{AA}^-)\mathbf{B}].
\end{aligned}$$

(b) According to Part (1) of Theorem 12.3.4, $(\mathbf{A}'\mathbf{A})^-\mathbf{A}'$ is a generalized inverse of $\mathbf{A}$. Substituting this generalized inverse for $\mathbf{A}^-$ in the result of Part (a) and making use of Lemma 12.5.2, we find that

$$\begin{aligned}
\mathcal{C}[(\mathbf{I}-\mathbf{P_A})\mathbf{B}] &= \mathcal{C}(\mathbf{I}-\mathbf{P_A}) \cap \mathcal{C}(\mathbf{A}, \mathbf{B}) \\
&= \mathcal{N}(\mathbf{A}') \cap \mathcal{C}(\mathbf{A}, \mathbf{B}).
\end{aligned}$$

EXERCISE 9. Let $\mathbf{A} = (\mathbf{T}, \mathbf{U})$ and $\mathbf{B} = (\mathbf{V}, \mathbf{0})$, where $\mathbf{T}$ is an $m \times p$ matrix, $\mathbf{U}$ an $m \times q$ matrix, and $\mathbf{V}$ an $n \times p$ matrix, and suppose that $\mathbf{U}$ is of full row rank. Show that $\mathcal{R}(\mathbf{A})$ and $\mathcal{R}(\mathbf{B})$ are essentially disjoint [even if $\mathcal{R}(\mathbf{T})$ and $\mathcal{R}(\mathbf{V})$ are not essentially disjoint].

Solution. Let $\mathbf{x}'$ represent an arbitrary $[1 \times (p+q)]$ vector in $\mathcal{R}(\mathbf{A}) \cap \mathcal{R}(\mathbf{B})$. Then, $\mathbf{x}' = \mathbf{r}'\mathbf{A}$ and $\mathbf{x}' = \mathbf{s}'\mathbf{B}$ for some (row) vectors $\mathbf{r}'$ and $\mathbf{s}'$. Partitioning $\mathbf{x}'$ as $\mathbf{x}' = (\mathbf{x}_1', \mathbf{x}_2')$ (where $\mathbf{x}_1'$ is of dimensions $1 \times p$), we find that

$$(\mathbf{x}_1', \mathbf{x}_2') = \mathbf{r}'(\mathbf{T}, \mathbf{U}) = (\mathbf{r}'\mathbf{T}, \mathbf{r}'\mathbf{U})$$

and similarly that

$$(\mathbf{x}_1', \mathbf{x}_2') = \mathbf{s}'(\mathbf{V}, \mathbf{0}) = (\mathbf{s}'\mathbf{V}, \mathbf{0}).$$

Thus, $\mathbf{r}'\mathbf{U} = \mathbf{x}_2' = \mathbf{0}$, implying (since the rows of $\mathbf{U}$ are linearly independent) that $\mathbf{r}' = \mathbf{0}$ and hence that $\mathbf{x}' = \mathbf{0}$. We conclude that $\mathcal{R}(\mathbf{A})$ and $\mathcal{R}(\mathbf{B})$ are essentially disjoint [even if $\mathcal{R}(\mathbf{T})$ and $\mathcal{R}(\mathbf{V})$ are not essentially disjoint].

EXERCISE 10. To what extent does the formula

$$\operatorname{rank}\begin{pmatrix} \mathbf{T} & \mathbf{U} \\ \mathbf{V} & \mathbf{0} \end{pmatrix} = \operatorname{rank}(\mathbf{U}) + \operatorname{rank}(\mathbf{V}) + \operatorname{rank}[(\mathbf{I}-\mathbf{UU}^-)\mathbf{T}(\mathbf{I}-\mathbf{V}^-\mathbf{V})] \qquad (*)$$

[where $\mathbf{T}$ is an $m \times p$ matrix, $\mathbf{U}$ an $m \times q$ matrix, and $\mathbf{V}$ an $n \times p$ matrix] simplify in (a) the special case where $\mathcal{C}(\mathbf{T})$ and $\mathcal{C}(\mathbf{U})$ are essentially disjoint [but $\mathcal{R}(\mathbf{T})$ and

$\mathcal{R}(\mathbf{V})$ are not necessarily essentially disjoint] and (b) the special case where $\mathcal{R}(\mathbf{T})$ and $\mathcal{R}(\mathbf{V})$ are essentially disjoint.

Solution. (a) If $\mathcal{C}(\mathbf{T})$ and $\mathcal{C}(\mathbf{U})$ are essentially disjoint, then {since $\mathcal{C}[\mathbf{T}(\mathbf{I}-\mathbf{V}^-\mathbf{V})] \subset \mathcal{C}(\mathbf{T})$} $\mathcal{C}[\mathbf{T}(\mathbf{I}-\mathbf{V}^-\mathbf{V})]$ and $\mathcal{C}(\mathbf{U})$ are essentially disjoint, and (in light of Corollary 17.2.10) formula (∗) [or, equivalently, formula (2.15)] simplifies to

$$\text{rank}\begin{pmatrix}\mathbf{T} & \mathbf{U}\\ \mathbf{V} & \mathbf{0}\end{pmatrix} = \text{rank}(\mathbf{U}) + \text{rank}(\mathbf{V}) + \text{rank}[\mathbf{T}(\mathbf{I} - \mathbf{V}^-\mathbf{V})].$$

(b) If $\mathcal{R}(\mathbf{T})$ and $\mathcal{R}(\mathbf{V})$ are essentially disjoint, then it follows from an analogous line of reasoning that formula (∗) [or, equivalently, formula (2.15)] simplifies to

$$\text{rank}\begin{pmatrix}\mathbf{T} & \mathbf{U}\\ \mathbf{V} & \mathbf{0}\end{pmatrix} = \text{rank}(\mathbf{U}) + \text{rank}(\mathbf{V}) + \text{rank}[(\mathbf{I} - \mathbf{U}\mathbf{U}^-)\mathbf{T}].$$

EXERCISE 11. Let $\mathbf{T}$ represent an $m \times p$ matrix, $\mathbf{U}$ an $m \times q$ matrix, and $\mathbf{V}$ an $n \times p$ matrix. Further, define $\mathbf{E}_T = \mathbf{I} - \mathbf{T}\mathbf{T}^-$, $\mathbf{F}_T = \mathbf{I} - \mathbf{T}^-\mathbf{T}$, $\mathbf{X} = \mathbf{E}_T\mathbf{U}$, and $\mathbf{Y} = \mathbf{V}\mathbf{F}_T$. Show (a) that the partitioned matrix $\begin{pmatrix}\mathbf{T}^- - \mathbf{T}^-\mathbf{U}\mathbf{X}^-\mathbf{E}_T\\ \mathbf{X}^-\mathbf{E}_T\end{pmatrix}$ is a generalized inverse of the partitioned matrix $(\mathbf{T}, \mathbf{U})$ and (b) that the partitioned matrix $(\mathbf{T}^- - \mathbf{F}_T\mathbf{Y}^-\mathbf{V}\mathbf{T}^-,\ \mathbf{F}_T\mathbf{Y}^-)$ is a generalized inverse of the partitioned matrix $\begin{pmatrix}\mathbf{T}\\ \mathbf{V}\end{pmatrix}$. Do so by applying formula (E.1) from Part (a) of Exercise 10.10 to the partitioned matrices $\begin{pmatrix}\mathbf{T} & \mathbf{U}\\ \mathbf{0} & \mathbf{0}\end{pmatrix}$ and $\begin{pmatrix}\mathbf{T} & \mathbf{0}\\ \mathbf{V} & \mathbf{0}\end{pmatrix}$ and by making use of the result that for any generalized inverse $\mathbf{G} = \begin{pmatrix}\mathbf{G}_1\\ \mathbf{G}_2\end{pmatrix}$ of the partitioned matrix $(\mathbf{A},\ \mathbf{B})$ and any generalized inverse $\mathbf{H} = (\mathbf{H}_1, \mathbf{H}_2)$ of the partitioned matrix $\begin{pmatrix}\mathbf{A}\\ \mathbf{C}\end{pmatrix}$ (where $\mathbf{A}$ is an $m \times n$ matrix, $\mathbf{B}$ an $m \times p$ matrix, and $\mathbf{C}$ a $q \times n$ matrix and where $\mathbf{G}_1$ has n rows and $\mathbf{H}_1$ m columns), (1) $\mathbf{G}_1$ is a generalized inverse of $\mathbf{A}$ and $\mathbf{G}_2$ a generalized inverse of $\mathbf{B}$ if and only if $\mathcal{C}(\mathbf{A})$ and $\mathcal{C}(\mathbf{B})$ are essentially disjoint, and, similarly, (2) $\mathbf{H}_1$ is a generalized inverse of $\mathbf{A}$ and $\mathbf{H}_2$ a generalized inverse of $\mathbf{C}$ if and only if $\mathcal{R}(\mathbf{A})$ and $\mathcal{R}(\mathbf{C})$ are essentially disjoint.

Solution. (a) Upon setting $\mathbf{V} = \mathbf{0}$ and $\mathbf{W} = \mathbf{0}$ (in which case $\mathbf{Y} = \mathbf{0}$, $\mathbf{Q} = \mathbf{0}$, and $\mathbf{Z} = \mathbf{0}$) and choosing $\mathbf{Y}^- = \mathbf{0}$ and $\mathbf{Z}^- = \mathbf{0}$ in formula (E.1) [from Part (a) of Exercise 10.10], we obtain as a generalized inverse for $\begin{pmatrix}\mathbf{T} & \mathbf{U}\\ \mathbf{0} & \mathbf{0}\end{pmatrix}$ the partitioned matrix

$$\mathbf{G} = \begin{pmatrix}\mathbf{T}^- - \mathbf{T}^-\mathbf{U}\mathbf{X}^-\mathbf{E}_T & \mathbf{0}\\ \mathbf{X}^-\mathbf{E}_T & \mathbf{0}\end{pmatrix}.$$

We conclude, on the basis of the cited result (or, equivalently, Theorem 17.3.3), that $\begin{pmatrix}\mathbf{T}^- - \mathbf{T}^-\mathbf{U}\mathbf{X}^-\mathbf{E}_T\\ \mathbf{X}^-\mathbf{E}_T\end{pmatrix}$ is a generalized inverse of $(\mathbf{T}, \mathbf{U})$.

(b) Upon setting $\mathbf{U} = \mathbf{0}$ and $\mathbf{W} = \mathbf{0}$ (in which case $\mathbf{X} = \mathbf{0}$, $\mathbf{Q} = \mathbf{0}$, and $\mathbf{Z} = \mathbf{0}$) and choosing $\mathbf{X}^- = \mathbf{0}$ and $\mathbf{Z}^- = \mathbf{0}$ in formula (E.1) [from Part (a) of Exercise 10.10], we obtain as a generalized inverse for $\begin{pmatrix} \mathbf{T} & \mathbf{0} \\ \mathbf{V} & \mathbf{0} \end{pmatrix}$ the partitioned matrix

$$\mathbf{G} = \begin{pmatrix} \mathbf{T}^- - \mathbf{F}_T\mathbf{Y}^-\mathbf{V}\mathbf{T}^- & \mathbf{F}_T\mathbf{Y}^- \\ \mathbf{0} & \mathbf{0} \end{pmatrix}.$$

We conclude, on the basis of the cited result (or, equivalently, Theorem 17.3.3), that $(\mathbf{T}^- - \mathbf{F}_T\mathbf{Y}^-\mathbf{V}\mathbf{T}^-,\ \mathbf{F}_T\mathbf{Y}^-)$ is a generalized inverse of $\begin{pmatrix} \mathbf{T} \\ \mathbf{V} \end{pmatrix}$.

EXERCISE 12. Let $\mathbf{T}$ represent an $m \times p$ matrix, $\mathbf{U}$ an $m \times q$ matrix, and $\mathbf{V}$ an $n \times p$ matrix. And, let $\begin{pmatrix} \mathbf{G}_{11} & \mathbf{G}_{12} \\ \mathbf{G}_{21} & \mathbf{G}_{22} \end{pmatrix}$ (where $\mathbf{G}_{11}$ is of dimensions $p \times m$) represent a generalized inverse of the partitioned matrix $\begin{pmatrix} \mathbf{T} & \mathbf{U} \\ \mathbf{V} & \mathbf{0} \end{pmatrix}$. Show that (a) if $\mathbf{G}_{11}$ is a generalized inverse of $\mathbf{T}$ and $\mathbf{G}_{12}$ a generalized inverse of $\mathbf{V}$, then $\mathcal{R}(\mathbf{T})$ and $\mathcal{R}(\mathbf{V})$ are essentially disjoint, and (b) if $\mathbf{G}_{11}$ is a generalized inverse of $\mathbf{T}$ and $\mathbf{G}_{21}$ a generalized inverse of $\mathbf{U}$, then $\mathcal{C}(\mathbf{T})$ and $\mathcal{C}(\mathbf{U})$ are essentially disjoint.

Solution. Clearly,

$$\begin{aligned} &\begin{pmatrix} \mathbf{T}\mathbf{G}_{11}\mathbf{T} + \mathbf{U}\mathbf{G}_{21}\mathbf{T} + \mathbf{T}\mathbf{G}_{12}\mathbf{V} + \mathbf{U}\mathbf{G}_{22}\mathbf{V} & \mathbf{T}\mathbf{G}_{11}\mathbf{U} + \mathbf{U}\mathbf{G}_{21}\mathbf{U} \\ \mathbf{V}\mathbf{G}_{11}\mathbf{T} + \mathbf{V}\mathbf{G}_{12}\mathbf{V} & \mathbf{V}\mathbf{G}_{11}\mathbf{U} \end{pmatrix} \\ &\qquad = \begin{pmatrix} \mathbf{T} & \mathbf{U} \\ \mathbf{V} & \mathbf{0} \end{pmatrix} \begin{pmatrix} \mathbf{G}_{11} & \mathbf{G}_{12} \\ \mathbf{G}_{21} & \mathbf{G}_{22} \end{pmatrix} \begin{pmatrix} \mathbf{T} & \mathbf{U} \\ \mathbf{V} & \mathbf{0} \end{pmatrix} = \begin{pmatrix} \mathbf{T} & \mathbf{U} \\ \mathbf{V} & \mathbf{0} \end{pmatrix}. \qquad \text{(S.1)} \end{aligned}$$

(a) Result (S.1) implies in particular that

$$\mathbf{V}\mathbf{G}_{11}\mathbf{T} = \mathbf{V} - \mathbf{V}\mathbf{G}_{12}\mathbf{V}. \qquad \text{(S.2)}$$

Now, suppose that $\mathbf{G}_{11}$ is a generalized inverse of $\mathbf{T}$ and $\mathbf{G}_{12}$ a generalized inverse of $\mathbf{V}$. Then, equality (S.2) reduces to

$$\mathbf{V}\mathbf{G}_{11}\mathbf{T} = \mathbf{0},$$

and it follows from Corollary 17.2.12 that $\mathcal{R}(\mathbf{T})$ and $\mathcal{R}(\mathbf{V})$ are essentially disjoint.

(b) The proof of Part (b) is analogous to that of Part (a).

EXERCISE 13. (a) Generalize the result that, for any two subspaces $\mathcal{U}$ and $\mathcal{V}$ of $\mathcal{R}^{m \times n}$,

$$\dim(\mathcal{U} + \mathcal{V}) = \dim(\mathcal{U}) + \dim(\mathcal{V}) - \dim(\mathcal{U} \cap \mathcal{V}), \qquad (*)$$

Do so by showing that, for any k subspaces $\mathcal{U}_1, \ldots, \mathcal{U}_k$,

$$\begin{aligned} \dim(\mathcal{U}_1 + \cdots + \mathcal{U}_k) &= \dim(\mathcal{U}_1) + \cdots + \dim(\mathcal{U}_k) \\ &\quad - \sum_{i=2}^{k} \dim[(\mathcal{U}_1 + \cdots + \mathcal{U}_{i-1}) \cap \mathcal{U}_i]. \qquad \text{(E.2)} \end{aligned}$$

(b) Generalize the result that, for any $m \times n$ matrix $\mathbf{A}$, $m \times p$ matrix $\mathbf{B}$, and $q \times n$ matrix $\mathbf{C}$,

$$\operatorname{rank}(\mathbf{A}, \mathbf{B}) = \operatorname{rank}(\mathbf{A}) + \operatorname{rank}(\mathbf{B}) - \dim[\mathcal{C}(\mathbf{A}) \cap \mathcal{C}(\mathbf{B})],$$
$$\operatorname{rank}\begin{pmatrix}\mathbf{A}\\ \mathbf{C}\end{pmatrix} = \operatorname{rank}(\mathbf{A}) + \operatorname{rank}(\mathbf{C}) - \dim[\mathcal{R}(\mathbf{A}) \cap \mathcal{R}(\mathbf{C})].$$

Do so by showing that, for any matrices $\mathbf{A}_1, \ldots, \mathbf{A}_k$ having the same number of rows,

$$\operatorname{rank}(\mathbf{A}_1, \ldots, \mathbf{A}_k) = \operatorname{rank}(\mathbf{A}_1) + \cdots + \operatorname{rank}(\mathbf{A}_k) - \sum_{i=2}^{k} \dim[\mathcal{C}(\mathbf{A}_1, \ldots \mathbf{A}_{i-1}) \cap \mathcal{C}(\mathbf{A}_i)]$$

and, for any matrices $\mathbf{B}_1, \ldots, \mathbf{B}_k$ having the same number of columns,

$$\operatorname{rank}\begin{pmatrix}\mathbf{B}_1\\ \vdots\\ \mathbf{B}_k\end{pmatrix} = \operatorname{rank}(\mathbf{B}_1) + \cdots + \operatorname{rank}(\mathbf{B}_k) - \sum_{i=2}^{k} \dim\left[\mathcal{R}\begin{pmatrix}\mathbf{B}_1\\ \vdots\\ \mathbf{B}_{i-1}\end{pmatrix} \cap \mathcal{R}(\mathbf{B}_i)\right].$$

Solution. (a) The proof is by mathematical induction. In the special case where $k = 2$, equality (E.2) reduces to the equality

$$\dim(\mathcal{U}_1 + \mathcal{U}_2) = \dim(\mathcal{U}_1) + \dim(\mathcal{U}_2) - \dim(\mathcal{U}_1 \cap \mathcal{U}_2),$$

which is equivalent to equality ($*$) and whose validity was established in Theorem 17.4.1.

Suppose now that equality (E.2) is valid for $k = k'$ (where $k' \geq 2$). Then, making use of Theorem 17.4.1, we find that

$$\begin{aligned}
\dim(\mathcal{U}_1 + \cdots + \mathcal{U}_{k'} + \mathcal{U}_{k'+1})
&= \dim(\mathcal{U}_1 + \cdots + \mathcal{U}_{k'}) + \dim(\mathcal{U}_{k'+1}) \\
&\qquad - \dim[(\mathcal{U}_1 + \cdots + \mathcal{U}_{k'}) \cap \mathcal{U}_{k'+1}] \\
&= \dim(\mathcal{U}_1) + \cdots + \dim(\mathcal{U}_{k'}) - \sum_{i=2}^{k'} \dim[(\mathcal{U}_1 + \cdots + \mathcal{U}_{i-1}) \cap \mathcal{U}_i] \\
&\qquad + \dim(\mathcal{U}_{k'+1}) - \dim[(\mathcal{U}_1 + \cdots + \mathcal{U}_{k'}) \cap \mathcal{U}_{k'+1}] \\
&= \dim(\mathcal{U}_1) + \cdots + \dim(\mathcal{U}_{k'+1}) - \sum_{i=2}^{k'+1} \dim[(\mathcal{U}_1 + \cdots + \mathcal{U}_{i-1}) \cap \mathcal{U}_i],
\end{aligned}$$

thereby completing the induction argument.

(b) Applying Part (a) with $\mathcal{U}_1 = \mathcal{C}(\mathbf{A}_1), \ldots, \mathcal{U}_k = \mathcal{C}(\mathbf{A}_k)$ [and recalling result (1.6)], we find that

$$\begin{aligned}\operatorname{rank}(\mathbf{A}_1, \ldots, \mathbf{A}_k) &= \dim[\mathcal{C}(\mathbf{A}_1, \ldots \mathbf{A}_k)] \\ &= \dim[\mathcal{C}(\mathbf{A}_1) + \cdots + \mathcal{C}(\mathbf{A}_k)] \\ &= \dim[\mathcal{C}(\mathbf{A}_1)] + \cdots + \dim[\mathcal{C}(\mathbf{A}_k)] \\ &\qquad - \sum_{i=2}^{k} \dim\{[\mathcal{C}(\mathbf{A}_1) + \cdots + \mathcal{C}(\mathbf{A}_{i-1})] \cap \mathcal{C}(\mathbf{A}_i)\} \\ &= \operatorname{rank}(\mathbf{A}_1) + \cdots + \operatorname{rank}(\mathbf{A}_k) \\ &\qquad - \sum_{i=2}^{k} \dim[\mathcal{C}(\mathbf{A}_1, \ldots, \mathbf{A}_{i-1}) \cap \mathcal{C}(\mathbf{A}_i)].\end{aligned}$$

And, similarly, applying Part (a) with $\mathcal{U}_1 = \mathcal{R}(\mathbf{B}_1), \ldots, \mathcal{U}_k = \mathcal{R}(\mathbf{B}_k)$ [and recalling result (1.7)], we find that

$$\begin{aligned}\operatorname{rank}\begin{pmatrix}\mathbf{B}_1 \\ \vdots \\ \mathbf{B}_k\end{pmatrix} &= \dim[\mathcal{R}\begin{pmatrix}\mathbf{B}_1 \\ \vdots \\ \mathbf{B}_k\end{pmatrix}] \\ &= \dim[\mathcal{R}(\mathbf{B}_1) + \cdots + \mathcal{R}(\mathbf{B}_k)] \\ &= \dim[\mathcal{R}(\mathbf{B}_1)] + \cdots + \dim[\mathcal{R}(\mathbf{B}_k)] \\ &\qquad - \sum_{i=2}^{k} \dim\{[\mathcal{R}(\mathbf{B}_1) + \cdots + \mathcal{R}(\mathbf{B}_{i-1})] \cap \mathcal{R}(\mathbf{B}_i)\} \\ &= \operatorname{rank}(\mathbf{B}_1) + \cdots + \operatorname{rank}(\mathbf{B}_k) - \sum_{i=2}^{k} \dim[\mathcal{R}\begin{pmatrix}\mathbf{B}_1 \\ \vdots \\ \mathbf{B}_{i-1}\end{pmatrix} \cap \mathcal{R}(\mathbf{B}_i)].\end{aligned}$$

EXERCISE 14. Show that, for any $m \times n$ matrix $\mathbf{A}$, $n \times q$ matrix $\mathbf{C}$, and $q \times p$ matrix $\mathbf{B}$,

$$\begin{aligned}\operatorname{rank}\{[\mathbf{I} - \mathbf{CB}(\mathbf{CB})^{-}]\mathbf{C}[\mathbf{I} - (\mathbf{AC})^{-}\mathbf{AC}]\} \\ = \operatorname{rank}(\mathbf{A}) + \operatorname{rank}(\mathbf{C}) - \operatorname{rank}(\mathbf{AC}) - n \\ + \operatorname{rank}\{[\mathbf{I} - \mathbf{CB}(\mathbf{CB})^{-}](\mathbf{I} - \mathbf{A}^{-}\mathbf{A})\}.\end{aligned}$$

Hint. Apply the equality

$$\operatorname{rank}(\mathbf{AC}) = \operatorname{rank}(\mathbf{A}) + \operatorname{rank}(\mathbf{C}) - n + \operatorname{rank}[(\mathbf{I} - \mathbf{CC}^{-})(\mathbf{I} - \mathbf{A}^{-}\mathbf{A})] \qquad (*)$$

to the product $\mathbf{A}(\mathbf{CB})$, and make use of the equality

$$\begin{aligned}\operatorname{rank}(\mathbf{ACB}) = \operatorname{rank}(\mathbf{AC}) + \operatorname{rank}(\mathbf{CB}) - \operatorname{rank}(\mathbf{C}) \\ + \operatorname{rank}\{[\mathbf{I} - \mathbf{CB}(\mathbf{CB})^{-}]\mathbf{C}[\mathbf{I} - (\mathbf{AC})^{-}\mathbf{AC}]\}. \qquad (**)\end{aligned}$$

Solution. Making use of equality ($*$) [or equivalently equality (5.8)], we find that

$$\begin{aligned}\text{rank}(\mathbf{ACB}) &= \text{rank}[\mathbf{A}(\mathbf{CB})] \\ &= \text{rank}(\mathbf{A}) + \text{rank}(\mathbf{CB}) - n \\ &\qquad + \text{rank}\{[\mathbf{I} - \mathbf{CB}(\mathbf{CB})^{-}](\mathbf{I} - \mathbf{A}^{-}\mathbf{A})\}. \qquad \text{(S.3)}\end{aligned}$$

And upon equating expression ($**$) [or equivalently expression (5.5)] to expression (S.3), we find that

$$\begin{aligned}&\text{rank}\{[\mathbf{I} - \mathbf{CB}(\mathbf{CB})^{-}]\mathbf{C}[\mathbf{I} - (\mathbf{AC})^{-}\mathbf{AC}]\} \\ &\qquad = \text{rank}(\mathbf{A}) + \text{rank}(\mathbf{C}) - \text{rank}(\mathbf{AC}) - n \\ &\qquad\qquad + \text{rank}\{[\mathbf{I} - \mathbf{CB}(\mathbf{CB})^{-}](\mathbf{I} - \mathbf{A}^{-}\mathbf{A})\}.\end{aligned}$$

EXERCISE 15. Show that if an $n \times n$ matrix $\mathbf{A}$ is the projection matrix for a subspace $\mathcal{U}$ of $\mathcal{R}^{n\times 1}$ along a subspace $\mathcal{V}$ of $\mathcal{R}^{n\times 1}$ (where $\mathcal{U} \oplus \mathcal{V} = \mathcal{R}^{n\times 1}$), then $\mathbf{A}'$ is the projection matrix for $\mathcal{V}^{\perp}$ along $\mathcal{U}^{\perp}$ [where $\mathcal{U}^{\perp}$ and $\mathcal{V}^{\perp}$ are the orthogonal complements (with respect to the usual inner product and relative to $\mathcal{R}^{n\times 1}$) of $\mathcal{U}$ and $\mathcal{V}$, respectively].

Solution. Suppose that $\mathbf{A}$ is the projection matrix for $\mathcal{U}$ along $\mathcal{V}$ (where $\mathcal{U} \oplus \mathcal{V} = \mathcal{R}^{n\times 1}$). Then, according to Theorem 17.6.14, $\mathbf{A}$ is idempotent, $\mathcal{U} = \mathcal{C}(\mathbf{A})$, and $\mathcal{V} = \mathcal{C}(\mathbf{I} - \mathbf{A})$. And, since (according to Lemma 10.1.2) $\mathbf{A}'$ is idempotent, it follows from Theorem 17.6.14 that $\mathbf{A}'$ is the projection matrix for $\mathcal{C}(\mathbf{A}')$ along $\mathcal{N}(\mathbf{A}')$. Moreover, making use of Corollary 11.7.2 and of Lemma 12.5.2, we find that

$$\mathcal{C}(\mathbf{A}') = \mathcal{N}(\mathbf{I} - \mathbf{A}') = \mathcal{C}^{\perp}(\mathbf{I} - \mathbf{A}) = \mathcal{V}^{\perp}$$

and that

$$\mathcal{N}(\mathbf{A}') = \mathcal{C}^{\perp}(\mathbf{A}) = \mathcal{U}^{\perp}.$$

EXERCISE 16. Show that, for any $n \times p$ matrix $\mathbf{X}$, $\mathbf{XX}^{-}$ is the projection matrix for $\mathcal{C}(\mathbf{X})$ along $\mathcal{N}(\mathbf{XX}^{-})$.

Solution. According to Lemma 10.2.5, $\mathbf{XX}^{-}$ is idempotent. Thus, it follows from Theorem 17.6.14 that $\mathbf{XX}^{-}$ is the projection matrix for $\mathcal{C}(\mathbf{XX}^{-})$ along $\mathcal{N}(\mathbf{XX}^{-})$. Moreover, according to Lemma 9.3.7, $\mathcal{C}(\mathbf{XX}^{-}) = \mathcal{C}(\mathbf{X})$.

EXERCISE 17. Let $\mathbf{Y}$ represent a matrix in a linear space $\mathcal{V}$ of $m \times n$ matrices, and let $\mathcal{U}_1, \ldots, \mathcal{U}_k$ represent subspaces of $\mathcal{V}$. Adopting the terminology and using the results of Exercise 6, show that if $\mathcal{U}_1, \ldots, \mathcal{U}_k$ are independent and if $\mathcal{U}_1 + \cdots + \mathcal{U}_k = \mathcal{V}$, then (a) there exist unique matrices $\mathbf{Z}_1, \ldots, \mathbf{Z}_k$ in $\mathcal{U}_1, \ldots, \mathcal{U}_k$, respectively, such that $\mathbf{Y} = \mathbf{Z}_1 + \cdots + \mathbf{Z}_k$ and (b) for $i = 1, \ldots, k$, $\mathbf{Z}_i$ equals the projection of $\mathbf{Y}$ on $\mathcal{U}_i$ along $\mathcal{U}_1 + \cdots + \mathcal{U}_{i-1} + \mathcal{U}_{i+1} + \cdots + \mathcal{U}_k$.

Solution. Suppose that $\mathcal{U}_1, \ldots, \mathcal{U}_k$ are independent and that $\mathcal{U}_1 + \cdots + \mathcal{U}_k = \mathcal{V}$.

(a) It follows from the very definition of a sum (of subspaces) that there exist matrices $\mathbf{Z}_1, \ldots, \mathbf{Z}_k$ in $\mathcal{U}_1, \ldots, \mathcal{U}_k$, respectively, such that $\mathbf{Y} = \mathbf{Z}_1 + \cdots + \mathbf{Z}_k$. For purposes of establishing the uniqueness of $\mathbf{Z}_1, \ldots, \mathbf{Z}_k$, let $\mathbf{Z}_1^*, \ldots, \mathbf{Z}_k^*$ represent matrices (potentially different from $\mathbf{Z}_1, \ldots, \mathbf{Z}_k$) in $\mathcal{U}_1, \ldots, \mathcal{U}_k$, respectively, such that $\mathbf{Y} = \mathbf{Z}_1^* + \cdots + \mathbf{Z}_k^*$. Then,

$$(\mathbf{Z}_1^* - \mathbf{Z}_1) + \cdots + (\mathbf{Z}_k^* - \mathbf{Z}_k) = \mathbf{Y} - \mathbf{Y} = \mathbf{0},$$

and (for $i = 1, \ldots, k$) $\mathbf{Z}_i^* - \mathbf{Z}_i \in \mathcal{U}_i$. Thus, $\mathbf{Z}_i^* - \mathbf{Z}_i = \mathbf{0}$ and hence $\mathbf{Z}_i^* = \mathbf{Z}_i$ $(i = 1, \ldots, k)$, thereby establishing the uniqueness of $\mathbf{Z}_1, \ldots, \mathbf{Z}_k$.

(b) That (for $i = 1, \ldots, k$) $\mathbf{Z}_i$ equals the projection of $\mathbf{Y}$ on $\mathcal{U}_i$ along $\mathcal{U}_1 + \cdots + \mathcal{U}_{i-1} + \mathcal{U}_{i+1} + \cdots + \mathcal{U}_k$ is evident upon observing that [as a consequence of Part (b) of Exercise 6] $\mathcal{U}_i$ and $\mathcal{U}_1 + \cdots + \mathcal{U}_{i-1} + \mathcal{U}_{i+1} + \cdots + \mathcal{U}_k$ are essentially disjoint and that

$$\mathbf{Y} - \mathbf{Z}_i = \mathbf{Z}_1 + \cdots + \mathbf{Z}_{i-1} + \mathbf{Z}_{i+1} + \cdots + \mathbf{Z}_k \in \mathcal{U}_1 + \cdots + \mathcal{U}_{i-1} + \mathcal{U}_{i+1} + \cdots + \mathcal{U}_k.$$

EXERCISE 18. Let $\mathcal{U}$ and $\mathcal{W}$ represent essentially disjoint subspaces (of $\mathcal{R}^{n \times 1}$) whose sum is $\mathcal{R}^{n \times 1}$, and let $\mathbf{U}$ represent any $n \times s$ matrix such that $\mathcal{C}(\mathbf{U}) = \mathcal{U}$ and $\mathbf{W}$ any $n \times t$ matrix such that $\mathcal{C}(\mathbf{W}) = \mathcal{W}$.

(a) Show that the $n \times (s + t)$ partitioned matrix $(\mathbf{U}, \mathbf{W})$ has a right inverse.

(b) Taking $\mathbf{R}$ to be an arbitrary right inverse of $(\mathbf{U}, \mathbf{W})$ and partitioning $\mathbf{R}$ as $\mathbf{R} = \begin{pmatrix} \mathbf{R}_1 \\ \mathbf{R}_2 \end{pmatrix}$ (where $\mathbf{R}_1$ has s rows), show that the projection matrix for $\mathcal{U}$ along $\mathcal{W}$ equals $\mathbf{U}\mathbf{R}_1$ and that the projection matrix for $\mathcal{W}$ along $\mathcal{U}$ equals $\mathbf{W}\mathbf{R}_2$.

Solution. (a) In light of result (1.4), we have that

$$\operatorname{rank}(\mathbf{U}, \mathbf{W}) = \dim[\mathcal{C}(\mathbf{U}, \mathbf{W})] = \dim(\mathcal{U} + \mathcal{W}) = \dim(\mathcal{R}^n) = n.$$

Thus, $(\mathbf{U}, \mathbf{W})$ is of full row rank, and it follows from Lemma 8.1.1 that $(\mathbf{U}, \mathbf{W})$ has a right inverse.

(b) For $j = 1, \ldots, n$, let $\mathbf{e}_j$ represent the jth column of $\mathbf{I}_n$; let $\mathbf{z}_j$ represent the projection of $\mathbf{e}_j$ on $\mathcal{U}$ along $\mathcal{W}$; let $\mathbf{r}_j$, $\mathbf{r}_{1j}$, and $\mathbf{r}_{2j}$ represent the jth columns of $\mathbf{R}$, $\mathbf{R}_1$, and $\mathbf{R}_2$, respectively, and observe that $\mathbf{r}_j = \begin{pmatrix} \mathbf{r}_{1j} \\ \mathbf{r}_{2j} \end{pmatrix}$.

By definition, $(\mathbf{U}, \mathbf{W})\mathbf{R} = \mathbf{I}_n$, implying that (for $j = 1, \ldots, n$) $(\mathbf{U}, \mathbf{W})\mathbf{r}_j = \mathbf{e}_j$. Thus, it follows from Corollary 17.6.5 that (for $j = 1, \ldots, n$) $\mathbf{z}_j = \mathbf{U}\mathbf{r}_{1j}$. We conclude (on the basis of Theorem 17.6.9) that the projection matrix for $\mathcal{U}$ along $\mathcal{W}$ equals

$$(\mathbf{z}_1, \ldots, \mathbf{z}_n) = (\mathbf{U}\mathbf{r}_{11}, \ldots, \mathbf{U}\mathbf{r}_{1n}) = \mathbf{U}\mathbf{R}_1.$$

And, since $\mathbf{U}\mathbf{R}_1 + \mathbf{W}\mathbf{R}_2 = \mathbf{I}_n$, we further conclude (on the basis of Theorem 17.6.10) that the projection matrix for $\mathcal{W}$ along $\mathcal{U}$ equals $\mathbf{I} - \mathbf{U}\mathbf{R}_1 = \mathbf{W}\mathbf{R}_2$.

EXERCISE 19. Let **A** represent the $(n \times n)$ projection matrix for a subspace $\mathcal{U}$ of $\mathcal{R}^{n\times 1}$ along a subspace $\mathcal{V}$ of $\mathcal{R}^{n\times 1}$ (where $\mathcal{U} \oplus \mathcal{V} = \mathcal{R}^{n\times 1}$), let **B** represent the $(n \times n)$ projection matrix for a subspace $\mathcal{W}$ of $\mathcal{R}^{n\times 1}$ along a subspace $\mathcal{X}$ of $\mathcal{R}^{n\times 1}$ (where $\mathcal{W} \oplus \mathcal{X} = \mathcal{R}^{n\times 1}$), and suppose that **A** and **B** commute (i.e., that **BA**= **AB**).

(a) Show that **AB** is the projection matrix for $\mathcal{U} \cap \mathcal{W}$ along $\mathcal{V} + \mathcal{X}$.

(b) Show that **A** + **B** – **AB** is the projection matrix for $\mathcal{U} + \mathcal{W}$ along $\mathcal{V} \cap \mathcal{X}$. [*Hint for Part (b).* Observe that $\mathbf{I} - (\mathbf{A} + \mathbf{B} - \mathbf{AB}) = (\mathbf{I} - \mathbf{A})(\mathbf{I} - \mathbf{B})$, and make use of Part (a).]

Solution. (a) According to Theorem 17.6.13, **A** and **B** are both idempotent, so that

$$(\mathbf{AB})^2 = \mathbf{A}(\mathbf{BA})\mathbf{B} = \mathbf{A}(\mathbf{AB})\mathbf{B} = \mathbf{A}^2\mathbf{B}^2 = \mathbf{AB}.$$

Thus, **AB** is idempotent, and it follows from Theorem 17.6.14 that **AB** is the projection matrix for $\mathcal{C}(\mathbf{AB})$ along $\mathcal{N}(\mathbf{AB})$.

It remains to show that $\mathcal{C}(\mathbf{AB}) = \mathcal{U} \cap \mathcal{W}$ and $\mathcal{N}(\mathbf{AB}) = \mathcal{V} + \mathcal{X}$ or equivalently (in light of Theorem 17.6.14) that $\mathcal{C}(\mathbf{AB}) = \mathcal{C}(\mathbf{A}) \cap \mathcal{C}(\mathbf{B})$ and $\mathcal{N}(\mathbf{AB}) = \mathcal{N}(\mathbf{A}) + \mathcal{N}(\mathbf{B})$. Clearly, $\mathcal{C}(\mathbf{AB}) \subset \mathcal{C}(\mathbf{A})$ and (since $\mathbf{AB} = \mathbf{BA}$) $\mathcal{C}(\mathbf{AB}) \subset \mathcal{C}(\mathbf{B})$, so that $\mathcal{C}(\mathbf{AB}) \subset \mathcal{C}(\mathbf{A}) \cap \mathcal{C}(\mathbf{B})$. And, for any vector **y** in $\mathcal{C}(\mathbf{A}) \cap \mathcal{C}(\mathbf{B})$, it follows from Lemma 17.6.7 that $\mathbf{y} = \mathbf{Ay}$ and $\mathbf{y} = \mathbf{By}$, implying that $\mathbf{y} = \mathbf{ABy}$ and hence that $\mathbf{y} \in \mathcal{C}(\mathbf{AB})$. Thus, $\mathcal{C}(\mathbf{A}) \cap \mathcal{C}(\mathbf{B}) \subset \mathcal{C}(\mathbf{AB})$, and hence [since $\mathcal{C}(\mathbf{AB}) \subset \mathcal{C}(\mathbf{A}) \cap \mathcal{C}(\mathbf{B})$] $\mathcal{C}(\mathbf{AB}) = \mathcal{C}(\mathbf{A}) \cap \mathcal{C}(\mathbf{B})$.

Further, for any vector **x** in $\mathcal{N}(\mathbf{A})$ and any vector **y** in $\mathcal{N}(\mathbf{B})$,

$$\mathbf{AB}(\mathbf{x} + \mathbf{y}) = \mathbf{ABx} + \mathbf{ABy} = \mathbf{BAx} + \mathbf{ABy} = \mathbf{0} + \mathbf{0} = \mathbf{0},$$

implying that $\mathbf{x} + \mathbf{y} \in \mathcal{N}(\mathbf{AB})$. Thus, $\mathcal{N}(\mathbf{A}) + \mathcal{N}(\mathbf{B}) \subset \mathcal{N}(\mathbf{AB})$. And, for any vector **z** in $\mathcal{N}(\mathbf{AB})$ (i.e., any vector **z** such that $\mathbf{ABz} = \mathbf{0}$), $\mathbf{Bz} \in \mathcal{N}(\mathbf{A})$, which since $\mathbf{z} = \mathbf{Bz} + (\mathbf{I} - \mathbf{B})\mathbf{z}$ and since $(\mathbf{I} - \mathbf{B})\mathbf{z} \in \mathcal{N}(\mathbf{B})$ [as is evident from Theorem 11.7.1 or upon observing that $\mathbf{B}(\mathbf{I} - \mathbf{B})\mathbf{z} = (\mathbf{B} - \mathbf{B}^2)\mathbf{z} = \mathbf{0}$] implies that $\mathbf{z} \in \mathcal{N}(\mathbf{A}) + \mathcal{N}(\mathbf{B})$. It follows that $\mathcal{N}(\mathbf{AB}) \subset \mathcal{N}(\mathbf{A}) + \mathcal{N}(\mathbf{B})$, and hence [since $\mathcal{N}(\mathbf{A}) + \mathcal{N}(\mathbf{B}) \subset \mathcal{N}(\mathbf{AB})$] that $\mathcal{N}(\mathbf{AB}) = \mathcal{N}(\mathbf{A}) + \mathcal{N}(\mathbf{B})$.

(b) As a consequence of Theorem 17.6.10, $\mathbf{I} - \mathbf{A}$ is the projection matrix for $\mathcal{V}$ along $\mathcal{U}$, and $\mathbf{I} - \mathbf{B}$ is the projection matrix for $\mathcal{X}$ along $\mathcal{W}$. Thus, it follows from Part (a) that $(\mathbf{I} - \mathbf{A})(\mathbf{I} - \mathbf{B})$ is the projection matrix for $\mathcal{V} \cap \mathcal{X}$ along $\mathcal{U} + \mathcal{W}$. Observing that $\mathbf{A} + \mathbf{B} - \mathbf{AB} = \mathbf{I} - (\mathbf{I} - \mathbf{A})(\mathbf{I} - \mathbf{B})$, we conclude, on the basis of Theorem 17.6.10, that $\mathbf{A} + \mathbf{B} - \mathbf{AB}$ is the projection matrix for $\mathcal{U} + \mathcal{W}$ along $\mathcal{V} \cap \mathcal{X}$.

EXERCISE 20. Let $\mathcal{V}$ represent a linear space of n–dimensional column vectors, and let $\mathcal{U}$ and $\mathcal{W}$ represent essentially disjoint subspaces whose sum is $\mathcal{V}$. Then, an $n \times n$ matrix **A** is said to be a *projection matrix for* $\mathcal{U}$ *along* $\mathcal{W}$ if **Ay** is the projection of **y** on $\mathcal{U}$ along $\mathcal{W}$ for every $\mathbf{y} \in \mathcal{V}$ — this represents an extension of the definition of a projection matrix for $\mathcal{U}$ along $\mathcal{W}$ in the special case where $\mathcal{V} = \mathcal{R}^n$. Further, let **U** represent an $n \times s$ matrix such that $\mathcal{C}(\mathbf{U}) = \mathcal{U}$, and let **W** represent an $n \times t$ matrix such that $\mathcal{C}(\mathbf{W}) = \mathcal{W}$.

(a) Show that an $n \times n$ matrix $\mathbf{A}$ is a projection matrix for $\mathcal{U}$ along $\mathcal{W}$ if and only if $\mathbf{AU} = \mathbf{U}$ and $\mathbf{AW} = \mathbf{0}$ or, equivalently, if and only if $\mathbf{A}'$ is a solution to the linear system $\begin{pmatrix} \mathbf{U}' \\ \mathbf{W}' \end{pmatrix} \mathbf{B} = \begin{pmatrix} \mathbf{U}' \\ \mathbf{0} \end{pmatrix}$ (in an $n \times n$ matrix $\mathbf{B}$).

(b) Establish the existence of a projection matrix for $\mathcal{U}$ along $\mathcal{W}$.

(c) Show that if $\mathbf{A}$ is a projection matrix for $\mathcal{U}$ along $\mathcal{W}$, then $\mathbf{I} - \mathbf{A}$ is a projection matrix for $\mathcal{W}$ along $\mathcal{U}$.

(d) Let $\mathbf{X}$ represent any $n \times p$ matrix whose columns span $\mathcal{N}(\mathbf{W}')$ or, equivalently, $\mathcal{W}^{\perp}$. Show that an $n \times n$ matrix $\mathbf{A}$ is a projection matrix for $\mathcal{U}$ along $\mathcal{W}$ if and only if $\mathbf{A}' = \mathbf{XR}_*$ for some solution $\mathbf{R}_*$ to the linear system $\mathbf{U}'\mathbf{XR} = \mathbf{U}'$ (in a $p \times n$ matrix $\mathbf{R}$).

Solution. (a) Clearly, an $n \times 1$ vector $\mathbf{y}$ is in $\mathcal{V}$ if and only if $\mathbf{y}$ is expressible as $\mathbf{y} = \mathbf{Ub} + \mathbf{Wc}$ for some vectors $\mathbf{b}$ and $\mathbf{c}$. Thus, an $n \times n$ matrix $\mathbf{A}$ is a projection matrix for $\mathcal{U}$ along $\mathcal{W}$ if and only if, for every $(s \times 1)$ vector $\mathbf{b}$ and every $(t \times 1)$ vector $\mathbf{c}$, $\mathbf{A}(\mathbf{Ub}+\mathbf{Wc})$ is the projection of $\mathbf{Ub}+\mathbf{Wc}$ on $\mathcal{U}$ along $\mathcal{W}$, or equivalently (in light of Corollary 17.6.2) if and only if, for every $\mathbf{b}$ and every $\mathbf{c}$, $\mathbf{A}(\mathbf{Ub} + \mathbf{Wc}) = \mathbf{Ub}$.

Now, if $\mathbf{AU} = \mathbf{U}$ and $\mathbf{AW} = \mathbf{0}$, then obviously $\mathbf{A}(\mathbf{Ub} + \mathbf{Wc}) = \mathbf{Ub}$ for every $\mathbf{b}$ and every $\mathbf{c}$. Conversely, suppose that $\mathbf{A}(\mathbf{Ub} + \mathbf{Wc}) = \mathbf{Ub}$ for every $\mathbf{b}$ and every $\mathbf{c}$. Then, $\mathbf{A}(\mathbf{Ub} + \mathbf{Wc}) = \mathbf{Ub}$ for every $\mathbf{b}$ and for $\mathbf{c} = \mathbf{0}$, or equivalently $\mathbf{AUb} = \mathbf{Ub}$ for every $\mathbf{b}$, implying (in light of Lemma 2.3.2) that $\mathbf{AU} = \mathbf{U}$. Similarly, $\mathbf{A}(\mathbf{Ub}+\mathbf{Wc}) = \mathbf{Ub}$ for $\mathbf{b} = \mathbf{0}$ and for every $\mathbf{c}$, or equivalently $\mathbf{AWc} = \mathbf{0}$ for every $\mathbf{c}$, implying that $\mathbf{AW} = \mathbf{0}$.

(b) Clearly, the linear systems $\mathbf{U}'\mathbf{B} = \mathbf{U}'$ and $\mathbf{W}'\mathbf{B} = \mathbf{0}$ (in $\mathbf{B}$) are both consistent. And, since (in light of Lemma 17.2.1) $\mathcal{R}(\mathbf{U}')$ and $\mathcal{R}(\mathbf{W}')$ are essentially disjoint, we have, as a consequence of Theorem 17.3.2, that the combined linear system $\begin{pmatrix} \mathbf{U}' \\ \mathbf{W}' \end{pmatrix} \mathbf{B} = \begin{pmatrix} \mathbf{U}' \\ \mathbf{0} \end{pmatrix}$ is consistent. Thus, the existence of a projection matrix for $\mathcal{U}$ along $\mathcal{W}$ follows from Part (a).

(c) Suppose that $\mathbf{A}$ is a projection matrix for $\mathcal{U}$ along $\mathcal{W}$. Then, according to Part (a), $\mathbf{AU} = \mathbf{U}$ and $\mathbf{AW} = \mathbf{0}$. Thus, $(\mathbf{I} - \mathbf{A})\mathbf{W} = \mathbf{W}$, and $(\mathbf{I} - \mathbf{A})\mathbf{U} = \mathbf{0}$. We conclude [on the basis of Part (a)] that $\mathbf{I} - \mathbf{A}$ is a projection matrix for $\mathcal{W}$ along $\mathcal{U}$.

(d) In light of Part (a), it suffices to show that $\mathbf{A}'$ is a solution to the linear system $\begin{pmatrix} \mathbf{U}' \\ \mathbf{W}' \end{pmatrix} \mathbf{B} = \begin{pmatrix} \mathbf{U}' \\ \mathbf{0} \end{pmatrix}$ (in $\mathbf{B}$) if and only if $\mathbf{A}' = \mathbf{XR}_*$ for some solution $\mathbf{R}_*$ to the linear system $\mathbf{U}'\mathbf{XR} = \mathbf{U}'$.

Suppose that $\mathbf{A}' = \mathbf{XR}_*$ for some solution $\mathbf{R}_*$ to $\mathbf{U}'\mathbf{XR} = \mathbf{U}'$. Then, $\mathbf{U}'\mathbf{A}' = \mathbf{U}'$, and (since clearly $\mathbf{W}'\mathbf{X} = \mathbf{0}$) $\mathbf{W}'\mathbf{A}' = \mathbf{0}$. Thus, $\mathbf{A}'$ is a solution to $\begin{pmatrix} \mathbf{U}' \\ \mathbf{W}' \end{pmatrix} \mathbf{B} = \begin{pmatrix} \mathbf{U}' \\ \mathbf{0} \end{pmatrix}$.

Conversely, suppose that $\mathbf{A}'$ is a solution to $\begin{pmatrix} \mathbf{U}' \\ \mathbf{W}' \end{pmatrix} \mathbf{B} = \begin{pmatrix} \mathbf{U}' \\ \mathbf{0} \end{pmatrix}$ or equivalently that $\mathbf{U}'\mathbf{A}' = \mathbf{U}'$ and $\mathbf{W}'\mathbf{A}' = \mathbf{0}$. Then, according to Lemma 11.4.1, $\mathcal{C}(\mathbf{A}') \subset \mathcal{N}(\mathbf{W}')$, or equivalently $\mathcal{C}(\mathbf{A}') \subset \mathcal{C}(\mathbf{X})$, and consequently $\mathbf{A}' = \mathbf{XR}_*$ for some matrix $\mathbf{R}_*$.

And, $\mathbf{U}'\mathbf{X}\mathbf{R}_* = \mathbf{U}'\mathbf{A}' = \mathbf{U}'$, so that $\mathbf{R}_*$ is a solution to $\mathbf{U}'\mathbf{X}\mathbf{R} = \mathbf{U}'$.

EXERCISE 21. Let $\mathcal{U}_1, \ldots, \mathcal{U}_k$ represent independent subspaces of $\mathcal{R}^{n\times 1}$ such that $\mathcal{U}_1 + \cdots + \mathcal{U}_k = \mathcal{R}^{n\times 1}$ (where the independence of subspaces is as defined in Exercise 6). Further, letting $s_i = \dim(\mathcal{U}_i)$ (and supposing that $s_i > 0$), take $\mathbf{U}_i$ to be any $n \times s_i$ matrix such that $\mathcal{C}(\mathbf{U}_i) = \mathcal{U}_i$ $(i = 1, \ldots, k)$. And, define $\mathbf{B} = (\mathbf{U}_1, \ldots, \mathbf{U}_k)^{-1}$, partition $\mathbf{B}$ as $\mathbf{B} = \begin{pmatrix} \mathbf{B}_1 \\ \vdots \\ \mathbf{B}_k \end{pmatrix}$ (where, for $i = 1, \ldots, k$, $\mathbf{B}_i$ has s_i rows), and let $\mathbf{H} = \mathbf{B}'\mathbf{B}$ or (more generally) let $\mathbf{H}$ represent any matrix of the form

$$\mathbf{H} = \mathbf{B}_1'\mathbf{A}_1\mathbf{B}_1 + \mathbf{B}_2'\mathbf{A}_2\mathbf{B}_2 + \cdots + \mathbf{B}_k'\mathbf{A}_k\mathbf{B}_k, \tag{E.3}$$

where $\mathbf{A}_1, \mathbf{A}_2, \ldots, \mathbf{A}_k$ are symmetric positive definite matrices.

(a) Using the result of Part (g)-(1) of Exercise 6 (or otherwise), verify that the partitioned matrix $(\mathbf{U}_1, \ldots, \mathbf{U}_k)$ is nonsingular (i.e., is square and of rank n).

(b) Show that $\mathbf{H}$ is positive definite.

(c) Show that (for $j \neq i = 1, \ldots, k$) $\mathcal{U}_i$ and $\mathcal{U}_j$ are orthogonal with respect to $\mathbf{H}$.

(d) Using the result of Part(a)-(2) of Exercise 3 (or otherwise), show that, for $i = 1, \ldots, k$, (1) $\mathcal{U}_1 + \cdots + \mathcal{U}_{i-1} + \mathcal{U}_{i+1} + \cdots + \mathcal{U}_k$ equals the orthogonal complement $\mathcal{U}_i^{\perp}$ of $\mathcal{U}_i$ (where the orthogonality in the orthogonal complement is with respect to the bilinear form $\mathbf{x}'\mathbf{H}\mathbf{y}$) and (2) the projection of any $n \times 1$ vector $\mathbf{y}$ on $\mathcal{U}_i$ along $\mathcal{U}_1 + \cdots + \mathcal{U}_{i-1} + \mathcal{U}_{i+1} + \cdots + \mathcal{U}_k$ equals the orthogonal projection of $\mathbf{y}$ on $\mathcal{U}_i$ with respect to $\mathbf{H}$.

(e) Show that if, for $j \neq i = 1, \ldots, k$, $\mathcal{U}_i$ and $\mathcal{U}_j$ are orthogonal with respect to some symmetric positive definite matrix $\mathbf{H}_*$, then $\mathbf{H}_*$ is expressible in the form (E.3).

Solution. (a) Clearly, $\dim(\mathcal{U}_1 + \cdots + \mathcal{U}_k) = \dim(\mathcal{R}^{n\times 1}) = n$. Thus, making use of Part (g)-(1) of Exercise 6, we find that

$$s_1 + \cdots + s_k = \dim(\mathcal{U}_1) + \cdots + \dim(\mathcal{U}_k) = \dim(\mathcal{U}_1 + \cdots + \mathcal{U}_k) = n.$$

And, making use of result (1.6), we find that

$$\begin{aligned} \operatorname{rank}(\mathbf{U}_1, \ldots, \mathbf{U}_k) = \dim[\mathcal{C}(\mathbf{U}_1, \ldots, \mathbf{U}_k)] &= \dim[\mathcal{C}(\mathbf{U}_1) + \cdots + \mathcal{C}(\mathbf{U}_k)] \\ &= \dim(\mathcal{U}_1 + \cdots + \mathcal{U}_k) = n. \end{aligned}$$

(b) Clearly, $\mathbf{H} = \mathbf{B}' \operatorname{diag}(\mathbf{A}_1, \ldots, \mathbf{A}_k)\mathbf{B}$. Thus, since (according to Lemma 14.8.3) $\operatorname{diag}(\mathbf{A}_1, \ldots, \mathbf{A}_k)$ is positive definite, it follows from Corollary 14.2.10 that $\mathbf{H}$ is positive definite.

(c) In light of Lemma 14.12.1, it suffices to show that (for $j \neq i$) $\mathbf{U}_i'\mathbf{H}\mathbf{U}_j = \mathbf{0}$.

By definition,

$$\begin{pmatrix} \mathbf{B}_1\mathbf{U}_1 & \mathbf{B}_1\mathbf{U}_2 & \dots & \mathbf{B}_1\mathbf{U}_k \\ \mathbf{B}_2\mathbf{U}_1 & \mathbf{B}_2\mathbf{U}_2 & \dots & \mathbf{B}_2\mathbf{U}_k \\ \vdots & \vdots & \ddots & \vdots \\ \mathbf{B}_k\mathbf{U}_1 & \mathbf{B}_k\mathbf{U}_2 & \dots & \mathbf{B}_k\mathbf{U}_k \end{pmatrix}$$

$$= \mathbf{B}(\mathbf{U}_1, \mathbf{U}_2, \dots, \mathbf{U}_k) = \mathbf{I}_n = \begin{pmatrix} \mathbf{I}_{s_1} & \mathbf{0} & \dots & \mathbf{0} \\ \mathbf{0} & \mathbf{I}_{s_2} & & \mathbf{0} \\ \vdots & & \ddots & \vdots \\ \mathbf{0} & \mathbf{0} & & \mathbf{I}_{s_k} \end{pmatrix},$$

implying in particular that (for $j \neq i$) $\mathbf{B}_j\mathbf{U}_i = \mathbf{0}$ and (for $r \neq j$) $\mathbf{B}_r\mathbf{U}_j = \mathbf{0}$. Thus, for $j \neq i$,

$$\mathbf{U}_i'\mathbf{H}\mathbf{U}_j = (\mathbf{B}_j\mathbf{U}_i)'\mathbf{A}_j\mathbf{B}_j\mathbf{U}_j + \sum_{r \neq j} \mathbf{U}_i'\mathbf{B}_r'\mathbf{A}_r\mathbf{B}_r\mathbf{U}_j = \mathbf{0} + \mathbf{0} = \mathbf{0}.$$

(d) (1) According to Part (c), $\mathcal{U}_i$ is orthogonal to $\mathcal{U}_1, \dots, \mathcal{U}_{i-1}, \mathcal{U}_{i+1}, \dots, \mathcal{U}_k$. Thus, making repeated ($k-2$ times) use of Part (a)-(2) of Exercise 3, we find that $\mathcal{U}_i$ is orthogonal to $\mathcal{U}_1 + \dots + \mathcal{U}_{i-1} + \mathcal{U}_{i+1} + \dots + \mathcal{U}_k$. We conclude (on the basis of Lemma 17.7.2) that $\mathcal{U}_1 + \dots + \mathcal{U}_{i-1} + \mathcal{U}_{i+1} + \dots + \mathcal{U}_k = \mathcal{U}_i^{\perp}$.

(2) That the projection of $\mathbf{y}$ on $\mathcal{U}_i$ along $\mathcal{U}_1 + \dots + \mathcal{U}_{i-1} + \mathcal{U}_{i+1} + \dots + \mathcal{U}_k$ equals the orthogonal projection of $\mathbf{y}$ on $\mathcal{U}_i$ (with respect to $\mathbf{H}$) is [in light of Part (1)] evident from Theorem 17.6.6.

(e) Suppose that, for $j \neq i = 1, \dots, k, \mathcal{U}_i$ and $\mathcal{U}_j$ are orthogonal with respect to some symmetric positive definite matrix $\mathbf{H}_*$. Then, according to Corollary 14.3.13, there exists an $n \times n$ nonsingular matrix $\mathbf{P}$ such that $\mathbf{H}_* = \mathbf{P}'\mathbf{P}$. Further,

$$\mathbf{P} = \mathbf{P}\mathbf{I}_n = \mathbf{P}(\mathbf{U}_1, \dots, \mathbf{U}_k)\begin{pmatrix} \mathbf{B}_1 \\ \vdots \\ \mathbf{B}_k \end{pmatrix} = \mathbf{L}_1\mathbf{B}_1 + \dots + \mathbf{L}_k\mathbf{B}_k,$$

where (for $i = 1, \dots, k$) $\mathbf{L}_i = \mathbf{P}\mathbf{U}_i$. And, making use of Lemma 14.12.1, we find that, for $j \neq i = 1, \dots, k$,

$$\mathbf{L}_i'\mathbf{L}_j = \mathbf{U}_i'\mathbf{P}'\mathbf{P}\mathbf{U}_j = \mathbf{U}_i'\mathbf{H}_*\mathbf{U}_j = \mathbf{0}.$$

Thus,

$$\begin{aligned} \mathbf{H}_* &= (\mathbf{L}_1\mathbf{B}_1 + \dots + \mathbf{L}_k\mathbf{B}_k)'(\mathbf{L}_1\mathbf{B}_1 + \dots + \mathbf{L}_k\mathbf{B}_k) \\ &= \mathbf{B}_1'\mathbf{L}_1'\mathbf{L}_1\mathbf{B}_1 + \mathbf{B}_2'\mathbf{L}_2'\mathbf{L}_2\mathbf{B}_2 + \dots + \mathbf{B}_k'\mathbf{L}_k'\mathbf{L}_k\mathbf{B}_k \\ &= \mathbf{B}_1'\mathbf{A}_1\mathbf{B}_1 + \mathbf{B}_2'\mathbf{A}_2\mathbf{B}_2 + \dots + \mathbf{B}_k'\mathbf{A}_k\mathbf{B}_k, \end{aligned}$$

where (for $i = 1, \dots, k$) $\mathbf{A}_i = \mathbf{L}_i'\mathbf{L}_i$ [which is a symmetric positive definite matrix, as is evident from Corollary 14.2.14 upon observing that $\text{rank}(\mathbf{L}_i) = \text{rank}(\mathbf{P}\mathbf{U}_i) = \text{rank}(\mathbf{U}_i) = s_i$].

18

Sums (and Differences) of Matrices

EXERCISE 1. Let $\mathbf{R}$ represent an $n \times n$ matrix, $\mathbf{S}$ an $n \times m$ matrix, $\mathbf{T}$ an $m \times m$ matrix, and $\mathbf{U}$ an $m \times n$ matrix. Derive (for the special case where $\mathbf{R}$ and $\mathbf{T}$ are nonsingular), the formula

$$|\mathbf{R} + \mathbf{STU}| = |\mathbf{R}|\ |\mathbf{T} + \mathbf{TUR}^{-1}\mathbf{ST}|/|\mathbf{T}|.$$

Do so by making two applications of the formula

$$\begin{vmatrix} \mathbf{T} & \mathbf{U} \\ \mathbf{V} & \mathbf{W} \end{vmatrix} = \begin{vmatrix} \mathbf{W} & \mathbf{V} \\ \mathbf{U} & \mathbf{T} \end{vmatrix} = |\mathbf{T}|\ |\mathbf{W} - \mathbf{VT}^{-1}\mathbf{U}|. \qquad (*)$$

(in which $\mathbf{V}$ is an $n \times m$ matrix and $\mathbf{W}$ an $n \times n$ matrix and in which $\mathbf{T}$ is assumed to be nonsingular) to the partitioned matrix $\begin{vmatrix} \mathbf{R} & -\mathbf{ST} \\ \mathbf{TU} & \mathbf{T} \end{vmatrix}$ — one with $\mathbf{W}$ set equal to $\mathbf{R}$, and the other with $\mathbf{T}$ set equal to $\mathbf{R}$.

Solution. Suppose that $\mathbf{R}$ and $\mathbf{T}$ are nonsingular. Then, making use of formula $(*)$ (or equivalently the formula of Theorem 13.3.8), we find that

$$\begin{vmatrix} \mathbf{R} & -\mathbf{ST} \\ \mathbf{TU} & \mathbf{T} \end{vmatrix} = |\mathbf{T}|\ |\mathbf{R} - (-\mathbf{ST})\mathbf{T}^{-1}\mathbf{TU}| = |\mathbf{T}|\ |\mathbf{R} + \mathbf{STU}|$$

and also that

$$\begin{vmatrix} \mathbf{R} & -\mathbf{ST} \\ \mathbf{TU} & \mathbf{T} \end{vmatrix} = |\mathbf{R}|\ |\mathbf{T} - (\mathbf{TU})\mathbf{R}^{-1}(-\mathbf{ST})| = |\mathbf{R}|\ |\mathbf{T} + \mathbf{TUR}^{-1}\mathbf{ST}|.$$

Thus,

$$|\mathbf{T}|\ |\mathbf{R} + \mathbf{STU}| = |\mathbf{R}|\ |\mathbf{T} + \mathbf{TUR}^{-1}\mathbf{ST}|,$$

or equivalently

$$|\mathbf{R}+\mathbf{STU}| = |\mathbf{R}|\,|\mathbf{T}+\mathbf{TUR}^{-1}\mathbf{ST}|/|\mathbf{T}|.$$

EXERCISE 2. Let **R** represent an $n \times n$ matrix, **S** an $n \times m$ matrix, **T** an $m \times m$ matrix, and **U** an $m \times n$ matrix. Show that if **R** is nonsingular, then

$$|\mathbf{R}+\mathbf{STU}| = |\mathbf{R}|\,|\mathbf{I}_m+\mathbf{UR}^{-1}\mathbf{ST}| = |\mathbf{R}|\,|\mathbf{I}_m+\mathbf{TUR}^{-1}\mathbf{S}|.$$

Do so by using the formula

$$|\mathbf{R}+\mathbf{STU}| = |\mathbf{R}|\,|\mathbf{T}|\,|\mathbf{T}^{-1}+\mathbf{UR}^{-1}\mathbf{S}|, \qquad (*)$$

(in which **R** and **T** are assumed to be nonsingular), or alternatively the formula $|\mathbf{I}_n+\mathbf{SU}| = |\mathbf{I}_m+\mathbf{US}|$ or the formula $|\mathbf{R}+\mathbf{STU}| = |\mathbf{R}|\,|\mathbf{T}+\mathbf{TUR}^{-1}\mathbf{ST}|/|\mathbf{T}|$.

Solution. Note that

$$\mathbf{R}+\mathbf{STU} = \mathbf{R}+(\mathbf{ST})\mathbf{I}_m\mathbf{U}, \qquad (\text{S}.1)$$

$$\mathbf{R}+\mathbf{STU} = \mathbf{R}+\mathbf{SI}_m(\mathbf{TU}). \qquad (\text{S}.2)$$

Now, suppose that **R** is nonsingular. By applying formula $(*)$ (or equivalently the formula of Theorem 18.1.1) to the right side of equality (S.1) [i.e., by applying formula $(*)$ with **ST** and $\mathbf{I}_m$ in place of **S** and **T**, respectively], we find that

$$|\mathbf{R}+\mathbf{STU}| = |\mathbf{R}|\,|\mathbf{I}_m|\,|\mathbf{I}_m^{-1}+\mathbf{UR}^{-1}\mathbf{ST}| = |\mathbf{R}|\,|\mathbf{I}_m+\mathbf{UR}^{-1}\mathbf{ST}|.$$

Similarly, by applying formula $(*)$ to the right side of equality (S.2) [i.e., by applying formula $(*)$ with $\mathbf{I}_m$ and **TU** in place of **T** and **U**, respectively], we find that

$$|\mathbf{R}+\mathbf{STU}| = |\mathbf{R}|\,|\mathbf{I}_m|\,|\mathbf{I}_m^{-1}+\mathbf{TUR}^{-1}\mathbf{S}| = |\mathbf{R}|\,|\mathbf{I}_m+\mathbf{TUR}^{-1}\mathbf{S}|.$$

EXERCISE 3. Let **A** represent an $n \times n$ symmetric nonnegative definite matrix. Show that if $\mathbf{I}-\mathbf{A}$ is nonnegative definite and if $|\mathbf{A}| = 1$, then $\mathbf{A} = \mathbf{I}$.

Solution. Suppose that $\mathbf{I}-\mathbf{A}$ is nonnegative definite and that $|\mathbf{A}| = 1$. Then, as a consequence of Corollary 14.3.12, **A** is positive definite. And,

$$|\mathbf{I}| = 1 = |\mathbf{A}|.$$

Thus, it follows from Corollary 18.1.7 (specifically from the special case of Corollary 18.1.7 where $\mathbf{C} = \mathbf{I}$) that $\mathbf{I} = \mathbf{A}$.

EXERCISE 4. Show that, for any $n \times n$ symmetric nonnegative definite matrix **B** and for any $n \times n$ symmetric matrix **C** such that $\mathbf{C}-\mathbf{B}$ is nonnegative definite,

$$|\mathbf{C}| \geq |\mathbf{C}-\mathbf{B}|,$$

with equality holding if and only if $\mathbf{C}$ is singular or $\mathbf{B} = \mathbf{0}$.

Solution. Let $\mathbf{A} = \mathbf{C} - \mathbf{B}$. Then, $\mathbf{C} - \mathbf{A} = \mathbf{B}$. So, by definition, $\mathbf{A}$ is a (symmetric) nonnegative definite matrix, and $\mathbf{C} - \mathbf{A}$ is nonnegative definite. Thus, it follows from Corollary 18.1.8 that

$$|\mathbf{C}| \geq |\mathbf{C} - \mathbf{B}|,$$

with equality holding if and only if $\mathbf{C}$ is singular or $\mathbf{C} = \mathbf{C} - \mathbf{B}$, or equivalently if and only if $\mathbf{C}$ is singular or $\mathbf{B} = \mathbf{0}$.

EXERCISE 5. Let $\mathbf{A}$ represent a symmetric nonnegative definite matrix that has been partitioned as

$$\mathbf{A} = \begin{pmatrix} \mathbf{T} & \mathbf{U} \\ \mathbf{U}' & \mathbf{W} \end{pmatrix},$$

where $\mathbf{T}$ is of dimensions $m \times m$ and $\mathbf{W}$ of dimensions $n \times n$ (and where $\mathbf{U}$ is of dimensions $m \times n$). And, define $\mathbf{Q} = \mathbf{W} - \mathbf{U}'\mathbf{T}^{-}\mathbf{U}$ (which is the Schur complement of $\mathbf{T}$).

(a) Using the result that the symmetry and nonnegative definiteness of $\mathbf{A}$ imply the nonnegative definiteness of $\mathbf{Q}$ and the result of Exercise 14.33 (or otherwise), show that

$$|\mathbf{W}| \geq |\mathbf{U}'\mathbf{T}^{-}\mathbf{U}|,$$

with equality holding if and only if $\mathbf{W}$ is singular or $\mathbf{Q} = \mathbf{0}$.

(b) Suppose that $n = m$ and that $\mathbf{T}$ is nonsingular. Show that

$$|\mathbf{W}|\ |\mathbf{T}| \geq |\mathbf{U}|^2,$$

with equality holding if and only if $\mathbf{W}$ is singular or $\operatorname{rank}(\mathbf{A}) = m$.

(c) Suppose that $n = m$ and that $\mathbf{A}$ is positive definite. Show that

$$|\mathbf{W}|\ |\mathbf{T}| > |\mathbf{U}|^2.$$

Solution. (a) According to the result of Exercise 14.33, $\mathbf{U}'\mathbf{T}^{-}\mathbf{U}$ is symmetric and nonnegative definite. Further, $\mathbf{W}$ is symmetric. And, in light of the result that the symmetry and nonnegative definiteness of $\mathbf{A}$ imply the nonnegative definiteness of $\mathbf{Q} = \mathbf{W} - \mathbf{U}'\mathbf{T}^{-}\mathbf{U}$ [a result that is implicit in Parts (1) and (2) of Theorem 14.8.4], it follows from Corollary 18.1.8 that

$$|\mathbf{W}| \geq |\mathbf{U}'\mathbf{T}^{-}\mathbf{U}|,$$

with equality holding if and only if $\mathbf{W}$ is singular or $\mathbf{W} = \mathbf{U}'\mathbf{T}^{-}\mathbf{U}$, or equivalently if and only if $\mathbf{W}$ is singular or $\mathbf{Q} = \mathbf{0}$.

(b) Since (in light of Corollary 14.2.12) $|\mathbf{T}| > 0$ and since

$$|\mathbf{U}'\mathbf{T}^{-1}\mathbf{U}| = |\mathbf{U}'|\ |\mathbf{T}^{-1}|\ |\mathbf{U}| = |\mathbf{U}|^2/|\mathbf{T}|,$$

$$|\mathbf{W}|\,|\mathbf{T}| \geq |\mathbf{U}|^2 \quad \Leftrightarrow \quad |\mathbf{W}| \geq |\mathbf{U}'\mathbf{T}^{-1}\mathbf{U}|$$

and

$$|\mathbf{W}|\,|\mathbf{T}| = |\mathbf{U}|^2 \quad \Leftrightarrow \quad |\mathbf{W}| = |\mathbf{U}'\mathbf{T}^{-1}\mathbf{U}|.$$

Moreover, in light of Theorem 8.5.10,

$$\text{rank}(\mathbf{A}) = m \quad \Leftrightarrow \quad \text{rank}(\mathbf{Q}) = 0 \quad \Leftrightarrow \quad \mathbf{Q} = \mathbf{0}.$$

Thus, it follows from Part (a) that

$$|\mathbf{W}|\,|\mathbf{T}| \geq |\mathbf{U}|^2,$$

with equality holding if and only if $\mathbf{W}$ is singular or $\text{rank}(\mathbf{A}) = m$.

(c) We have (in light of Lemma 14.2.8 and Corollary 14.2.12) that $\text{rank}(\mathbf{A}) = 2m > m$ and that $\mathbf{W}$ (and $\mathbf{T}$) are nonsingular. Thus, it follows from Part (b) that

$$|\mathbf{W}|\,|\mathbf{T}| > |\mathbf{U}|^2.$$

EXERCISE 6. Show that, for any $n \times p$ matrix $\mathbf{X}$ and any symmetric positive definite matrix $\mathbf{W}$,

$$|\mathbf{X}'\mathbf{W}\mathbf{X}|\,|\mathbf{X}'\mathbf{W}^{-1}\mathbf{X}| \geq |\mathbf{X}'\mathbf{X}|^2 . \tag{E.1}$$

[*Hint*. Begin by showing that the matrices $\mathbf{X}'\mathbf{X}(\mathbf{X}'\mathbf{W}\mathbf{X})^-\mathbf{X}'\mathbf{X}$ and $\mathbf{X}'\mathbf{W}^{-1}\mathbf{X} - \mathbf{X}'\mathbf{X}(\mathbf{X}'\mathbf{W}\mathbf{X})^-\mathbf{X}'\mathbf{X}$ are symmetric and nonnegative definite.]

Solution. Let $\mathbf{A} = \mathbf{X}'\mathbf{X}(\mathbf{X}'\mathbf{W}\mathbf{X})^-\mathbf{X}'\mathbf{X}$ and $\mathbf{C} = \mathbf{X}'\mathbf{W}^{-1}\mathbf{X}$. Then, making use of Part (6$'$) of Theorem 14.12.11, we find that

$$\begin{aligned}\mathbf{A} = \mathbf{X}'\mathbf{W}^{-1}\mathbf{W}\mathbf{P}_{\mathbf{X},\mathbf{W}}\mathbf{W}^{-1}\mathbf{X} &= \mathbf{X}'\mathbf{W}^{-1}\mathbf{P}_{\mathbf{X},\mathbf{W}}'\mathbf{W}\mathbf{P}_{\mathbf{X},\mathbf{W}}\mathbf{W}^{-1}\mathbf{X} \\ &= (\mathbf{P}_{\mathbf{X},\mathbf{W}}\mathbf{W}^{-1}\mathbf{X})'\mathbf{W}(\mathbf{P}_{\mathbf{X},\mathbf{W}}\mathbf{W}^{-1}\mathbf{X}),\end{aligned}$$

so that (in light of Theorem 14.2.9) $\mathbf{A}$ is symmetric and nonnegative definite. Further, $\mathbf{C}$ is symmetric, and, making use of Part (9$'$) of Theorem 14.12.11, we find that

$$\begin{aligned}\mathbf{C} - \mathbf{A} &= \mathbf{X}'\mathbf{W}^{-1}\mathbf{W}(\mathbf{I} - \mathbf{P}_{\mathbf{X},\mathbf{W}})\mathbf{W}^{-1}\mathbf{X} \\ &= \mathbf{X}'\mathbf{W}^{-1}(\mathbf{I} - \mathbf{P}_{\mathbf{X},\mathbf{W}})'\mathbf{W}(\mathbf{I} - \mathbf{P}_{\mathbf{X},\mathbf{W}})\mathbf{W}^{-1}\mathbf{X} \\ &= [(\mathbf{I} - \mathbf{P}_{\mathbf{X},\mathbf{W}})\mathbf{W}^{-1}\mathbf{X}]'\mathbf{W}[(\mathbf{I} - \mathbf{P}_{\mathbf{X},\mathbf{W}})\mathbf{W}^{-1}\mathbf{X}],\end{aligned}$$

so that $\mathbf{C} - \mathbf{A}$ is nonnegative definite. Thus, it follows from Corollary 18.1.8 that

$$|\mathbf{X}'\mathbf{W}^{-1}\mathbf{X}| \geq |\mathbf{X}'\mathbf{X}(\mathbf{X}'\mathbf{W}\mathbf{X})^-\mathbf{X}'\mathbf{X}|. \tag{S.3}$$

If $\text{rank}(\mathbf{X}) = p$, then (in light of Theorem 14.2.9 and Lemma 14.9.1) $|\mathbf{X}'\mathbf{W}\mathbf{X}| > 0$ and (in light of Theorems 13.3.4 and 13.3.7)

$$|\mathbf{X}'\mathbf{X}(\mathbf{X}'\mathbf{W}\mathbf{X})^-\mathbf{X}'\mathbf{X}| = |\mathbf{X}'\mathbf{X}|\,|(\mathbf{X}'\mathbf{W}\mathbf{X})^{-1}|\,|\mathbf{X}'\mathbf{X}| = |\mathbf{X}'\mathbf{X}|^2/|\mathbf{X}'\mathbf{W}\mathbf{X}|,$$

in which case inequality (S.3) is equivalent to inequality (S.1). Alternatively, if $\text{rank}(\mathbf{X}) < p$, then [since (according to Corollary 14.11.3) $\text{rank}(\mathbf{X}'\mathbf{W}\mathbf{X}) = \text{rank}(\mathbf{X})$ and $\text{rank}(\mathbf{X}'\mathbf{X}) = \text{rank}(\mathbf{X})$] both sides of inequality (E.1) equal 0 and hence inequality (E.1) holds as an equality.

EXERCISE 7. (a) Show that, for any $n \times n$ skew-symmetric matrix $\mathbf{C}$,

$$|\mathbf{I}_n + \mathbf{C}| \geq 1,$$

with equality holding if and only if $\mathbf{C} = \mathbf{0}$.

(b) Generalize the result of Part (a) by showing that, for any $n \times n$ symmetric positive definite matrix $\mathbf{A}$ and any $n \times n$ skew-symmetric matrix $\mathbf{B}$,

$$|\mathbf{A} + \mathbf{B}| \geq |\mathbf{A}|,$$

with equality holding if and only if $\mathbf{B} = \mathbf{0}$.

Solution. (a) Clearly,

$$|\mathbf{I} + \mathbf{C}| = |(\mathbf{I} + \mathbf{C})'| = |\mathbf{I} + \mathbf{C}'| = |\mathbf{I} - \mathbf{C}|,$$

so that

$$|\mathbf{I} + \mathbf{C}|^2 = |\mathbf{I} + \mathbf{C}|\ |\mathbf{I} - \mathbf{C}| = |(\mathbf{I} + \mathbf{C})(\mathbf{I} - \mathbf{C})| = |\mathbf{I} - \mathbf{C}\mathbf{C}| = |\mathbf{I} + \mathbf{C}'\mathbf{C}|.$$

Moreover, since $\mathbf{C}'\mathbf{C}$ is symmetric and nonnegative definite, it follows from Theorem 18.1.6 that

$$|\mathbf{I} + \mathbf{C}'\mathbf{C}| \geq |\mathbf{I}|,$$

with equality holding if and only if $\mathbf{C}'\mathbf{C} = \mathbf{0}$ or equivalently if and only if $\mathbf{C} = \mathbf{0}$. Since $|\mathbf{I}| = 1$, we conclude that

$$|\mathbf{I} + \mathbf{C}|^2 \geq 1,$$

with equality holding if and only if $\mathbf{C} = \mathbf{0}$.

To complete the proof, it suffices to show that $|\mathbf{I} + \mathbf{C}| > 0$. According to Lemma 14.6.4, $\mathbf{C}$ is nonnegative definite. Thus, we have (in light of Lemma 14.2.4) that $\mathbf{I} + \mathbf{C}$ is positive definite and hence (in light of Corollary 14.9.4) that $|\mathbf{I} + \mathbf{C}| > 0$.

(b) According to Corollary 14.3.13, there exists a nonsingular matrix $\mathbf{P}$ such that $\mathbf{A} = \mathbf{P}'\mathbf{P}$. Then,

$$\mathbf{A} + \mathbf{B} = \mathbf{P}'(\mathbf{I} + \mathbf{C})\mathbf{P},$$

where $\mathbf{C} = (\mathbf{P}^{-1})'\mathbf{B}\mathbf{P}^{-1}$. Moreover, since (according to Lemma 14.6.2) $\mathbf{C}$ is skew-symmetric, we have [as a consequence of Part (a)] that

$$|\mathbf{I} + \mathbf{C}| \geq 1,$$

with equality holding if and only if $\mathbf{C} = \mathbf{0}$ or equivalently if and only if $\mathbf{B} = \mathbf{0}$. The proof is complete upon observing that, since $|\mathbf{A} + \mathbf{B}| = |\mathbf{P}|^2|\mathbf{I} + \mathbf{C}|$ and $|\mathbf{A}| = |\mathbf{P}|^2$

(and since $|\mathbf{P}| \neq 0$), $|\mathbf{A}+\mathbf{B}| \geq |\mathbf{A}| \Leftrightarrow |\mathbf{I}+\mathbf{C}| \geq 1$, and $|\mathbf{A}+\mathbf{B}| = |\mathbf{A}| \Leftrightarrow |\mathbf{I}+\mathbf{C}| = 1$.

EXERCISE 8. (a) Let $\mathbf{R}$ represent an $n \times n$ nonsingular matrix, and let $\mathbf{B}$ represent an $n \times n$ matrix of rank one. Show that $\mathbf{R}+\mathbf{B}$ is nonsingular if and only if $\text{tr}(\mathbf{R}^{-1}\mathbf{B}) \neq -1$, in which case

$$(\mathbf{R}+\mathbf{B})^{-1} = \mathbf{R}^{-1} - [1+\text{tr}(\mathbf{R}^{-1}\mathbf{B})]^{-1}\mathbf{R}^{-1}\mathbf{B}\mathbf{R}^{-1}.$$

(b) To what does the result of Part (a) simplify in the special case where $\mathbf{R} = \mathbf{I}_n$?

Solution. (a) It follows from Theorem 4.4.8 that there exist n-dimensional column vectors $\mathbf{s}$ and $\mathbf{u}$ such that $\mathbf{B} = \mathbf{s}\mathbf{u}'$. Then, as a consequence of Corollary 18.2.10, we find that $\mathbf{R}+\mathbf{B}$ is nonsingular if and only if $\mathbf{u}'\mathbf{R}^{-1}\mathbf{s} \neq -1$. Moreover, upon applying result (5.2.6) (with $\mathbf{b} = \mathbf{u}$ and $\mathbf{a} = \mathbf{R}^{-1}\mathbf{s}$), we obtain

$$\mathbf{u}'\mathbf{R}^{-1}\mathbf{s} = \text{tr}(\mathbf{R}^{-1}\mathbf{s}\mathbf{u}') = \text{tr}(\mathbf{R}^{-1}\mathbf{B}).$$

Thus, $\mathbf{R}+\mathbf{B}$ is nonsingular if and only if $\text{tr}(\mathbf{R}^{-1}\mathbf{B}) \neq -1$. And, if $\text{tr}(\mathbf{R}^{-1}\mathbf{B}) \neq -1$, then we have, as a further consequence of Corollary 18.2.10, that

$$\begin{aligned}(\mathbf{R}+\mathbf{B})^{-1} &= \mathbf{R}^{-1} - (1+\mathbf{u}'\mathbf{R}^{-1}\mathbf{s})\mathbf{R}^{-1}\mathbf{s}\mathbf{u}'\mathbf{R}^{-1} \\ &= \mathbf{R}^{-1} - [1+\text{tr}(\mathbf{R}^{-1}\mathbf{B})]^{-1}\mathbf{R}^{-1}\mathbf{B}\mathbf{R}^{-1}.\end{aligned}$$

(b) In the special case where $\mathbf{R} = \mathbf{I}_n$, the result of Part (a) can be restated as follows: $\mathbf{I}_n + \mathbf{B}$ is nonsingular if and only if $\text{tr}(\mathbf{B}) \neq -1$, in which case

$$(\mathbf{I}_n+\mathbf{B})^{-1} = \mathbf{I}_n - [1+\text{tr}(\mathbf{B})]^{-1}\mathbf{B}.$$

EXERCISE 9. Let $\mathbf{R}$ represent an $n \times n$ matrix, $\mathbf{S}$ an $n \times m$ matrix, $\mathbf{T}$ an $m \times m$ matrix, and $\mathbf{U}$ an $m \times n$ matrix. Suppose that $\mathbf{R}$ is nonsingular. Show (a) that $\mathbf{R}+\mathbf{STU}$ is nonsingular if and only if $\mathbf{I}_m + \mathbf{U}\mathbf{R}^{-1}\mathbf{S}\mathbf{T}$ is nonsingular, in which case

$$(\mathbf{R}+\mathbf{STU})^{-1} = \mathbf{R}^{-1} - \mathbf{R}^{-1}\mathbf{S}\mathbf{T}(\mathbf{I}_m + \mathbf{U}\mathbf{R}^{-1}\mathbf{S}\mathbf{T})^{-1}\mathbf{U}\mathbf{R}^{-1},$$

and (b) that $\mathbf{R}+\mathbf{STU}$ is nonsingular if and only if $\mathbf{I}_m + \mathbf{T}\mathbf{U}\mathbf{R}^{-1}\mathbf{S}$ is nonsingular, in which case

$$(\mathbf{R}+\mathbf{STU})^{-1} = \mathbf{R}^{-1} - \mathbf{R}^{-1}\mathbf{S}(\mathbf{I}_m + \mathbf{T}\mathbf{U}\mathbf{R}^{-1}\mathbf{S})^{-1}\mathbf{T}\mathbf{U}\mathbf{R}^{-1}.$$

Do so by using the result—applicable when $\mathbf{T}$ (as well as $\mathbf{R}$) is nonsingular—that $\mathbf{R}+\mathbf{STU}$ is nonsingular if and only if $\mathbf{T}^{-1}+\mathbf{U}\mathbf{R}^{-1}\mathbf{S}$ is nonsingular, or equivalently if and only if $\mathbf{T}+\mathbf{T}\mathbf{U}\mathbf{R}^{-1}\mathbf{S}\mathbf{T}$ is nonsingular, in which case

$$\begin{aligned}(\mathbf{R}+\mathbf{STU})^{-1} &= \mathbf{R}^{-1} - \mathbf{R}^{-1}\mathbf{S}(\mathbf{T}^{-1}+\mathbf{U}\mathbf{R}^{-1}\mathbf{S})^{-1}\mathbf{U}\mathbf{R}^{-1} \\ &= \mathbf{R}^{-1} - \mathbf{R}^{-1}\mathbf{S}\mathbf{T}(\mathbf{T}+\mathbf{T}\mathbf{U}\mathbf{R}^{-1}\mathbf{S}\mathbf{T})^{-1}\mathbf{T}\mathbf{U}\mathbf{R}^{-1}.\end{aligned}$$

[*Hint*. Reexpress $\mathbf{R}+\mathbf{STU}$ as $\mathbf{R}+\mathbf{STU} = \mathbf{R}+(\mathbf{ST})\mathbf{I}_m\mathbf{U}$ and as $\mathbf{R}+\mathbf{STU} = \mathbf{R}+\mathbf{SI}_m\mathbf{TU}$.]

Solution. (a) Reexpress $\mathbf{R}+\mathbf{STU}$ as

$$\mathbf{R}+\mathbf{STU} = \mathbf{R}+(\mathbf{ST})\mathbf{I}_m\mathbf{U}.$$

Then, applying the cited result (or equivalently Theorem 18.2.8) with $\mathbf{ST}$ and $\mathbf{I}_m$ in place of $\mathbf{S}$ and $\mathbf{T}$, respectively, we find that $\mathbf{R}+\mathbf{STU}$ is nonsingular if and only if $\mathbf{I}_m+\mathbf{UR}^{-1}\mathbf{ST}$ is nonsingular, in which case

$$(\mathbf{R}+\mathbf{STU})^{-1} = \mathbf{R}^{-1} - \mathbf{R}^{-1}\mathbf{ST}(\mathbf{I}_m+\mathbf{UR}^{-1}\mathbf{ST})^{-1}\mathbf{UR}^{-1}.$$

(b) Reexpress $\mathbf{R}+\mathbf{STU}$ as

$$\mathbf{R}+\mathbf{STU} = \mathbf{R}+\mathbf{SI}_m(\mathbf{TU}).$$

Then, applying the cited result (or equivalently Theorem 18.2.8) with $\mathbf{I}_m$ and $\mathbf{TU}$ in place of $\mathbf{T}$ and $\mathbf{U}$, respectively, we find that $\mathbf{R}+\mathbf{STU}$ is nonsingular if and only if $\mathbf{I}_m+\mathbf{TUR}^{-1}\mathbf{S}$ is nonsingular, in which case

$$(\mathbf{R}+\mathbf{STU})^{-1} = \mathbf{R}^{-1} - \mathbf{R}^{-1}\mathbf{S}(\mathbf{I}_m+\mathbf{TUR}^{-1}\mathbf{S})^{-1}\mathbf{TUR}^{-1}.$$

EXERCISE 10. Let $\mathbf{R}$ represent an $n \times q$ matrix, $\mathbf{S}$ an $n \times m$ matrix, $\mathbf{T}$ an $m \times p$ matrix, and $\mathbf{U}$ a $p \times q$ matrix. Extend the results of Exercise 9 by showing that if $\mathcal{R}(\mathbf{STU}) \subset \mathcal{R}(\mathbf{R})$ and $\mathcal{C}(\mathbf{STU}) \subset \mathcal{C}(\mathbf{R})$, then the matrix

$$\mathbf{R}^- - \mathbf{R}^-\mathbf{ST}(\mathbf{I}_p+\mathbf{UR}^-\mathbf{ST})^-\mathbf{UR}^-$$

and the matrix

$$\mathbf{R}^- - \mathbf{R}^-\mathbf{S}(\mathbf{I}_m+\mathbf{TUR}^-\mathbf{S})^-\mathbf{TUR}^-$$

are both generalized inverses of the matrix $\mathbf{R}+\mathbf{STU}$.

Solution. Observe that $\mathbf{R}+\mathbf{STU}$ can be reexpressed as

$$\mathbf{R}+\mathbf{STU} = \mathbf{R}+(\mathbf{ST})\mathbf{I}_p\mathbf{U}$$

and also as

$$\mathbf{R}+\mathbf{STU} = \mathbf{R}+\mathbf{SI}_m(\mathbf{TU}).$$

Suppose now that $\mathcal{R}(\mathbf{STU}) \subset \mathcal{R}(\mathbf{R})$ and $\mathcal{C}(\mathbf{STU}) \subset \mathcal{C}(\mathbf{R})$. Then, upon applying Theorem 18.2.14 with $\mathbf{ST}$ and $\mathbf{I}_p$ in place of $\mathbf{S}$ and $\mathbf{T}$, respectively, we find that

$$\mathbf{R}^- - \mathbf{R}^-\mathbf{ST}(\mathbf{I}_p+\mathbf{UR}^-\mathbf{ST})^-\mathbf{UR}^-$$

is a generalized inverse of the matrix $\mathbf{R}+\mathbf{STU}$. And, upon applying Theorem 18.2.14 with $\mathbf{I}_m$ and $\mathbf{TU}$ in place of $\mathbf{T}$ and $\mathbf{U}$, respectively, we find that

$$\mathbf{R}^- - \mathbf{R}^-\mathbf{S}(\mathbf{I}_m+\mathbf{TUR}^-\mathbf{S})^-\mathbf{TUR}^-$$

is also a generalized inverse of $\mathbf{R} + \mathbf{STU}$.

EXERCISE 11. Let $\mathbf{R}$ represent an $n \times q$ matrix, $\mathbf{S}$ an $n \times m$ matrix, $\mathbf{T}$ an $m \times p$ matrix, and $\mathbf{U}$ a $p \times q$ matrix.

(a) Take $\mathbf{G}$ to be a generalized inverse of the partitioned matrix $\begin{pmatrix} \mathbf{R} & -\mathbf{ST} \\ \mathbf{TU} & \mathbf{T} \end{pmatrix}$, and partition $\mathbf{G}$ as $\mathbf{G} = \begin{pmatrix} \mathbf{G}_{11} & \mathbf{G}_{12} \\ \mathbf{G}_{21} & \mathbf{G}_{22} \end{pmatrix}$ (where $\mathbf{G}_{11}$ is of dimensions $q \times n$). Show that $\mathbf{G}_{11}$ is a generalized inverse of the matrix $\mathbf{R} + \mathbf{STU}$. Do so by using the result that, for any partitioned matrix $\mathbf{A} = \begin{pmatrix} \mathbf{A}_{11} & \mathbf{A}_{12} \\ \mathbf{A}_{21} & \mathbf{A}_{22} \end{pmatrix}$ such that $\mathcal{C}(\mathbf{A}_{21}) \subset \mathcal{C}(\mathbf{A}_{22})$ and $\mathcal{R}(\mathbf{A}_{12}) \subset \mathcal{R}(\mathbf{A}_{22})$ and for any generalized inverse $\begin{pmatrix} \mathbf{C}_{11} & \mathbf{C}_{12} \\ \mathbf{C}_{21} & \mathbf{C}_{22} \end{pmatrix}$ of $\mathbf{A}$ (where $\mathbf{C}_{11}$ is of the same dimensions as $\mathbf{A}'_{11}$), $\mathbf{C}_{11}$ is a generalized inverse of the matrix $\mathbf{A}_{11} - \mathbf{A}_{12}\mathbf{A}_{22}^{-}\mathbf{A}_{21}$.

(b) Let $\mathbf{E}_R = \mathbf{I} - \mathbf{RR}^{-}$, $\mathbf{F}_R = \mathbf{I} - \mathbf{R}^{-}\mathbf{R}$, $\mathbf{X} = \mathbf{E}_R\mathbf{ST}$, $\mathbf{Y} = \mathbf{TUF}_R$, $\mathbf{E}_Y = \mathbf{I} - \mathbf{YY}^{-}$, $\mathbf{F}_X = \mathbf{I} - \mathbf{X}^{-}\mathbf{X}$, $\mathbf{Q} = \mathbf{T} + \mathbf{TUR}^{-}\mathbf{ST}$, $\mathbf{Z} = \mathbf{E}_Y\mathbf{QF}_X$, and $\mathbf{Q}^* = \mathbf{F}_X\mathbf{Z}^{-}\mathbf{E}_Y$. Use the result of Part (a) of Exercise 10.10 to show that the matrix

$$\begin{aligned} &\mathbf{R}^{-} - \mathbf{R}^{-}\mathbf{STQ}^*\mathbf{TUR}^{-} - \mathbf{R}^{-}\mathbf{ST}(\mathbf{I} - \mathbf{Q}^*\mathbf{Q})\mathbf{X}^{-}\mathbf{E}_R \\ &\quad - \mathbf{F}_R\mathbf{Y}^{-}(\mathbf{I} - \mathbf{QQ}^*)\mathbf{TUR}^{-} + \mathbf{F}_R\mathbf{Y}^{-}(\mathbf{I} - \mathbf{QQ}^*)\mathbf{QX}^{-}\mathbf{E}_R \end{aligned} \tag{E.2}$$

is a generalized inverse of the matrix $\mathbf{R} + \mathbf{STU}$.

(c) Show that if $\mathcal{R}(\mathbf{TU}) \subset \mathcal{R}(\mathbf{R})$ and $\mathcal{C}(\mathbf{ST}) \subset \mathcal{C}(\mathbf{R})$, then the formula

$$\mathbf{R}^{-} - \mathbf{R}^{-}\mathbf{STQ}^{-}\mathbf{TUR}^{-} \tag{$*$}$$

for a generalized inverse of $\mathbf{R} + \mathbf{STU}$ can be obtained as a special case of formula (E.2).

Solution. (a) It follows from the cited result (or equivalently from the second part of Theorem 9.6.5) that $\mathbf{G}_{11}$ is a generalized inverse of the matrix

$$\mathbf{R} - (-\mathbf{ST})\mathbf{T}^{-}\mathbf{TU} = \mathbf{R} + \mathbf{STT}^{-}\mathbf{TU} = \mathbf{R} + \mathbf{STU}.$$

(b) Let $\mathbf{G}$ represent the generalized inverse of the matrix $\begin{pmatrix} \mathbf{R} & -\mathbf{ST} \\ \mathbf{TU} & \mathbf{T} \end{pmatrix}$ obtained by applying formula (10.E.1). Partition $\mathbf{G}$ as $\mathbf{G} = \begin{pmatrix} \mathbf{G}_{11} & \mathbf{G}_{12} \\ \mathbf{G}_{21} & \mathbf{G}_{22} \end{pmatrix}$ (where $\mathbf{G}_{11}$ is of dimensions $q \times n$), and assume that [in applying formula (10.E.1)] the generalized inverse of $-\mathbf{X}$ is set equal to $-\mathbf{X}^{-}$ [in which case $\mathbf{F}_X = \mathbf{I} - (-\mathbf{X}^{-})(-\mathbf{X})$]. Then, $\mathbf{G}_{11}$ equals the matrix (E.2), and we conclude on the basis of Part (a) (of the current exercise) that the matrix (E.2) is a generalized inverse of the matrix $\mathbf{R} + \mathbf{STU}$.

(c) Suppose that $\mathcal{R}(\mathbf{TU}) \subset \mathcal{R}(\mathbf{R})$ and $\mathcal{C}(\mathbf{ST}) \subset \mathcal{C}(\mathbf{R})$. Then, it follows from Lemma 9.3.5 that $\mathbf{X} = \mathbf{0}$ and $\mathbf{Y} = \mathbf{0}$ (so that $\mathbf{F}_X = \mathbf{I}$ and $\mathbf{E}_Y = \mathbf{I}$ and consequently

$\mathbf{Q}^*$ is an arbitrary generalized inverse of $\mathbf{Q}$). Thus, formula $(*)$ [or equivalently formula (2.27)] can be obtained as a special case of formula (E.2) by setting $\mathbf{X}^- = \mathbf{0}$ and $\mathbf{Y}^- = \mathbf{0}$.

EXERCISE 12. Let $\mathbf{A}_1, \mathbf{A}_2, \ldots$ represent a sequence of $m \times n$ matrices, and let $\mathbf{A}$ represent another $m \times n$ matrix.

(a) Using the result of Exercise 6.1 (i.e., the triangle inequality), show that if $\|\mathbf{A}_k - \mathbf{A}\| \to 0$, then $\|\mathbf{A}_k\| \to \|\mathbf{A}\|$.

(b) Show that if $\mathbf{A}_k \to \mathbf{A}$, then $\|\mathbf{A}_k\| \to \|\mathbf{A}\|$ (where the norms are the usual norms).

Solution. (a) Making use of the triangle inequality, we find that

$$\|\mathbf{A}_k\| = \|(\mathbf{A}_k - \mathbf{A}) + \mathbf{A}\| \leq \|\mathbf{A}_k - \mathbf{A}\| + \|\mathbf{A}\|$$

and that

$$\|\mathbf{A}\| = \|\mathbf{A}_k - (\mathbf{A}_k - \mathbf{A})\| \leq \|\mathbf{A}_k\| + \|\mathbf{A}_k - \mathbf{A}\|.$$

Thus,

$$\|\mathbf{A}_k\| - \|\mathbf{A}\| \leq \|\mathbf{A}_k - \mathbf{A}\|,$$

and

$$-(\|\mathbf{A}_k\| - \|\mathbf{A}\|) = \|\mathbf{A}\| - \|\mathbf{A}_k\| \leq \|\mathbf{A}_k - \mathbf{A}\|,$$

implying that

$$|\ \|\mathbf{A}_k\| - \|\mathbf{A}\|\ | \leq \|\mathbf{A}_k - \mathbf{A}\|.$$

Suppose now that $\|\mathbf{A}_k - \mathbf{A}\| \to 0$. Then, corresponding to each positive scaler ϵ, there exists a positive integer p such that, for $k > p$, $\|\mathbf{A}_k - \mathbf{A}\| < \epsilon$ and hence such that, for $k > p$, $|\ \|\mathbf{A}_k\| - \|\mathbf{A}\|\ | < \epsilon$. We conclude that $\|\mathbf{A}_k\| \to \|\mathbf{A}\|$.

(b) In light of Lemma 18.2.20, Part (b) follows from Part (a).

EXERCISE 13. Let $\mathbf{A}$ represent an $n \times n$ matrix. Using the results of Exercise 6.1 and of Part (b) of Exercise 12, show that if $\|\mathbf{A}\| < 1$, then (for $k = 0, 1, 2, \ldots$)

$$\|(\mathbf{I} - \mathbf{A})^{-1} - (\mathbf{I} + \mathbf{A} + \mathbf{A}^2 + \cdots + \mathbf{A}^k)\| \leq \|\mathbf{A}\|^{k+1}/(1 - \|\mathbf{A}\|)$$

(where the norms are the usual norms). (*Note.* If $\|\mathbf{A}\| < 1$, then $\mathbf{I} - \mathbf{A}$ is nonsingular.)

Solution. Suppose that $\|\mathbf{A}\| < 1$, and (for $p = 0, 1, 2, \ldots$) let $\mathbf{S}_p = \sum_{m=0}^{p} \mathbf{A}^m$ (where $\mathbf{A}^0 = \mathbf{I}$). Then, as a consequence of Theorems 18.2.16 and 18.2.19, we have that$(\mathbf{I} - \mathbf{A})^{-1} = \lim_{p \to \infty} \mathbf{S}_p$, implying that

$$\begin{aligned}(\mathbf{I} - \mathbf{A})^{-1} - \mathbf{S}_k = (\lim_{p \to \infty} \mathbf{S}_p) - \mathbf{S}_k &= \lim_{p \to \infty} (\mathbf{S}_p - \mathbf{S}_k) \\ &= \lim_{p \to \infty} \sum_{m=k+1}^{p} \mathbf{A}^m,\end{aligned}$$

and it follows from the result of Part (b) of Exercise 12 that

$$\|(\mathbf{I}-\mathbf{A})^{-1}-\mathbf{S}_k\| = \lim_{p\to\infty} \|\sum_{m=k+1}^{p} \mathbf{A}^m\|. \tag{S.4}$$

Moreover, making repeated use of the result of Exercise 6.1 (i.e., of the triangle inequality) and of Lemma 18.2.21, we find that (for $p \geq k+1$)

$$\|\sum_{m=k+1}^{p} \mathbf{A}^m\| \leq \sum_{m=k+1}^{p} \|\mathbf{A}^m\| \leq \sum_{m=k+1}^{p} \|\mathbf{A}\|^m = \|\mathbf{A}\|^{k+1} \sum_{m=0}^{p-k-1} \|\mathbf{A}\|^m. \tag{S.5}$$

It follows from a basic result on geometric series [which is example 34.8(c) in Bartle's (1976) book] that $\sum_{m=0}^{\infty} \|\mathbf{A}\|^m = 1/(1-\|\mathbf{A}\|)$. Thus, combining result (S.5) with result (S.4), we find that

$$\begin{aligned}\|(\mathbf{I}-\mathbf{A})^{-1}-\mathbf{S}_k\| &\leq \lim_{p\to\infty} [\|\mathbf{A}\|^{k+1} \sum_{m=0}^{p-k-1} \|\mathbf{A}\|^m] \\ &= \|\mathbf{A}\|^{k+1} \sum_{m=0}^{\infty} \|\mathbf{A}\|^m = \|\mathbf{A}\|^{k+1}/(1-\|\mathbf{A}\|).\end{aligned}$$

EXERCISE 14. Let **A** and **B** represent $n \times n$ matrices. Suppose that **B** is nonsingular, and define $\mathbf{F} = \mathbf{B}^{-1}\mathbf{A}$. Using the result of Exercise 13, show that if $\|\mathbf{F}\| < 1$, then (for $k = 0, 1, 2, \ldots$)

$$\begin{aligned}&\|(\mathbf{B}-\mathbf{A})^{-1} - (\mathbf{B}^{-1} + \mathbf{F}\mathbf{B}^{-1} + \mathbf{F}^2\mathbf{B}^{-1} + \cdots + \mathbf{F}^k\mathbf{B}^{-1})\| \\ &\qquad\qquad \leq \|\mathbf{B}^{-1}\| \, \|\mathbf{F}\|^{k+1}/(1-\|\mathbf{F}\|)\end{aligned}$$

(where the norms are the usual norms). (*Note.* If $\|\mathbf{F}\| < 1$, then $\mathbf{B}-\mathbf{A}$ is nonsingular.)

Solution. Suppose that $\|\mathbf{F}\| < 1$. Then, since $\mathbf{B}-\mathbf{A} = \mathbf{B}(\mathbf{I}-\mathbf{F})$ (and since $\mathbf{B}-\mathbf{A}$ is nonsingular), $\mathbf{I}-\mathbf{F}$ is nonsingular, and

$$(\mathbf{B}-\mathbf{A})^{-1} = (\mathbf{I}-\mathbf{F})^{-1}\mathbf{B}^{-1}.$$

Thus, making use of Lemma 18.2.21 and the result of Exercise 13, we find that

$$\begin{aligned}&\|(\mathbf{B}-\mathbf{A})^{-1} - (\mathbf{B}^{-1} + \mathbf{F}\mathbf{B}^{-1} + \mathbf{F}^2\mathbf{B}^{-1} + \cdots + \mathbf{F}^k\mathbf{B}^{-1})\| \\ &\qquad = \|[(\mathbf{I}-\mathbf{F})^{-1} - (\mathbf{I}+\mathbf{F}+\mathbf{F}^2+\cdots+\mathbf{F}^k)]\mathbf{B}^{-1}\| \\ &\qquad \leq \|(\mathbf{I}-\mathbf{F})^{-1} - (\mathbf{I}+\mathbf{F}+\mathbf{F}^2+\cdots+\mathbf{F}^k)\| \, \|\mathbf{B}^{-1}\| \\ &\qquad \leq \|\mathbf{B}^{-1}\| \, \|\mathbf{F}\|^{k+1}/(1-\|\mathbf{F}\|).\end{aligned}$$

EXERCISE 15. Let **A** represent an $n \times n$ symmetric nonnegative definite matrix, and let **B** represent an $n \times n$ matrix. Show that if $\mathbf{B}-\mathbf{A}$ is nonnegative definite (in which case **B** is also nonnegative definite), then $\mathcal{R}(\mathbf{A}) \subset \mathcal{R}(\mathbf{B})$ and $\mathcal{C}(\mathbf{A}) \subset \mathcal{C}(\mathbf{B})$.

Solution. Define $\mathbf{C} = (\mathbf{I} - \mathbf{B}^-\mathbf{B})'(\mathbf{B} - \mathbf{A})(\mathbf{I} - \mathbf{B}^-\mathbf{B})$. Since $\mathbf{A}$ is symmetric and nonnegative definite, there exists a matrix $\mathbf{R}$ such that $\mathbf{A} = \mathbf{R}'\mathbf{R}$. Clearly,

$$\mathbf{C} = -(\mathbf{I} - \mathbf{B}^-\mathbf{B})'\mathbf{A}(\mathbf{I} - \mathbf{B}^-\mathbf{B}) = -[\mathbf{R}(\mathbf{I} - \mathbf{B}^-\mathbf{B})]'\mathbf{R}(\mathbf{I} - \mathbf{B}^-\mathbf{B}). \quad \text{(S.6)}$$

Suppose now that $\mathbf{B} - \mathbf{A}$ is nonnegative definite. Then, according to Theorem 14.2.9, $\mathbf{C}$ is nonnegative definite. Moreover, it is clear from expression (S.6) that $\mathbf{C}$ is nonpositive definite and symmetric. Consequently, it follows from Lemma 14.2.2 that $\mathbf{C} = \mathbf{0}$ or equivalently that $[\mathbf{R}(\mathbf{I} - \mathbf{B}^-\mathbf{B}]'\mathbf{R}(\mathbf{I} - \mathbf{B}^-\mathbf{B}) = \mathbf{0}$, implying (in light of Corollary 5.3.2) that $\mathbf{R}(\mathbf{I} - \mathbf{B}^-\mathbf{B}) = \mathbf{0}$ and hence (since $\mathbf{A} = \mathbf{R}'\mathbf{R}$) that $\mathbf{A}(\mathbf{I} - \mathbf{B}^-\mathbf{B}) = \mathbf{0}$. We conclude (in light of Lemma 9.3.5) that $\mathcal{R}(\mathbf{A}) \subset \mathcal{R}(\mathbf{B})$.

Further, since $\mathbf{B} - \mathbf{A}$ is nonnegative definite, $(\mathbf{B} - \mathbf{A})' = \mathbf{B}' - \mathbf{A}'$ is also nonnegative definite. Thus, by employing an argument analogous to that employed in establishing that $\mathcal{R}(\mathbf{A}) \subset \mathcal{R}(\mathbf{B})$, it can be shown that $\mathcal{R}(\mathbf{A}') \subset \mathcal{R}(\mathbf{B}')$ or equivalently (in light of Corollary 4.2.5) that $\mathcal{C}(\mathbf{A}) \subset \mathcal{C}(\mathbf{B})$.

An alternative solution to Exercise 15 can be obtained by making use of Corollary 12.5.6. Suppose that $\mathbf{B} - \mathbf{A}$ is nonnegative definite. And, let $\mathbf{x}$ represent an arbitrary vector in $\mathcal{C}^{\perp}(\mathbf{B})$. Then,

$$0 \leq \mathbf{x}'(\mathbf{B} - \mathbf{A})\mathbf{x} = -\mathbf{x}'\mathbf{A}\mathbf{x} \leq 0,$$

implying that $\mathbf{x}'\mathbf{A}\mathbf{x} = 0$ and hence (in light of Corollary 14.3.11) that $\mathbf{A}'\mathbf{x} = \mathbf{A}\mathbf{x} = \mathbf{0}$ or equivalently that $\mathbf{x} \in \mathcal{C}^{\perp}(\mathbf{A})$. Thus, $\mathcal{C}^{\perp}(\mathbf{B}) \subset \mathcal{C}^{\perp}(\mathbf{A})$, and it follows from Corollary 12.5.6 that $\mathcal{C}(\mathbf{A}) \subset \mathcal{C}(\mathbf{B})$. That $\mathcal{R}(\mathbf{A}) \subset \mathcal{R}(\mathbf{B})$ can be established via an analogous argument.

EXERCISE 16. Let $\mathbf{A}$ represent an $n \times n$ symmetric idempotent matrix, and let $\mathbf{B}$ represent an $n \times n$ symmetric nonnegative definite matrix. Show that if $\mathbf{I} - \mathbf{A} - \mathbf{B}$ is nonnegative definite, then $\mathbf{BA} = \mathbf{AB} = \mathbf{0}$. (*Hint*. Show that $\mathbf{A}'(\mathbf{I} - \mathbf{A} - \mathbf{B})\mathbf{A} = -\mathbf{A}'\mathbf{BA}$, and then consider the implications of this equality.)

Solution. Clearly,

$$\mathbf{A}'(\mathbf{I} - \mathbf{A} - \mathbf{B})\mathbf{A} = \mathbf{A}'(\mathbf{A} - \mathbf{A}^2 - \mathbf{BA}) = \mathbf{A}'(\mathbf{A} - \mathbf{A} - \mathbf{BA}) = -\mathbf{A}'\mathbf{BA}. \quad \text{(S.7)}$$

Suppose now that $\mathbf{I} - \mathbf{A} - \mathbf{B}$ is nonnegative definite. Then, as a consequence of Theorem 14.2.9, $\mathbf{A}'(\mathbf{I} - \mathbf{A} - \mathbf{B})\mathbf{A}$ is nonnegative definite, in which case it follows from result (S.7) that $\mathbf{A}'\mathbf{BA}$ is nonpositive definite. Moreover, as a further consequence of Theorem 14.2.9, $\mathbf{A}'\mathbf{BA}$ is nonnegative definite. Thus, in light of Lemma 14.2.2, we have that

$$\mathbf{A}'\mathbf{BA} = \mathbf{0}. \quad \text{(S.8)}$$

And, since $\mathbf{B}$ is symmetric as well as nonnegative definite, we conclude (on the basis of Corollary 14.3.11) that $\mathbf{BA} = \mathbf{0}$ and also [upon observing that $\mathbf{AB} = \mathbf{A}'\mathbf{B}' = (\mathbf{BA})'$] that $\mathbf{AB} = \mathbf{0}$.

EXERCISE 17. Let $\mathbf{A}_1, \ldots, \mathbf{A}_k$ represent $n \times n$ symmetric matrices, and define $\mathbf{A} = \mathbf{A}_1 + \cdots + \mathbf{A}_k$. Suppose that $\mathbf{A}$ is idempotent. Suppose further that

$\mathbf{A}_1, \ldots, \mathbf{A}_{k-1}$ are idempotent and that $\mathbf{A}_k$ is nonnegative definite. Using the result of Exercise 16 (or otherwise), show that $\mathbf{A}_i\mathbf{A}_j = \mathbf{0}$ (for $j \neq i = 1, \ldots, k$), that $\mathbf{A}_k$ is idempotent, and that $\text{rank}(\mathbf{A}_k) = \text{rank}(\mathbf{A}) - \sum_{i=1}^{k-1} \text{rank}(\mathbf{A}_i)$.

Solution. Let $\mathbf{A}_0 = \mathbf{I} - \mathbf{A}$. Then, $\sum_{i=0}^{k} \mathbf{A}_i = \mathbf{I}$. Further, $\mathbf{A}_0$ (like $\mathbf{A}_1, \ldots, \mathbf{A}_{k-1}$) is symmetric and idempotent, and (in light of Lemma 14.2.17) $\mathbf{A}_0, \mathbf{A}_1, \ldots, \mathbf{A}_{k-1}$ (like $\mathbf{A}_k$) are nonnegative definite.

Thus, for $i = 1, \ldots, k-1$ and $j = i+1, \ldots, k$, $\mathbf{A}_i$ is idempotent, $\mathbf{A}_j$ is nonnegative definite, and (since $\mathbf{I} - \mathbf{A}_i - \mathbf{A}_j = \sum_{m=0\ (m \neq i,j)}^{k} \mathbf{A}_m$) $\mathbf{I} - \mathbf{A}_i - \mathbf{A}_j$ is nonnegative definite. And, it follows from the result of Exercise 16 that (for $i = 1, \ldots, k-1$ and $j = i+1, \ldots, k$) $\mathbf{A}_i\mathbf{A}_j = \mathbf{0}$ and $\mathbf{A}_j\mathbf{A}_i = \mathbf{0}$ or equivalently that, for $j \neq i = 1, \ldots, k$, $\mathbf{A}_i\mathbf{A}_j = \mathbf{0}$. Moreover, since $\mathbf{A}_1, \ldots, \mathbf{A}_k$ are symmetric, we conclude from Theorem 18.4.1 that $\mathbf{A}_k$ (like $\mathbf{A}_1, \ldots, \mathbf{A}_{k-1}$) is idempotent and that $\sum_{i=1}^{k} \text{rank}(\mathbf{A}_i) = \text{rank}(\mathbf{A})$ or, equivalently, $\text{rank}(\mathbf{A}_k) = \text{rank}(\mathbf{A}) - \sum_{i=1}^{k-1} \text{rank}(\mathbf{A}_i)$.

EXERCISE 18. Let $\mathbf{A}_1, \ldots, \mathbf{A}_k$ represent $n \times n$ symmetric matrices, and define $\mathbf{A} = \mathbf{A}_1 + \cdots + \mathbf{A}_k$. Suppose that $\mathbf{A}$ is idempotent. Show that if $\mathbf{A}_1, \ldots, \mathbf{A}_k$ are nonnegative definite and if $\text{tr}(\mathbf{A}) \leq \sum_{i=1}^{k} \text{tr}(\mathbf{A}_i^2)$, then $\mathbf{A}_i\mathbf{A}_j = \mathbf{0}$ (for $j \neq i = 1, \ldots, k$) and $\mathbf{A}_1, \ldots, \mathbf{A}_k$ are idempotent. *Hint*. Show that $\sum_{i,j \neq i} \text{tr}(\mathbf{A}_i\mathbf{A}_j) \leq 0$ and then make use of the result that, for any two symmetric nonnegative definite matrices $\mathbf{B}$ and $\mathbf{C}$ (of the same order), $\text{tr}(\mathbf{BC}) \geq 0$, with equality holding if and only if $\mathbf{BC} = \mathbf{0}$.

Solution. Clearly,

$$\mathbf{A} = \mathbf{A}^2 = (\sum_i \mathbf{A}_i)^2 = \sum_i \mathbf{A}_i^2 + \sum_{i,j \neq i} \mathbf{A}_i\mathbf{A}_j,$$

so that

$$\text{tr}(\mathbf{A}) = \text{tr}(\sum_i \mathbf{A}_i^2 + \sum_{i,j \neq i} \mathbf{A}_i\mathbf{A}_j) = \sum_i \text{tr}(\mathbf{A}_i^2) + \sum_{i,j \neq i} \text{tr}(\mathbf{A}_i\mathbf{A}_j)$$

and hence

$$\sum_{i,j \neq i} \text{tr}(\mathbf{A}_i\mathbf{A}_j) = \text{tr}(\mathbf{A}) - \sum_i \text{tr}(\mathbf{A}_i^2). \tag{S.9}$$

Suppose now that $\mathbf{A}_1, \ldots, \mathbf{A}_k$ are nonnegative definite and also that $\text{tr}(\mathbf{A}) \leq \sum_i \text{tr}(\mathbf{A}_i^2)$. Then, it follows from result (S.9) that

$$\sum_{i,j \neq i} \text{tr}(\mathbf{A}_i\mathbf{A}_j) \leq 0.$$

And, since (according to Corollary 14.7.7, which is the result cited in the hint) $\text{tr}(\mathbf{A}_i\mathbf{A}_j) \geq 0$ (for all i and $j \neq i$), we have that $\text{tr}(\mathbf{A}_i\mathbf{A}_j) = 0$ (for all i and $j \neq i$). We conclude (on the basis of Corollary 14.7.7) that $\mathbf{A}_i\mathbf{A}_j = \mathbf{0}$ (for all i

and $j \neq i$). And, in light of Theorem 18.4.1 (and the symmetry of $\mathbf{A}_1, \ldots, \mathbf{A}_k$), we further conclude that $\mathbf{A}_1, \ldots, \mathbf{A}_k$ are idempotent.

EXERCISE 19. Let $\mathbf{A}_1, \ldots, \mathbf{A}_k$ represent $n \times n$ symmetric matrices such that $\mathbf{A}_1 + \cdots + \mathbf{A}_k = \mathbf{I}$. Show that if $\text{rank}(\mathbf{A}_1) + \cdots + \text{rank}(\mathbf{A}_k) = n$, then, for any (strictly) positive scalars $c_1, \ldots, c_k$, the matrix $c_1\mathbf{A}_1 + \cdots + c_k\mathbf{A}_k$ is positive definite.

Solution. Suppose that $\text{rank}(\mathbf{A}_1) + \cdots + \text{rank}(\mathbf{A}_k) = n$. Then, it follows from Theorem 18.4.5 that $\mathbf{A}_1, \ldots, \mathbf{A}_k$ are idempotent. Thus,

$$c_1\mathbf{A}_1 + \cdots + c_k\mathbf{A}_k = \begin{pmatrix} \sqrt{c_1}\mathbf{A}_1 \\ \vdots \\ \sqrt{c_k}\mathbf{A}_k \end{pmatrix}' \begin{pmatrix} \sqrt{c_1}\mathbf{A}_1 \\ \vdots \\ \sqrt{c_k}\mathbf{A}_k \end{pmatrix},$$

implying (in light of Corollary 14.2.14) that $c_1\mathbf{A}_1 + \cdots + c_k\mathbf{A}_k$ is nonnegative definite and (in light of Corollaries 7.4.5 and 4.5.6) that

$$\begin{aligned} \text{rank}(c_1\mathbf{A}_1 + \cdots + c_k\mathbf{A}_k) = \text{rank}\begin{pmatrix} \sqrt{c_1}\mathbf{A}_1 \\ \vdots \\ \sqrt{c_k}\mathbf{A}_k \end{pmatrix} &= \text{rank}\begin{pmatrix} \mathbf{A}_1 \\ \vdots \\ \mathbf{A}_k \end{pmatrix} \\ &= \text{rank}\left[\begin{pmatrix} \mathbf{A}_1 \\ \vdots \\ \mathbf{A}_k \end{pmatrix}' \begin{pmatrix} \mathbf{A}_1 \\ \vdots \\ \mathbf{A}_k \end{pmatrix}\right] \\ &= \text{rank}(\mathbf{A}_1'\mathbf{A}_1 + \cdots + \mathbf{A}_k'\mathbf{A}_k) \\ &= \text{rank}(\mathbf{A}_1 + \cdots + \mathbf{A}_k) \\ &= \text{rank}(\mathbf{I}_n) \\ &= n. \end{aligned}$$

We conclude (on the basis of Corollary 14.3.12) that $c_1\mathbf{A}_1 + \cdots + c_k\mathbf{A}_k$ is positive definite.

EXERCISE 20. Let $\mathbf{A}_1, \ldots, \mathbf{A}_k$ represent $n \times n$ symmetric idempotent matrices such that $\mathbf{A}_i\mathbf{A}_j = \mathbf{0}$ for $j \neq i = 1, \ldots, k$. Show that, for any (strictly) positive scalar c_0 and any nonnegative scalars $c_1, \ldots, c_k$, the matrix $c_0\mathbf{I} + \sum_{i=1}^k c_i\mathbf{A}_i$ is positive definite (and hence nonsingular), and

$$(c_0\mathbf{I} + \sum_{i=1}^k c_i\mathbf{A}_i)^{-1} = d_0\mathbf{I} + \sum_{i=1}^k d_i\mathbf{A}_i \,,$$

where $d_0 = 1/c_0$ and (for $i = 1, \ldots, k$) $d_i = -c_i/[c_0(c_0 + c_i)]$.

Solution. Clearly, $c_0\mathbf{I}$ is positive definite. Moreover, as a consequence of Lemma 14.2.17, $\mathbf{A}_1, \ldots, \mathbf{A}_k$ are nonnegative definite, and hence $c_1\mathbf{A}_1, \ldots, c_k\mathbf{A}_k$ are nonnegative definite. Thus, it follows from Corollary 14.2.5 that $c_0\mathbf{I} + \sum_{i=1}^k c_i\mathbf{A}_i$ is positive definite (and hence, in light of Lemma 14.2.8, nonsingular).

That $(c_0\mathbf{I} + \sum_{i=1}^k c_i\mathbf{A}_i)^{-1} = d_0\mathbf{I} + \sum_{i=1}^k d_i\mathbf{A}_i$ is clear upon observing that

$$\begin{aligned}(c_0\mathbf{I} + \sum_i c_i\mathbf{A}_i)(d_0\mathbf{I} + \sum_i d_i\mathbf{A}_i) &= c_0d_0\mathbf{I} + c_0\sum_i d_i\mathbf{A}_i + d_0\sum_i c_i\mathbf{A}_i + \sum_i c_id_i\mathbf{A}_i^2 + \sum_{i,j\neq i} c_id_j\mathbf{A}_i\mathbf{A}_j \\ &= \mathbf{I} - \sum_i \frac{c_i}{c_0+c_i}\mathbf{A}_i + \sum_i \frac{c_i}{c_0}\mathbf{A}_i - \sum_i \frac{c_i^2}{c_0(c_0+c_i)}\mathbf{A}_i + \mathbf{0} \\ &= \mathbf{I} - \sum_i \frac{c_0c_i - c_i(c_0+c_i) + c_i^2}{c_0(c_0+c_i)}\mathbf{A}_i \\ &= \mathbf{I}.\end{aligned}$$

EXERCISE 21. Let $\mathbf{A}_1, \ldots, \mathbf{A}_k$ represent $n \times n$ symmetric idempotent matrices such that (for $j \neq i = 1, \ldots, k$) $\mathbf{A}_i\mathbf{A}_j = \mathbf{0}$, and let $\mathbf{A}$ represent an $n \times n$ symmetric idempotent matrix such that (for $i = 1, \ldots, k$) $\mathcal{C}(\mathbf{A}_i) \subset \mathcal{C}(\mathbf{A})$. Show that if $\text{rank}(\mathbf{A}_1) + \cdots + \text{rank}(\mathbf{A}_k) = \text{rank}(\mathbf{A})$, then $\mathbf{A}_1 + \cdots + \mathbf{A}_k = \mathbf{A}$.

Solution. Suppose that $\text{rank}(\mathbf{A}_1) + \cdots + \text{rank}(\mathbf{A}_k) = \text{rank}(\mathbf{A})$. Then, in light of Corollary 10.2.2, we have that

$$\begin{aligned}\text{tr}(\mathbf{A}_1 + \cdots + \mathbf{A}_k) &= \text{tr}(\mathbf{A}_1) + \cdots + \text{tr}(\mathbf{A}_k) \\ &= \text{rank}(\mathbf{A}_1) + \cdots + \text{rank}(\mathbf{A}_k) = \text{rank}(\mathbf{A}) = \text{tr}(\mathbf{A}).\end{aligned}$$

Moreover, [since $\mathcal{C}(\mathbf{A}_i) \subset \mathcal{C}(\mathbf{A})$] there exists a matrix $\mathbf{L}_i$ such that $\mathbf{A}_i = \mathbf{A}\mathbf{L}_i$, so that $\mathbf{A}\mathbf{A}_i = \mathbf{A}^2\mathbf{L}_i = \mathbf{A}\mathbf{L}_i = \mathbf{A}_i$ and $\mathbf{A}_i\mathbf{A} = \mathbf{A}_i'\mathbf{A}' = (\mathbf{A}\mathbf{A}_i)' = \mathbf{A}_i' = \mathbf{A}_i$ $(i = 1, \ldots, k)$, implying that

$$\begin{aligned}(\mathbf{A} - \sum_i \mathbf{A}_i)'(\mathbf{A} - \sum_i \mathbf{A}_i) &= (\mathbf{A} - \sum_i \mathbf{A}_i)(\mathbf{A} - \sum_i \mathbf{A}_i) \\ &= \mathbf{A}^2 - \sum_i \mathbf{A}_i\mathbf{A} - \sum_i \mathbf{A}\mathbf{A}_i + \sum_i \mathbf{A}_i^2 + \sum_{i,j\neq i} \mathbf{A}_i\mathbf{A}_j \\ &= \mathbf{A} - \sum_i \mathbf{A}_i - \sum_i \mathbf{A}_i + \sum_i \mathbf{A}_i + \mathbf{0} \\ &= \mathbf{A} - \sum_i \mathbf{A}_i.\end{aligned}$$

Thus,

$$\text{tr}[(\mathbf{A} - \sum_i \mathbf{A}_i)'(\mathbf{A} - \sum_i \mathbf{A}_i)] = \text{tr}(\mathbf{A} - \sum_i \mathbf{A}_i) = \text{tr}(\mathbf{A}) - \text{tr}(\sum_i \mathbf{A}_i) = 0.$$

We conclude (on the basis of Lemma 5.3.1) that $\mathbf{A} - \sum_i \mathbf{A}_i = \mathbf{0}$ or equivalently that $\mathbf{A}_1 + \cdots + \mathbf{A}_k = \mathbf{A}$.

EXERCISE 22. Let $\mathbf{A}$ represent an $m \times n$ matrix and $\mathbf{B}$ an $n \times m$ matrix. If $\mathbf{B}$ is a generalized inverse of $\mathbf{A}$, then $\text{rank}(\mathbf{I} - \mathbf{BA}) = n - \text{rank}(\mathbf{A})$. Show that the converse is also true; that is, show that if $\text{rank}(\mathbf{I} - \mathbf{BA}) = n - \text{rank}(\mathbf{A})$, then $\mathbf{B}$ is a generalized inverse of $\mathbf{A}$.

Solution. Suppose that $\text{rank}(\mathbf{I} - \mathbf{BA}) = n - \text{rank}(\mathbf{A})$. Then, since $\text{rank}(\mathbf{BA}) \leq \text{rank}(\mathbf{A})$, we have that

$$n - \text{rank}(\mathbf{BA}) \geq n - \text{rank}(\mathbf{A}) = \text{rank}(\mathbf{I} - \mathbf{BA}). \tag{S.10}$$

Moreover, making use of Corollary 4.5.9, we find that

$$\text{rank}(\mathbf{I} - \mathbf{BA}) + \text{rank}(\mathbf{BA}) \geq \text{rank}[(\mathbf{I} - \mathbf{BA}) + \mathbf{BA}] = \text{rank}(\mathbf{I}_n) = n$$

and hence that

$$\text{rank}(\mathbf{I} - \mathbf{BA}) \geq n - \text{rank}(\mathbf{BA}). \tag{S.11}$$

Together, results (S.10) and (S.11) imply that

$$\text{rank}(\mathbf{I} - \mathbf{BA}) = n - \text{rank}(\mathbf{BA})$$

or equivalently that

$$\text{rank}(\mathbf{BA}) + \text{rank}(\mathbf{I} - \mathbf{BA}) = n.$$

Thus, it follows from Lemma 18.4.2 that $\mathbf{BA}$ is idempotent. Further,

$$n - \text{rank}(\mathbf{A}) = \text{rank}(\mathbf{I} - \mathbf{BA}) = n - \text{rank}(\mathbf{BA}),$$

implying that $\text{rank}(\mathbf{BA}) = \text{rank}(\mathbf{A})$. We conclude (on the basis of Theorem 10.2.7) that $\mathbf{B}$ is a generalized inverse of $\mathbf{A}$.

EXERCISE 23. Let $\mathbf{A}$ represent the $(n \times n)$ projection matrix for a subspace $\mathcal{U}$ of $\mathcal{R}^{n\times 1}$ along a subspace $\mathcal{V}$ of $\mathcal{R}^{n\times 1}$ (where $\mathcal{U} \oplus \mathcal{V} = \mathcal{R}^{n\times 1}$), and let $\mathbf{B}$ represent the $(n \times n)$ projection matrix for a subspace $\mathcal{W}$ of $\mathcal{R}^{n\times 1}$ along a subspace $\mathcal{X}$ of $\mathcal{R}^{n\times 1}$ (where $\mathcal{W} \oplus \mathcal{X} = \mathcal{R}^{n\times 1}$).

(a) Show that $\mathbf{A}+\mathbf{B}$ is the projection matrix for some subspace $\mathcal{L}$ of $\mathcal{R}^{n\times 1}$ along some subspace $\mathcal{M}$ of $\mathcal{R}^{n\times 1}$ (where $\mathcal{L}\oplus\mathcal{M} = \mathcal{R}^{n\times 1}$) if and only if $\mathbf{BA} = \mathbf{AB} = \mathbf{0}$, in which case $\mathcal{L} = \mathcal{U} \oplus \mathcal{W}$ and $\mathcal{M} = \mathcal{V} \cap \mathcal{X}$.

(b) Show that $\mathbf{A}-\mathbf{B}$ is the projection matrix for some subspace $\mathcal{L}$ of $\mathcal{R}^{n\times 1}$ along some subspace $\mathcal{M}$ of $\mathcal{R}^{n\times 1}$ (where $\mathcal{L}\oplus\mathcal{M} = \mathcal{R}^{n\times 1}$) if and only if $\mathbf{BA} = \mathbf{AB} = \mathbf{B}$, in which case $\mathcal{L} = \mathcal{U} \cap \mathcal{X}$ and $\mathcal{M} = \mathcal{V} \oplus \mathcal{W}$. [*Hint*. Observe (in light of the result that a matrix is a projection matrix for one subspace along another if and only if it is idempotent and the result that a matrix, say $\mathbf{K}$, is idempotent if and only if $\mathbf{I}-\mathbf{K}$ is idempotent) that $\mathbf{A}-\mathbf{B}$ is the projection matrix for some subspace $\mathcal{L}$ along some subspace $\mathcal{M}$ if and only if $\mathbf{I} - (\mathbf{A} - \mathbf{B}) = (\mathbf{I} - \mathbf{A}) + \mathbf{B}$ is the projection matrix for some subspace $\mathcal{L}^*$ along some subspace $\mathcal{M}^*$, and then make use of Part (a) and the result that the projection matrix for a subspace $\mathcal{M}^*$ along a subspace $\mathcal{L}^*$

(where $\mathcal{M}^* \oplus \mathcal{L}^* = \mathcal{R}^{n \times 1}$) equals $\mathbf{I} - \mathbf{H}$, where $\mathbf{H}$ is the projection matrix for $\mathcal{L}^*$ along $\mathcal{M}^*$.]

Solution. (a) According to Theorem 17.6.14, $\mathbf{A}$ and $\mathbf{B}$ are idempotent, $\mathcal{U} = \mathcal{C}(\mathbf{A})$, $\mathcal{W} = \mathcal{C}(\mathbf{B})$, $\mathcal{V} = \mathcal{N}(\mathbf{A})$, and $\mathcal{X} = \mathcal{N}(\mathbf{B})$.

Suppose now that $\mathbf{A} + \mathbf{B}$ is the projection matrix for some subspace $\mathcal{L}$ along some subspace $\mathcal{M}$. Then, as a consequence of Theorem 17.6.13 (or 17.6.14), $\mathbf{A}+\mathbf{B}$ is idempotent. And, it follows from Lemma 18.4.3 that $\mathbf{BA} = \mathbf{AB} = \mathbf{0}$.

Conversely, suppose that $\mathbf{BA} = \mathbf{AB} = \mathbf{0}$. Then, as a consequence of Lemma 18.4.3, $\mathbf{A}+\mathbf{B}$ is idempotent. And, it follows from Theorem 17.6.13 that $\mathbf{A}+\mathbf{B}$ is the projection matrix for some subspace $\mathcal{L}$ along some subspace $\mathcal{M}$ and from Theorem 17.6.12 that $\mathcal{L} = \mathcal{C}(\mathbf{A}+\mathbf{B})$ and (in light of Theorem 11.7.1) that $\mathcal{M} = \mathcal{N}(\mathbf{A}+\mathbf{B})$. Moreover, since $\mathbf{A}$, $\mathbf{B}$, and $\mathbf{A} + \mathbf{B}$ are all idempotent, it follows from Theorem 18.4.1 that

$$\operatorname{rank}(\mathbf{A} + \mathbf{B}) = \operatorname{rank}(\mathbf{A}) + \operatorname{rank}(\mathbf{B}). \tag{S.12}$$

Since (according to Lemmas 4.5.8 and 4.5.7)

$$\operatorname{rank}(\mathbf{A} + \mathbf{B}) \leq \operatorname{rank}(\mathbf{A}, \mathbf{B}) \leq \operatorname{rank}(\mathbf{A}) + \operatorname{rank}(\mathbf{B}),$$

we have [as a consequence of result (S.12)] that

$$\operatorname{rank}(\mathbf{A} + \mathbf{B}) = \operatorname{rank}(\mathbf{A}, \mathbf{B}) = \operatorname{rank}(\mathbf{A}) + \operatorname{rank}(\mathbf{B}),$$

implying [in light of result (4.5.5)] that $\mathcal{C}(\mathbf{A} + \mathbf{B}) = \mathcal{C}(\mathbf{A}, \mathbf{B})$ and (in light of Theorem 17.2.4) that $\mathcal{C}(\mathbf{A})$ and $\mathcal{C}(\mathbf{B})$ are essentially disjoint. Thus, in light of result (17.1.4), it follows that

$$\mathcal{C}(\mathbf{A} + \mathbf{B}) = \mathcal{C}(\mathbf{A}) \oplus \mathcal{C}(\mathbf{B})$$

or equivalently that $\mathcal{L} = \mathcal{U} \oplus \mathcal{W}$.

It remains to show that $\mathcal{M} = \mathcal{V} \cap \mathcal{X}$ or equivalently that $\mathcal{N}(\mathbf{A}+\mathbf{B}) = \mathcal{N}(\mathbf{A}) \cap \mathcal{N}(\mathbf{B})$. Let $\mathbf{x}$ represent an arbitrary vector in $\mathcal{N}(\mathbf{A} + \mathbf{B})$. Then, $\mathbf{Ax} + \mathbf{Bx} = \mathbf{0}$, and consequently (since $\mathbf{A}^2 = \mathbf{A}$, $\mathbf{B}^2 = \mathbf{B}$, and $\mathbf{BA} = \mathbf{AB} = \mathbf{0}$)

$$\mathbf{Ax} = \mathbf{A}^2\mathbf{x} = \mathbf{A}^2\mathbf{x} + \mathbf{ABx} = \mathbf{A}(\mathbf{Ax} + \mathbf{Bx}) = \mathbf{0},$$

$$\mathbf{Bx} = \mathbf{B}^2\mathbf{x} = \mathbf{B}^2\mathbf{x} + \mathbf{BAx} = \mathbf{B}(\mathbf{Ax} + \mathbf{Bx}) = \mathbf{0}.$$

Thus, $\mathbf{x} \in \mathcal{N}(\mathbf{A}) \cap \mathcal{N}(\mathbf{B})$. We conclude that $\mathcal{N}(\mathbf{A}+\mathbf{B}) \subset \mathcal{N}(\mathbf{A}) \cap \mathcal{N}(\mathbf{B})$ and hence [since clearly $\mathcal{N}(\mathbf{A}) \cap \mathcal{N}(\mathbf{B}) \subset \mathcal{N}(\mathbf{A} + \mathbf{B})$] that $\mathcal{N}(\mathbf{A} + \mathbf{B}) = \mathcal{N}(\mathbf{A}) \cap \mathcal{N}(\mathbf{B})$.

(b) Since (according to Lemma 10.1.2) $\mathbf{I} - (\mathbf{A} - \mathbf{B})$ is idempotent if and only if $\mathbf{A} - \mathbf{B}$ is idempotent, it follows from Theorem 17.6.13 that $\mathbf{A} - \mathbf{B}$ is the projection matrix for some subspace $\mathcal{L}$ along some subspace $\mathcal{M}$ if and only if $\mathbf{I} - (\mathbf{A} - \mathbf{B}) = (\mathbf{I} - \mathbf{A}) + \mathbf{B}$ is the projection matrix for some subspace $\mathcal{L}^*$ along some subspace $\mathcal{M}^*$ — Lemma 10.1.2 and Theorem 17.6.13 are the results mentioned parenthetically in the hint. And, since (according to Theorem 17.6.10) $\mathbf{I} - \mathbf{A}$ is the projection matrix for $\mathcal{V}$ along $\mathcal{U}$, it follows from Part (a) that $(\mathbf{I} - \mathbf{A}) + \mathbf{B}$ is the projection matrix for some subspace $\mathcal{L}^*$ along some subspace $\mathcal{M}^*$ if and only if

$$\mathbf{B}(\mathbf{I} - \mathbf{A}) = (\mathbf{I} - \mathbf{A})\mathbf{B} = \mathbf{0},$$

or equivalently if and only if

$$\mathbf{BA} = \mathbf{AB} = \mathbf{B},$$

in which case $\mathcal{L}^* = \mathcal{V} \oplus \mathcal{W}$ and $\mathcal{M}^* = \mathcal{U} \cap \mathcal{X}$. The proof is complete upon observing (in light of Theorem 17.6.10, which is the result whose use is prescribed in the hint) that if $(\mathbf{I} - \mathbf{A}) + \mathbf{B}$ is the projection matrix for $\mathcal{V} \oplus \mathcal{W}$ along $\mathcal{U} \cap \mathcal{X}$, then $\mathbf{A} - \mathbf{B} = \mathbf{I} - [(\mathbf{I} - \mathbf{A}) + \mathbf{B}]$ is the projection matrix for $\mathcal{U} \cap \mathcal{X}$ along $\mathcal{V} \oplus \mathcal{W}$.

EXERCISE 24. (a) Let $\mathbf{B}$ represent an $n \times n$ symmetric matrix, and let $\mathbf{W}$ represent an $n \times n$ symmetric nonnegative definite matrix. Show that

$$\begin{aligned}\mathbf{WBWBW} = \mathbf{WBW} \;&\Leftrightarrow\; (\mathbf{BW})^3 = (\mathbf{BW})^2 \\ &\Leftrightarrow\; \mathrm{tr}[(\mathbf{BW})^2] = \mathrm{tr}[(\mathbf{BW})^3] = \mathrm{tr}[(\mathbf{BW})^4].\end{aligned}$$

(b) Let $\mathbf{A}_1, \ldots, \mathbf{A}_k$ represent $n \times n$ matrices, let $\mathbf{V}$ represent an $n \times n$ symmetric nonnegative definite matrix, and define $\mathbf{A} = \mathbf{A}_1 + \cdots \mathbf{A}_k$. If $\mathbf{VA}_i\mathbf{VA}_i\mathbf{V} = \mathbf{VA}_i\mathbf{V}$ for all i and if $\mathbf{VA}_i\mathbf{VA}_j\mathbf{V} = \mathbf{0}$ for all i and $j \neq i$, then $\mathbf{VAVAV} = \mathbf{VAV}$ and $\mathrm{rank}(\mathbf{VA}_1\mathbf{V}) + \cdots + \mathrm{rank}(\mathbf{VA}_k\mathbf{V}) = \mathrm{rank}(\mathbf{VAV})$. Conversely, if $\mathbf{VAVAV} = \mathbf{VAV}$, then each of the following three conditions implies the other two: (1) $\mathbf{VA}_i\mathbf{VA}_j\mathbf{V} = \mathbf{0}$ (for $j \neq i = 1, \ldots, k$) and $\mathrm{rank}(\mathbf{VA}_i\mathbf{VA}_i\mathbf{V}) = \mathrm{rank}(\mathbf{VA}_i\mathbf{V})$ (for $i = 1, \ldots, k$); (2) $\mathbf{VA}_i\mathbf{VA}_i\mathbf{V} = \mathbf{VA}_i\mathbf{V}$ (for $i = 1, \ldots, k$); (3) $\mathrm{rank}(\mathbf{VA}_1\mathbf{V}) + \cdots + \mathrm{rank}(\mathbf{VA}_k\mathbf{V}) = \mathrm{rank}(\mathbf{VAV})$. Indicate how, in the special case where $\mathbf{A}_1, \ldots, \mathbf{A}_k$ are symmetric, the conditions $\mathbf{VAVAV} = \mathbf{VAV}$ and $\mathbf{VA}_i\mathbf{VA}_i\mathbf{V} = \mathbf{VA}_i\mathbf{V}$ can be reexpressed by applying the results of Part (a) (of the current exercise).

Solution. (a) Let $\mathbf{S}$ represent any matrix such that $\mathbf{W} = \mathbf{S}'\mathbf{S}$ — the existence of such a matrix follows from Corollary 14.3.8.

If $\mathbf{WBWBW} = \mathbf{WBW}$, then clearly $\mathbf{BWBWBW} = \mathbf{BWBW}$, or equivalently $(\mathbf{BW})^3 = (\mathbf{BW})^2$. Conversely, suppose that $(\mathbf{BW})^3 = (\mathbf{BW})^2$. Then,

$$(\mathbf{SB})'\mathbf{SBWBW} = (\mathbf{SB})'\mathbf{SBW},$$

implying (in light of Corollary 5.3.3) that $\mathbf{SBWBW} = \mathbf{SBW}$, so that

$$\mathbf{WBWBW} = \mathbf{S}'\mathbf{SBWBW} = \mathbf{S}'\mathbf{SBW} = \mathbf{WBW}.$$

It remains to show that

$$(\mathbf{BW})^3 = (\mathbf{BW})^2 \;\Leftrightarrow\; \mathrm{tr}[(\mathbf{BW})^2] = \mathrm{tr}[(\mathbf{BW})^3] = \mathrm{tr}[(\mathbf{BW})^4].$$

Suppose that $(\mathbf{BW})^3 = (\mathbf{BW})^2$. Then, clearly,

$$(\mathbf{BW})^4 = \mathbf{BW}(\mathbf{BW})^3 = \mathbf{BW}(\mathbf{BW})^2 = (\mathbf{BW})^3.$$

Thus, $\mathrm{tr}[(\mathbf{BW})^2] = \mathrm{tr}[(\mathbf{BW})^3] = \mathrm{tr}[(\mathbf{BW})^4]$.

Conversely, suppose that $\mathrm{tr}[(\mathbf{BW})^2] = \mathrm{tr}[(\mathbf{BW})^3] = \mathrm{tr}[(\mathbf{BW})^4]$. Then, making use of Lemma 5.2.1, we find that

$$\begin{aligned}
&\mathrm{tr}[(\mathbf{SBS}' - \mathbf{SBWBS}')'(\mathbf{SBS}' - \mathbf{SBWBS}')] \\
&\qquad = \mathrm{tr}(\mathbf{SBWBS}' - 2\mathbf{SBWBWBS}' + \mathbf{SBWBWBWBS}') \\
&\qquad = \mathrm{tr}(\mathbf{BWBS}'\mathbf{S}) - 2\mathrm{tr}(\mathbf{BWBWBS}'\mathbf{S}) + \mathrm{tr}(\mathbf{BWBWBWBS}'\mathbf{S}) \\
&\qquad = \mathrm{tr}[(\mathbf{BW})^2] - 2\mathrm{tr}[(\mathbf{BW})^3] + \mathrm{tr}[(\mathbf{BW})^4] \\
&\qquad = 0.
\end{aligned}$$

Thus, it follows from Lemma 5.3.1 that $\mathbf{SBS}' - \mathbf{SBWBS}' = \mathbf{0}$, or equivalently that $\mathbf{SBWBS}' = \mathbf{SBS}'$, so that

$$(\mathbf{BW})^3 = \mathbf{BS}'(\mathbf{SBWBS}')\mathbf{S} = \mathbf{BS}'(\mathbf{SBS}')\mathbf{S} = (\mathbf{BW})^2.$$

(b) Suppose that $\mathbf{A}_1, \ldots, \mathbf{A}_k$ are symmetric (in which case $\mathbf{A}$ is also symmetric). Then, applying the results of Part (a) (with $\mathbf{A}$ in place of $\mathbf{B}$ and $\mathbf{V}$ in place of $\mathbf{W}$), we find that

$$\mathbf{VAVAV} = \mathbf{VAV} \;\Leftrightarrow\; (\mathbf{AV})^3 = (\mathbf{AV})^2 \;\Leftrightarrow\; \mathrm{tr}[(\mathbf{AV})^2] = \mathrm{tr}[(\mathbf{AV})^3] = \mathrm{tr}[(\mathbf{AV})^4].$$

Similarly, applying the results of Part (a) (with $\mathbf{A}_i$ in place of $\mathbf{B}$ and $\mathbf{V}$ in place of $\mathbf{W}$), we find that

$$\begin{aligned}
\mathbf{VA}_i\mathbf{VA}_i\mathbf{V} = \mathbf{VA}_i\mathbf{V} \;&\Leftrightarrow\; (\mathbf{A}_i\mathbf{V})^3 = (\mathbf{A}_i\mathbf{V})^2 \\
&\Leftrightarrow\; \mathrm{tr}[(\mathbf{A}_i\mathbf{V})^2] = \mathrm{tr}[(\mathbf{A}_i\mathbf{V})^3] = \mathrm{tr}[(\mathbf{A}_i\mathbf{V})^4].
\end{aligned}$$

EXERCISE 25. Let $\mathbf{R}$ represent an $n \times q$ matrix, $\mathbf{S}$ an $n \times m$ matrix, $\mathbf{T}$ an $m \times p$ matrix, and $\mathbf{U}$ a $p \times q$ matrix.

(a) Show that

$$\mathrm{rank}(\mathbf{R} + \mathbf{STU}) = \mathrm{rank}\begin{pmatrix} \mathbf{R} & -\mathbf{ST} \\ \mathbf{TU} & \mathbf{T} \end{pmatrix} - \mathrm{rank}(\mathbf{T}). \tag{E.3}$$

(b) Let $\mathbf{E}_R = \mathbf{I} - \mathbf{RR}^-$, $\mathbf{F}_R = \mathbf{I} - \mathbf{R}^-\mathbf{R}$, $\mathbf{X} = \mathbf{E}_R\mathbf{ST}$, $\mathbf{Y} = \mathbf{TUF}_R$, $\mathbf{E}_Y = \mathbf{I} - \mathbf{YY}^-$, $\mathbf{F}_X = \mathbf{I} - \mathbf{X}^-\mathbf{X}$, $\mathbf{Q} = \mathbf{T} + \mathbf{TUR}^-\mathbf{ST}$, and $\mathbf{Z} = \mathbf{E}_Y\mathbf{QF}_X$. Use the result of Part (b) of Exercise 10.10 to show that

$$\mathrm{rank}(\mathbf{R} + \mathbf{STU}) = \mathrm{rank}(\mathbf{R}) + \mathrm{rank}(\mathbf{X}) + \mathrm{rank}(\mathbf{Y}) + \mathrm{rank}(\mathbf{Z}) - \mathrm{rank}(\mathbf{T}).$$

Solution. (a) Observing that

$$\begin{pmatrix} \mathbf{R} & -\mathbf{ST} \\ \mathbf{TU} & \mathbf{T} \end{pmatrix}\begin{pmatrix} \mathbf{I} & \mathbf{0} \\ -\mathbf{U} & \mathbf{I} \end{pmatrix} = \begin{pmatrix} \mathbf{R} + \mathbf{STU} & -\mathbf{ST} \\ \mathbf{0} & \mathbf{T} \end{pmatrix}$$

and making use of Lemma 8.5.2 and Corollary 9.6.2, we find that

$$\operatorname{rank}\begin{pmatrix} \mathbf{R} & -\mathbf{ST} \\ \mathbf{TU} & \mathbf{T} \end{pmatrix} = \operatorname{rank}\begin{pmatrix} \mathbf{R}+\mathbf{STU} & -\mathbf{ST} \\ \mathbf{0} & \mathbf{T} \end{pmatrix} = \operatorname{rank}(\mathbf{R}+\mathbf{STU}) + \operatorname{rank}(\mathbf{T})$$

and hence that

$$\operatorname{rank}(\mathbf{R}+\mathbf{STU}) = \operatorname{rank}\begin{pmatrix} \mathbf{R} & -\mathbf{ST} \\ \mathbf{TU} & \mathbf{T} \end{pmatrix} - \operatorname{rank}(\mathbf{T}).$$

Or, alternatively, equality (E.3) can be validated by making use of result (9.6.1) — observing that $\mathcal{C}(\mathbf{TU}) \subset \mathcal{C}(\mathbf{T})$ and $\mathcal{R}(-\mathbf{ST}) \subset \mathcal{R}(\mathbf{T})$, we find that

$$\begin{aligned}\operatorname{rank}\begin{pmatrix} \mathbf{R} & -\mathbf{ST} \\ \mathbf{TU} & \mathbf{T} \end{pmatrix} &= \operatorname{rank}(\mathbf{T}) + \operatorname{rank}[\mathbf{R} - (-\mathbf{ST})\mathbf{T}^{-}\mathbf{TU}] \\ &= \operatorname{rank}(\mathbf{T}) + \operatorname{rank}(\mathbf{R}+\mathbf{STU})\end{aligned}$$

and hence that

$$\operatorname{rank}(\mathbf{R}+\mathbf{STU}) = \operatorname{rank}\begin{pmatrix} \mathbf{R} & -\mathbf{ST} \\ \mathbf{TU} & \mathbf{T} \end{pmatrix} - \operatorname{rank}(\mathbf{T}).$$

(b) Upon observing that $\operatorname{rank}(\mathbf{X}) = \operatorname{rank}(-\mathbf{X})$, that $-\mathbf{X}^{-}$ is a generalized inverse of $-\mathbf{X}$, and that $\mathbf{F}_X = \mathbf{I} - (-\mathbf{X}^{-})(-\mathbf{X})$, it follows from the result of Part (b) of Exercise 10.10 that

$$\operatorname{rank}\begin{pmatrix} \mathbf{R} & -\mathbf{ST} \\ \mathbf{TU} & \mathbf{T} \end{pmatrix} = \operatorname{rank}(\mathbf{R}) + \operatorname{rank}(\mathbf{X}) + \operatorname{rank}(\mathbf{Y}) + \operatorname{rank}(\mathbf{Z}).$$

We conclude, on the basis of Part (a) (of the current exercise) that

$$\operatorname{rank}(\mathbf{R}+\mathbf{STU}) = \operatorname{rank}(\mathbf{R}) + \operatorname{rank}(\mathbf{X}) + \operatorname{rank}(\mathbf{Y}) + \operatorname{rank}(\mathbf{Z}) - \operatorname{rank}(\mathbf{T}).$$

EXERCISE 26. Show that, for any $m \times n$ matrices $\mathbf{A}$ and $\mathbf{B}$,

$$\operatorname{rank}(\mathbf{A}+\mathbf{B}) \geq |\operatorname{rank}(\mathbf{A}) - \operatorname{rank}(\mathbf{B})|.$$

Solution. Making use of results (4.5.7) and (4.4.3), we find that

$$\operatorname{rank}(\mathbf{A}) = \operatorname{rank}[(\mathbf{A}+\mathbf{B})-\mathbf{B}] \leq \operatorname{rank}(\mathbf{A}+\mathbf{B})+\operatorname{rank}(-\mathbf{B}) = \operatorname{rank}(\mathbf{A}+\mathbf{B})+\operatorname{rank}(\mathbf{B})$$

and hence that

$$\operatorname{rank}(\mathbf{A}+\mathbf{B}) \geq \operatorname{rank}(\mathbf{A}) - \operatorname{rank}(\mathbf{B}). \tag{S.13}$$

Similarly, we find that

$$\operatorname{rank}(\mathbf{B}) = \operatorname{rank}[(\mathbf{A}+\mathbf{B})-\mathbf{A}] \leq \operatorname{rank}(\mathbf{A}+\mathbf{B})+\operatorname{rank}(-\mathbf{A}) = \operatorname{rank}(\mathbf{A}+\mathbf{B})+\operatorname{rank}(\mathbf{A})$$

and hence that

$$\text{rank}(\mathbf{A}+\mathbf{B}) \geq \text{rank}(\mathbf{B}) - \text{rank}(\mathbf{A}) = -[\text{rank}(\mathbf{A}) - \text{rank}(\mathbf{B})]. \tag{S.14}$$

Together, results (S.13) and (S.14) imply that

$$\text{rank}(\mathbf{A}+\mathbf{B}) \geq |\text{rank}(\mathbf{A}) - \text{rank}(\mathbf{B})|.$$

EXERCISE 27. Show that, for any $n \times n$ symmetric nonnegative definite matrices $\mathbf{A}$ and $\mathbf{B}$,

$$\mathcal{C}(\mathbf{A}+\mathbf{B}) = \mathcal{C}(\mathbf{A}, \mathbf{B}), \qquad \mathcal{R}(\mathbf{A}+\mathbf{B}) = \mathcal{R}\begin{pmatrix}\mathbf{A}\\ \mathbf{B}\end{pmatrix},$$

$$\text{rank}(\mathbf{A}+\mathbf{B}) = \text{rank}(\mathbf{A}, \mathbf{B}) = \text{rank}\begin{pmatrix}\mathbf{A}\\ \mathbf{B}\end{pmatrix}.$$

Solution. According to Corollary 14.3.8, there exist matrices $\mathbf{R}$ and $\mathbf{S}$ such that $\mathbf{A} = \mathbf{R}'\mathbf{R}$ and $\mathbf{B} = \mathbf{S}'\mathbf{S}$. And, upon observing that

$$\mathbf{A}+\mathbf{B} = \begin{pmatrix}\mathbf{R}\\ \mathbf{S}\end{pmatrix}'\begin{pmatrix}\mathbf{R}\\ \mathbf{S}\end{pmatrix}$$

and recalling Corollaries 7.4.5 and 4.5.6, we find that

$$\mathcal{C}(\mathbf{A}, \mathbf{B}) = \mathcal{C}(\mathbf{R}'\mathbf{R},\ \mathbf{S}'\mathbf{S}) = \mathcal{C}(\mathbf{R}', \mathbf{S}') = \mathcal{C}\left[\begin{pmatrix}\mathbf{R}\\ \mathbf{S}\end{pmatrix}'\right] = \mathcal{C}(\mathbf{A}+\mathbf{B})$$

[which implies that $\text{rank}(\mathbf{A}, \mathbf{B}) = \text{rank}(\mathbf{A}+\mathbf{B})$] and similarly that

$$\mathcal{R}\begin{pmatrix}\mathbf{A}\\ \mathbf{B}\end{pmatrix} = \mathcal{R}\begin{pmatrix}\mathbf{R}'\mathbf{R}\\ \mathbf{S}'\mathbf{S}\end{pmatrix} = \mathcal{R}\begin{pmatrix}\mathbf{R}\\ \mathbf{S}\end{pmatrix} = \mathcal{R}(\mathbf{A}+\mathbf{B})$$

[which implies that $\text{rank}\begin{pmatrix}\mathbf{A}\\ \mathbf{B}\end{pmatrix} = \text{rank}(\mathbf{A}+\mathbf{B})$].

EXERCISE 28. Let $\mathbf{A}$ and $\mathbf{B}$ represent $m \times n$ matrices.

(a) Show that (1) $\mathcal{C}(\mathbf{A}) \subset \mathcal{C}(\mathbf{A}+\mathbf{B})$ if and only if $\text{rank}(\mathbf{A}, \mathbf{B}) = \text{rank}(\mathbf{A}+\mathbf{B})$ and (2) $\mathcal{R}(\mathbf{A}) \subset \mathcal{R}(\mathbf{A}+\mathbf{B})$ if and only if $\text{rank}\begin{pmatrix}\mathbf{A}\\ \mathbf{B}\end{pmatrix} = \text{rank}(\mathbf{A}+\mathbf{B})$.

(b) Show that (1) if $\mathcal{R}(\mathbf{A})$ and $\mathcal{R}(\mathbf{B})$ are essentially disjoint, then $\mathcal{C}(\mathbf{A}) \subset \mathcal{C}(\mathbf{A}+\mathbf{B})$ and (2) if $\mathcal{C}(\mathbf{A})$ and $\mathcal{C}(\mathbf{B})$ are essentially disjoint, then $\mathcal{R}(\mathbf{A}) \subset \mathcal{R}(\mathbf{A}+\mathbf{B})$.

Solution. (a) (1) Suppose that $\text{rank}(\mathbf{A}, \mathbf{B}) = \text{rank}(\mathbf{A}+\mathbf{B})$. Then, since (according to Lemma 4.5.8) $\mathcal{C}(\mathbf{A}+\mathbf{B}) \subset \mathcal{C}(\mathbf{A}, \mathbf{B})$, it follows from Theorem 4.4.6 that $\mathcal{C}(\mathbf{A}, \mathbf{B}) = \mathcal{C}(\mathbf{A}+\mathbf{B})$. Then, since $\mathcal{C}(\mathbf{A}) \subset \mathcal{C}(\mathbf{A}, \mathbf{B})$, we have that $\mathcal{C}(\mathbf{A}) \subset \mathcal{C}(\mathbf{A}+\mathbf{B})$.

Conversely, suppose that $\mathcal{C}(\mathbf{A}) \subset \mathcal{C}(\mathbf{A}+\mathbf{B})$. Then, according to Lemma 4.2.2, there exists a matrix $\mathbf{F}$ such that $\mathbf{A} = (\mathbf{A}+\mathbf{B})\mathbf{F}$. Further, $\mathbf{B} = (\mathbf{A}+\mathbf{B}) - \mathbf{A} =$

$(\mathbf{A}+\mathbf{B})(\mathbf{I}-\mathbf{F})$. Thus $(\mathbf{A}, \mathbf{B}) = (\mathbf{A}+\mathbf{B})(\mathbf{F}, \mathbf{I}-\mathbf{F})$, implying that $\mathcal{C}(\mathbf{A}, \mathbf{B}) \subset \mathcal{C}(\mathbf{A}+\mathbf{B})$. Since (according to Lemma 4.5.8) $\mathcal{C}(\mathbf{A}+\mathbf{B}) \subset \mathcal{C}(\mathbf{A}, \mathbf{B})$, we conclude that $\mathcal{C}(\mathbf{A}, \mathbf{B}) = \mathcal{C}(\mathbf{A}+\mathbf{B})$ and hence that $\text{rank}(\mathbf{A}, \mathbf{B}) = \text{rank}(\mathbf{A}+\mathbf{B})$.

(2) The proof of Part (2) is analogous to that of Part (1).

(b) Let $c = \dim[\mathcal{C}(\mathbf{A}) \cap \mathcal{C}(\mathbf{B})]$, $d = \dim[\mathcal{R}(\mathbf{A}) \cap \mathcal{R}(\mathbf{B})]$, and

$$\mathbf{H} = \left[\mathbf{I} - \begin{pmatrix}\mathbf{A}\\ \mathbf{B}\end{pmatrix}\begin{pmatrix}\mathbf{A}\\ \mathbf{B}\end{pmatrix}^{-}\right]\begin{pmatrix}\mathbf{A} & \mathbf{0}\\ \mathbf{0} & \mathbf{B}\end{pmatrix}[\mathbf{I} - (\mathbf{A}, \mathbf{B})^{-}(\mathbf{A}, \mathbf{B})].$$

(1) If $\mathcal{R}(\mathbf{A})$ and $\mathcal{R}(\mathbf{B})$ are essentially disjoint (or equivalently if $d = 0$), then (as a consequence of Theorem 18.5.6) $\text{rank}(\mathbf{H}) = 0 = d$, implying [in light of result (5.14)] that

$$\text{rank}(\mathbf{A}+\mathbf{B}) = \text{rank}(\mathbf{A}, \mathbf{B}),$$

and hence [in light of part (a)-(1)] that $\mathcal{C}(\mathbf{A}) \subset \mathcal{C}(\mathbf{A}+\mathbf{B})$.

(2) Similarly, if $\mathcal{C}(\mathbf{A}$ and $\mathcal{C}(\mathbf{B})$ are essentially disjoint (or equivalently if $c = 0$), then (as a consequence of Theorem 18.5.6) $\text{rank}(\mathbf{H}) = 0 = c$, implying [in light of result (5.18)] that

$$\text{rank}(\mathbf{A}+\mathbf{B}) = \text{rank}\begin{pmatrix}\mathbf{A}\\ \mathbf{B}\end{pmatrix}$$

and hence [in light of Part (a)-(2)] that $\mathcal{R}(\mathbf{A}) \subset \mathcal{R}(\mathbf{A}+\mathbf{B})$.

EXERCISE 29. Let $\mathbf{A}$ and $\mathbf{B}$ represent $m \times n$ matrices. Show that *each* of the following five conditions is necessary and sufficient for rank additivity [i.e., for $\text{rank}(\mathbf{A}+\mathbf{B}) = \text{rank}(\mathbf{A}) + \text{rank}(\mathbf{B})$]:

(a) $\text{rank}(\mathbf{A}, \mathbf{B}) = \text{rank}\begin{pmatrix}\mathbf{A}\\ \mathbf{B}\end{pmatrix} = \text{rank}(\mathbf{A}) + \text{rank}(\mathbf{B})$;

(b) $\text{rank}(\mathbf{A}) = \text{rank}[\mathbf{A}(\mathbf{I}-\mathbf{B}^{-}\mathbf{B})] = \text{rank}[(\mathbf{I}-\mathbf{B}\mathbf{B}^{-})\mathbf{A}]$;

(c) $\text{rank}(\mathbf{B}) = \text{rank}[\mathbf{B}(\mathbf{I}-\mathbf{A}^{-}\mathbf{A})] = \text{rank}[(\mathbf{I}-\mathbf{A}\mathbf{A}^{-})\mathbf{B}]$;

(d) $\text{rank}(\mathbf{A}) = \text{rank}[\mathbf{A}(\mathbf{I}-\mathbf{B}^{-}\mathbf{B})]$ and $\text{rank}(\mathbf{B}) = \text{rank}[(\mathbf{I}-\mathbf{A}\mathbf{A}^{-})\mathbf{B}]$;

(e) $\text{rank}(\mathbf{A}) = \text{rank}[(\mathbf{I}-\mathbf{B}\mathbf{B}^{-})\mathbf{A}]$ and $\text{rank}(\mathbf{B}) = \text{rank}[\mathbf{B}(\mathbf{I}-\mathbf{A}^{-}\mathbf{A})]$.

Solution. Let $r = \dim[\mathcal{C}(\mathbf{A}) \cap \mathcal{C}(\mathbf{B})]$ and $s = \dim[\mathcal{R}(\mathbf{A}) \cap \mathcal{R}(\mathbf{B})]$. In light of Theorem 18.5.7, it suffices to show that the condition $r = s = 0$ is equivalent to each of Conditions (a)-(e).

That $r = s = 0$ implies Condition (a) and conversely is an immediate consequence of results (5.15) and (5.19). That $r = s = 0$ is equivalent to each of Conditions (b) - (e) becomes clear upon observing that (as a consequence of Corollary 17.2.10)

$$r = 0 \Leftrightarrow \text{rank}(\mathbf{A}) = \text{rank}[(\mathbf{I}-\mathbf{B}\mathbf{B}^{-})\mathbf{A}] \Leftrightarrow \text{rank}(\mathbf{B}) = \text{rank}[(\mathbf{I}-\mathbf{A}\mathbf{A}^{-})\mathbf{B}],$$

$$s = 0 \Leftrightarrow \text{rank}(\mathbf{A}) = \text{rank}[\mathbf{A}(\mathbf{I} - \mathbf{B}^-\mathbf{B})] \Leftrightarrow \text{rank}(\mathbf{B}) = \text{rank}[\mathbf{B}(\mathbf{I} - \mathbf{A}^-\mathbf{A})].$$

EXERCISE 30. Let $\mathbf{A}$ and $\mathbf{B}$ represent $m \times n$ matrices. And, let

$$\mathbf{H} = \left[\mathbf{I} - \begin{pmatrix}\mathbf{A}\\ \mathbf{B}\end{pmatrix}\begin{pmatrix}\mathbf{A}\\ \mathbf{B}\end{pmatrix}^-\right]\begin{pmatrix}\mathbf{A} & \mathbf{0}\\ \mathbf{0} & -\mathbf{B}\end{pmatrix}[\mathbf{I} - (\mathbf{A}, \mathbf{B})^-(\mathbf{A}, \mathbf{B})].$$

(a) Show that

$$\text{rank}(\mathbf{A} - \mathbf{B}) = \text{rank}(\mathbf{A}) - \text{rank}(\mathbf{B}) + [\text{rank}(\mathbf{A}, \mathbf{B}) - \text{rank}(\mathbf{A})] + \left[\text{rank}\begin{pmatrix}\mathbf{A}\\ \mathbf{B}\end{pmatrix} - \text{rank}(\mathbf{A})\right] + \text{rank}(\mathbf{H}).$$

Do so by applying (with $-\mathbf{B}$ in place of $\mathbf{B}$) the formula

$$\text{rank}(\mathbf{A} + \mathbf{B}) = \text{rank}(\mathbf{A}, \mathbf{B}) + \text{rank}\begin{pmatrix}\mathbf{A}\\ \mathbf{B}\end{pmatrix} - \text{rank}(\mathbf{A}) - \text{rank}(\mathbf{B}) + \text{rank}(\mathbf{K}), \quad (*)$$

where

$$\mathbf{K} = \left[\mathbf{I} - \begin{pmatrix}\mathbf{A}\\ \mathbf{B}\end{pmatrix}\begin{pmatrix}\mathbf{A}\\ \mathbf{B}\end{pmatrix}^-\right]\begin{pmatrix}\mathbf{A} & \mathbf{0}\\ \mathbf{0} & \mathbf{B}\end{pmatrix}[\mathbf{I} - (\mathbf{A}, \mathbf{B})^-(\mathbf{A}, \mathbf{B})].$$

(b) Show that $\mathbf{A}$ and $\mathbf{B}$ are rank subtractive [in the sense that $\text{rank}(\mathbf{A} - \mathbf{B}) = \text{rank}(\mathbf{A}) - \text{rank}(\mathbf{B})$] if and only if $\text{rank}(\mathbf{A}, \mathbf{B}) = \text{rank}\begin{pmatrix}\mathbf{A}\\ \mathbf{B}\end{pmatrix} = \text{rank}(\mathbf{A})$ and $\mathbf{H} = \mathbf{0}$.

(c) Show that if $\text{rank}(\mathbf{A}, \mathbf{B}) = \text{rank}\begin{pmatrix}\mathbf{A}\\ \mathbf{B}\end{pmatrix} = \text{rank}(\mathbf{A})$, then (1) $(\mathbf{A}^-, \mathbf{0})$ and $\begin{pmatrix}\mathbf{A}^-\\ \mathbf{0}\end{pmatrix}$ are generalized inverses of $\begin{pmatrix}\mathbf{A}\\ \mathbf{B}\end{pmatrix}$ and $(\mathbf{A}, \mathbf{B})$, respectively, and (2) for $\begin{pmatrix}\mathbf{A}\\ \mathbf{B}\end{pmatrix}^- = (\mathbf{A}^-, \mathbf{0})$ and $(\mathbf{A}, \mathbf{B})^- = \begin{pmatrix}\mathbf{A}^-\\ \mathbf{0}\end{pmatrix}$, $\mathbf{H} = \begin{pmatrix}\mathbf{0} & \mathbf{0}\\ \mathbf{0} & \mathbf{B}\mathbf{A}^-\mathbf{B} - \mathbf{B}\end{pmatrix}$.

(d) Show that *each* of the following three conditions is necessary and sufficient for rank subtractivity [i.e., for $\text{rank}(\mathbf{A} - \mathbf{B}) = \text{rank}(\mathbf{A}) - \text{rank}(\mathbf{B})$]:

(1) $\text{rank}(\mathbf{A}, \mathbf{B}) = \text{rank}\begin{pmatrix}\mathbf{A}\\ \mathbf{B}\end{pmatrix} = \text{rank}(\mathbf{A})$ and $\mathbf{B}\mathbf{A}^-\mathbf{B} = \mathbf{B}$;

(2) $\mathcal{C}(\mathbf{B}) \subset \mathcal{C}(\mathbf{A})$, $\mathcal{R}(\mathbf{B}) \subset \mathcal{R}(\mathbf{A})$, and $\mathbf{B}\mathbf{A}^-\mathbf{B} = \mathbf{B}$;

(3) $\mathbf{A}\mathbf{A}^-\mathbf{B} = \mathbf{B}\mathbf{A}^-\mathbf{A} = \mathbf{B}\mathbf{A}^-\mathbf{B} = \mathbf{B}$.

(e) Using the result of Exercise 29 (or otherwise), show that $\text{rank}(\mathbf{A} - \mathbf{B}) = \text{rank}(\mathbf{A}) - \text{rank}(\mathbf{B})$ if and only if $\text{rank}(\mathbf{A} - \mathbf{B}) = \text{rank}[\mathbf{A}(\mathbf{I} - \mathbf{B}^-\mathbf{B})] = \text{rank}[(\mathbf{I} - \mathbf{B}\mathbf{B}^-)\mathbf{A}]$.

Solution. (a) Clearly,

$$\begin{pmatrix}\mathbf{A}\\ -\mathbf{B}\end{pmatrix} = \begin{pmatrix}\mathbf{I}_m & \mathbf{0}\\ \mathbf{0} & -\mathbf{I}_m\end{pmatrix}\begin{pmatrix}\mathbf{A}\\ \mathbf{B}\end{pmatrix}, \qquad (\mathbf{A}, -\mathbf{B}) = (\mathbf{A}, \mathbf{B})\begin{pmatrix}\mathbf{I}_n & \mathbf{0}\\ \mathbf{0} & -\mathbf{I}_n\end{pmatrix}.$$

Thus, it follows from Parts (1) and (2) of Lemma 9.2.4 that $\begin{pmatrix} \mathbf{A} \\ \mathbf{B} \end{pmatrix}^{-} \begin{pmatrix} \mathbf{I}_m & \mathbf{0} \\ \mathbf{0} & -\mathbf{I}_m \end{pmatrix}^{-1}$ is a generalized inverse of $\begin{pmatrix} \mathbf{A} \\ -\mathbf{B} \end{pmatrix}$ and that $\begin{pmatrix} \mathbf{I}_n & \mathbf{0} \\ \mathbf{0} & -\mathbf{I}_n \end{pmatrix}^{-1} (\mathbf{A}, \mathbf{B})^{-}$ is a generalized inverse of $(\mathbf{A}, -\mathbf{B})$. Further,

$$\begin{aligned}
&\begin{pmatrix} \mathbf{I}_m & \mathbf{0} \\ \mathbf{0} & -\mathbf{I}_m \end{pmatrix} \mathbf{H} \begin{pmatrix} \mathbf{I}_n & \mathbf{0} \\ \mathbf{0} & -\mathbf{I}_n \end{pmatrix} \\
&= \left[\mathbf{I} - \begin{pmatrix} \mathbf{A} \\ -\mathbf{B} \end{pmatrix} \begin{pmatrix} \mathbf{A} \\ \mathbf{B} \end{pmatrix}^{-} \begin{pmatrix} \mathbf{I}_m & \mathbf{0} \\ \mathbf{0} & -\mathbf{I}_m \end{pmatrix}^{-1} \right] \begin{pmatrix} \mathbf{I}_m & \mathbf{0} \\ \mathbf{0} & -\mathbf{I}_m \end{pmatrix} \\
&\qquad \times \begin{pmatrix} \mathbf{A} & \mathbf{0} \\ \mathbf{0} & -\mathbf{B} \end{pmatrix} \begin{pmatrix} \mathbf{I}_n & \mathbf{0} \\ \mathbf{0} & -\mathbf{I}_n \end{pmatrix} \left[\mathbf{I} - \begin{pmatrix} \mathbf{I}_n & \mathbf{0} \\ \mathbf{0} & -\mathbf{I}_n \end{pmatrix}^{-1} (\mathbf{A}, \mathbf{B})^{-} (\mathbf{A}, -\mathbf{B}) \right] \\
&= \left[\mathbf{I} - \begin{pmatrix} \mathbf{A} \\ -\mathbf{B} \end{pmatrix} \begin{pmatrix} \mathbf{A} \\ \mathbf{B} \end{pmatrix}^{-} \begin{pmatrix} \mathbf{I}_m & \mathbf{0} \\ \mathbf{0} & -\mathbf{I}_m \end{pmatrix}^{-1} \right] \\
&\qquad \times \begin{pmatrix} \mathbf{A} & \mathbf{0} \\ \mathbf{0} & -\mathbf{B} \end{pmatrix} \left[\mathbf{I} - \begin{pmatrix} \mathbf{I}_n & \mathbf{0} \\ \mathbf{0} & -\mathbf{I}_n \end{pmatrix}^{-1} (\mathbf{A}, \mathbf{B})^{-} (\mathbf{A}, -\mathbf{B}) \right].
\end{aligned}$$

Now, applying result (∗) [which is equivalent to result (5.7) or, when combined with result (17.4.13) or (17.4.12), to result (5.8) or (5.9)] and recalling Corollary 4.5.6, we find that

$$\begin{aligned}
\operatorname{rank}(\mathbf{A} - \mathbf{B}) &= \operatorname{rank}[\mathbf{A} + (-\mathbf{B})] \\
&= \operatorname{rank}(\mathbf{A}, -\mathbf{B}) + \operatorname{rank}\begin{pmatrix} \mathbf{A} \\ -\mathbf{B} \end{pmatrix} - \operatorname{rank}(\mathbf{A}) - \operatorname{rank}(-\mathbf{B}) \\
&\qquad + \operatorname{rank}\left[\begin{pmatrix} \mathbf{I}_m & \mathbf{0} \\ \mathbf{0} & -\mathbf{I}_m \end{pmatrix} \mathbf{H} \begin{pmatrix} \mathbf{I}_n & \mathbf{0} \\ \mathbf{0} & -\mathbf{I}_n \end{pmatrix} \right] \\
&= \operatorname{rank}(\mathbf{A}, \mathbf{B}) + \operatorname{rank}\begin{pmatrix} \mathbf{A} \\ \mathbf{B} \end{pmatrix} - \operatorname{rank}(\mathbf{A}) - \operatorname{rank}(\mathbf{B}) + \operatorname{rank}(\mathbf{H}) \\
&= \operatorname{rank}(\mathbf{A}) - \operatorname{rank}(\mathbf{B}) + [\operatorname{rank}(\mathbf{A}, \mathbf{B}) - \operatorname{rank}(\mathbf{A})] \\
&\qquad + \left[\operatorname{rank}\begin{pmatrix} \mathbf{A} \\ \mathbf{B} \end{pmatrix} - \operatorname{rank}(\mathbf{A})\right] + \operatorname{rank}(\mathbf{H}).
\end{aligned}$$

(b) It follows from Part (a) that $\operatorname{rank}(\mathbf{A} - \mathbf{B}) = \operatorname{rank}(\mathbf{A}) - \operatorname{rank}(\mathbf{B})$ if and only if

$$[\operatorname{rank}(\mathbf{A}, \mathbf{B}) - \operatorname{rank}(\mathbf{A})] + \left[\operatorname{rank}\begin{pmatrix} \mathbf{A} \\ \mathbf{B} \end{pmatrix} - \operatorname{rank}(\mathbf{A})\right] + \operatorname{rank}(\mathbf{H}) = 0. \qquad \text{(S.15)}$$

Since all three terms of the left side of equality (S.15) are nonnegative, we conclude that $\operatorname{rank}(\mathbf{A} - \mathbf{B}) = \operatorname{rank}(\mathbf{A}) - \operatorname{rank}(\mathbf{B})$ if and only if $\operatorname{rank}(\mathbf{A}, \mathbf{B}) - \operatorname{rank}(\mathbf{A}) = 0$, $\operatorname{rank}\begin{pmatrix} \mathbf{A} \\ \mathbf{B} \end{pmatrix} - \operatorname{rank}(\mathbf{A}) = 0$, and $\operatorname{rank}(\mathbf{H}) = 0$ or, equivalently, if and only if

$$\operatorname{rank}(\mathbf{A}, \mathbf{B}) = \operatorname{rank}\begin{pmatrix} \mathbf{A} \\ \mathbf{B} \end{pmatrix} = \operatorname{rank}(\mathbf{A}) \quad \text{and} \quad \mathbf{H} = \mathbf{0}.$$

(c) Suppose that $\operatorname{rank}(\mathbf{A}, \mathbf{B}) = \operatorname{rank}\begin{pmatrix}\mathbf{A}\\ \mathbf{B}\end{pmatrix} = \operatorname{rank}(\mathbf{A})$. Then, according to Corollary 4.5.2, $\mathcal{C}(\mathbf{B}) \subset \mathcal{C}(\mathbf{A})$ and $\mathcal{R}(\mathbf{B}) \subset \mathcal{R}(\mathbf{A})$, and it follows from Lemma 9.3.5 that $\mathbf{A}\mathbf{A}^-\mathbf{B} = \mathbf{B}$ and $\mathbf{B}\mathbf{A}^-\mathbf{A} = \mathbf{B}$.

Thus, (1)

$$\begin{pmatrix}\mathbf{A}\\ \mathbf{B}\end{pmatrix}(\mathbf{A}^-, \mathbf{0})\begin{pmatrix}\mathbf{A}\\ \mathbf{B}\end{pmatrix} = \begin{pmatrix}\mathbf{A}\\ \mathbf{B}\end{pmatrix}\mathbf{A}^-\mathbf{A} = \begin{pmatrix}\mathbf{A}\mathbf{A}^-\mathbf{A}\\ \mathbf{B}\mathbf{A}^-\mathbf{A}\end{pmatrix} = \begin{pmatrix}\mathbf{A}\\ \mathbf{B}\end{pmatrix},$$

$$(\mathbf{A}, \mathbf{B})\begin{pmatrix}\mathbf{A}^-\\ \mathbf{0}\end{pmatrix}(\mathbf{A}, \mathbf{B}) = \mathbf{A}\mathbf{A}^-(\mathbf{A}, \mathbf{B}) = (\mathbf{A}\mathbf{A}^-\mathbf{A},\ \mathbf{A}\mathbf{A}^-\mathbf{B}) = (\mathbf{A}, \mathbf{B});$$

and (2) upon setting $\begin{pmatrix}\mathbf{A}\\ \mathbf{B}\end{pmatrix}^-$ and $(\mathbf{A}, \mathbf{B})^-$ equal to $(\mathbf{A}^-, \mathbf{0})$ and $\begin{pmatrix}\mathbf{A}^-\\ \mathbf{0}\end{pmatrix}$, respectively, we obtain

$$\begin{aligned}
\mathbf{H} &= \left[\mathbf{I} - \begin{pmatrix}\mathbf{A}\\ \mathbf{B}\end{pmatrix}(\mathbf{A}^-, \mathbf{0})\right]\begin{pmatrix}\mathbf{A} & \mathbf{0}\\ \mathbf{0} & -\mathbf{B}\end{pmatrix}\left[\mathbf{I} - \begin{pmatrix}\mathbf{A}^-\\ \mathbf{0}\end{pmatrix}(\mathbf{A}, \mathbf{B})\right]\\
&= \begin{pmatrix}\mathbf{I} - \mathbf{A}\mathbf{A}^- & \mathbf{0}\\ -\mathbf{B}\mathbf{A}^- & \mathbf{I}\end{pmatrix}\begin{pmatrix}\mathbf{A} & \mathbf{0}\\ \mathbf{0} & -\mathbf{B}\end{pmatrix}\begin{pmatrix}\mathbf{I} - \mathbf{A}^-\mathbf{A} & -\mathbf{A}^-\mathbf{B}\\ \mathbf{0} & \mathbf{I}\end{pmatrix}\\
&= \begin{pmatrix}\mathbf{0} & \mathbf{0}\\ -\mathbf{B}\mathbf{A}^-\mathbf{A} & -\mathbf{B}\end{pmatrix}\begin{pmatrix}\mathbf{I} - \mathbf{A}^-\mathbf{A} & -\mathbf{A}^-\mathbf{B}\\ \mathbf{0} & \mathbf{I}\end{pmatrix}\\
&= \begin{pmatrix}\mathbf{0} & \mathbf{0}\\ \mathbf{0} & \mathbf{B}\mathbf{A}^-\mathbf{A}\mathbf{A}^-\mathbf{B} - \mathbf{B}\end{pmatrix} = \begin{pmatrix}\mathbf{0} & \mathbf{0}\\ \mathbf{0} & \mathbf{B}\mathbf{A}^-\mathbf{B} - \mathbf{B}\end{pmatrix}.
\end{aligned}$$

(d) (1) Suppose that $\operatorname{rank}(\mathbf{A}, \mathbf{B}) = \operatorname{rank}\begin{pmatrix}\mathbf{A}\\ \mathbf{B}\end{pmatrix} = \operatorname{rank}(\mathbf{A})$ and that $\mathbf{B}\mathbf{A}^-\mathbf{B} = \mathbf{B}$. Then, it follows from Part (c) that $(\mathbf{A}^-, \mathbf{0})$ and $\begin{pmatrix}\mathbf{A}^-\\ \mathbf{0}\end{pmatrix}$ are generalized inverses of $\begin{pmatrix}\mathbf{A}\\ \mathbf{B}\end{pmatrix}$ and $(\mathbf{A}, \mathbf{B})$, respectively, and that, for $\begin{pmatrix}\mathbf{A}\\ \mathbf{B}\end{pmatrix}^- = (\mathbf{A}^-, \mathbf{0})$ and $(\mathbf{A}, \mathbf{B})^- = \begin{pmatrix}\mathbf{A}^-\\ \mathbf{0}\end{pmatrix}$, $\mathbf{H} = \mathbf{0}$. Thus, as a consequence of Part (b), we have that $\operatorname{rank}(\mathbf{A} - \mathbf{B}) = \operatorname{rank}(\mathbf{A}) - \operatorname{rank}(\mathbf{B})$.

Conversely, suppose that $\operatorname{rank}(\mathbf{A} - \mathbf{B}) = \operatorname{rank}(\mathbf{A}) - \operatorname{rank}(\mathbf{B})$. Then, according to Part (b), $\operatorname{rank}(\mathbf{A}, \mathbf{B}) = \operatorname{rank}\begin{pmatrix}\mathbf{A}\\ \mathbf{B}\end{pmatrix} = \operatorname{rank}(\mathbf{A})$ and $\mathbf{H} = \mathbf{0}$ [for any choice of $\begin{pmatrix}\mathbf{A}\\ \mathbf{B}\end{pmatrix}^-$ and $(\mathbf{A}, \mathbf{B})^-$]. And, observing [in light of Part (c)] that $(\mathbf{A}^-, \mathbf{0})$ and $\begin{pmatrix}\mathbf{A}^-\\ \mathbf{0}\end{pmatrix}$ are generalized inverses of $\begin{pmatrix}\mathbf{A}\\ \mathbf{B}\end{pmatrix}$ and $(\mathbf{A}, \mathbf{B})$, respectively, and that, for $\begin{pmatrix}\mathbf{A}\\ \mathbf{B}\end{pmatrix}^- = (\mathbf{A}^-, \mathbf{0})$ and $(\mathbf{A}, \mathbf{B})^- = \begin{pmatrix}\mathbf{A}^-\\ \mathbf{0}\end{pmatrix}$, $\mathbf{H} = \begin{pmatrix}\mathbf{0} & \mathbf{0}\\ \mathbf{0} & \mathbf{B}\mathbf{A}^-\mathbf{B} - \mathbf{B}\end{pmatrix}$, we find that $\mathbf{B}\mathbf{A}^-\mathbf{B} - \mathbf{B} = \mathbf{0}$ or equivalently that $\mathbf{B}\mathbf{A}^-\mathbf{B} = \mathbf{B}$.

(2) Since (according to Corollary 4.5.2) $\mathcal{C}(\mathbf{B}) \subset \mathcal{C}(\mathbf{A}) \Leftrightarrow \text{rank}(\mathbf{A}, \mathbf{B}) = \text{rank}(\mathbf{A})$ and $\mathcal{R}(\mathbf{B}) \subset \mathcal{R}(\mathbf{A}) \Leftrightarrow \text{rank}\begin{pmatrix}\mathbf{A}\\ \mathbf{B}\end{pmatrix} = \text{rank}(\mathbf{A})$, Condition (2) is equivalent to Condition (1) and hence is necessary and sufficient for rank subtractivity.

(3) Since (according to Lemma 9.3.5) $\mathbf{AA^-B} = \mathbf{B} \Leftrightarrow \mathcal{C}(\mathbf{B}) \subset \mathcal{C}(\mathbf{A})$ and $\mathbf{BA^-A} = \mathbf{B} \Leftrightarrow \mathcal{R}(\mathbf{B}) \subset \mathcal{R}(\mathbf{A})$, Condition (3) is equivalent to Condition (2) and hence is necessary and sufficient for rank subtractivity.

(e) Clearly, $\text{rank}(\mathbf{A}-\mathbf{B}) = \text{rank}(\mathbf{A}) - \text{rank}(\mathbf{B})$ if and only if $\text{rank}[(\mathbf{A}-\mathbf{B})+\mathbf{B}] = \text{rank}(\mathbf{A}-\mathbf{B}) + \text{rank}(\mathbf{B})$, that is, if and only if $\mathbf{A}-\mathbf{B}$ and $\mathbf{B}$ are rank additive. Moreover, it follows from the result of Exercise 29 [specifically, Condition (b)] that $\mathbf{A}-\mathbf{B}$ and $\mathbf{B}$ are rank additive if and only if

$$\text{rank}(\mathbf{A}-\mathbf{B}) = \text{rank}[(\mathbf{A}-\mathbf{B})(\mathbf{I}-\mathbf{B^-B})] = \text{rank}[(\mathbf{I}-\mathbf{BB^-})(\mathbf{A}-\mathbf{B})].$$

Since $(\mathbf{A}-\mathbf{B})(\mathbf{I}-\mathbf{B^-B}) = \mathbf{A}(\mathbf{I}-\mathbf{B^-B})$ and $(\mathbf{I}-\mathbf{BB^-})(\mathbf{A}-\mathbf{B}) = (\mathbf{I}-\mathbf{BB^-})\mathbf{A}$, we conclude that $\text{rank}(\mathbf{A}-\mathbf{B}) = \text{rank}(\mathbf{A}) - \text{rank}(\mathbf{B})$ if and only if $\text{rank}(\mathbf{A}-\mathbf{B}) = \text{rank}[\mathbf{A}(\mathbf{I}-\mathbf{B^-B})] = \text{rank}[(\mathbf{I}-\mathbf{BB^-})\mathbf{A}]$.

EXERCISE 31. Let $\mathbf{A}_1, \ldots, \mathbf{A}_k$ represent $m \times n$ matrices. Adopting the terminology of Exercise 17.6, use Part (a) of that exercise to show that if $\mathcal{R}(\mathbf{A}_1), \ldots, \mathcal{R}(\mathbf{A}_k)$ are independent and $\mathcal{C}(\mathbf{A}_1), \ldots, \mathcal{C}(\mathbf{A}_k)$ are independent, then $\text{rank}(\mathbf{A}_1 + \cdots + \mathbf{A}_k) = \text{rank}(\mathbf{A}_1) + \cdots + \text{rank}(\mathbf{A}_k)$.

Solution. Suppose that $\mathcal{R}(\mathbf{A}_1), \ldots, \mathcal{R}(\mathbf{A}_k)$ are independent and also that $\mathcal{C}(\mathbf{A}_1), \ldots, \mathcal{C}(\mathbf{A}_k)$ are independent. Then, as a consequence of Part (a) of Exercise 17.6, we have that, for $i = 2, \ldots, k$, $\mathcal{R}(\mathbf{A}_i)$ and $\mathcal{R}(\mathbf{A}_1) + \cdots + \mathcal{R}(\mathbf{A}_{i-1})$ are essentially disjoint and $\mathcal{C}(\mathbf{A}_i)$ and $\mathcal{C}(\mathbf{A}_1) + \cdots + \mathcal{C}(\mathbf{A}_{i-1})$ are essentially disjoint or equivalently [in light of results (17.1.7) and (17.1.6)] that, for $i = 2, \ldots, k$, $\mathcal{R}(\mathbf{A}_i)$ and $\mathcal{R}\begin{pmatrix}\mathbf{A}_1\\ \vdots \\ \mathbf{A}_{i-1}\end{pmatrix}$ are essentially disjoint and $\mathcal{C}(\mathbf{A}_i)$ and $\mathcal{C}(\mathbf{A}_1, \ldots, \mathbf{A}_{i-1})$ are essentially disjoint. Moreover, it follows from results (4.5.9) and (4.5.8) that (for $i = 2, \ldots, k$)

$$\mathcal{R}(\mathbf{A}_1 + \cdots + \mathbf{A}_{i-1}) \subset \mathcal{R}\begin{pmatrix}\mathbf{A}_1\\ \vdots \\ \mathbf{A}_{i-1}\end{pmatrix} \text{ and } \mathcal{C}(\mathbf{A}_1 + \cdots + \mathbf{A}_{i-1}) \subset \mathcal{C}(\mathbf{A}_1, \ldots, \mathbf{A}_{i-1}).$$

Thus, for $i = 2, \ldots, k$, $\mathcal{R}(\mathbf{A}_i)$ and $\mathcal{R}(\mathbf{A}_1 + \cdots + \mathbf{A}_{i-1})$ are essentially disjoint and $\mathcal{C}(\mathbf{A}_i)$ and $\mathcal{C}(\mathbf{A}_1 + \cdots + \mathbf{A}_{i-1})$ are essentially disjoint, implying (in light of Theorem 18.5.7) that (for $i = 2, \ldots, k$)

$$\text{rank}(\mathbf{A}_1 + \cdots + \mathbf{A}_{i-1} + \mathbf{A}_i) = \text{rank}(\mathbf{A}_1 + \cdots + \mathbf{A}_{i-1}) + \text{rank}(\mathbf{A}_i).$$

We conclude that

$$\text{rank}(\mathbf{A}_1 + \cdots + \mathbf{A}_k) = \text{rank}(\mathbf{A}_1 + \cdots + \mathbf{A}_{k-1}) + \text{rank}(\mathbf{A}_k)$$

$$= \text{rank}(\mathbf{A}_1 + \cdots + \mathbf{A}_{k-2}) + \text{rank}(\mathbf{A}_{k-1}) + \text{rank}(\mathbf{A}_k)$$
$$\vdots$$
$$= \text{rank}(\mathbf{A}_1) + \cdots + \text{rank}(\mathbf{A}_k).$$

EXERCISE 32. Let $\mathbf{T}$ represent an $m \times p$ matrix, $\mathbf{U}$ an $m \times q$ matrix, $\mathbf{V}$ an $n \times p$ matrix, and $\mathbf{W}$ an $n \times q$ matrix, and define $\mathbf{Q} = \mathbf{W} - \mathbf{VT}^-\mathbf{U}$. Further, let $\mathbf{E}_T = \mathbf{I} - \mathbf{TT}^-$, $\mathbf{F}_T = \mathbf{I} - \mathbf{T}^-\mathbf{T}$, $\mathbf{X} = \mathbf{E}_T\mathbf{U}$, and $\mathbf{Y} = \mathbf{VF}_T$.

(a) Show that

$$\text{rank}\begin{pmatrix} \mathbf{T} & \mathbf{U} \\ \mathbf{V} & \mathbf{W} \end{pmatrix} = \text{rank}(\mathbf{T}) + \text{rank}\begin{pmatrix} \mathbf{0} & \mathbf{X} \\ \mathbf{Y} & \mathbf{Q} \end{pmatrix}. \tag{E.4}$$

(b) Show that

$$\text{rank}\begin{pmatrix} \mathbf{0} & \mathbf{V} \\ \mathbf{U} & \mathbf{T} \end{pmatrix} = \text{rank}(\mathbf{T}) + \text{rank}\begin{pmatrix} -\mathbf{VT}^-\mathbf{U} & \mathbf{Y} \\ \mathbf{X} & \mathbf{0} \end{pmatrix}.$$

[*Hint*. Observe that (since the rank of a matrix is not affected by a permutation of rows or columns) $\text{rank}\begin{pmatrix} \mathbf{0} & \mathbf{V} \\ \mathbf{U} & \mathbf{T} \end{pmatrix} = \text{rank}\begin{pmatrix} \mathbf{T} & \mathbf{U} \\ \mathbf{V} & \mathbf{0} \end{pmatrix}$, and make use of Part (a).]

(c) Show that

$$\text{rank}\begin{pmatrix} \mathbf{0} & \mathbf{V} \\ \mathbf{U} & \mathbf{T} \end{pmatrix} = \text{rank}(\mathbf{T}) + \text{rank}(\mathbf{X}) + \text{rank}(\mathbf{Y}) + \text{rank}(\mathbf{E}_Y\mathbf{VT}^-\mathbf{UF}_X),$$

where $\mathbf{E}_Y = \mathbf{I} - \mathbf{YY}^-$ and $\mathbf{F}_X = \mathbf{I} - \mathbf{X}^-\mathbf{X}$. *Hint*. Use Part (b) in combination with the result that, for any $r \times s$ matrix $\mathbf{A}$, $r \times v$ matrix $\mathbf{B}$, and $u \times s$ matrix $\mathbf{C}$,

$$\text{rank}\begin{pmatrix} \mathbf{A} & \mathbf{B} \\ \mathbf{C} & \mathbf{0} \end{pmatrix} = \text{rank}(\mathbf{B}) + \text{rank}(\mathbf{C}) + \text{rank}[(\mathbf{I} - \mathbf{BB}^-)\mathbf{A}(\mathbf{I} - \mathbf{C}^-\mathbf{C})]. \tag{$*$}$$

(d) Show that

$$\begin{aligned}\text{rank}\begin{pmatrix} \mathbf{T} & \mathbf{U} \\ \mathbf{V} & \mathbf{W} \end{pmatrix} = \text{rank}(\mathbf{T}) &+ \text{rank}(\mathbf{Q}) + \text{rank}(\mathbf{A}) + \text{rank}(\mathbf{B}) \\ &+ \text{rank}[(\mathbf{I} - \mathbf{AA}^-)\mathbf{XQ}^-\mathbf{Y}(\mathbf{I} - \mathbf{B}^-\mathbf{B})],\end{aligned}$$

where $\mathbf{A} = \mathbf{X}(\mathbf{I} - \mathbf{Q}^-\mathbf{Q})$ and $\mathbf{B} = (\mathbf{I} - \mathbf{QQ}^-)\mathbf{Y}$. [*Hint*. Use Part (c) in combination with Part (a).]

Solution. (a) Let $\mathbf{K} = \mathbf{Q} - \mathbf{YT}^-\mathbf{U}$. Then, clearly,

$$\begin{aligned}\begin{pmatrix} \mathbf{I} & \mathbf{0} \\ -\mathbf{VT}^- & \mathbf{I} \end{pmatrix}\begin{pmatrix} \mathbf{T} & \mathbf{U} \\ \mathbf{V} & \mathbf{W} \end{pmatrix}\begin{pmatrix} \mathbf{I} & -\mathbf{T}^-\mathbf{U} \\ \mathbf{0} & \mathbf{I} \end{pmatrix} &= \begin{pmatrix} \mathbf{T} & \mathbf{U} \\ \mathbf{Y} & \mathbf{Q} \end{pmatrix}\begin{pmatrix} \mathbf{I} & -\mathbf{T}^-\mathbf{U} \\ \mathbf{0} & \mathbf{I} \end{pmatrix} \\ &= \begin{pmatrix} \mathbf{T} & \mathbf{X} \\ \mathbf{Y} & \mathbf{K} \end{pmatrix},\end{aligned}$$

so that (in light of Lemma 8.5.2)

$$\operatorname{rank}\begin{pmatrix} \mathbf{T} & \mathbf{U} \\ \mathbf{V} & \mathbf{W} \end{pmatrix} = \operatorname{rank}\begin{pmatrix} \mathbf{T} & \mathbf{X} \\ \mathbf{Y} & \mathbf{K} \end{pmatrix}. \tag{S.16}$$

Further,

$$\begin{pmatrix} \mathbf{T} & \mathbf{X} \\ \mathbf{Y} & \mathbf{K} \end{pmatrix} = \begin{pmatrix} \mathbf{T} & \mathbf{0} \\ \mathbf{0} & \mathbf{0} \end{pmatrix} + \begin{pmatrix} \mathbf{0} & \mathbf{X} \\ \mathbf{Y} & \mathbf{K} \end{pmatrix}.$$

And, since $\mathcal{R}\begin{pmatrix} \mathbf{T} \\ \mathbf{0} \end{pmatrix} = \mathcal{R}(\mathbf{T})$ and $\mathcal{R}\begin{pmatrix} \mathbf{0} \\ \mathbf{Y} \end{pmatrix} = \mathcal{R}(\mathbf{Y})$ and since (according to Corollary 17.2.8) $\mathcal{R}(\mathbf{T})$ and $\mathcal{R}(\mathbf{Y})$ are essentially disjoint, $\mathcal{R}\begin{pmatrix} \mathbf{T} \\ \mathbf{0} \end{pmatrix}$ and $\mathcal{R}\begin{pmatrix} \mathbf{0} \\ \mathbf{Y} \end{pmatrix}$ are essentially disjoint, and hence it follows from Corollary 17.2.16 that $\mathcal{R}\begin{pmatrix} \mathbf{T} & \mathbf{0} \\ \mathbf{0} & \mathbf{0} \end{pmatrix}$ and $\mathcal{R}\begin{pmatrix} \mathbf{0} & \mathbf{X} \\ \mathbf{Y} & \mathbf{K} \end{pmatrix}$ are essentially disjoint. Similarly, $\mathcal{C}\begin{pmatrix} \mathbf{T} & \mathbf{0} \\ \mathbf{0} & \mathbf{0} \end{pmatrix}$ and $\mathcal{C}\begin{pmatrix} \mathbf{0} & \mathbf{X} \\ \mathbf{Y} & \mathbf{K} \end{pmatrix}$ are essentially disjoint. Thus, making use of Theorem 18.5.7, we find that

$$\begin{aligned} \operatorname{rank}\begin{pmatrix} \mathbf{T} & \mathbf{X} \\ \mathbf{Y} & \mathbf{K} \end{pmatrix} &= \operatorname{rank}\begin{pmatrix} \mathbf{T} & \mathbf{0} \\ \mathbf{0} & \mathbf{0} \end{pmatrix} + \operatorname{rank}\begin{pmatrix} \mathbf{0} & \mathbf{X} \\ \mathbf{Y} & \mathbf{K} \end{pmatrix} \\ &= \operatorname{rank}(\mathbf{T}) + \operatorname{rank}\begin{pmatrix} \mathbf{0} & \mathbf{X} \\ \mathbf{Y} & \mathbf{K} \end{pmatrix}. \end{aligned} \tag{S.17}$$

Now, observe that

$$\begin{pmatrix} \mathbf{0} & \mathbf{X} \\ \mathbf{Y} & \mathbf{K} \end{pmatrix} \begin{pmatrix} \mathbf{I} & \mathbf{T}^{-}\mathbf{U} \\ \mathbf{0} & \mathbf{I} \end{pmatrix} = \begin{pmatrix} \mathbf{0} & \mathbf{X} \\ \mathbf{Y} & \mathbf{Q} \end{pmatrix}$$

and hence that

$$\operatorname{rank}\begin{pmatrix} \mathbf{0} & \mathbf{X} \\ \mathbf{Y} & \mathbf{K} \end{pmatrix} = \operatorname{rank}\begin{pmatrix} \mathbf{0} & \mathbf{X} \\ \mathbf{Y} & \mathbf{Q} \end{pmatrix}. \tag{S.18}$$

Finally, combining results (S.16)–(S.18), we obtain

$$\operatorname{rank}\begin{pmatrix} \mathbf{T} & \mathbf{U} \\ \mathbf{V} & \mathbf{W} \end{pmatrix} = \operatorname{rank}(\mathbf{T}) + \operatorname{rank}\begin{pmatrix} \mathbf{0} & \mathbf{X} \\ \mathbf{Y} & \mathbf{K} \end{pmatrix} = \operatorname{rank}(\mathbf{T}) + \operatorname{rank}\begin{pmatrix} \mathbf{0} & \mathbf{X} \\ \mathbf{Y} & \mathbf{Q} \end{pmatrix}.$$

(b) Since (as indicated by Lemma 8.5.1) the rank of a matrix is not affected by a permutation of rows or columns,

$$\operatorname{rank}\begin{pmatrix} \mathbf{0} & \mathbf{V} \\ \mathbf{U} & \mathbf{T} \end{pmatrix} = \operatorname{rank}\begin{pmatrix} \mathbf{T} & \mathbf{U} \\ \mathbf{V} & \mathbf{0} \end{pmatrix}. \tag{S.19}$$

Moreover, applying result (E.4) (in the special case where $\mathbf{W} = \mathbf{0}$) and again making use of Lemma 8.5.1, we find that

$$\begin{aligned} \operatorname{rank}\begin{pmatrix} \mathbf{T} & \mathbf{U} \\ \mathbf{V} & \mathbf{0} \end{pmatrix} &= \operatorname{rank}(\mathbf{T}) + \operatorname{rank}\begin{pmatrix} \mathbf{0} & \mathbf{X} \\ \mathbf{Y} & -\mathbf{V}\mathbf{T}^{-}\mathbf{U} \end{pmatrix} \\ &= \operatorname{rank}(\mathbf{T}) + \operatorname{rank}\begin{pmatrix} -\mathbf{V}\mathbf{T}^{-}\mathbf{U} & \mathbf{Y} \\ \mathbf{X} & \mathbf{0} \end{pmatrix}. \end{aligned} \tag{S.20}$$

And, upon combining result (S.20) with result (S.19), we obtain

$$\operatorname{rank}\begin{pmatrix} \mathbf{0} & \mathbf{V} \\ \mathbf{U} & \mathbf{T} \end{pmatrix} = \operatorname{rank}(\mathbf{T}) + \operatorname{rank}\begin{pmatrix} -\mathbf{V}\mathbf{T}^{-}\mathbf{U} & \mathbf{Y} \\ \mathbf{X} & \mathbf{0} \end{pmatrix}.$$

(c) Applying result (∗) [or equivalently result (17.2.15), which is part of Theorem 17.2.17] with $-\mathbf{V}\mathbf{T}^{-}\mathbf{U}$, $\mathbf{Y}$, and $\mathbf{X}$ in place of $\mathbf{A}$, $\mathbf{B}$, and $\mathbf{C}$, respectively (or $\mathbf{T}$, $\mathbf{U}$, and $\mathbf{V}$, respectively), we find that

$$\begin{aligned} \operatorname{rank}\begin{pmatrix} -\mathbf{V}\mathbf{T}^{-}\mathbf{U} & \mathbf{Y} \\ \mathbf{X} & \mathbf{0} \end{pmatrix} &= \operatorname{rank}(\mathbf{Y}) + \operatorname{rank}(\mathbf{X}) \\ &\quad + \operatorname{rank}[(\mathbf{I} - \mathbf{Y}\mathbf{Y}^{-})(-\mathbf{V}\mathbf{T}^{-}\mathbf{U})(\mathbf{I} - \mathbf{X}^{-}\mathbf{X})] \\ &= \operatorname{rank}(\mathbf{Y}) + \operatorname{rank}(\mathbf{X}) + \operatorname{rank}(\mathbf{E}_Y\mathbf{V}\mathbf{T}^{-}\mathbf{U}\mathbf{F}_X). \end{aligned} \tag{S.21}$$

And, upon combining result (S.21) with Part (b), we obtain

$$\operatorname{rank}\begin{pmatrix} \mathbf{0} & \mathbf{V} \\ \mathbf{U} & \mathbf{T} \end{pmatrix} = \operatorname{rank}(\mathbf{T}) + \operatorname{rank}(\mathbf{Y}) + \operatorname{rank}(\mathbf{X}) + \operatorname{rank}(\mathbf{E}_Y\mathbf{V}\mathbf{T}^{-}\mathbf{U}\mathbf{F}_X).$$

(d) Applying Part (c) (with $\mathbf{Q}$, $\mathbf{Y}$, and $\mathbf{X}$ in place of $\mathbf{T}$, $\mathbf{U}$, and $\mathbf{V}$, respectively, and hence with $\mathbf{A}$ and $\mathbf{B}$ in place of $\mathbf{Y}$ and $\mathbf{X}$, respectively), we find that

$$\begin{aligned} \operatorname{rank}\begin{pmatrix} \mathbf{0} & \mathbf{X} \\ \mathbf{Y} & \mathbf{Q} \end{pmatrix} &= \operatorname{rank}(\mathbf{Q}) + \operatorname{rank}(\mathbf{B}) + \operatorname{rank}(\mathbf{A}) \\ &\quad + \operatorname{rank}[(\mathbf{I} - \mathbf{A}\mathbf{A}^{-})\mathbf{X}\mathbf{Q}^{-}\mathbf{Y}(\mathbf{I} - \mathbf{B}^{-}\mathbf{B})]. \end{aligned} \tag{S.22}$$

And, upon combining result (S.22) with Part (a), we obtain

$$\begin{aligned} \operatorname{rank}\begin{pmatrix} \mathbf{T} & \mathbf{U} \\ \mathbf{V} & \mathbf{W} \end{pmatrix} &= \operatorname{rank}(\mathbf{T}) + \operatorname{rank}(\mathbf{Q}) + \operatorname{rank}(\mathbf{A}) + \operatorname{rank}(\mathbf{B}) \\ &\quad + \operatorname{rank}[(\mathbf{I} - \mathbf{A}\mathbf{A}^{-})\mathbf{X}\mathbf{Q}^{-}\mathbf{Y}(\mathbf{T} - \mathbf{B}^{-}\mathbf{B})]. \end{aligned}$$

EXERCISE 33. Let $\mathbf{R}$ represent an $n \times q$ matrix, $\mathbf{S}$ an $n \times m$ matrix, $\mathbf{T}$ an $m \times p$ matrix, and $\mathbf{U}$ a $p \times q$ matrix, and define $\mathbf{Q} = \mathbf{T} + \mathbf{T}\mathbf{U}\mathbf{R}^{-}\mathbf{S}\mathbf{T}$. Further, let $\mathbf{E}_R = \mathbf{I} - \mathbf{R}\mathbf{R}^{-}$, $\mathbf{F}_R = \mathbf{I} - \mathbf{R}^{-}\mathbf{R}$, $\mathbf{X} = \mathbf{E}_R\mathbf{S}\mathbf{T}$, $\mathbf{Y} = \mathbf{T}\mathbf{U}\mathbf{F}_R$, $\mathbf{A} = \mathbf{X}(\mathbf{I} - \mathbf{Q}^{-}\mathbf{Q})$, $\mathbf{B} = (\mathbf{I} - \mathbf{Q}\mathbf{Q}^{-})\mathbf{Y}$. Use the result of Part (d) of Exercise 32 in combination with the result of Part (a) of Exercise 25 to show that

$$\begin{aligned} \operatorname{rank}(\mathbf{R} + \mathbf{S}\mathbf{T}\mathbf{U}) &= \operatorname{rank}(\mathbf{R}) + \operatorname{rank}(\mathbf{Q}) \\ &\quad - \operatorname{rank}(\mathbf{T}) + \operatorname{rank}(\mathbf{A}) + \operatorname{rank}(\mathbf{B}) \\ &\quad + \operatorname{rank}[(\mathbf{I} - \mathbf{A}\mathbf{A}^{-})\mathbf{X}\mathbf{Q}^{-}\mathbf{Y}(\mathbf{I} - \mathbf{B}^{-}\mathbf{B})]. \end{aligned}$$

Solution. Upon observing that $\operatorname{rank}(\mathbf{A}) = \operatorname{rank}(-\mathbf{A})$, that $-\mathbf{A}^{-}$ is a generalized inverse of $-\mathbf{A}$, and that

$$\begin{aligned} &\operatorname{rank}[(\mathbf{I} - \mathbf{A}\mathbf{A}^{-})\mathbf{X}\mathbf{Q}^{-}\mathbf{Y}(\mathbf{I} - \mathbf{B}^{-}\mathbf{B})] \\ &\qquad = \operatorname{rank}[-(\mathbf{I} - \mathbf{A}\mathbf{A}^{-})\mathbf{X}\mathbf{Q}^{-}\mathbf{Y}(\mathbf{I} - \mathbf{B}^{-}\mathbf{B})] \\ &\qquad = \operatorname{rank}\{[\mathbf{I} - (-\mathbf{A})(-\mathbf{A}^{-})](-\mathbf{X})\mathbf{Q}^{-}\mathbf{Y}(\mathbf{I} - \mathbf{B}^{-}\mathbf{B})\}, \end{aligned}$$

it follows from the result of Part (d) of Exercise 32 that

$$\operatorname{rank}\begin{pmatrix} \mathbf{R} & -\mathbf{ST} \\ \mathbf{TU} & \mathbf{T} \end{pmatrix} = \operatorname{rank}(\mathbf{R}) + \operatorname{rank}(\mathbf{Q}) + \operatorname{rank}(\mathbf{A}) + \operatorname{rank}(\mathbf{B}) + \operatorname{rank}[(\mathbf{I} - \mathbf{AA}^-)\mathbf{XQ}^-\mathbf{Y}(\mathbf{I} - \mathbf{B}^-\mathbf{B})].$$

We conclude, on the basis of Part (a) of Exercise 25, that

$$\begin{aligned}\operatorname{rank}(\mathbf{R} + \mathbf{STU}) = \operatorname{rank}(\mathbf{R}) &+ \operatorname{rank}(\mathbf{Q}) - \operatorname{rank}(\mathbf{T}) \\ &+ \operatorname{rank}(\mathbf{A}) + \operatorname{rank}(\mathbf{B}) \\ &+ \operatorname{rank}[(\mathbf{I} - \mathbf{AA}^-)\mathbf{XQ}^-\mathbf{Y}(\mathbf{I} - \mathbf{B}^-\mathbf{B})].\end{aligned}$$

19

Minimization of a Second-Degree Polynomial (in n Variables) Subject to Linear Constraints

EXERCISE 1. Let $\mathbf{a}$ represent an $n \times 1$ vector of (unconstrained) variables, and define $f(\mathbf{a}) = \mathbf{a}'\mathbf{V}\mathbf{a} - 2\mathbf{b}'\mathbf{a}$, where $\mathbf{V}$ is an $n \times n$ matrix and $\mathbf{b}$ an $n \times 1$ vector. Show that if $\mathbf{V}$ is not nonnegative definite or if $\mathbf{b} \notin \mathcal{C}(\mathbf{V})$, then $f(\mathbf{a})$ is unbounded from below, that is, corresponding to any scalar c, there exists a vector $\mathbf{a}_*$ such that $f(\mathbf{a}_*) < c$.

Solution. Let c represent an arbitrary scalar.

Suppose that $\mathbf{V}$ is not nonnegative definite. Then, there exists an $n \times 1$ vector $\mathbf{x}$ such that $\mathbf{x}'\mathbf{V}\mathbf{x} < 0$. Moreover, for any scalar k,

$$f(k\mathbf{x}) = k^2(\mathbf{x}'\mathbf{V}\mathbf{x}) - 2k(\mathbf{b}'\mathbf{x}) = k[k(\mathbf{x}'\mathbf{V}\mathbf{x}) - 2(\mathbf{b}'\mathbf{x})].$$

Thus, $\lim_{k\to\pm\infty} f(k\mathbf{x}) = -\infty$, and hence there exists a scalar k_* such that $f(k_*\mathbf{x}) < c$, so that, for $\mathbf{a}_* = k_*\mathbf{x}$, $f(\mathbf{a}_*) < c$.

Or, suppose that $\mathbf{b} \notin \mathcal{C}(\mathbf{V})$. According to Theorem 12.5.11, $\mathcal{C}(\mathbf{V})$ contains a (unique) vector $\mathbf{b}_1$ and $\mathcal{C}^{\perp}(\mathbf{V})$ [or equivalently $\mathcal{N}(\mathbf{V}')$] contains a (unique) vector $\mathbf{b}_2$ such that $\mathbf{b} = \mathbf{b}_1 + \mathbf{b}_2$. Clearly, $\mathbf{b}_2 \neq \mathbf{0}$ [since otherwise we would arrive at a contradiction of the supposition that $\mathbf{b} \notin \mathcal{C}(\mathbf{V})$]. Moreover, for any scalar k,

$$f(k\mathbf{b}_2) = k^2(\mathbf{V}'\mathbf{b}_2)'\mathbf{b}_2 - 2k\mathbf{b}'\mathbf{b}_2 = -2k(\mathbf{b}_1' + \mathbf{b}_2')\mathbf{b}_2 = -2k(\mathbf{b}_2'\mathbf{b}_2).$$

Thus, $\lim_{k\to\infty} f(k\mathbf{b}_2) = -\infty$, and hence there exists a scalar k_* such that $f(k_*\mathbf{b}_2) < c$, so that, for $\mathbf{a}_* = k_*\mathbf{b}_2$, $f(\mathbf{a}_*) < c$.

EXERCISE 2. Let $\mathbf{V}$ represent an $n \times n$ symmetric matrix and $\mathbf{X}$ an $n \times p$ matrix. Show that, for any $p \times p$ matrix $\mathbf{U}$ such that $\mathcal{C}(\mathbf{X}) \subset \mathcal{C}(\mathbf{V} + \mathbf{X}\mathbf{U}\mathbf{X}')$,

(1) $(\mathbf{V}+\mathbf{XUX}')(\mathbf{V}+\mathbf{XUX}')^-\mathbf{V}=\mathbf{V}$;

(2) $\mathbf{V}(\mathbf{V}+\mathbf{XUX}')^-(\mathbf{V}+\mathbf{XUX}')=\mathbf{V}$.

Solution. (1) According to Lemma 19.3.4, $\mathcal{C}(\mathbf{V},\mathbf{X})=\mathcal{C}(\mathbf{V}+\mathbf{XUX}')$, implying (in light of Lemma 4.5.1) that $\mathcal{C}(\mathbf{V})\subset\mathcal{C}(\mathbf{V}+\mathbf{XUX}')$ and hence (in light of Lemma 9.3.5) that $(\mathbf{V}+\mathbf{XUX}')(\mathbf{V}+\mathbf{XUX}')^-\mathbf{V}=\mathbf{V}$.

(2) According to Lemma 19.3.4, $\mathcal{C}(\mathbf{V},\mathbf{X})=\mathcal{C}(\mathbf{V}+\mathbf{XU}'\mathbf{X}')$, implying that $\mathcal{C}(\mathbf{V})\subset\mathcal{C}(\mathbf{V}+\mathbf{XU}'\mathbf{X}')$ and hence (in light of Lemma 4.2.5) that

$$\mathcal{R}(\mathbf{V})\subset\mathcal{R}[(\mathbf{V}+\mathbf{XU}'\mathbf{X}')']=\mathcal{R}(\mathbf{V}+\mathbf{XUX}').$$

Thus, it follows from Lemma 9.3.5 that

$$\mathbf{V}(\mathbf{V}+\mathbf{XUX}')^-(\mathbf{V}+\mathbf{XUX}')=\mathbf{V}.$$

EXERCISE 3. Let $\mathbf{V}$ represent an $n\times n$ symmetric nonnegative definite matrix, $\mathbf{X}$ an $n\times p$ matrix, $\mathbf{B}$ an $n\times s$ matrix such that $\mathcal{C}(\mathbf{B})\subset\mathcal{C}(\mathbf{V},\mathbf{X})$, and $\mathbf{D}$ a $p\times s$ matrix such that $\mathcal{C}(\mathbf{D})\subset\mathcal{C}(\mathbf{X}')$. Further, let $\mathbf{U}$ represent any $p\times p$ matrix such that $\mathcal{C}(\mathbf{X})\subset\mathcal{C}(\mathbf{V}+\mathbf{XUX}')$, and let $\mathbf{W}$ represent an arbitrary generalized inverse of $\mathbf{V}+\mathbf{XUX}'$. Devise a short proof of the result that $\mathbf{A}_*$ and $\mathbf{R}_*$ are respectively the first $(n\times s)$ and second $(p\times s)$ parts of a solution to the (consistent) linear system

$$\begin{pmatrix}\mathbf{V} & \mathbf{X}\\ \mathbf{X}' & \mathbf{0}\end{pmatrix}\begin{pmatrix}\mathbf{A}\\ \mathbf{R}\end{pmatrix}=\begin{pmatrix}\mathbf{B}\\ \mathbf{D}\end{pmatrix} \qquad (*)$$

(in an $n\times s$ matrix $\mathbf{A}$ and a $p\times s$ matrix $\mathbf{R}$) if and only if

$$\mathbf{R}_*=\mathbf{T}_*+\mathbf{UD}$$

and

$$\mathbf{A}_*=\mathbf{WB}-\mathbf{WXT}_*+[\mathbf{I}-\mathbf{W}(\mathbf{V}+\mathbf{XUX}')]\mathbf{L}$$

for some solution $\mathbf{T}_*$ to the (consistent) linear system

$$\mathbf{X}'\mathbf{WXT}=\mathbf{X}'\mathbf{WB}-\mathbf{D} \qquad (**)$$

(in a $p\times s$ matrix $\mathbf{T}$) and for some $n\times s$ matrix $\mathbf{L}$. Do so by taking advantage of the result that if the coefficient matrix, right side, and matrix of unknowns in a linear system $\mathbf{HY}=\mathbf{S}$ (in $\mathbf{Y}$) are partitioned (conformally) as

$$\mathbf{H}=\begin{pmatrix}\mathbf{H}_{11} & \mathbf{H}_{12}\\ \mathbf{H}_{21} & \mathbf{H}_{22}\end{pmatrix},\quad \mathbf{S}=\begin{pmatrix}\mathbf{S}_1\\ \mathbf{S}_2\end{pmatrix},\quad\text{and}\quad \mathbf{Y}=\begin{pmatrix}\mathbf{Y}_1\\ \mathbf{Y}_2\end{pmatrix},$$

and if

$$\mathcal{C}(\mathbf{H}_{12})\subset\mathcal{C}(\mathbf{H}_{11}),\quad \mathcal{C}(\mathbf{S}_1)\subset\mathcal{C}(\mathbf{H}_{11}),\quad\text{and}\quad \mathcal{R}(\mathbf{H}_{21})\subset\mathcal{R}(\mathbf{H}_{11}),$$

then the matrix $\mathbf{Y}^* = \begin{pmatrix} \mathbf{Y}_1^* \\ \mathbf{Y}_2^* \end{pmatrix}$ is a solution to the linear system $\mathbf{HY} = \mathbf{S}$ if and only if $\mathbf{Y}_2^*$ is a solution to the linear system

$$(\mathbf{H}_{22} - \mathbf{H}_{21}\mathbf{H}_{11}^{-}\mathbf{H}_{12})\mathbf{Y}_2 = \mathbf{S}_2 - \mathbf{H}_{21}\mathbf{H}_{11}^{-}\mathbf{S}_1 \quad (\text{in } \mathbf{Y}_2)$$

and $\mathbf{Y}_1^*$ and $\mathbf{Y}_2^*$ are a solution to the linear system

$$\mathbf{H}_{11}\mathbf{Y}_1 + \mathbf{H}_{12}\mathbf{Y}_2 = \mathbf{S}_1 \ (\text{in } \mathbf{Y}_1 \text{ and } \mathbf{Y}_2).$$

Solution. According to Lemma 19.3.2, $\mathbf{A}_*$ and $\mathbf{R}_*$ are the first and second parts of a solution to linear system $(*)$ [or equivalently linear system (3.14)] if and only if $\mathbf{A}_*$ and $\mathbf{R}_* - \mathbf{UD}$ are the first and second parts of a solution to the linear system

$$\begin{pmatrix} \mathbf{V} + \mathbf{XUX}' & \mathbf{X} \\ \mathbf{X}' & \mathbf{0} \end{pmatrix} \begin{pmatrix} \mathbf{A} \\ \mathbf{T} \end{pmatrix} = \begin{pmatrix} \mathbf{B} \\ \mathbf{D} \end{pmatrix} \tag{S.1}$$

(in $\mathbf{A}$ and $\mathbf{T}$).

Now, observing (in light of Lemma 19.3.4) that $\mathcal{R}(\mathbf{X}') \subset \mathcal{R}(\mathbf{V} + \mathbf{XUX}')$ and $\mathcal{C}(\mathbf{B}) \subset \mathcal{C}(\mathbf{V} + \mathbf{XUX}')$, it follows from the cited result [or equivalently from Part (1) of Theorem 11.11.1] that $\mathbf{A}_*$ and $\mathbf{T}_*$ are the first and second parts of a solution to linear system (S.1) if and only if $\mathbf{T}_*$ is a solution to the linear system

$$(\mathbf{0} - \mathbf{X}'\mathbf{WX})\mathbf{T} = \mathbf{D} - \mathbf{X}'\mathbf{WB} \tag{S.2}$$

(in $\mathbf{T}$) and

$$(\mathbf{V} + \mathbf{XUX}')\mathbf{A}_* + \mathbf{XT}_* = \mathbf{B}. \tag{S.3}$$

Note that linear system (S.2) is equivalent to linear system $(**)$ [which is identical to linear system (3.17)]. Note also that condition (S.3) is equivalent to the condition

$$(\mathbf{V} + \mathbf{XUX}')\mathbf{A}_* = \mathbf{B} - \mathbf{XT}_*. \tag{S.4}$$

And, since (in light of Lemma 19.3.4) $\mathcal{C}(\mathbf{B} - \mathbf{XT}_*) \subset \mathcal{C}(\mathbf{V} + \mathbf{XUX}')$, it follows from Theorem 11.2.4 that condition (S.4) is equivalent to the condition that

$$\mathbf{A}_* = \mathbf{WB} - \mathbf{WXT}_* + [\mathbf{I} - \mathbf{W}(\mathbf{V} + \mathbf{XUX}')]\mathbf{L}$$

for some matrix $\mathbf{L}$.

Thus, $\mathbf{A}_*$ and $\mathbf{R}_* - \mathbf{UD}$ are the first and second parts of a solution to linear system (S.1) if and only if $\mathbf{R}_* - \mathbf{UD} = \mathbf{T}_*$ (or equivalently $\mathbf{R}_* = \mathbf{T}_* + \mathbf{UD}$) and

$$\mathbf{A}_* = \mathbf{WB} - \mathbf{WXT}_* + [\mathbf{I} - \mathbf{W}(\mathbf{V} + \mathbf{XUX}')]\mathbf{L}$$

for some solution $\mathbf{T}_*$ to linear system $(**)$ [or equivalently linear system (3.17)] and for some matrix $\mathbf{L}$.

EXERCISE 4. Let $\mathbf{V}$ represent an $n \times n$ symmetric matrix and $\mathbf{X}$ an $n \times p$ matrix. Further, let $\mathbf{U} = \mathbf{X}'\mathbf{TT}'\mathbf{X}$, where $\mathbf{T}$ is any matrix whose columns span the null space of $\mathbf{V}$. Show that $\mathcal{C}(\mathbf{X}) \subset \mathcal{C}(\mathbf{V} + \mathbf{XUX}')$ and that $\mathcal{C}(\mathbf{V})$ and $\mathcal{C}(\mathbf{XUX}')$ are essentially disjoint and $\mathcal{R}(\mathbf{V})$ and $\mathcal{R}(\mathbf{XUX}')$ are essentially disjoint (even in the absence of any assumption that $\mathbf{V}$ is nonnegative definite).

Solution. Making use of Corollaries 7.4.5, 4.5.6, and 17.2.14, we find that

$$\mathcal{C}(\mathbf{V}, \mathbf{X}) = \mathcal{C}(\mathbf{V}, \mathbf{XX}') = \mathcal{C}(\mathbf{V}, \mathbf{XX}'\mathbf{T}).$$

Further,

$$\mathcal{C}(\mathbf{XX}'\mathbf{T}) = \mathcal{C}[(\mathbf{XX}'\mathbf{T})(\mathbf{XX}'\mathbf{T})'] = \mathcal{C}(\mathbf{XUX}').$$

Thus, again making use of Corollary 4.5.6, we have that

$$\text{rank}(\mathbf{V}, \mathbf{X}) = \text{rank}(\mathbf{V}, \mathbf{XX}'\mathbf{T}) = \text{rank}(\mathbf{V}, \mathbf{XUX}').$$

And, in light of Corollary 17.2.14, $\mathcal{C}(\mathbf{V})$ and $\mathcal{C}(\mathbf{XUX}')$ are essentially disjoint. Moreover, since $\mathbf{XUX}'$ is clearly symmetric, it follows from Lemma 17.2.1 that $\mathcal{R}(\mathbf{V})$ and $\mathcal{R}(\mathbf{XUX}')$ are also essentially disjoint.

Finally, in light of Theorem 18.5.6, it follows from result (5.13) or (5.14) that $\text{rank}(\mathbf{V} + \mathbf{XUX}') = \text{rank}(\mathbf{V}, \mathbf{XUX}')$, so that $\text{rank}(\mathbf{V} + \mathbf{XUX}') = \text{rank}(\mathbf{V}, \mathbf{X})$ or equivalently (in light of Lemma 19.3.4) $\mathcal{C}(\mathbf{X}) \subset \mathcal{C}(\mathbf{V} + \mathbf{XUX}')$.

EXERCISE 5. Let $\mathbf{V}$ represent an $n \times n$ symmetric nonnegative definite matrix and $\mathbf{X}$ an $n \times p$ matrix. Further, let $\mathbf{Z}$ represent any matrix whose columns span $\mathcal{N}(\mathbf{X}')$ or, equivalently, $\mathcal{C}^{\perp}(\mathbf{X})$. And, adopt the same terminology as in Exercise 17.20.

(a) Using the result of Part (a) of Exercise 17.20 (or otherwise) show that an $n \times n$ matrix $\mathbf{H}$ is a projection matrix for $\mathcal{C}(\mathbf{X})$ along $\mathcal{C}(\mathbf{VZ})$ if and only if $\mathbf{H}'$ is the first $(n \times n)$ part of a solution to the consistent linear system

$$\begin{pmatrix} \mathbf{V} & \mathbf{X} \\ \mathbf{X}' & \mathbf{0} \end{pmatrix} \begin{pmatrix} \mathbf{A} \\ \mathbf{R} \end{pmatrix} = \begin{pmatrix} \mathbf{0} \\ \mathbf{X}' \end{pmatrix} \tag{E.1}$$

(in an $n \times n$ matrix $\mathbf{A}$ and a $p \times n$ matrix $\mathbf{R}$).

(b) Letting $\mathbf{U}$ represent any $p \times p$ matrix such that $\mathcal{C}(\mathbf{X}) \subset \mathcal{C}(\mathbf{V} + \mathbf{XUX}')$ and letting $\mathbf{W}$ represent an arbitrary generalized inverse of $\mathbf{V} + \mathbf{XUX}'$, show that an $n \times n$ matrix $\mathbf{H}$ is a projection matrix for $\mathcal{C}(\mathbf{X})$ along $\mathcal{C}(\mathbf{VZ})$ if and only if

$$\mathbf{H} = \mathbf{P}_{\mathbf{X},\mathbf{W}} + \mathbf{K}[\mathbf{I} - (\mathbf{V} + \mathbf{XUX}')\mathbf{W}]$$

for some $n \times n$ matrix $\mathbf{K}$.

Solution. (a) In light of the result of Part (a) of Exercise 17.20, it suffices to show that $\mathbf{HX} = \mathbf{X}$ and $\mathbf{HVZ} = \mathbf{0}$ (or equivalently that $\mathbf{X}'\mathbf{H}' = \mathbf{X}'$ and $\mathbf{Z}'\mathbf{VH}' = \mathbf{0}$) if and only if $\mathbf{H}'$ is the first part of a solution to linear system (E.1).

Now, suppose that $\mathbf{X}'\mathbf{H}' = \mathbf{X}'$ and $\mathbf{Z}'\mathbf{VH}' = \mathbf{0}$. Then, in light of Corollary 12.1.2, it follows from Corollary 12.5.5 that $\mathcal{C}(\mathbf{VH}') \subset \mathcal{C}(\mathbf{X})$ and hence (in light of Lemma 4.2.2) that $\mathbf{VH}' = \mathbf{XT}$ for some matrix $\mathbf{T}$. Thus,

$$\begin{pmatrix} \mathbf{V} & \mathbf{X} \\ \mathbf{X}' & \mathbf{0} \end{pmatrix} \begin{pmatrix} \mathbf{H}' \\ -\mathbf{T} \end{pmatrix} = \begin{pmatrix} \mathbf{0} \\ \mathbf{X}' \end{pmatrix},$$

so that $\mathbf{H}'$ is the first part of a solution to linear system (E.1).

Conversely, suppose that $\mathbf{H}'$ is the first part of a solution to linear system (E.1) and hence that

$$\begin{pmatrix} \mathbf{V} & \mathbf{X} \\ \mathbf{X}' & \mathbf{0} \end{pmatrix} \begin{pmatrix} \mathbf{H}' \\ \mathbf{R}_* \end{pmatrix} = \begin{pmatrix} \mathbf{0} \\ \mathbf{X}' \end{pmatrix}$$

for some $p \times n$ matrix $\mathbf{R}_*$. Then, $\mathbf{X}'\mathbf{H}' = \mathbf{X}'$. And, $\mathbf{VH}' = \mathbf{X}(-\mathbf{R}_*)$, implying that $\mathcal{C}(\mathbf{VH}') \subset \mathcal{C}(\mathbf{X})$ and hence (in light of Corollaries 12.5.5 and 12.1.2) that $\mathbf{Z}'\mathbf{VH}' = \mathbf{0}$.

(b) In light of Part (a), it suffices to show that $\mathbf{H}'$ is the first part of a solution to linear system (E.1) if and only if

$$\mathbf{H} = \mathbf{P}_{\mathbf{X},\mathbf{W}} + \mathbf{K}[\mathbf{I} - (\mathbf{V} + \mathbf{XUX}')\mathbf{W}]$$

for some matrix $\mathbf{K}$.

According to Lemma 19.3.4, $\mathcal{C}(\mathbf{X}) \subset \mathcal{C}(\mathbf{V} + \mathbf{XU}'\mathbf{X}')$. Moreover, since $\mathbf{V} + \mathbf{XU}'\mathbf{X}' = (\mathbf{V} + \mathbf{XUX}')'$ and since $\mathbf{X}'\mathbf{W}'\mathbf{X} = (\mathbf{X}'\mathbf{WX})'$, $\mathbf{W}'$ is a generalized inverse of $\mathbf{V} + \mathbf{XU}'\mathbf{X}'$, and $[(\mathbf{X}'\mathbf{WX})^-]'$ is a generalized inverse of $\mathbf{X}'\mathbf{W}'\mathbf{X}$. Thus, it follows from the results of Section 19.3c that $\mathbf{H}'$ is the first part of a solution to linear system (E.1) if and only if

$$\mathbf{H}' = \mathbf{W}'\mathbf{X}[(\mathbf{X}'\mathbf{WX})^-]'\mathbf{X}' + [\mathbf{I} - \mathbf{W}'(\mathbf{V} + \mathbf{XU}'\mathbf{X}')]\mathbf{K}'$$

for some $(n \times n)$ matrix $\mathbf{K}$, or equivalently if and only if

$$\mathbf{H} = \mathbf{P}_{\mathbf{X},\mathbf{W}} + \mathbf{K}[\mathbf{I} - (\mathbf{V} + \mathbf{XUX}')\mathbf{W}]$$

for some matrix $\mathbf{K}$.

EXERCISE 6. Let $\mathbf{V}$ represent an $n \times n$ symmetric nonnegative definite matrix, $\mathbf{W}$ an $n \times n$ matrix, and $\mathbf{X}$ an $n \times p$ matrix. Show that, for the matrix $\mathbf{WX}(\mathbf{X}'\mathbf{WX})^-\mathbf{X}'$ to be the first $(n \times n)$ part of some solution to the (consistent) linear system

$$\begin{pmatrix} \mathbf{V} & \mathbf{X} \\ \mathbf{X}' & \mathbf{0} \end{pmatrix} \begin{pmatrix} \mathbf{A} \\ \mathbf{R} \end{pmatrix} = \begin{pmatrix} \mathbf{0} \\ \mathbf{X}' \end{pmatrix} \tag{E.2}$$

(in an $n \times n$ matrix $\mathbf{A}$ and a $p \times n$ matrix $\mathbf{R}$), it is necessary and sufficient that $\mathcal{C}(\mathbf{VWX}) \subset \mathcal{C}(\mathbf{X})$ and $\operatorname{rank}(\mathbf{X}'\mathbf{WX}) = \operatorname{rank}(\mathbf{X})$.

Solution. Clearly, $\mathbf{WX}(\mathbf{X}'\mathbf{WX})^-\mathbf{X}'$ is the first part of a solution to linear system (E.2) if and only if $\mathbf{VWX}(\mathbf{X}'\mathbf{WX})^-\mathbf{X}' + \mathbf{XR} = \mathbf{0}$ for some matrix $\mathbf{R}$ and $\mathbf{X}'\mathbf{WX}(\mathbf{X}'\mathbf{WX})^-\mathbf{X}' = \mathbf{X}'$, or equivalently if and only if

$$\mathcal{C}[\mathbf{VWX}(\mathbf{X}'\mathbf{WX})^-\mathbf{X}'] \subset \mathcal{C}(\mathbf{X}) \tag{S.5}$$

and

$$\mathbf{X}'\mathbf{W}\mathbf{X}(\mathbf{X}'\mathbf{W}\mathbf{X})^-\mathbf{X}' = \mathbf{X}'. \tag{S.6}$$

And, by following the same line of reasoning as in the latter part of the proof of Theorem 19.5.1, we find that conditions (S.5) and (S.6) are equivalent to the conditions that $\mathcal{C}(\mathbf{VWX}) \subset \mathcal{C}(\mathbf{X})$ and $\text{rank}(\mathbf{X}'\mathbf{W}\mathbf{X}) = \text{rank}(\mathbf{X})$.

EXERCISE 7. Let $\mathbf{a}$ represent an $n \times 1$ vector of variables, and impose on $\mathbf{a}$ the constraint $\mathbf{X}'\mathbf{a} = \mathbf{d}$, where $\mathbf{X}$ is an $n \times p$ matrix and $\mathbf{d}$ is a $p \times 1$ vector such that $\mathbf{d} \in \mathcal{C}(\mathbf{X}')$. Define $f(\mathbf{a}) = \mathbf{a}'\mathbf{V}\mathbf{a} - 2\mathbf{b}'\mathbf{a}$, where $\mathbf{V}$ is an $n \times n$ symmetric nonnegative definite matrix and $\mathbf{b}$ is an $n \times 1$ vector such that $\mathbf{b} \in \mathcal{C}(\mathbf{V}, \mathbf{X})$. Further, define $g(\mathbf{a}) = \mathbf{a}'(\mathbf{V} + \mathbf{W})\mathbf{a} - 2(\mathbf{b} + \mathbf{c})'\mathbf{a}$, where $\mathbf{W}$ is any $n \times n$ matrix such that $\mathcal{C}(\mathbf{W}) \subset \mathcal{C}(\mathbf{X})$ and $\mathcal{R}(\mathbf{W}) \subset \mathcal{R}(\mathbf{X}')$ and where $\mathbf{c}$ is any $n \times 1$ vector in $\mathcal{C}(\mathbf{X})$. Show that the constrained (by $\mathbf{X}'\mathbf{a} = \mathbf{d}$) minimization of $g(\mathbf{a})$ is equivalent to the constrained minimization of $f(\mathbf{a})$ [in the sense that $g(\mathbf{a})$ and $f(\mathbf{a})$ attain their minimum values at the same points].

Solution. Clearly, $\mathbf{c} = \mathbf{X}\mathbf{r}$ for some $p \times 1$ vector $\mathbf{r}$. Further, in light of Lemma 9.3.5, we have that $\mathbf{W} = \mathbf{X}\mathbf{X}^-\mathbf{W}$ and $\mathbf{W} = \mathbf{W}(\mathbf{X}')^-\mathbf{X}'$ and hence that

$$\mathbf{W} = \mathbf{X}\mathbf{X}^-\mathbf{W}(\mathbf{X}')^-\mathbf{X}' = \mathbf{X}\mathbf{U}\mathbf{X}',$$

where $\mathbf{U} = \mathbf{X}^-\mathbf{W}(\mathbf{X}')^-$. Thus,

$$\begin{aligned} g(\mathbf{a}) &= f(\mathbf{a}) + \mathbf{a}'\mathbf{W}\mathbf{a} - 2\mathbf{c}'\mathbf{a} \\ &= f(\mathbf{a}) + (\mathbf{X}'\mathbf{a})'\mathbf{U}\mathbf{X}'\mathbf{a} - 2\mathbf{r}'\mathbf{X}'\mathbf{a}, \end{aligned}$$

so that, for $\mathbf{a}$ such that $\mathbf{X}'\mathbf{a} = \mathbf{d}$,

$$g(\mathbf{a}) = f(\mathbf{a}) + \mathbf{d}'\mathbf{U}\mathbf{d} - 2\mathbf{r}'\mathbf{d}.$$

We conclude that, for $\mathbf{a}$ such that $\mathbf{X}'\mathbf{a} = \mathbf{d}$, $g(\mathbf{a})$ differs from $f(\mathbf{a})$ only by an additive constant and hence that $g(\mathbf{a})$ and $f(\mathbf{a})$ attain their minimum values (under the constraint $\mathbf{X}'\mathbf{a} = \mathbf{d}$) at the same points.

EXERCISE 8. Let $\mathbf{V}$ represent an $n \times n$ symmetric nonnegative definite matrix, $\mathbf{W}$ an $n \times n$ matrix, $\mathbf{X}$ an $n \times p$ matrix, $\mathbf{f}$ an $n \times 1$ vector, and $\mathbf{d}$ a $p \times 1$ vector. Further, let $\mathbf{b}$ represent an $n \times 1$ vector such that $\mathbf{b} \in \mathcal{C}(\mathbf{V}, \mathbf{X})$. Show that, for the vector $\mathbf{W}(\mathbf{I} - \mathbf{P}_{\mathbf{X},\mathbf{W}})\mathbf{f} + \mathbf{W}\mathbf{X}(\mathbf{X}'\mathbf{W}\mathbf{X})^-\mathbf{d}$ to be a solution, for every $\mathbf{d} \in \mathcal{C}(\mathbf{X}')$, to the problem of minimizing the second-degree polynomial $\mathbf{a}'\mathbf{V}\mathbf{a} - 2\mathbf{b}'\mathbf{a}$ (in $\mathbf{a}$) subject to $\mathbf{X}'\mathbf{a} = \mathbf{d}$, it is necessary and sufficient that

$$\mathbf{V}\mathbf{W}\mathbf{f} - \mathbf{b} \in \mathcal{C}(\mathbf{X}), \tag{E.3}$$

$$\mathcal{C}(\mathbf{V}\mathbf{W}\mathbf{X}) \subset \mathcal{C}(\mathbf{X}), \tag{E.4}$$

and

$$\text{rank}(\mathbf{X}'\mathbf{W}\mathbf{X}) = \text{rank}(\mathbf{X}). \tag{E.5}$$

Solution. It follows from Theorem 19.2.1 that $\mathbf{a}'\mathbf{V}\mathbf{a} - 2\mathbf{b}'\mathbf{a}$ has a minimum at $\mathbf{W}(\mathbf{I}-\mathbf{P}_{\mathbf{X},\mathbf{W}})\mathbf{f}+\mathbf{W}\mathbf{X}(\mathbf{X}'\mathbf{W}\mathbf{X})^{-}\mathbf{d}$ under the constraint $\mathbf{X}'\mathbf{a} = \mathbf{d}$ [where $\mathbf{d} \in \mathcal{C}(\mathbf{X}')$] if and only if $\mathbf{V}\mathbf{W}(\mathbf{I}-\mathbf{P}_{\mathbf{X},\mathbf{W}})\mathbf{f}+\mathbf{V}\mathbf{W}\mathbf{X}(\mathbf{X}'\mathbf{W}\mathbf{X})^{-}\mathbf{d}+\mathbf{X}\mathbf{r} = \mathbf{b}$ for some vector $\mathbf{r}$ and $\mathbf{X}'\mathbf{W}(\mathbf{I}-\mathbf{P}_{\mathbf{X},\mathbf{W}})\mathbf{f}+\mathbf{X}'\mathbf{W}\mathbf{X}(\mathbf{X}'\mathbf{W}\mathbf{X})^{-}\mathbf{d} = \mathbf{d}$, or equivalently if and only if $\mathbf{V}\mathbf{W}(\mathbf{I}-\mathbf{P}_{\mathbf{X},\mathbf{W}})\mathbf{f}-\mathbf{b}+\mathbf{V}\mathbf{W}\mathbf{X}(\mathbf{X}'\mathbf{W}\mathbf{X})^{-}\mathbf{d} \in \mathcal{C}(\mathbf{X})$ and $\mathbf{X}'\mathbf{W}(\mathbf{I}-\mathbf{P}_{\mathbf{X},\mathbf{W}})\mathbf{f}+\mathbf{X}'\mathbf{W}\mathbf{X}(\mathbf{X}'\mathbf{W}\mathbf{X})^{-}\mathbf{d} = \mathbf{d}$. Thus, for $\mathbf{W}(\mathbf{I}-\mathbf{P}_{\mathbf{X},\mathbf{W}})\mathbf{f}+\mathbf{W}\mathbf{X}(\mathbf{X}'\mathbf{W}\mathbf{X})^{-}\mathbf{d}$ to be a solution, for every $\mathbf{d} \in \mathcal{C}(\mathbf{X}')$, to the problem of minimizing $\mathbf{a}'\mathbf{V}\mathbf{a} - 2\mathbf{b}'\mathbf{a}$ subject to $\mathbf{X}'\mathbf{a} = \mathbf{d}$, it is necesary and sufficient that, for every $n \times 1$ vector $\mathbf{u}$, $\mathbf{V}\mathbf{W}(\mathbf{I}-\mathbf{P}_{\mathbf{X},\mathbf{W}})\mathbf{f}-\mathbf{b}+\mathbf{V}\mathbf{W}\mathbf{X}(\mathbf{X}'\mathbf{W}\mathbf{X})^{-}\mathbf{X}'\mathbf{u} \in \mathcal{C}(\mathbf{X})$ and $\mathbf{X}'\mathbf{W}(\mathbf{I}-\mathbf{P}_{\mathbf{X},\mathbf{W}})\mathbf{f}+\mathbf{X}'\mathbf{W}\mathbf{X}(\mathbf{X}'\mathbf{W}\mathbf{X})^{-}\mathbf{X}'\mathbf{u} = \mathbf{X}'\mathbf{u}$, a requirement equivalent to a requirement that

$$\mathbf{V}\mathbf{W}(\mathbf{I}-\mathbf{P}_{\mathbf{X},\mathbf{W}})\mathbf{f}-\mathbf{b} \in \mathcal{C}(\mathbf{X}), \tag{S.7}$$

$$\mathcal{C}[\mathbf{V}\mathbf{W}\mathbf{X}(\mathbf{X}'\mathbf{W}\mathbf{X})^{-}\mathbf{X}'] \subset \mathcal{C}(\mathbf{X}), \tag{S.8}$$

and

$$\mathbf{X}'\mathbf{W}\mathbf{X}(\mathbf{X}'\mathbf{W}\mathbf{X})^{-}\mathbf{X}' = \mathbf{X}', \tag{S.9}$$

as we now show.

Suppose that conditions (S.7) – (S.9) are satisfied. Then, observing that

$$\mathbf{X}'\mathbf{W}\mathbf{P}_{\mathbf{X},\mathbf{W}} = \mathbf{X}'\mathbf{W}\mathbf{X}(\mathbf{X}'\mathbf{W}\mathbf{X})^{-}\mathbf{X}'\mathbf{W},$$

we find that, for every $\mathbf{u}$,

$$\mathbf{V}\mathbf{W}(\mathbf{I}-\mathbf{P}_{\mathbf{X},\mathbf{W}})\mathbf{f}-\mathbf{b}+\mathbf{V}\mathbf{W}\mathbf{X}(\mathbf{X}'\mathbf{W}\mathbf{X})^{-}\mathbf{X}'\mathbf{u} \in \mathcal{C}(\mathbf{X})$$

and

$$\mathbf{X}'\mathbf{W}(\mathbf{I}-\mathbf{P}_{\mathbf{X},\mathbf{W}})\mathbf{f}+\mathbf{X}'\mathbf{W}\mathbf{X}(\mathbf{X}'\mathbf{W}\mathbf{X})^{-}\mathbf{X}'\mathbf{u} = (\mathbf{X}'\mathbf{W}-\mathbf{X}'\mathbf{W})\mathbf{f}+\mathbf{X}'\mathbf{u} = \mathbf{X}'\mathbf{u}.$$

Conversely, suppose that, for every $\mathbf{u}$,

$$\mathbf{V}\mathbf{W}(\mathbf{I}-\mathbf{P}_{\mathbf{X},\mathbf{W}})\mathbf{f}-\mathbf{b}+\mathbf{V}\mathbf{W}\mathbf{X}(\mathbf{X}'\mathbf{W}\mathbf{X})^{-}\mathbf{X}'\mathbf{u} \in \mathcal{C}(\mathbf{X}) \tag{S.10}$$

and

$$\mathbf{X}'\mathbf{W}(\mathbf{I}-\mathbf{P}_{\mathbf{X},\mathbf{W}})\mathbf{f}+\mathbf{X}'\mathbf{W}\mathbf{X}(\mathbf{X}'\mathbf{W}\mathbf{X})^{-}\mathbf{X}'\mathbf{u} = \mathbf{X}'\mathbf{u}. \tag{S.11}$$

Then, since conditions (S.10) and (S.11) are satisfied in particular for $\mathbf{u} = \mathbf{0}$, we have that

$$\mathbf{V}\mathbf{W}(\mathbf{I}-\mathbf{P}_{\mathbf{X},\mathbf{W}})\mathbf{f}-\mathbf{b} \in \mathcal{C}(\mathbf{X})$$

and

$$\mathbf{X}'\mathbf{W}(\mathbf{I}-\mathbf{P}_{\mathbf{X},\mathbf{W}})\mathbf{f} = \mathbf{0}.$$

Further, for every $\mathbf{u}$, $\mathbf{V}\mathbf{W}\mathbf{X}(\mathbf{X}'\mathbf{W}\mathbf{X})^{-}\mathbf{X}'\mathbf{u} \in \mathcal{C}(\mathbf{X})$ and $\mathbf{X}'\mathbf{W}\mathbf{X}(\mathbf{X}'\mathbf{W}\mathbf{X})^{-}\mathbf{X}'\mathbf{u} = \mathbf{X}'\mathbf{u}$, implying that

$$\mathcal{C}[\mathbf{V}\mathbf{W}\mathbf{X}(\mathbf{X}'\mathbf{W}\mathbf{X})^{-}\mathbf{X}'] \subset \mathcal{C}(\mathbf{X})$$

and

$$\mathbf{X}'\mathbf{W}\mathbf{X}(\mathbf{X}'\mathbf{W}\mathbf{X})^{-}\mathbf{X}' = \mathbf{X}'.$$

Now, when condition (S.8) is satisfied, condition (E.3) is equivalent to condition (S.7), as is evident from Lemma 4.1.2 upon observing that $\mathbf{VWP}_{\mathbf{X},\mathbf{W}} = \mathbf{VWX}(\mathbf{X}'\mathbf{WX})^{-}\mathbf{X}'\mathbf{W}$ and hence that $\mathbf{VWP}_{\mathbf{X},\mathbf{W}}\mathbf{f} \in \mathcal{C}[\mathbf{VWX}(\mathbf{X}'\mathbf{WX})^{-}\mathbf{X}']$. Moreover, by employing the same line of reasoning as in the latter part of the proof of Theorem 19.5.1, we find that conditions (E.4) and (E.5) are equivalent to conditions (S.8) and (S.9). Thus, conditions (E.3) – (E.5) are equivalent to conditions (S.7) – (S.9). And, we conclude that, for $\mathbf{W}(\mathbf{I} - \mathbf{P}_{\mathbf{X},\mathbf{W}})\mathbf{b} + \mathbf{WX}(\mathbf{X}'\mathbf{WX})^{-}\mathbf{d}$ to be a solution, for every $\mathbf{d} \in \mathcal{C}(\mathbf{X}')$, to the problem of minimizing $\mathbf{a}'\mathbf{Va} - 2\mathbf{b}'\mathbf{a}$ subject to $\mathbf{X}'\mathbf{a} = \mathbf{d}$, it is necessary and sufficient that conditions (E.3) – (E.5) be satisfied.

EXERCISE 9. Let $\mathbf{V}$ and $\mathbf{W}$ represent $n \times n$ matrices, and let $\mathbf{X}$ represent an $n \times p$ matrix. Show that if $\mathbf{V}$ and $\mathbf{W}$ are nonsingular, then the condition $\mathcal{C}(\mathbf{VWX}) \subset \mathcal{C}(\mathbf{X})$ is equivalent to the condition $\mathcal{C}(\mathbf{V}^{-1}\mathbf{X}) \subset \mathcal{C}(\mathbf{WX})$ and is also equivalent to the condition $\mathcal{C}(\mathbf{W}^{-1}\mathbf{V}^{-1}\mathbf{X}) \subset \mathcal{C}(\mathbf{X})$.

Solution. Assume that $\mathbf{V}$ and $\mathbf{W}$ are nonsingular. Then, in light of Corollary 8.3.3, $\text{rank}(\mathbf{VWX}) = \text{rank}(\mathbf{X})$, $\text{rank}(\mathbf{V}^{-1}\mathbf{X}) = \text{rank}(\mathbf{X}) = \text{rank}(\mathbf{WX})$, and $\text{rank}(\mathbf{W}^{-1}\mathbf{V}^{-1}\mathbf{X}) = \text{rank}(\mathbf{X})$. Thus, as a consequence of Theorem 4.4.6,

$$\begin{aligned} \mathcal{C}(\mathbf{VWX}) \subset \mathcal{C}(\mathbf{X}) \quad &\Leftrightarrow \quad \mathcal{C}(\mathbf{VWX}) = \mathcal{C}(\mathbf{X}), \\ \mathcal{C}(\mathbf{V}^{-1}\mathbf{X}) \subset \mathcal{C}(\mathbf{WX}) \quad &\Leftrightarrow \quad \mathcal{C}(\mathbf{V}^{-1}\mathbf{X}) = \mathcal{C}(\mathbf{WX}), \quad \text{and} \\ \mathcal{C}(\mathbf{W}^{-1}\mathbf{V}^{-1}\mathbf{X}) \subset \mathcal{C}(\mathbf{X}) \quad &\Leftrightarrow \quad \mathcal{C}(\mathbf{W}^{-1}\mathbf{V}^{-1}\mathbf{X}) = \mathcal{C}(\mathbf{X}). \end{aligned}$$

Now, if $\mathcal{C}(\mathbf{VWX}) \subset \mathcal{C}(\mathbf{X})$, then $\mathcal{C}(\mathbf{X}) = \mathcal{C}(\mathbf{VWX})$, so that $\mathbf{X} = \mathbf{VWXQ}$ for some matrix $\mathbf{Q}$, in which case $\mathbf{V}^{-1}\mathbf{X} = \mathbf{WXQ}$ and $\mathbf{W}^{-1}\mathbf{V}^{-1}\mathbf{X} = \mathbf{XQ}$, implying that $\mathcal{C}(\mathbf{V}^{-1}\mathbf{X}) \subset \mathcal{C}(\mathbf{WX})$ and $\mathcal{C}(\mathbf{W}^{-1}\mathbf{V}^{-1}\mathbf{X}) \subset \mathcal{C}(\mathbf{X})$. Conversely, if $\mathcal{C}(\mathbf{V}^{-1}\mathbf{X}) \subset \mathcal{C}(\mathbf{WX})$, then $\mathcal{C}(\mathbf{WX}) = \mathcal{C}(\mathbf{V}^{-1}\mathbf{X})$, so that $\mathbf{WX} = \mathbf{V}^{-1}\mathbf{XQ}$ for some matrix $\mathbf{Q}$, in which case $\mathbf{VWX} = \mathbf{XQ}$, implying that $\mathcal{C}(\mathbf{VWX}) \subset \mathcal{C}(\mathbf{X})$. And, similarly, if $\mathcal{C}(\mathbf{W}^{-1}\mathbf{V}^{-1}\mathbf{X}) \subset \mathcal{C}(\mathbf{X})$, then $\mathcal{C}(\mathbf{X}) = \mathcal{C}(\mathbf{W}^{-1}\mathbf{V}^{-1}\mathbf{X})$, so that $\mathbf{X} = \mathbf{W}^{-1}\mathbf{V}^{-1}\mathbf{XQ}$ for some matrix $\mathbf{Q}$, in which case $\mathbf{VWX} = \mathbf{XQ}$, implying that $\mathcal{C}(\mathbf{VWX}) \subset \mathcal{C}(\mathbf{X})$. We conclude that $\mathcal{C}(\mathbf{VWX}) \subset \mathcal{C}(\mathbf{X}) \Leftrightarrow \mathcal{C}(\mathbf{V}^{-1}\mathbf{X}) \subset \mathcal{C}(\mathbf{WX})$ and that $\mathcal{C}(\mathbf{VWX}) \subset \mathcal{C}(\mathbf{X}) \Leftrightarrow \mathcal{C}(\mathbf{W}^{-1}\mathbf{V}^{-1}\mathbf{X}) \subset \mathcal{C}(\mathbf{X})$.

EXERCISE 10. Let $\mathbf{V}$ represent an $n \times n$ symmetric positive definite matrix, $\mathbf{W}$ an $n \times n$ matrix, $\mathbf{X}$ an $n \times p$ matrix, and $\mathbf{d}$ a $p \times 1$ vector. Show that, for the vector $\mathbf{WX}(\mathbf{X}'\mathbf{WX})^{-}\mathbf{d}$ to be a solution, for every $\mathbf{d} \in \mathcal{C}(\mathbf{X}')$, to the problem of minimizing the quadratic form $\mathbf{a}'\mathbf{Va}$ (in $\mathbf{a}$) subject to $\mathbf{X}'\mathbf{a} = \mathbf{d}$, it is necessary and sufficient that $\mathbf{V}^{-1}\mathbf{P}_{\mathbf{X},\mathbf{W}'}$ be symmetric and $\text{rank}(\mathbf{X}'\mathbf{WX}) = \text{rank}(\mathbf{X})$. Show that it is also necessary and sufficient that $(\mathbf{I} - \mathbf{P}'_{\mathbf{X},\mathbf{W}'})\mathbf{V}^{-1}\mathbf{P}_{\mathbf{X},\mathbf{W}'} = \mathbf{0}$ and $\text{rank}(\mathbf{X}'\mathbf{WX}) = \text{rank}(\mathbf{X})$.

Solution. Since $(\mathbf{V}^{-1}\mathbf{P}_{\mathbf{X},\mathbf{W}'})' = \mathbf{P}'_{\mathbf{X},\mathbf{W}'}\mathbf{V}^{-1}$, $\mathbf{V}^{-1}\mathbf{P}_{\mathbf{X},\mathbf{W}'}$ is symmetric if and only if $\mathbf{V}^{-1}\mathbf{P}_{\mathbf{X},\mathbf{W}'} = \mathbf{P}'_{\mathbf{X},\mathbf{W}'}\mathbf{V}^{-1}$. Moreover, if $\mathbf{V}^{-1}\mathbf{P}_{\mathbf{X},\mathbf{W}'} = \mathbf{P}'_{\mathbf{X},\mathbf{W}'}\mathbf{V}^{-1}$, then

$$\mathbf{P}_{\mathbf{X},\mathbf{W}'}\mathbf{V} = \mathbf{V}(\mathbf{V}^{-1}\mathbf{P}_{\mathbf{X},\mathbf{W}'})\mathbf{V} = \mathbf{V}(\mathbf{P}'_{\mathbf{X},\mathbf{W}'}\mathbf{V}^{-1})\mathbf{V} = \mathbf{VP}'_{\mathbf{X},\mathbf{W}'}.$$

And, conversely, if $\mathbf{P_{X,W'}V} = \mathbf{VP'_{X,W'}}$, then

$$\mathbf{V}^{-1}\mathbf{P_{X,W'}} = \mathbf{V}^{-1}(\mathbf{P_{X,W'}V})\mathbf{V}^{-1} = \mathbf{V}^{-1}(\mathbf{VP'_{X,W'}})\mathbf{V}^{-1} = \mathbf{P'_{X,W'}}\mathbf{V}^{-1}.$$

Thus, $\mathbf{V}^{-1}\mathbf{P_{X,W'}}$ is symmetric if and only if $\mathbf{P_{X,W'}V} = \mathbf{VP'_{X,W'}}$, and the necessity and sufficiency of $\mathbf{V}^{-1}\mathbf{P_{X,W'}}$ being symmetric and $\text{rank}(\mathbf{X'WX}) = \text{rank}(\mathbf{X})$ follows from Theorem 19.5.4.

To complete the proof, it suffices to show that $(\mathbf{I} - \mathbf{P'_{X,W'}})\mathbf{V}^{-1}\mathbf{P_{X,W'}} = \mathbf{0}$ and $\text{rank}(\mathbf{X'WX}) = \text{rank}(\mathbf{X})$, or equivalently that $\mathbf{V}^{-1}\mathbf{P_{X,W'}} = \mathbf{P'_{X,W'}}\mathbf{V}^{-1}\mathbf{P_{X,W'}}$ and $\text{rank}(\mathbf{X'WX}) = \text{rank}(\mathbf{X})$, if and only if $\mathbf{V}^{-1}\mathbf{P_{X,W'}} = \mathbf{P'_{X,W'}}\mathbf{V}^{-1}$ and $\text{rank}(\mathbf{X'WX}) = \text{rank}(\mathbf{X})$.

Suppose that $\mathbf{V}^{-1}\mathbf{P_{X,W'}} = \mathbf{P'_{X,W'}}\mathbf{V}^{-1}$ and $\text{rank}(\mathbf{X'WX}) = \text{rank}(\mathbf{X})$. Then, since $\mathbf{X'W'X} = (\mathbf{X'WX})'$, $\text{rank}(\mathbf{X'W'X}) = \text{rank}(\mathbf{X})$, and it follows from Part (3) of Lemma 19.5.5 that $\mathbf{P}^2_{\mathbf{X,W'}} = \mathbf{P_{X,W'}}$. Thus,

$$\mathbf{V}^{-1}\mathbf{P_{X,W'}} = (\mathbf{V}^{-1}\mathbf{P_{X,W'}})\mathbf{P_{X,W'}} = \mathbf{P'_{X,W'}}\mathbf{V}^{-1}\mathbf{P_{X,W'}}.$$

Conversely, if $\mathbf{V}^{-1}\mathbf{P_{X,W'}} = \mathbf{P'_{X,W'}}\mathbf{V}^{-1}\mathbf{P_{X,W'}}$, then (since clearly the matrix $\mathbf{P'_{X,W'}}\mathbf{V}^{-1}\mathbf{P_{X,W'}}$ is symmetric)

$$\mathbf{V}^{-1}\mathbf{P_{X,W'}} = (\mathbf{P'_{X,W'}}\mathbf{V}^{-1}\mathbf{P_{X,W'}})' = (\mathbf{V}^{-1}\mathbf{P_{X,W'}})' = \mathbf{P'_{X,W'}}\mathbf{V}^{-1}.$$

We conclude that $\mathbf{V}^{-1}\mathbf{P_{X,W'}} = \mathbf{P'_{X,W'}}\mathbf{V}^{-1}\mathbf{P_{X,W'}}$ and $\text{rank}(\mathbf{X'WX}) = \text{rank}(\mathbf{X})$ if and only if $\mathbf{V}^{-1}\mathbf{P_{X,W'}} = \mathbf{P'_{X,W'}}\mathbf{V}^{-1}$ and $\text{rank}(\mathbf{X'WX}) = \text{rank}(\mathbf{X})$.

EXERCISE 11. Let $\mathbf{V}$ represent an $n \times n$ symmetric nonnegative definite matrix, $\mathbf{X}$ an $n \times p$ matrix, and $\mathbf{d}$ a $p \times 1$ vector. Show that each of the following six conditions is necessary and sufficient for the vector $\mathbf{X}(\mathbf{X'X})^-\mathbf{d}$ to be a solution, for every $\mathbf{d} \in \mathcal{C}(\mathbf{X'})$, to the problem of minimizing the quadratic form $\mathbf{a'Va}$ (in $\mathbf{a}$) subject to $\mathbf{X'a} = \mathbf{d}$:

(a) $\mathcal{C}(\mathbf{VX}) \subset \mathcal{C}(\mathbf{X})$ (or, equivalently, $\mathbf{VX} = \mathbf{XQ}$ for some matrix $\mathbf{Q}$);

(b) $\mathbf{P_X V}(\mathbf{I} - \mathbf{P_X}) = \mathbf{0}$ (or, equivalently, $\mathbf{P_X V} = \mathbf{P_X V P_X}$);

(c) $\mathbf{P_X V} = \mathbf{V P_X}$ (or, equivalently, $\mathbf{P_X V}$ is symmetric);

(d) $\mathcal{C}(\mathbf{VP_X}) \subset \mathcal{C}(\mathbf{P_X})$;

(e) $\mathcal{C}(\mathbf{VP_X}) = \mathcal{C}(\mathbf{V}) \cap \mathcal{C}(\mathbf{P_X})$;

(f) $\mathcal{C}(\mathbf{VX}) = \mathcal{C}(\mathbf{V}) \cap \mathcal{C}(\mathbf{X})$.

Solution. (a), (b), and (c) Upon applying Theorems 19.5.1 and 19.5.4 (with $\mathbf{W} = \mathbf{I}$) and recalling (from Corollary 7.4.5) that $\text{rank}(\mathbf{X'X}) = \text{rank}(\mathbf{X})$ and (from Theorem 12.3.4) that $\mathbf{P_X}$ is symmetric, we find that each of Conditions (a) – (c) is necessary and sufficient for $\mathbf{X}(\mathbf{X'X})^-\mathbf{d}$ to be a solution to the problem of minimizing $\mathbf{a'Va}$ subject to $\mathbf{X'a} = \mathbf{d}$.

(d) According to Theorem 12.3.4, $\mathcal{C}(\mathbf{P_X}) = \mathcal{C}(\mathbf{X})$. And, in light of Corollary 4.2.4, $\mathcal{C}(\mathbf{VP_X}) = \mathcal{C}(\mathbf{VX})$. Thus, Condition (d) is equivalent to Condition (a).

(e) Let $\mathbf{y}$ represent an arbitrary vector in $\mathcal{C}(\mathbf{V}) \cap \mathcal{C}(\mathbf{P_X})$. Then, $\mathbf{y} = \mathbf{Va}$ for some vector $\mathbf{a}$ and $\mathbf{y} = \mathbf{P_X b}$ for some vector $\mathbf{b}$, implying (since, according to Theorem 12.3.4, $\mathbf{P_X}$ is idempotent) that

$$\mathbf{y} = \mathbf{P_X P_X b} = \mathbf{P_X y} = \mathbf{P_X Va} \in \mathcal{C}(\mathbf{P_X V}).$$

Thus,

$$\mathcal{C}(\mathbf{V}) \cap \mathcal{C}(\mathbf{P_X}) \subset \mathcal{C}(\mathbf{P_X V}). \tag{S.12}$$

Now, suppose that $\mathbf{X}(\mathbf{X'X})^{-}\mathbf{d}$ is a solution, for every $\mathbf{d} \in \mathcal{C}(\mathbf{X'})$, to the problem of minimizing $\mathbf{a'Va}$ subject to $\mathbf{X'a} = \mathbf{d}$. Then, Condition (c) is satisfied (i.e., $\mathbf{P_X V} = \mathbf{VP_X}$), implying [since, clearly, $\mathcal{C}(\mathbf{P_X V}) \subset \mathcal{C}(\mathbf{P_X})$ and $\mathcal{C}(\mathbf{VP_X}) \subset \mathcal{C}(\mathbf{V})$] that

$$\mathcal{C}(\mathbf{VP_X}) \subset \mathcal{C}(\mathbf{V}) \cap \mathcal{C}(\mathbf{P_X})$$

and also [in light of result (S.12)] that

$$\mathcal{C}(\mathbf{V}) \cap \mathcal{C}(\mathbf{P_X}) \subset \mathcal{C}(\mathbf{VP_X}).$$

Thus, $\mathcal{C}(\mathbf{VP_X}) = \mathcal{C}(\mathbf{V}) \cap \mathcal{C}(\mathbf{P_X})$ [i.e., Condition (e) is satisfied].

Conversely, suppose that $\mathcal{C}(\mathbf{VP_X}) = \mathcal{C}(\mathbf{V}) \cap \mathcal{C}(\mathbf{P_X})$. Then, obviously, $\mathcal{C}(\mathbf{VP_X}) \subset \mathcal{C}(\mathbf{P_X})$ [i.e., Condition (d) is satisfied], implying that $\mathbf{X}(\mathbf{X'X})^{-}\mathbf{d}$ is a solution, for every $\mathbf{d} \in \mathcal{C}(\mathbf{X'})$, to the problem of minimizing $\mathbf{a'Va}$ subject to $\mathbf{X'a} = \mathbf{d}$.

(f) Since [as noted in the proof of the necessity and sufficiency of Condition (d)] $\mathcal{C}(\mathbf{P_X}) = \mathcal{C}(\mathbf{X})$ and $\mathcal{C}(\mathbf{VP_X}) = \mathcal{C}(\mathbf{VX})$, Condition (f) is equivalent to Condition (e).

EXERCISE 12. Let $\mathbf{V}$ represent an $n \times n$ symmetric nonnegative definite matrix, $\mathbf{W}$ an $n \times n$ matrix, $\mathbf{X}$ an $n \times p$ matrix, and $\mathbf{d}$ a $p \times 1$ vector. Further, let $\mathbf{K}$ represent any $n \times q$ matrix such that $\mathcal{C}(\mathbf{K}) = \mathcal{C}(\mathbf{I} - \mathbf{P_{X,W'}})$. Show that if $\text{rank}(\mathbf{X'WX}) = \text{rank}(\mathbf{X})$, then each of the following two conditions is necessary and sufficient for the vector $\mathbf{WX}(\mathbf{X'WX})^{-}\mathbf{d}$ to be a solution, for every $\mathbf{d} \in \mathcal{C}(\mathbf{X'})$, to the problem of minimizing the quadratic form $\mathbf{a'Va}$ (in $\mathbf{a}$) subject to $\mathbf{X'a} = \mathbf{d}$:

(a) $\mathbf{V} = \mathbf{XR_1X'} + (\mathbf{I} - \mathbf{P_{X,W'}})\mathbf{R_2}(\mathbf{I} - \mathbf{P_{X,W'}})'$
for some $p \times p$ matrix $\mathbf{R_1}$ and some $n \times n$ matrix $\mathbf{R_2}$;

(b) $\mathbf{V} = \mathbf{XS_1X'} + \mathbf{KS_2K'}$
for some $p \times p$ matrix $\mathbf{S_1}$ and some $q \times q$ matrix $\mathbf{S_2}$.

And, show that if $\text{rank}(\mathbf{X'WX}) = \text{rank}(\mathbf{X})$ and $\mathbf{W}$ is nonsingular, then another necessary and sufficient condition is:

(c) $\mathbf{V} = t\mathbf{W}^{-1} + \mathbf{XT_1X'} + \mathbf{KT_2K'}$
for some scalar t, some $p \times p$ matrix $\mathbf{T_1}$, and some $q \times q$ matrix $\mathbf{T_2}$.

[*Hint*. To establish the necessity of Condition (a), begin by observing that $\mathbf{V} = \mathbf{CC'}$ for some matrix $\mathbf{C}$ and by expressing $\mathbf{C}$ as $\mathbf{C} = \mathbf{P_{X,W'}C} + (\mathbf{I} - \mathbf{P_{X,W'}})\mathbf{C}$.]

Solution. Assume that $\text{rank}(\mathbf{X}'\mathbf{W}\mathbf{X}) = \text{rank}(\mathbf{X})$. Then, since $\mathbf{X}'\mathbf{W}'\mathbf{X} = (\mathbf{X}'\mathbf{W}\mathbf{X})'$, $\text{rank}(\mathbf{X}'\mathbf{W}'\mathbf{X}) = \text{rank}(\mathbf{X})$. Thus, applying Parts (1) and (3) of Lemma 19.5.5 (with $\mathbf{W}'$ in place of $\mathbf{W}$), we find that $\mathbf{P}_{\mathbf{X},\mathbf{W}'}\mathbf{X} = \mathbf{X}$ and $\mathbf{P}^2_{\mathbf{X},\mathbf{W}'} = \mathbf{P}_{\mathbf{X},\mathbf{W}'}$. And, applying Part (2) of Lemma 19.5.5, we find that $\mathbf{X}'\mathbf{W}'\mathbf{P}_{\mathbf{X},\mathbf{W}'} = \mathbf{X}'\mathbf{W}'$ and hence that

$$\mathbf{P}'_{\mathbf{X},\mathbf{W}'}\mathbf{W}\mathbf{X} = [\mathbf{X}'\mathbf{W}'\mathbf{P}_{\mathbf{X},\mathbf{W}'}]' = (\mathbf{X}'\mathbf{W}')' = \mathbf{W}\mathbf{X}.$$

To establish the necesity and sufficiency of Condition (a), it suffices (in light of Theorem 19.5.1) to show that Condition (a) is equivalent to the condition that $\mathcal{C}(\mathbf{V}\mathbf{W}\mathbf{X}) \subset \mathcal{C}(\mathbf{X})$. If Condition (a) is satisfied, then

$$\begin{aligned}\mathbf{V}\mathbf{W}\mathbf{X} &= \mathbf{X}\mathbf{R}_1\mathbf{X}'\mathbf{W}\mathbf{X} + (\mathbf{I} - \mathbf{P}_{\mathbf{X},\mathbf{W}'})\mathbf{R}_2(\mathbf{W}\mathbf{X} - \mathbf{P}'_{\mathbf{X},\mathbf{W}'}\mathbf{W}\mathbf{X})\\ &= \mathbf{X}\mathbf{R}_1\mathbf{X}'\mathbf{W}\mathbf{X} + (\mathbf{I} - \mathbf{P}_{\mathbf{X},\mathbf{W}'})\mathbf{R}_2(\mathbf{W}\mathbf{X} - \mathbf{W}\mathbf{X}) = \mathbf{X}\mathbf{R}_1\mathbf{X}'\mathbf{W}\mathbf{X},\end{aligned}$$

and consequently $\mathcal{C}(\mathbf{V}\mathbf{W}\mathbf{X}) \subset \mathcal{C}(\mathbf{X})$.

Conversely, suppose that $\mathcal{C}(\mathbf{V}\mathbf{W}\mathbf{X}) \subset \mathcal{C}(\mathbf{X})$. Then, $\mathbf{V}\mathbf{W}\mathbf{X} = \mathbf{X}\mathbf{Q}$ for some matrix $\mathbf{Q}$, so that

$$(\mathbf{I} - \mathbf{P}_{\mathbf{X},\mathbf{W}'})\mathbf{V}\mathbf{W}\mathbf{X} = (\mathbf{I} - \mathbf{P}_{\mathbf{X},\mathbf{W}'})\mathbf{X}\mathbf{Q} = \mathbf{0}. \tag{S.13}$$

Now, observe (in light of Corollary 14.3.8) that there exists a matrix $\mathbf{C}$ such that $\mathbf{V} = \mathbf{C}\mathbf{C}'$. Thus,

$$\begin{aligned}\mathbf{V} &= [\mathbf{P}_{\mathbf{X},\mathbf{W}'}\mathbf{C} + (\mathbf{I} - \mathbf{P}_{\mathbf{X},\mathbf{W}'})\mathbf{C}][\mathbf{P}_{\mathbf{X},\mathbf{W}'}\mathbf{C} + (\mathbf{I} - \mathbf{P}_{\mathbf{X},\mathbf{W}'})\mathbf{C}]'\\ &= \mathbf{P}_{\mathbf{X},\mathbf{W}'}\mathbf{C}\mathbf{C}'\mathbf{P}'_{\mathbf{X},\mathbf{W}'} + \mathbf{P}_{\mathbf{X},\mathbf{W}'}\mathbf{C}\mathbf{C}'(\mathbf{I} - \mathbf{P}'_{\mathbf{X},\mathbf{W}'})\\ &\quad + (\mathbf{I} - \mathbf{P}_{\mathbf{X},\mathbf{W}'})\mathbf{C}\mathbf{C}'\mathbf{P}'_{\mathbf{X},\mathbf{W}'} + (\mathbf{I} - \mathbf{P}_{\mathbf{X},\mathbf{W}'})\mathbf{C}\mathbf{C}'(\mathbf{I} - \mathbf{P}'_{\mathbf{X},\mathbf{W}'}).\end{aligned}$$

Moreover,

$$\begin{aligned}(\mathbf{I} - \mathbf{P}_{\mathbf{X},\mathbf{W}'})\mathbf{V}\mathbf{W}\mathbf{X} &= \mathbf{0}\mathbf{C}\mathbf{C}'\mathbf{W}\mathbf{X} + \mathbf{0}\mathbf{C}\mathbf{C}'\mathbf{0}\\ &\qquad + (\mathbf{I} - \mathbf{P}_{\mathbf{X},\mathbf{W}'})\mathbf{C}\mathbf{C}'\mathbf{W}\mathbf{X} + (\mathbf{I} - \mathbf{P}_{\mathbf{X},\mathbf{W}'})\mathbf{C}\mathbf{C}'\mathbf{0}\\ &= (\mathbf{I} - \mathbf{P}_{\mathbf{X},\mathbf{W}'})\mathbf{C}\mathbf{C}'\mathbf{W}\mathbf{X}.\end{aligned} \tag{S.14}$$

Together, results (S.13) and (S.14) imply that

$$(\mathbf{I} - \mathbf{P}_{\mathbf{X},\mathbf{W}'})\mathbf{C}\mathbf{C}'\mathbf{W}\mathbf{X} = \mathbf{0},$$

so that

$$(\mathbf{I} - \mathbf{P}_{\mathbf{X},\mathbf{W}'})\mathbf{C}\mathbf{C}'\mathbf{P}'_{\mathbf{X},\mathbf{W}'} = (\mathbf{I} - \mathbf{P}_{\mathbf{X},\mathbf{W}'})\mathbf{C}\mathbf{C}'\mathbf{W}\mathbf{X}[(\mathbf{X}'\mathbf{W}'\mathbf{X})^-]'\mathbf{X}' = \mathbf{0}$$

and

$$\mathbf{P}_{\mathbf{X},\mathbf{W}'}\mathbf{C}\mathbf{C}'(\mathbf{I} - \mathbf{P}'_{\mathbf{X},\mathbf{W}'}) = [(\mathbf{I} - \mathbf{P}_{\mathbf{X},\mathbf{W}'})\mathbf{C}\mathbf{C}'\mathbf{P}'_{\mathbf{X},\mathbf{W}'}]' = \mathbf{0}.$$

We conclude that

$$\begin{aligned}\mathbf{V} &= \mathbf{P}_{\mathbf{X},\mathbf{W}'}\mathbf{C}\mathbf{C}'\mathbf{P}'_{\mathbf{X},\mathbf{W}'} + (\mathbf{I} - \mathbf{P}_{\mathbf{X},\mathbf{W}'})\mathbf{C}\mathbf{C}'(\mathbf{I} - \mathbf{P}'_{\mathbf{X},\mathbf{W}'})\\ &= \mathbf{X}\mathbf{R}_1\mathbf{X}' + (\mathbf{I} - \mathbf{P}_{\mathbf{X},\mathbf{W}'})\mathbf{R}_2(\mathbf{I} - \mathbf{P}_{\mathbf{X},\mathbf{W}'})',\end{aligned}$$

where $\mathbf{R}_1 = (\mathbf{X}'\mathbf{W}'\mathbf{X})^-\mathbf{X}'\mathbf{W}'\mathbf{C}\mathbf{C}'\mathbf{W}\mathbf{X}[(\mathbf{X}'\mathbf{W}'\mathbf{X})^-]'$ and $\mathbf{R}_2 = \mathbf{C}\mathbf{C}'$.

Thus, Condition (a) is equivalent to the condition that $\mathcal{C}(\mathbf{VWX}) \subset \mathcal{C}(\mathbf{X})$.

To establish the necessity and sufficiency of Condition (b), it suffices to show that Conditions (a) and (b) are equivalent. According to Lemma 4.2.2, there exist matrices $\mathbf{A}$ and $\mathbf{B}$ such that $\mathbf{I} - \mathbf{P}_{\mathbf{X},\mathbf{W}'} = \mathbf{KA}$ and $\mathbf{K} = (\mathbf{I} - \mathbf{P}_{\mathbf{X},\mathbf{W}'})\mathbf{B}$. Thus, if Condition (a) is satisfied, then

$$\mathbf{V} = \mathbf{X}\mathbf{S}_1\mathbf{X}' + \mathbf{K}\mathbf{S}_2\mathbf{K}',$$

where $\mathbf{S}_1 = \mathbf{R}_1$ and $\mathbf{S}_2 = \mathbf{A}\mathbf{R}_2\mathbf{A}'$. Conversely, if Condition (b) is satisfied, then

$$\mathbf{V} = \mathbf{X}\mathbf{R}_1\mathbf{X}' + (\mathbf{I} - \mathbf{P}_{\mathbf{X},\mathbf{W}'})\mathbf{R}_2(\mathbf{I} - \mathbf{P}_{\mathbf{X},\mathbf{W}'})',$$

where $\mathbf{R}_1 = \mathbf{S}_1$ and $\mathbf{R}_2 = \mathbf{B}\mathbf{S}_2\mathbf{B}'$. Thus, Conditions (a) and (b) are equivalent.

Assume now that $\mathbf{W}$ is nonsingular [and continue to assume that $\text{rank}(\mathbf{X}'\mathbf{W}\mathbf{X}) = \text{rank}(\mathbf{X})$]. And [for purposes of establishing the necessity of Condition (c)] suppose that $\mathbf{W}\mathbf{X}(\mathbf{X}'\mathbf{W}\mathbf{X})^-\mathbf{d}$ is a solution, for every $\mathbf{d} \in \mathcal{C}(\mathbf{X}')$, to the problem of minimizing $\mathbf{a}'\mathbf{V}\mathbf{a}$ subject to $\mathbf{X}'\mathbf{a} = \mathbf{d}$. Then, Condition (b) is satisfied, in which case

$$\mathbf{V} = t\mathbf{W}^{-1} + \mathbf{X}\mathbf{T}_1\mathbf{X}' + \mathbf{K}\mathbf{T}_2\mathbf{K}',$$

where $t = 0$, $\mathbf{T}_1 = \mathbf{S}_1$, and $\mathbf{T}_2 = \mathbf{S}_2$.

Conversely, suppose that Condition (c) is satisfied. Then, recalling that $\mathbf{K} = (\mathbf{I} - \mathbf{P}_{\mathbf{X},\mathbf{W}'})\mathbf{B}$ for some matrix $\mathbf{B}$, we find that

$$\begin{aligned}
\mathbf{VWX} &= t\mathbf{W}^{-1}\mathbf{W}\mathbf{X} + \mathbf{X}\mathbf{T}_1\mathbf{X}'\mathbf{W}\mathbf{X} + \mathbf{K}\mathbf{T}_2\mathbf{K}'\mathbf{W}\mathbf{X} \\
&= t\mathbf{X} + \mathbf{X}\mathbf{T}_1\mathbf{X}'\mathbf{W}\mathbf{X} + \mathbf{K}\mathbf{T}_2\mathbf{B}'(\mathbf{I} - \mathbf{P}'_{\mathbf{X},\mathbf{W}'})\mathbf{W}\mathbf{X} \\
&= t\mathbf{X} + \mathbf{X}\mathbf{T}_1\mathbf{X}'\mathbf{W}\mathbf{X} + \mathbf{K}\mathbf{T}_2\mathbf{B}'\mathbf{0} \\
&= \mathbf{X}(t\mathbf{I} + \mathbf{T}_1\mathbf{X}'\mathbf{W}\mathbf{X}).
\end{aligned}$$

Thus, $\mathcal{C}(\mathbf{VWX}) \subset \mathcal{C}(\mathbf{X})$, and it follows from Theorem 19.5.1 that Condition (c) is sufficient (as well as necessary) for $\mathbf{W}\mathbf{X}(\mathbf{X}'\mathbf{W}\mathbf{X})^-\mathbf{d}$ to be a solution, for every $\mathbf{d} \in \mathcal{C}(\mathbf{X}')$, to the problem of minimizing $\mathbf{a}'\mathbf{V}\mathbf{a}$ subject to $\mathbf{X}'\mathbf{a} = \mathbf{d}$.

20

The Moore-Penrose Inverse

EXERCISE 1. Show that, for any $m \times n$ matrix $\mathbf{B}$ of full column rank and for any $n \times p$ matrix $\mathbf{C}$ of full row rank,

$$(\mathbf{BC})^+ = \mathbf{C}^+\mathbf{B}^+.$$

Solution. As a consequence of result (1.2), we have that

$$(\mathbf{BC})^+ = \mathbf{C}'(\mathbf{CC}')^{-1}(\mathbf{B}'\mathbf{B})^{-1}\mathbf{B}'.$$

And, in light of results (2.1) and (2.2), it follows that

$$(\mathbf{BC})^+ = \mathbf{C}^+\mathbf{B}^+.$$

EXERCISE 2. Show that, for any $m \times n$ matrix $\mathbf{A}$, $\mathbf{A}^+ = \mathbf{A}'$ if and only if $\mathbf{A}'\mathbf{A}$ is idempotent.

Solution. Suppose that $\mathbf{A}'\mathbf{A}$ is idempotent or equivalently that

$$\mathbf{A}'\mathbf{A} = \mathbf{A}'\mathbf{A}\mathbf{A}'\mathbf{A}. \tag{S.1}$$

Then, premultiplying both sides of equality (S.1) by $\mathbf{A}^+(\mathbf{A}^+)'$ and postmultiplying both sides by $\mathbf{A}^+$, we find that

$$\mathbf{A}^+(\mathbf{A}^+)'\mathbf{A}'\mathbf{A}\mathbf{A}^+ = \mathbf{A}^+(\mathbf{A}^+)'\mathbf{A}'\mathbf{A}\mathbf{A}'\mathbf{A}\mathbf{A}^+.$$

Moreover,

$$\mathbf{A}^+(\mathbf{A}^+)'\mathbf{A}'\mathbf{A}\mathbf{A}^+ = \mathbf{A}^+(\mathbf{A}\mathbf{A}^+)'\mathbf{A}\mathbf{A}^+ = \mathbf{A}^+\mathbf{A}\mathbf{A}^+\mathbf{A}\mathbf{A}^+ = \mathbf{A}^+\mathbf{A}\mathbf{A}^+ = \mathbf{A}^+,$$

and

$$\begin{aligned}\mathbf{A}^+(\mathbf{A}^+)'\mathbf{A}'\mathbf{A}\mathbf{A}'\mathbf{A}\mathbf{A}^+ &= \mathbf{A}^+(\mathbf{A}\mathbf{A}^+)'\mathbf{A}\mathbf{A}'(\mathbf{A}\mathbf{A}^+)' \\ &= \mathbf{A}^+\mathbf{A}\mathbf{A}^+\mathbf{A}(\mathbf{A}\mathbf{A}^+\mathbf{A})' \\ &= \mathbf{A}^+\mathbf{A}\mathbf{A}' = (\mathbf{A}^+\mathbf{A})'\mathbf{A}' = (\mathbf{A}\mathbf{A}^+\mathbf{A})' = \mathbf{A}'.\end{aligned}$$

Thus, $\mathbf{A}^+ = \mathbf{A}'$.

Conversely, suppose that $\mathbf{A}^+ = \mathbf{A}'$. Then, clearly, $\mathbf{A}'\mathbf{A} = \mathbf{A}^+\mathbf{A}$, implying (in light of Lemma 10.2.5) that $\mathbf{A}'\mathbf{A}$ is idempotent.

EXERCISE 3. Let $\mathbf{T}$ represent an $m \times p$ matrix, $\mathbf{U}$ an $m \times q$ matrix, $\mathbf{V}$ an $n \times p$ matrix, and $\mathbf{W}$ an $n \times q$ matrix, and define $\mathbf{Q} = \mathbf{W} - \mathbf{V}\mathbf{T}^-\mathbf{U}$. If $\mathcal{C}(\mathbf{U}) \subset \mathcal{C}(\mathbf{T})$ and $\mathcal{R}(\mathbf{V}) \subset \mathcal{R}(\mathbf{T})$, then

$$\begin{pmatrix} \mathbf{T}^- + \mathbf{T}^-\mathbf{U}\mathbf{Q}^-\mathbf{V}\mathbf{T}^- & -\mathbf{T}^-\mathbf{U}\mathbf{Q}^- \\ -\mathbf{Q}^-\mathbf{V}\mathbf{T}^- & \mathbf{Q}^- \end{pmatrix} \tag{$*$}$$

is a generalized inverse of the partitioned matrix $\begin{pmatrix} \mathbf{T} & \mathbf{U} \\ \mathbf{V} & \mathbf{W} \end{pmatrix}$, and

$$\begin{pmatrix} \mathbf{Q}^- & -\mathbf{Q}^-\mathbf{V}\mathbf{T}^- \\ -\mathbf{T}^-\mathbf{U}\mathbf{Q}^- & \mathbf{T}^- + \mathbf{T}^-\mathbf{U}\mathbf{Q}^-\mathbf{V}\mathbf{T}^- \end{pmatrix} \tag{$**$}$$

a generalized inverse of $\begin{pmatrix} \mathbf{W} & \mathbf{V} \\ \mathbf{U} & \mathbf{T} \end{pmatrix}$. Show that if the generalized inverses $\mathbf{T}^-$ and $\mathbf{Q}^-$ (of $\mathbf{T}$ and $\mathbf{Q}$, respectively) are both reflexive [and if $\mathcal{C}(\mathbf{U}) \subset \mathcal{C}(\mathbf{T})$ and $\mathcal{R}(\mathbf{V}) \subset \mathcal{R}(\mathbf{T})$], then generalized inverses ($*$) and ($**$) are also reflexive.

Solution. Suppose that $\mathcal{C}(\mathbf{U}) \subset \mathcal{C}(\mathbf{T})$ and $\mathcal{R}(\mathbf{V}) \subset \mathcal{R}(\mathbf{T})$. [That these conditions are sufficient to insure that partitioned matrices ($*$) and ($**$) are generalized inverses of $\begin{pmatrix} \mathbf{T} & \mathbf{U} \\ \mathbf{V} & \mathbf{W} \end{pmatrix}$ and $\begin{pmatrix} \mathbf{W} & \mathbf{V} \\ \mathbf{U} & \mathbf{T} \end{pmatrix}$, respectively, is the content of Theorem 9.6.1.] Then, $\mathbf{T}\mathbf{T}^-\mathbf{U} = \mathbf{U}$ and $\mathbf{V}\mathbf{T}^-\mathbf{T} = \mathbf{V}$ (as is evident from Lemma 9.3.5). Further,

$$\begin{aligned}&\begin{pmatrix} \mathbf{T}^- + \mathbf{T}^-\mathbf{U}\mathbf{Q}^-\mathbf{V}\mathbf{T}^- & -\mathbf{T}^-\mathbf{U}\mathbf{Q}^- \\ -\mathbf{Q}^-\mathbf{V}\mathbf{T}^- & \mathbf{Q}^- \end{pmatrix}\begin{pmatrix} \mathbf{T} & \mathbf{U} \\ \mathbf{V} & \mathbf{W} \end{pmatrix} \\ &\qquad\qquad \times \begin{pmatrix} \mathbf{T}^- + \mathbf{T}^-\mathbf{U}\mathbf{Q}^-\mathbf{V}\mathbf{T}^- & -\mathbf{T}^-\mathbf{U}\mathbf{Q}^- \\ -\mathbf{Q}^-\mathbf{V}\mathbf{T}^- & \mathbf{Q}^- \end{pmatrix} \\ &= \begin{pmatrix} \mathbf{T}^- + \mathbf{T}^-\mathbf{U}\mathbf{Q}^-\mathbf{V}\mathbf{T}^- & -\mathbf{T}^-\mathbf{U}\mathbf{Q}^- \\ -\mathbf{Q}^-\mathbf{V}\mathbf{T}^- & \mathbf{Q}^- \end{pmatrix}\begin{pmatrix} \mathbf{T}\mathbf{T}^- & \mathbf{0} \\ (\mathbf{I} - \mathbf{Q}\mathbf{Q}^-)\mathbf{V}\mathbf{T}^- & \mathbf{Q}\mathbf{Q}^- \end{pmatrix} \\ &= \begin{pmatrix} \begin{matrix}\mathbf{T}^-\mathbf{T}\mathbf{T}^- + \mathbf{T}\mathbf{U}\mathbf{Q}^-\mathbf{V}\mathbf{T}^-\mathbf{T}\mathbf{T}^- \\ -\mathbf{T}^-\mathbf{U}\mathbf{Q}^-(\mathbf{I} - \mathbf{Q}\mathbf{Q}^-)\mathbf{V}\mathbf{T}^-\end{matrix} & -\mathbf{T}^-\mathbf{U}\mathbf{Q}^-\mathbf{Q}\mathbf{Q}^- \\ -\mathbf{Q}^-\mathbf{V}\mathbf{T}^-\mathbf{T}\mathbf{T}^- + \mathbf{Q}^-(\mathbf{I} - \mathbf{Q}\mathbf{Q}^-)\mathbf{V}\mathbf{T}^- & \mathbf{Q}^-\mathbf{Q}\mathbf{Q}^- \end{pmatrix}.\end{aligned} \tag{S.2}$$

Now, if the generalized inverses $\mathbf{T}^-$ and $\mathbf{Q}^-$ are both reflexive (i.e., if $\mathbf{T}^-\mathbf{T}\mathbf{T}^- = \mathbf{T}^-$ and $\mathbf{Q}^-\mathbf{Q}\mathbf{Q}^- = \mathbf{Q}^-$), then partitioned matrix (S.2) simplifies to partitioned matrix (∗). We conclude that if the generalized inverses $\mathbf{T}^-$ and $\mathbf{Q}^-$ are both reflexive [and if if $\mathcal{C}(\mathbf{U}) \subset \mathcal{C}(\mathbf{T})$ and $\mathcal{R}(\mathbf{V}) \subset \mathcal{R}(\mathbf{T})$], then the generalized inverse (∗) of $\begin{pmatrix} \mathbf{T} & \mathbf{U} \\ \mathbf{V} & \mathbf{W} \end{pmatrix}$ [or equivalently the generalized inverse given by expression (9.6.2)] is reflexive. And, it can be shown in similar fashion that if the generalized inverses $\mathbf{T}^-$ and $\mathbf{Q}^-$ are both reflexive [and if if $\mathcal{C}(\mathbf{U}) \subset \mathcal{C}(\mathbf{T})$ and $\mathcal{R}(\mathbf{V}) \subset \mathcal{R}(\mathbf{T})$], then the generalized inverse (∗∗) of $\begin{pmatrix} \mathbf{W} & \mathbf{V} \\ \mathbf{U} & \mathbf{T} \end{pmatrix}$ [or equivalently the generalized inverse given by expression (9.6.3)] is reflexive.

EXERCISE 4. Determine which of Penrose Conditions (1) – (4) [also known as Moore-Penrose Conditions (1) – (4)] are necessarily satisfied by a left inverse of an $m \times n$ matrix $\mathbf{A}$ (when a left inverse exists). Which of the Penrose conditions are necessarily satisfied by a right inverse of an $m \times n$ matrix $\mathbf{A}$ (when a right inverse exists)?

Solution. Suppose that $\mathbf{A}$ has a left inverse $\mathbf{L}$. Then, by definition, $\mathbf{LA} = \mathbf{I}_n$. And, as previously indicated (in Section 9.2d), $\mathbf{ALA} = \mathbf{AI} = \mathbf{A}$. Thus, $\mathbf{L}$ necessarily satisfies Penrose Condition (1). Further, $\mathbf{LAL} = \mathbf{IL} = \mathbf{L}$ and $(\mathbf{LA})' = \mathbf{I}' = \mathbf{I} = \mathbf{LA}$, so that $\mathbf{L}$ also necessarily satisfies Penrose Conditions (2) and (4).

However, there exist matrices that have left inverses that do not satisfy Penrose Condition (3). Suppose, for example, that $\mathbf{A} = \begin{pmatrix} \mathbf{I}_n \\ \mathbf{0} \end{pmatrix}$ (where $m > n$). And, take $\mathbf{L} = (\mathbf{I}_n, \mathbf{K})$, where $\mathbf{K}$ is an arbitrary $n \times (m - n)$ matrix. Then, $\mathbf{LA} = \mathbf{I}_n$, and $\mathbf{AL} = \begin{pmatrix} \mathbf{I}_n & \mathbf{K} \\ \mathbf{0} & \mathbf{0} \end{pmatrix}$, so that $\mathbf{L}$ is a left inverse of $\mathbf{A}$ that (unless $\mathbf{K} = \mathbf{0}$) does not satisfy Penrose Condition (3).

Similarly, if $\mathbf{A}$ has a right inverse $\mathbf{R}$, then $\mathbf{R}$ necessarily satisfies Penrose Conditions (1), (2), and (3). However, there exist matrices that have right inverses that do not satisfy Penrose Condition (4).

EXERCISE 5. Let $\mathbf{A}$ represent an $m \times n$ matrix and $\mathbf{G}$ an $n \times m$ matrix.

(a) Show that $\mathbf{G}$ is the Moore-Penrose inverse of $\mathbf{A}$ if and only if $\mathbf{G}$ is a minimum norm generalized inverse of $\mathbf{A}$ and $\mathbf{A}$ is a minimum norm generalized inverse of $\mathbf{G}$.

(b) Show that $\mathbf{G}$ is the Moore-Penrose inverse of $\mathbf{A}$ if and only if $\mathbf{GAA}' = \mathbf{A}'$ and $\mathbf{AGG}' = \mathbf{G}'$.

(c) Show that $\mathbf{G}$ is the Moore-Penrose inverse of $\mathbf{A}$ if and only if $\mathbf{GA} = \mathbf{P}_{\mathbf{A}'}$ and $\mathbf{AG} = \mathbf{P}_{\mathbf{G}'}$.

Solution. (a) By definition, $\mathbf{G}$ is a minimum norm generalized inverse of $\mathbf{A}$ if and only if $\mathbf{AGA} = \mathbf{A}$ and $(\mathbf{GA})' = \mathbf{GA}$ [which are Penrose Conditions (1) and (4)], and $\mathbf{A}$ is a minimum norm generalized inverse of $\mathbf{G}$ if and only if $\mathbf{GAG} = \mathbf{G}$ and $(\mathbf{AG})' = \mathbf{AG}$ [which, in the relevant context, are Penrose Conditions (2) and (3)]. Thus, $\mathbf{G}$ is the Moore-Penrose inverse of $\mathbf{A}$ if and only if $\mathbf{G}$ is a minimum norm generalized inverse of $\mathbf{A}$ and $\mathbf{A}$ is a minimum norm generalized inverse of $\mathbf{G}$.

(b) Part (b) follows from Part (a) upon observing (in light of Theorem 20.3.7) that $\mathbf{G}$ is a minimum norm generalized inverse of $\mathbf{A}$ if and only if $\mathbf{GAA}' = \mathbf{A}'$ and that $\mathbf{A}$ is a minimum norm generalized inverse of $\mathbf{G}$ if and only if $\mathbf{AGG}' = \mathbf{G}'$.

(c) Part (c) follows from Part (a) upon observing (in light of Corollary 20.3.8) that $\mathbf{G}$ is a minimum norm generalized inverse of $\mathbf{A}$ if and only if $\mathbf{GA} = \mathbf{P}_{\mathbf{A}'}$ and that $\mathbf{A}$ is a minimum norm generalized inverse of $\mathbf{G}$ if and only if $\mathbf{AG} = \mathbf{P}_{\mathbf{G}'}$.

EXERCISE 6. (a) Show that, for any $m \times n$ matrices $\mathbf{A}$ and $\mathbf{B}$ such that $\mathbf{A}'\mathbf{B} = \mathbf{0}$ and $\mathbf{BA}' = \mathbf{0}$, $(\mathbf{A}+\mathbf{B})^+ = \mathbf{A}^+ + \mathbf{B}^+$.

(b) Let $\mathbf{A}_1, \mathbf{A}_2, \ldots, \mathbf{A}_k$ represent $m \times n$ matrices such that, for $j > i = 1, \ldots, k-1$, $\mathbf{A}_i'\mathbf{A}_j = \mathbf{0}$ and $\mathbf{A}_j\mathbf{A}_i' = \mathbf{0}$. Generalize the result of Part (a) by showing that $(\mathbf{A}_1 + \mathbf{A}_2 + \cdots + \mathbf{A}_k)^+ = \mathbf{A}_1^+ + \mathbf{A}_2^+ + \cdots + \mathbf{A}_k^+$.

Solution. (a) Let $\mathbf{X}$ represent any $n \times m$ matrix such that $(\mathbf{A}+\mathbf{B})'(\mathbf{A}+\mathbf{B})\mathbf{X} = (\mathbf{A}+\mathbf{B})'$ and $\mathbf{Y}$ any $m \times n$ matrix such that $(\mathbf{A}+\mathbf{B})(\mathbf{A}+\mathbf{B})'\mathbf{Y} = \mathbf{A}+\mathbf{B}$. Then, since $\mathbf{B}'\mathbf{A} = (\mathbf{A}'\mathbf{B})' = \mathbf{0}$ and $\mathbf{AB}' = (\mathbf{BA}')' = \mathbf{0}$, we have that

$$(\mathbf{A}'\mathbf{A} + \mathbf{B}'\mathbf{B})\mathbf{X} = \mathbf{A}' + \mathbf{B}' \qquad \text{and} \qquad (\mathbf{AA}' + \mathbf{BB}')\mathbf{Y} = \mathbf{A} + \mathbf{B}.$$

Moreover, as a consequence of Corollary 12.1.2, we have that $\mathcal{C}(\mathbf{A}) \perp \mathcal{C}(\mathbf{B})$ and [since $(\mathbf{A}')'\mathbf{B}' = (\mathbf{BA}')' = \mathbf{0}$] that $\mathcal{C}(\mathbf{A}') \perp \mathcal{C}(\mathbf{B}')$, implying (in light of Lemma 17.1.9) that

$$\mathcal{C}(\mathbf{A}) \cap \mathcal{C}(\mathbf{B}) = \{\mathbf{0}\} \qquad \text{and} \qquad \mathcal{C}(\mathbf{A}') \cap \mathcal{C}(\mathbf{B}') = \{\mathbf{0}\}.$$

Thus, upon observing that $\mathcal{C}(\mathbf{A}') = \mathcal{C}(\mathbf{A}'\mathbf{A})$, $\mathcal{C}(\mathbf{B}') = \mathcal{C}(\mathbf{B}'\mathbf{B})$, $\mathcal{C}(\mathbf{A}) = \mathcal{C}(\mathbf{AA}')$, and $\mathcal{C}(\mathbf{B}) = \mathcal{C}(\mathbf{BB}')$, it follows from Theorem 18.2.7 that

$$\mathbf{A}'\mathbf{AX} = \mathbf{A}', \quad \mathbf{B}'\mathbf{BX} = \mathbf{B}', \quad \mathbf{AA}'\mathbf{Y} = \mathbf{A}, \quad \text{and} \quad \mathbf{BB}'\mathbf{Y} = \mathbf{B}.$$

Now, making use of Theorem 20.4.4, we find that

$$(\mathbf{A}+\mathbf{B})^+ = \mathbf{Y}'(\mathbf{A}+\mathbf{B})\mathbf{X} = \mathbf{Y}'\mathbf{AX} + \mathbf{Y}'\mathbf{BX} = \mathbf{A}^+ + \mathbf{B}^+.$$

(b) The proof is by mathematical induction. The result of Part (b) is valid for $k = 2$, as is evident from Part (a).

Suppose now that the result of Part (b) is valid for $k = k^* - 1$. And, let $\mathbf{A}_1, \ldots, \mathbf{A}_{k^*-1}, \mathbf{A}_{k^*}$ represent $m \times n$ matrices such that, for $j > i = 1, \ldots, k^*-1$, $\mathbf{A}_i'\mathbf{A}_j = \mathbf{0}$ and $\mathbf{A}_j\mathbf{A}_i' = \mathbf{0}$. Then, observing that

$$(\mathbf{A}_1 + \cdots + \mathbf{A}_{k^*-1})'\mathbf{A}_{k^*} = \mathbf{A}_1'\mathbf{A}_{k^*} + \cdots + \mathbf{A}_{k^*-1}'\mathbf{A}_{k^*} = \mathbf{0}$$

and that

$$\mathbf{A}_{k^*}(\mathbf{A}_1 + \cdots + \mathbf{A}_{k^*-1})' = \mathbf{A}_{k^*}\mathbf{A}_1' + \cdots + \mathbf{A}_{k^*}\mathbf{A}_{k^*-1}' = \mathbf{0}$$

and using the result of Part (a), we find that

$$\begin{aligned}(\mathbf{A}_1 + \cdots + \mathbf{A}_{k^*-1} + \mathbf{A}_{k^*})^+ &= [(\mathbf{A}_1 + \cdots + \mathbf{A}_{k^*-1}) + \mathbf{A}_{k^*}]^+ \\ &= (\mathbf{A}_1 + \cdots + \mathbf{A}_{k^*-1})^+ + \mathbf{A}_{k^*}^+ \\ &= \mathbf{A}_1^+ + \ldots + \mathbf{A}_{k^*-1}^+ + \mathbf{A}_{k^*}^+ ,\end{aligned}$$

which establishes the validity of the result of Part (b) for $k = k^*$ and completes the induction argument.

EXERCISE 7. Show that, for any $m \times n$ matrix $\mathbf{A}$, $(\mathbf{A}^+\mathbf{A})^+ = \mathbf{A}^+\mathbf{A}$, and $(\mathbf{A}\mathbf{A}^+)^+ = \mathbf{A}\mathbf{A}^+$.

Solution. According to Corollary 20.5.2, $\mathbf{A}^+\mathbf{A}$ and $\mathbf{A}\mathbf{A}^+$ are symmetric and idempotent. Thus, it follows from Lemma 20.2.1 that $(\mathbf{A}^+\mathbf{A})^+ = \mathbf{A}^+\mathbf{A}$ and $(\mathbf{A}\mathbf{A}^+)^+ = \mathbf{A}\mathbf{A}^+$.

EXERCISE 8. Show that, for any $n \times n$ symmetric matrix $\mathbf{A}$, $\mathbf{A}\mathbf{A}^+ = \mathbf{A}^+\mathbf{A}$.

Solution. That $\mathbf{A}\mathbf{A}^+ = \mathbf{A}^+\mathbf{A}$ is an immediate consequence of Part (2) of Theorem 20.5.1.

Or, alternatively, this equality can be verified by making use of Part (2) of Theorem 20.5.3 (and of the very definition of the Moore-Penrose inverse). We find that

$$\mathbf{A}\mathbf{A}^+ = (\mathbf{A}\mathbf{A}^+)' = (\mathbf{A}^+)'\mathbf{A}' = (\mathbf{A}^+)'\mathbf{A} = \mathbf{A}^+\mathbf{A}.$$

EXERCISE 9. Let $\mathbf{V}$ represent an $n \times n$ symmetric nonnegative definite matrix, $\mathbf{X}$ an $n \times p$ matrix, and $\mathbf{d}$ a $p \times 1$ vector. Using the results of Exercises 8 and 19.11 (or otherwise), show that, for the vector $\mathbf{X}(\mathbf{X}'\mathbf{X})^-\mathbf{d}$ to be a solution, for every $\mathbf{d} \in \mathcal{C}(\mathbf{X}')$, to the problem of minimizing the quadratic form $\mathbf{a}'\mathbf{V}\mathbf{a}$ (in $\mathbf{a}$) subject to $\mathbf{X}'\mathbf{a} = \mathbf{d}$, it is necessary and sufficient that $\mathcal{C}(\mathbf{V}^+\mathbf{X}) \subset \mathcal{C}(\mathbf{X})$.

Solution. In light of the results of Exercise 19.11, it suffices to show that $\mathcal{C}(\mathbf{V}\mathbf{X}) \subset \mathcal{C}(\mathbf{X}) \Leftrightarrow \mathcal{C}(\mathbf{V}^+\mathbf{X}) \subset \mathcal{C}(\mathbf{X})$.

Suppose that $\mathcal{C}(\mathbf{V}\mathbf{X}) \subset \mathcal{C}(\mathbf{X})$. Then, $\mathbf{V}\mathbf{X} = \mathbf{X}\mathbf{Q}$ for some matrix $\mathbf{Q}$. And, using the result of Exercise 8, we find that

$$\mathbf{V}\mathbf{X} = \mathbf{V}\mathbf{V}^+\mathbf{V}\mathbf{X} = \mathbf{V}\mathbf{V}^+\mathbf{X}\mathbf{Q} = \mathbf{V}^+\mathbf{V}\mathbf{X}\mathbf{Q} = \mathbf{V}^+\mathbf{X}\mathbf{Q}^2$$

and hence that

$$\mathcal{C}(\mathbf{V}\mathbf{X}) \subset \mathcal{C}(\mathbf{V}^+\mathbf{X}). \tag{S.3}$$

Moreover, since (according to Theorem 20.5.3) $\mathbf{V}^+$ is symmetric and nonnegative definite, we have (in light of Lemma 14.11.2 and the result of Exercise 8) that

$$\begin{aligned}\operatorname{rank}(\mathbf{V}\mathbf{X}) \geq \operatorname{rank}(\mathbf{V}^+\mathbf{V}\mathbf{X}) = \operatorname{rank}(\mathbf{V}\mathbf{V}^+\mathbf{X}) &\geq \operatorname{rank}(\mathbf{X}'\mathbf{V}^+\mathbf{V}\mathbf{V}^+\mathbf{X}) \\ &= \operatorname{rank}(\mathbf{X}'\mathbf{V}^+\mathbf{X}) = \operatorname{rank}(\mathbf{V}^+\mathbf{X}),\end{aligned}$$

implying [since, in light of result (S.3), $\text{rank}(\mathbf{VX}) \leq \text{rank}(\mathbf{V}^+\mathbf{X})$] that $\text{rank}(\mathbf{VX}) = \text{rank}(\mathbf{V}^+\mathbf{X})$. Thus, it follows from Theorem 4.4.6 that $\mathcal{C}(\mathbf{VX}) = \mathcal{C}(\mathbf{V}^+\mathbf{X})$. We conclude that $\mathcal{C}(\mathbf{V}^+\mathbf{X}) \subset \mathcal{C}(\mathbf{X})$.

Conversely, suppose that $\mathcal{C}(\mathbf{V}^+\mathbf{X}) \subset \mathcal{C}(\mathbf{X})$. Then, $\mathbf{V}^+\mathbf{X} = \mathbf{XR}$ for some matrix $\mathbf{R}$. And, using the result of Exercise 8, we find that

$$\mathbf{V}^+\mathbf{X} = \mathbf{V}^+\mathbf{V}\mathbf{V}^+\mathbf{X} = \mathbf{V}^+\mathbf{VXR} = \mathbf{V}\mathbf{V}^+\mathbf{XR} = \mathbf{VXR}^2$$

and hence that

$$\mathcal{C}(\mathbf{V}^+\mathbf{X}) \subset \mathcal{C}(\mathbf{VX}). \tag{S.4}$$

Moreover, in light of Lemma 14.11.2 and the result of Exercise 8, we have that

$$\begin{aligned} \text{rank}(\mathbf{V}^+\mathbf{X}) \geq \text{rank}(\mathbf{V}\mathbf{V}^+\mathbf{X}) = \text{rank}(\mathbf{V}^+\mathbf{VX}) &\geq \text{rank}(\mathbf{X}'\mathbf{V}\mathbf{V}^+\mathbf{VX}) \\ &= \text{rank}(\mathbf{X}'\mathbf{VX}) = \text{rank}(\mathbf{VX}), \end{aligned}$$

implying [since, in light of result (S.4), $\text{rank}(\mathbf{V}^+\mathbf{X}) \leq \text{rank}(\mathbf{VX})$] that $\text{rank}(\mathbf{V}^+\mathbf{X}) = \text{rank}(\mathbf{VX})$. Thus, it follows from Theorem 4.4.6 that $\mathcal{C}(\mathbf{V}^+\mathbf{X}) = \mathcal{C}(\mathbf{VX})$. We conclude that $\mathcal{C}(\mathbf{VX}) \subset \mathcal{C}(\mathbf{X})$.

EXERCISE 10. Let $\mathbf{A}$ represent an $n \times n$ matrix. Show that if $\mathbf{A}$ is symmetric and positive semidefinite, then $\mathbf{A}^+$ is symmetric and positive semidefinite and that if $\mathbf{A}$ is symmetric and positive definite, then $\mathbf{A}^+$ is symmetric and positive definite. Do so by taking advantage of the result that if $\mathbf{A}$ is symmetric and nonnegative definite (and nonnull), then $\mathbf{A}^+ = \mathbf{T}^+(\mathbf{T}^+)'$ for any matrix $\mathbf{T}$ of full row rank (and with n columns) such that $\mathbf{A} = \mathbf{T}'\mathbf{T}$.

Solution. Suppose that $\mathbf{A}$ is symmetric and nonnegative definite. Further, assume that $\mathbf{A}$ is nonnull — if $\mathbf{A} = \mathbf{0}$, then $\mathbf{A}$ is positive semidefinite, and $\mathbf{A}^+ = \mathbf{0}$, so that $\mathbf{A}^+$ is also positive semidefinite (and symmetric). Then, it follows from the result cited in the exercise [which is taken from Theorem 20.4.5] that

$$\mathbf{A}^+ = \mathbf{T}^+(\mathbf{T}^+)'$$

for any matrix $\mathbf{T}$ of full row rank (and with n columns) such that $\mathbf{A} = \mathbf{T}'\mathbf{T}$.

Thus, $\mathbf{A}^+$ is symmetric and (in light of Corollary 14.2.14) nonnegative definite. And, since $(\mathbf{T}^+)'$ has n columns and since [in light of Part (1) of Theorem 20.5.1] $\text{rank}\,(\mathbf{T}^+)' = \text{rank}\,\mathbf{T}^+ = \text{rank}\,\mathbf{T}$, it follows from Corollary 14.2.14 that $\mathbf{A}^+$ is positive semidefinite if $\text{rank}\,(\mathbf{T}) < n$ or equivalently if $\mathbf{A}$ is positive semidefinite and that $\mathbf{A}^+$ is positive definite if $\text{rank}(\mathbf{T}) = n$ or equivalently if $\mathbf{A}$ is positive definite.

EXERCISE 11. Let $\mathbf{C}$ represent an $m \times n$ matrix. Show that, for any $m \times m$ idempotent matrix $\mathbf{A}$, $(\mathbf{AC})^+\mathbf{A}' = (\mathbf{AC})^+$ and that, for any $n \times n$ idempotent matrix $\mathbf{B}$, $\mathbf{B}'(\mathbf{CB})^+ = (\mathbf{CB})^+$.

Solution. According to Corollary 20.5.5,

$$(\mathbf{AC})^+ = [(\mathbf{AC})'\mathbf{AC}]^+(\mathbf{AC})' = [(\mathbf{AC})'\mathbf{AC}]^+\mathbf{C}'\mathbf{A}',$$

and

$$(\mathbf{CB})^+ = (\mathbf{CB})'[\mathbf{CB}(\mathbf{CB})']^+ = \mathbf{B}'\mathbf{C}'[\mathbf{CB}(\mathbf{CB})']^+.$$

Thus,

$$\begin{aligned}(\mathbf{AC})^+\mathbf{A}' = [(\mathbf{AC})'\mathbf{AC}]^+\mathbf{C}'\mathbf{A}'\mathbf{A}' &= [(\mathbf{AC})'\mathbf{AC}]^+\mathbf{C}'(\mathbf{AA})' \\ &= [(\mathbf{AC})'\mathbf{AC}]^+\mathbf{C}'\mathbf{A}' = (\mathbf{AC})^+,\end{aligned}$$

and

$$\begin{aligned}\mathbf{B}'(\mathbf{CB})^+ = \mathbf{B}'\mathbf{B}'\mathbf{C}'[\mathbf{CB}(\mathbf{CB})']^+ &= (\mathbf{BB})'\mathbf{C}'[\mathbf{CB}(\mathbf{CB})']^+ \\ &= \mathbf{B}'\mathbf{C}'[\mathbf{CB}(\mathbf{CB})']^+ = (\mathbf{CB})^+.\end{aligned}$$

EXERCISE 12. Let $\mathbf{a}$ represent an $n \times 1$ vector of variables, and impose on $\mathbf{a}$ the constraint $\mathbf{X}'\mathbf{a} = \mathbf{d}$, where $\mathbf{X}$ is an $n \times p$ matrix and $\mathbf{d}$ a $p \times 1$ vector such that $\mathbf{d} \in \mathcal{C}(\mathbf{X}')$. And, define $f(\mathbf{a}) = \mathbf{a}'\mathbf{V}\mathbf{a} - 2\mathbf{b}'\mathbf{a}$, where $\mathbf{V}$ is an $n \times n$ symmetric nonnegative definite matrix and $\mathbf{b}$ is an $n \times 1$ vector such that $\mathbf{b} \in \mathcal{C}(\mathbf{V}, \mathbf{X})$. Further, let $\mathbf{R}$ represent any matrix such that $\mathbf{V} = \mathbf{R}'\mathbf{R}$, let $\mathbf{a}_0$ represent any $n \times 1$ vector such that $\mathbf{X}'\mathbf{a}_0 = \mathbf{d}$, and take $\mathbf{s}$ to be any $n \times 1$ vector such that $\mathbf{b} = \mathbf{V}\mathbf{s} + \mathbf{X}\mathbf{t}$ for some $p \times 1$ vector $\mathbf{t}$. Show that $f(\mathbf{a})$ attains its minimum value (under the constraint $\mathbf{X}'\mathbf{a} = \mathbf{d}$) at a point $\mathbf{a}_*$ if and only if

$$\mathbf{a}_* = \mathbf{a}_0 + [\mathbf{R}(\mathbf{I} - \mathbf{P_X})]^+\mathbf{R}(\mathbf{s} - \mathbf{a}_0) + \{\mathbf{I} - [\mathbf{R}(\mathbf{I} - \mathbf{P_X})]^+\mathbf{R}\}(\mathbf{I} - \mathbf{P_X})\mathbf{w}$$

for some $n \times 1$ vector $\mathbf{w}$. Do so by, for instance, using the results of Exercise 11 in combination with the result that, for any $n \times k$ matrix $\mathbf{Z}$ whose columns span $\mathcal{N}(\mathbf{X}')$, $f(\mathbf{a})$ attains its minimum value (subject to the constraint $\mathbf{X}'\mathbf{a} = \mathbf{d}$) at a point $\mathbf{a}_*$ if and only if

$$\mathbf{a}_* = \mathbf{a}_0 + \mathbf{Z}(\mathbf{Z}'\mathbf{V}\mathbf{Z})^-\mathbf{Z}'(\mathbf{b} - \mathbf{V}\mathbf{a}_0) + \mathbf{Z}[\mathbf{I} - (\mathbf{Z}'\mathbf{V}\mathbf{Z})^-\mathbf{Z}'\mathbf{V}\mathbf{Z}]\mathbf{w}$$

for some $k \times 1$ vector $\mathbf{w}$.

Solution. Take $\mathbf{Z} = \mathbf{I} - \mathbf{P_X}$. Then, according to Lemma 12.5.2, $\mathcal{C}(\mathbf{Z}) = \mathcal{N}(\mathbf{X}')$. And, it follows from the cited result on constrained minimization (which is taken from Section 19.6) that $f(\mathbf{a})$ attains its minimum value (under the constraint $\mathbf{X}'\mathbf{a} = \mathbf{d}$) at a point $\mathbf{a}_*$ if and only if

$$\mathbf{a}_* = \mathbf{a}_0 + \mathbf{Z}(\mathbf{Z}'\mathbf{V}\mathbf{Z})^+\mathbf{Z}'(\mathbf{b} - \mathbf{V}\mathbf{a}_0) + [\mathbf{I} - \mathbf{Z}(\mathbf{Z}'\mathbf{V}\mathbf{Z})^+\mathbf{Z}'\mathbf{V}]\mathbf{Z}\mathbf{w}$$

for some $n \times 1$ vector $\mathbf{w}$.

Moreover, according to Part (9) of Theorem 12.3.4, $\mathbf{Z}$ is symmetric and idempotent. Thus, making use of Corollary 20.5.5 and of the results of Exercise 11, we find that

$$\mathbf{Z}(\mathbf{Z}'\mathbf{V}\mathbf{Z})^+\mathbf{Z}'\mathbf{V} = \mathbf{Z}[(\mathbf{RZ})'\mathbf{RZ}]^+(\mathbf{RZ})'\mathbf{R} = \mathbf{Z}(\mathbf{RZ})^+\mathbf{R} = (\mathbf{RZ})^+\mathbf{R}.$$

And, since [in light of Part (1) of Theorem 12.3.4] $\mathbf{Z}'\mathbf{X} = \mathbf{Z}\mathbf{X} = \mathbf{0}$,

$$\begin{aligned}\mathbf{Z}(\mathbf{Z}'\mathbf{V}\mathbf{Z})^{+}\mathbf{Z}'(\mathbf{b} - \mathbf{V}\mathbf{a}_0) &= \mathbf{Z}(\mathbf{Z}'\mathbf{V}\mathbf{Z})^{+}\mathbf{Z}'(\mathbf{V}\mathbf{s} + \mathbf{X}\mathbf{t} - \mathbf{V}\mathbf{a}_0) \\ &= \mathbf{Z}(\mathbf{Z}'\mathbf{V}\mathbf{Z})^{+}\mathbf{Z}'\mathbf{V}(\mathbf{s} - \mathbf{a}_0) = (\mathbf{R}\mathbf{Z})^{+}\mathbf{R}(\mathbf{s} - \mathbf{a}_0).\end{aligned}$$

We conclude that $f(\mathbf{a})$ attains its minimum value (under the constraint $\mathbf{X}'\mathbf{a} = \mathbf{d}$) at a point $\mathbf{a}_*$ if and only if

$$\mathbf{a}_* = \mathbf{a}_0 + (\mathbf{R}\mathbf{Z})^{+}\mathbf{R}(\mathbf{s} - \mathbf{a}_0) + [\mathbf{I} - (\mathbf{R}\mathbf{Z})^{+}\mathbf{R}]\mathbf{Z}\mathbf{w}$$

for some $n \times 1$ vector $\mathbf{w}$.

EXERCISE 13. Let $\mathbf{A}$ represent an $n \times n$ symmetric nonnegative definite matrix, and let $\mathbf{B}$ represent an $n \times n$ matrix. Suppose that $\mathbf{B} - \mathbf{A}$ is symmetric and nonnegative definite (in which case $\mathbf{B}$ is symmetric and nonnegative definite). Show that $\mathbf{A}^{+} - \mathbf{B}^{+}$ is nonnegative definite if and only if $\text{rank}(\mathbf{A}) = \text{rank}(\mathbf{B})$. Do so by, for instance, using the results of Exercises 1 and 18.15, the result that $\mathbf{W}^{-1} - \mathbf{V}^{-1}$ is nonnegative definite for any $m \times m$ symmetric positive definite matrices $\mathbf{W}$ and $\mathbf{V}$ such that $\mathbf{V} - \mathbf{W}$ is nonnegative definite, and the result that the Moore-Penrose inverse $\mathbf{H}^{+}$ of a $k \times k$ symmetric nonnegative definite matrix $\mathbf{H}$ equals $\mathbf{T}^{+}(\mathbf{T}^{+})'$, where $\mathbf{T}$ is any matrix of full row rank (and with k columns) such that $\mathbf{H} = \mathbf{T}'\mathbf{T}$.

Solution. Let $r = \text{rank}(\mathbf{B})$. And, assume that $r > 0$ — if $r = 0$, then $\mathbf{B} = \mathbf{0}$ and (in light of Lemma 4.2.2) $\mathbf{A} = \mathbf{0}$, in which case $\mathbf{A}^{+} - \mathbf{B}^{+} = \mathbf{0} - \mathbf{0} = \mathbf{0}$ and $\text{rank}(\mathbf{A}) = 0 = \text{rank}(\mathbf{B})$. Then, according to Theorem 14.3.7, there exists an $r \times n$ matrix $\mathbf{P}$ such that $\mathbf{B} = \mathbf{P}'\mathbf{P}$. Similarly, according to Corollary 14.3.8, there exists a matrix $\mathbf{Q}$ such that $\mathbf{A} = \mathbf{Q}'\mathbf{Q}$. And, according to the result of Exercise 18.15,

$$\mathcal{R}(\mathbf{A}) \subset \mathcal{R}(\mathbf{B}), \tag{S.5}$$

implying [since $\mathcal{R}(\mathbf{Q}) = \mathcal{R}(\mathbf{A})$ and $\mathcal{R}(\mathbf{P}) = \mathcal{R}(\mathbf{B})$] that

$$\mathcal{R}(\mathbf{Q}) \subset \mathcal{R}(\mathbf{P})$$

and hence that there exists a matrix $\mathbf{K}$ (having r columns) such that

$$\mathbf{Q} = \mathbf{K}\mathbf{P}.$$

Thus,

$$\mathbf{B} - \mathbf{A} = \mathbf{P}'\mathbf{P} - \mathbf{Q}'\mathbf{Q} = \mathbf{P}'(\mathbf{I} - \mathbf{K}'\mathbf{K})\mathbf{P}.$$

Moreover, according to Lemma 8.1.1, $\mathbf{P}$ has a right inverse $\mathbf{R}$, so that

$$\mathbf{I} - \mathbf{K}'\mathbf{K} = (\mathbf{P}\mathbf{R})'(\mathbf{I} - \mathbf{K}'\mathbf{K})\mathbf{P}\mathbf{R} = \mathbf{R}'(\mathbf{B} - \mathbf{A})\mathbf{R}.$$

And, as a consequence, $\mathbf{I} - \mathbf{K}'\mathbf{K}$ is nonnegative definite.

Now, suppose that $\text{rank}(\mathbf{A}) = \text{rank}(\mathbf{B})$ $(= r)$. Then,

$$r = \text{rank}(\mathbf{Q}) = \text{rank}(\mathbf{K}\mathbf{P}) \le \text{rank}(\mathbf{K}),$$

implying (since clearly rank $\mathbf{K} \le r$) that $\text{rank}(\mathbf{K}) = r$ and hence (in light of Corollary 14.2.14) that $\mathbf{K}'\mathbf{K}$ is positive definite. Thus, it follows from one of the cited results (a result encompassed in Theorem 18.3.4) that $(\mathbf{K}'\mathbf{K})^{-1} - \mathbf{I}$ is nonnegative definite. Moreover, upon observing that $\mathbf{A} = \mathbf{P}'(\mathbf{K}'\mathbf{K})\mathbf{P}$, it follows from the result of Exercise 1 that

$$\mathbf{A}^{+} = \mathbf{P}^{+}(\mathbf{K}'\mathbf{K})^{-1}(\mathbf{P}')^{+} = \mathbf{P}^{+}(\mathbf{K}'\mathbf{K})^{-1}(\mathbf{P}^{+})'$$

and from another of the cited results (a result covered by Theorem 20.4.5) that

$$\mathbf{B}^{+} = \mathbf{P}^{+}(\mathbf{P}^{+})',$$

so that

$$\mathbf{A}^{+} - \mathbf{B}^{+} = \mathbf{P}^{+}[(\mathbf{K}'\mathbf{K})^{-1} - \mathbf{I}](\mathbf{P}^{+})'.$$

And, in light of Theorem 14.2.9, we conclude that $\mathbf{A}^{+} - \mathbf{B}^{+}$ is nonnegative definite.

Conversely, suppose that $\mathbf{A}^{+} - \mathbf{B}^{+}$ is nonnegative definite. Then, it follows from the result of Exercise 18.15 that $\mathcal{R}(\mathbf{B}^{+}) \subset \mathcal{R}(\mathbf{A}^{+})$, implying that $\text{rank}(\mathbf{B}^{+}) \le \text{rank}(\mathbf{A}^{+})$ and hence [in light of Part (1) of Theorem 20.5.1] that $\text{rank}(\mathbf{B}) \le \text{rank}(\mathbf{A})$. Since [in light of result (S.5)] $\text{rank}(\mathbf{A}) \le \text{rank}(\mathbf{B})$, we conclude that $\text{rank}(\mathbf{A}) = \text{rank}(\mathbf{B})$.

21

Eigenvalues and Eigenvectors

EXERCISE 1. Show that an $n \times n$ skew-symmetric matrix $\mathbf{A}$ has no nonzero eigenvalues.

Solution. Let λ represent any eigenvalue of $\mathbf{A}$ and let $\mathbf{x}$ represent an eigenvector that corresponds to λ. Then, $-\mathbf{A}'\mathbf{x} = \mathbf{Ax} = \lambda\mathbf{x}$, implying that

$$-\mathbf{A}'\mathbf{Ax} = -\mathbf{A}'(\lambda\mathbf{x}) = \lambda(-\mathbf{A}'\mathbf{x}) = \lambda(\lambda\mathbf{x}) = \lambda^2\mathbf{x}$$

and hence that $-\mathbf{x}'\mathbf{A}'\mathbf{Ax} = \lambda^2\mathbf{x}'\mathbf{x}$. Thus, observing that $\mathbf{x} \neq \mathbf{0}$ and that $\mathbf{A}'\mathbf{A}$ is nonnegative definite, we find that

$$0 \leq \lambda^2 = -\mathbf{x}'\mathbf{A}'\mathbf{Ax}/\mathbf{x}'\mathbf{x} \leq 0,$$

leading to the conclusion that $\lambda^2 = 0$ or equivalently that $\lambda = 0$.

EXERCISE 2. Let $\mathbf{A}$ represent an $n \times n$ matrix, $\mathbf{B}$ a $k \times k$ matrix, and $\mathbf{X}$ an $n \times k$ matrix such that $\mathbf{AX} = \mathbf{XB}$.

(a) Show that $\mathcal{C}(\mathbf{X})$ is an invariant subspace (of $\mathcal{R}^{n\times 1}$) relative to $\mathbf{A}$.

(b) Show that if $\mathbf{X}$ is of full column rank, then every eigenvalue of $\mathbf{B}$ is an eigenvalue of $\mathbf{A}$.

Solution. (a) Corresponding to any $(n \times 1)$ vector $\mathbf{u}$ in $\mathcal{C}(\mathbf{X})$, there exists a $k \times 1$ vector $\mathbf{r}$ such that $\mathbf{u} = \mathbf{Xr}$, so that

$$\mathbf{Au} = \mathbf{AXr} = \mathbf{XBr} \in \mathcal{C}(\mathbf{X}).$$

Thus, $\mathcal{C}(\mathbf{X})$ is an invariant subspace relative to $\mathbf{A}$.

(b) Let λ represent an eigenvalue of $\mathbf{B}$, and let $\mathbf{y}$ represent an eigenvector of $\mathbf{B}$ corresponding to λ. By definition, $\mathbf{By} = \lambda\mathbf{y}$, so that

$$\mathbf{A}(\mathbf{Xy}) = \mathbf{XBy} = \mathbf{X}(\lambda\mathbf{y}) = \lambda(\mathbf{Xy}).$$

Now, suppose that $\mathbf{X}$ is of full column rank. Then (since $\mathbf{y} \neq \mathbf{0}$) $\mathbf{Xy} \neq \mathbf{0}$, leading us to conclude that λ is an eigenvalue of $\mathbf{A}$ (and that $\mathbf{Xy}$ is an eigenvector of $\mathbf{A}$ corresponding to λ).

EXERCISE 3. Let $p(\lambda)$ represent the characteristic polynomial of an $n \times n$ matrix $\mathbf{A}$, and let $c_0, c_1, c_2, \ldots, c_n$ represent the respective coefficients of the characteristic polynomial, so that

$$p(\lambda) = c_0\lambda^0 + c_1\lambda + c_2\lambda^2 + \cdots + c_n\lambda^n = \sum_{s=0}^{n} c_s\lambda^s$$

(for $\lambda \in \mathcal{R}$). Further, let $\mathbf{P}$ represent the $n \times n$ matrix obtained from $p(\lambda)$ by formally replacing the scalar λ with the $n \times n$ matrix $\mathbf{A}$ (and by setting $\mathbf{A}^0 = \mathbf{I}_n$). That is, let

$$\mathbf{P} = c_0\mathbf{I} + c_1\mathbf{A} + c_2\mathbf{A}^2 + \cdots + c_n\mathbf{A}^n = \sum_{s=0}^{n} c_s\mathbf{A}^s.$$

Show that $\mathbf{P} = \mathbf{0}$ (a result that is known as the Cayley-Hamilton theorem) by carrying out the following four steps.

(a) Letting $\mathbf{B}(\lambda) = \mathbf{A} - \lambda\mathbf{I}$ and letting $\mathbf{H}(\lambda)$ represent the adjoint matrix of $\mathbf{B}(\lambda)$, show that (for $\lambda \in \mathcal{R}$)

$$\mathbf{H}(\lambda) = \mathbf{K}_0 + \lambda\mathbf{K}_1 + \lambda^2\mathbf{K}_2 + \cdots + \lambda^{n-1}\mathbf{K}_{n-1},$$

where $\mathbf{K}_0, \mathbf{K}_1, \mathbf{K}_2, \ldots, \mathbf{K}_{n-1}$ are $n \times n$ matrices (that do not vary with λ).

(b) Letting $\mathbf{T}_0 = \mathbf{AK}_0$, $\mathbf{T}_n = -\mathbf{K}_{n-1}$, and (for $s = 1, \ldots, n-1$) $\mathbf{T}_s = \mathbf{AK}_s - \mathbf{K}_{s-1}$, show that (for $\lambda \in \mathcal{R}$)

$$\mathbf{T}_0 + \lambda\mathbf{T}_1 + \lambda^2\mathbf{T}_2 + \cdots + \lambda^n\mathbf{T}_n = p(\lambda)\mathbf{I}_n.$$

[*Hint*. It follows from a fundamental result on adjoint matrices that (for $\lambda \in \mathcal{R}$) $\mathbf{B}(\lambda)\mathbf{H}(\lambda) = |\mathbf{B}(\lambda)|\mathbf{I}_n = p(\lambda)\mathbf{I}_n$.]

(c) Show that, for $s = 0, 1, \ldots, n$, $\mathbf{T}_s = c_s\mathbf{I}$.

(d) Show that

$$\mathbf{P} = \mathbf{T}_0 + \mathbf{AT}_1 + \mathbf{A}^2\mathbf{T}_2 + \cdots + \mathbf{A}^n\mathbf{T}_n = \mathbf{0}.$$

Solution. (a) Let $h_{ij}(\lambda)$ represent the ijth element of $\mathbf{H}(\lambda)$. Then, $h_{ij}(\lambda)$ is the cofactor of the jith element of $\mathbf{B}(\lambda)$. And, it is apparent from the definition of a cofactor and from the definition of a determinant [given by formula (13.1.2)] that $h_{ij}(\lambda)$ is a polynomial (in λ) of degree $n-1$ or $n-2$. Thus,

$$h_{ij}(\lambda) = k_{ij}^{(0)} + k_{ij}^{(1)}\lambda + k_{ij}^{(2)}\lambda^2 + \cdots + k_{ij}^{(n-1)}\lambda^{n-1}$$

for some scalars $k_{ij}^{(0)}, k_{ij}^{(1)}, k_{ij}^{(2)}, \ldots, k_{ij}^{(n-1)}$ (that do not vary with λ). And, it follows that

$$\mathbf{H}(\lambda) = \mathbf{K}_0 + \lambda\mathbf{K}_1 + \lambda^2\mathbf{K}_2 + \cdots + \lambda^{n-1}\mathbf{K}_{n-1},$$

where (for $s = 0, 1, 2, \ldots, n-1$) $\mathbf{K}_s$ is the $n \times n$ matrix whose ijth element is $k_{ij}^{(s)}$.

(b) In light of Part (a), we have that (for $\lambda \in \mathcal{R}$)

$$\begin{aligned}\mathbf{B}(\lambda)\mathbf{H}(\lambda) &= (\mathbf{A} - \lambda\mathbf{I})(\mathbf{K}_0 + \lambda\mathbf{K}_1 + \lambda^2\mathbf{K}_2 + \cdots + \lambda^{n-1}\mathbf{K}_{n-1})\\ &= \mathbf{T}_0 + \lambda\mathbf{T}_1 + \lambda^2\mathbf{T}_2 + \cdots + \lambda^n\mathbf{T}_n.\end{aligned}$$

And, making use of the hint, we find that (for $\lambda \in \mathcal{R}$)

$$\mathbf{T}_0 + \lambda\mathbf{T}_1 + \lambda^2\mathbf{T}_2 + \cdots + \lambda^n\mathbf{T}_n = p(\lambda)\mathbf{I}_n\,.$$

(c) For $s = 0, 1, \ldots, n$, let $t_{ij}^{(s)}$ represent the ijth element of $\mathbf{T}_s$. Then, it follows from Part (b) that (for $\lambda \in \mathcal{R}$)

$$t_{ij}^{(0)} + \lambda t_{ij}^{(1)} + \lambda^2 t_{ij}^{(2)} + \cdots + \lambda^n t_{ij}^{(n)} = \begin{cases} p(\lambda), & \text{if } j = i,\\ 0, & \text{if } j \neq i.\end{cases}$$

Consequently,

$$t_{ij}^{(s)} = \begin{cases} c_s, & \text{if } j = i,\\ 0, & \text{if } j \neq i,\end{cases}$$

and hence $\mathbf{T}_s = c_s\mathbf{I}$.

(d) Making use of Part (c), we find that

$$\begin{aligned}&\mathbf{T}_0 + \mathbf{A}\mathbf{T}_1 + \mathbf{A}^2\mathbf{T}_2 + \cdots + \mathbf{A}^n\mathbf{T}_n\\ &\qquad = c_0\mathbf{I} + \mathbf{A}(c_1\mathbf{I}) + \mathbf{A}^2(c_2\mathbf{I}) + \cdots + \mathbf{A}^n(c_n\mathbf{I}) = \mathbf{P}.\end{aligned}$$

Moreover,

$$\begin{aligned}&\mathbf{T}_0 + \mathbf{A}\mathbf{T}_1 + \mathbf{A}^2\mathbf{T}_2 + \cdots + \mathbf{A}^n\mathbf{T}_n\\ &\qquad = (\mathbf{A}-\mathbf{A})\mathbf{K}_0 + (\mathbf{A}-\mathbf{A})\mathbf{A}\mathbf{K}_1 + (\mathbf{A}-\mathbf{A})\mathbf{A}^2\mathbf{K}_2\\ &\qquad\qquad + \cdots + (\mathbf{A}-\mathbf{A})\mathbf{A}^{n-1}\mathbf{K}_{n-1}\\ &\qquad = \mathbf{0}.\end{aligned}$$

EXERCISE 4. Let $c_0, c_1, \ldots, c_{n-1}, c_n$ represent the respective coefficients of the characteristic polynomial $p(\lambda)$ of an $n \times n$ matrix $\mathbf{A}$ [so that $p(\lambda) = c_0 +$

$c_1\lambda + \cdots + c_{n-1}\lambda^{n-1} + c_n\lambda^n$ (for $\lambda \in \mathcal{R}$)]. Using the result of Exercise 3 (the Cayley-Hamilton theorem), show that if **A** is nonsingular, then $c_0 \neq 0$, and

$$\mathbf{A}^{-1} = -(1/c_0)(c_1\mathbf{I} + c_2\mathbf{A} + \cdots + c_n\mathbf{A}^{n-1}).$$

Solution. According to result (1.8), $c_0 = |\mathbf{A}|$, and, according to the result of Exercise 3,

$$c_0\mathbf{I} + c_1\mathbf{A} + c_2\mathbf{A}^2 + \cdots + c_n\mathbf{A}^n = \mathbf{0}. \tag{S.1}$$

Now, suppose that **A** is nonsingular. Then, it follows from Theorem 13.3.7 that $c_0 \neq 0$. Moreover, upon premultiplying both sides of equality (S.1) by $\mathbf{A}^{-1}$, we find that

$$c_0\mathbf{A}^{-1} + c_1\mathbf{I} + c_2\mathbf{A} + c_3\mathbf{A}^2 + \cdots + c_n\mathbf{A}^{n-1} = \mathbf{0}$$

and hence that

$$\mathbf{A}^{-1} = -(1/c_0)(c_1\mathbf{I} + c_2\mathbf{A} + c_3\mathbf{A}^2 + \cdots + c_n\mathbf{A}^{n-1}).$$

EXERCISE 5. Show that if an $n \times n$ matrix **B** is similar to an $n \times n$ matrix **A**, then (1) $\mathbf{B}^k$ is similar to $\mathbf{A}^k$ ($k = 2, 3, \ldots$) and (2) $\mathbf{B}'$ is similar to $\mathbf{A}'$.

Solution. Suppose that **B** is similar to **A**. Then, there exists an $n \times n$ nonsingular matrix **C** such that $\mathbf{B} = \mathbf{C}^{-1}\mathbf{AC}$.

(1) Clearly, it suffices to show that (for $k = 1, 2, 3, \ldots$) $\mathbf{B}^k = \mathbf{C}^{-1}\mathbf{A}^k\mathbf{C}$. Let us proceed by mathematical induction. Obviously, $\mathbf{B}^1 = \mathbf{C}^{-1}\mathbf{A}^1\mathbf{C}$. Now, suppose that $\mathbf{B}^{k-1} = \mathbf{C}^{-1}\mathbf{A}^{k-1}\mathbf{C}$ (where $k \geq 2$). Then,

$$\mathbf{B}^k = \mathbf{B}\mathbf{B}^{k-1} = \mathbf{C}^{-1}\mathbf{ACC}^{-1}\mathbf{A}^{k-1}\mathbf{C} = \mathbf{C}^{-1}\mathbf{A}^k\mathbf{C}.$$

(2) We find that

$$\mathbf{B}' = (\mathbf{C}^{-1}\mathbf{AC})' = \mathbf{C}'\mathbf{A}'(\mathbf{C}^{-1})' = [(\mathbf{C}')^{-1}]^{-1}\mathbf{A}'(\mathbf{C}')^{-1}.$$

Thus, $\mathbf{B}'$ is similar to $\mathbf{A}'$.

EXERCISE 6. Show that if an $n \times n$ matrix **B** is similar to an ($n \times n$) idempotent matrix, then **B** is idempotent.

Solution. Let **A** represent an $n \times n$ idempotent matrix, and suppose that **B** is similar to **A**. Then, there exists an $n \times n$ nonsingular matrix **C** such that $\mathbf{B} = \mathbf{C}^{-1}\mathbf{AC}$. And, it follows that

$$\mathbf{B}^2 = \mathbf{C}^{-1}\mathbf{ACC}^{-1}\mathbf{AC} = \mathbf{C}^{-1}\mathbf{A}^2\mathbf{C} = \mathbf{C}^{-1}\mathbf{AC} = \mathbf{B}.$$

EXERCISE 7. Let $\mathbf{A} = \begin{pmatrix} 1 & 0 \\ 0 & 1 \end{pmatrix}$ and $\mathbf{B} = \begin{pmatrix} 1 & 1 \\ 0 & 1 \end{pmatrix}$. Show that **B** has the same rank, determinant, trace, and characteristic polynomial as **A**, but that, nevertheless, **B** is not similar to **A**.

Solution. Clearly, $|\mathbf{B}| = 1 = |\mathbf{A}|$, $\text{rank}(\mathbf{B}) = 2 = \text{rank}(\mathbf{A})$, $\text{tr}(\mathbf{B}) = 2 = \text{tr}(\mathbf{A})$, and the characteristic polynomial of both $\mathbf{B}$ and $\mathbf{A}$ is $p(\lambda) = (\lambda - 1)^2$.

Now, suppose that $\mathbf{CB} = \mathbf{AC}$ for some 2×2 matrix $\mathbf{C} = \{c_{ij}\}$. Then, since

$$\mathbf{CB} = \begin{pmatrix} c_{11} & c_{11} + c_{12} \\ c_{21} & c_{21} + c_{22} \end{pmatrix} \quad \text{and} \quad \mathbf{AC} = \mathbf{C} = \begin{pmatrix} c_{11} & c_{12} \\ c_{21} & c_{22} \end{pmatrix},$$

$c_{11} + c_{12} = c_{12}$ and $c_{21} + c_{22} = c_{22}$, implying that $c_{11} = 0$ and $c_{21} = 0$ and hence that $\mathbf{C}$ is singular.

Thus, there exists no 2×2 *nonsingular* matrix $\mathbf{C}$ such that $\mathbf{CB} = \mathbf{AC}$. And, we conclude that $\mathbf{B}$ is not similar to $\mathbf{A}$.

EXERCISE 8. Expand on the result of Exercise 7 by showing (for an arbitrary positive integer n) that for an $n \times n$ matrix $\mathbf{B}$ to be similar to an $n \times n$ matrix $\mathbf{A}$ it is not sufficient for $\mathbf{B}$ to have the same rank, determinant, trace, and characteristic polynomial as $\mathbf{A}$.

Solution. Suppose that $\mathbf{A} = \mathbf{I}_n$, and suppose that $\mathbf{B}$ is a triangular matrix, all of whose diagonal elements equal 1. Then, in light of Lemma 13.1.1 and Corollary 8.5.6, $\mathbf{B}$ has the same determinant, rank, trace, and characteristic polynomial as $\mathbf{A}$. However, for $\mathbf{B}$ to be similar to $\mathbf{A}$, it is necessary (and sufficient) that there exist an $n \times n$ nonsingular matrix $\mathbf{C}$ such that $\mathbf{B} = \mathbf{C}^{-1}\mathbf{AC}$ or equivalently (since $\mathbf{A} = \mathbf{I}_n$) that $\mathbf{B} = \mathbf{I}_n$. Thus, it is only in the special case where $\mathbf{B} = \mathbf{I}_n$ that $\mathbf{B}$ is similar to $\mathbf{A}$.

EXERCISE 9. Let $\mathbf{A}$ represent an $n \times n$ matrix, $\mathbf{B}$ a $k \times k$ matrix, and $\mathbf{X}$ an $n \times k$ matrix such that $\mathbf{AX} = \mathbf{XB}$. Show that if $\mathbf{X}$ is of full column rank, then there exists an orthogonal matrix $\mathbf{Q}$ such that $\mathbf{Q}'\mathbf{AQ} = \begin{pmatrix} \mathbf{T}_{11} & \mathbf{T}_{12} \\ \mathbf{0} & \mathbf{T}_{22} \end{pmatrix}$, where $\mathbf{T}_{11}$ is a $k \times k$ matrix that is similar to $\mathbf{B}$.

Solution. Suppose that $\text{rank}(\mathbf{X}) = k$. Then, according to Theorem 6.4.3, there exists an $n \times k$ matrix $\mathbf{U}$ whose columns are orthonormal (with respect to the usual inner product) vectors that form a basis for $\mathcal{C}(\mathbf{X})$. And, $\mathbf{X} = \mathbf{UC}$ for some $k \times k$ matrix $\mathbf{C}$. Moreover, $\mathbf{C}$ is nonsingular [since $\text{rank}(\mathbf{X}) = k$].

Now, observe that $\mathbf{AUC} = \mathbf{AX} = \mathbf{XB} = \mathbf{UCB}$ and hence that $\mathbf{AU} = \mathbf{AUCC}^{-1} = \mathbf{UCBC}^{-1}$. Then, in light of Theorem 21.3.2, there exists an orthogonal matrix $\mathbf{Q}$ such that $\mathbf{Q}'\mathbf{AQ} = \begin{pmatrix} \mathbf{T}_{11} & \mathbf{T}_{12} \\ \mathbf{0} & \mathbf{T}_{22} \end{pmatrix}$, where $\mathbf{T}_{11} = \mathbf{CBC}^{-1}$. Moreover, $\mathbf{T}_{11}$ is similar to $\mathbf{B}$.

EXERCISE 10. Show that if 0 is an eigenvalue of an $n \times n$ matrix $\mathbf{A}$, then its algebraic multiplicity is greater than or equal to $n - \text{rank}(\mathbf{A})$.

Solution. Suppose that 0 is an eigenvalue of $\mathbf{A}$. Then, according to Theorem 21.3.4, its algebraic multiplicity is greater than or equal to its geometric multiplicity, and, according to Lemma 11.3.1, its geometric multiplicity equals $n - \text{rank}(\mathbf{A})$. Thus, the algebraic multiplicity of the eigenvalue 0 is greater than or equal to $n - \text{rank}(\mathbf{A})$.

EXERCISE 11. Let $\mathbf{A}$ represent an $n \times n$ matrix. Show that if a scalar λ is an eigenvalue of $\mathbf{A}$ of algebraic multiplicity γ, then $\text{rank}(\mathbf{A} - \lambda\mathbf{I}) \geq n - \gamma$.

Solution. Suppose that λ is an eigenvalue of $\mathbf{A}$ of algebraic multiplicity γ. Then, making use of Theorem 21.3.4 and result (1.1) {and recalling that, by definition, the geometric multiplicity of λ equals $\dim[\mathcal{N}(\mathbf{A} - \lambda\mathbf{I})]$}, we find that

$$\gamma \geq \dim[\mathcal{N}(\mathbf{A} - \lambda\mathbf{I})] = n - \text{rank}(\mathbf{A} - \lambda\mathbf{I})$$

and hence that

$$\text{rank}(\mathbf{A} - \lambda\mathbf{I}) \geq n - \gamma.$$

EXERCISE 12. Let γ_1 represent the algebraic multiplicity and ν_1 the geometric multiplicity of 0 when 0 is regarded as an eigenvalue of an $n \times n$ (singular) matrix $\mathbf{A}$. And let γ_2 represent the algebraic multiplicity and ν_2 the geometric multiplicity of 0 when 0 is regarded as an eigenvalue of $\mathbf{A}^2$. Show that if $\nu_1 = \gamma_1$, then $\nu_2 = \gamma_2 = \nu_1$.

Solution. For any $n \times 1$ vector $\mathbf{x}$ such that $\mathbf{Ax} = \mathbf{0}$, we find that $\mathbf{A}^2\mathbf{x} = \mathbf{AAx} = \mathbf{A0} = \mathbf{0}$. Thus,

$$\nu_2 = \dim[\mathcal{N}(\mathbf{A}^2)] \geq \dim[\mathcal{N}(\mathbf{A})] = \nu_1. \tag{S.2}$$

There exists an $n \times \nu_1$ matrix $\mathbf{U}$ whose columns form an orthonormal (with respect to the usual inner product) basis for $\mathcal{N}(\mathbf{A})$. Then, $\mathbf{AU} = \mathbf{0} = \mathbf{U\,0}$. And, it follows from Theorem 21.3.2 that there exists an $n \times (n - \nu_1)$ matrix $\mathbf{V}$ such that the $n \times n$ matrix $(\mathbf{U}, \mathbf{V})$ is orthogonal and, taking $\mathbf{V}$ to be any such matrix, that

$$(\mathbf{U}, \mathbf{V})'\mathbf{A}(\mathbf{U}, \mathbf{V}) = \begin{pmatrix} \mathbf{0} & \mathbf{U}'\mathbf{AV} \\ \mathbf{0} & \mathbf{V}'\mathbf{AV} \end{pmatrix}$$

[so that $\mathbf{A}$ is similar to $\begin{pmatrix} \mathbf{0} & \mathbf{U}'\mathbf{AV} \\ \mathbf{0} & \mathbf{V}'\mathbf{AV} \end{pmatrix}$]. Moreover, it follows from Theorem 21.3.1 that γ_1 equals the algebraic multiplicity of 0 when 0 is regarded as an eigenvalue of $\begin{pmatrix} \mathbf{0} & \mathbf{U}'\mathbf{AV} \\ \mathbf{0} & \mathbf{V}'\mathbf{AV} \end{pmatrix}$ and hence (in light of Lemma 21.2.1) that γ_1 equals ν_1 plus the algebraic multiplicity of 0 when 0 is regarded as an eigenvalue of $\mathbf{V}'\mathbf{AV}$.

Now, suppose that $\nu_1 = \gamma_1$. Then, the algebraic multiplicity of 0 when 0 is regarded as an eigenvalue of $\mathbf{V}'\mathbf{AV}$ equals 0; that is, 0 is not an eigenvalue of $\mathbf{V}'\mathbf{AV}$. Thus, it follows from Lemma 11.3.1 that $\mathbf{V}'\mathbf{AV}$ is nonsingular.

Further,

$$\begin{aligned}(\mathbf{U}, \mathbf{V})'\mathbf{A}^2(\mathbf{U}, \mathbf{V}) &= (\mathbf{U}, \mathbf{V})'\mathbf{A}(\mathbf{U}, \mathbf{V})(\mathbf{U}, \mathbf{V})'\mathbf{A}(\mathbf{U}, \mathbf{V}) \\ &= \begin{pmatrix} \mathbf{0} & \mathbf{U}'\mathbf{AV} \\ \mathbf{0} & \mathbf{V}'\mathbf{AV} \end{pmatrix}^2 = \begin{pmatrix} \mathbf{0} & \mathbf{U}'\mathbf{AVV}'\mathbf{AV} \\ \mathbf{0} & (\mathbf{V}'\mathbf{AV})^2 \end{pmatrix},\end{aligned}$$

so that $\mathbf{A}^2$ is similar to $\begin{pmatrix} \mathbf{0} & \mathbf{U}'\mathbf{AVV}'\mathbf{AV} \\ \mathbf{0} & (\mathbf{V}'\mathbf{AV})^2 \end{pmatrix}$. And, since $(\mathbf{V}'\mathbf{AV})^2$ is nonsingular, it follows from Lemma 21.2.1 that $\gamma_2 = \nu_1$. Recalling inequality (S.2), we conclude, on the basis of Theorem 21.3.4, that $\nu_1 = \gamma_2 \geq \nu_2 \geq \nu_1$ and hence that $\nu_2 = \gamma_2 = \nu_1$.

EXERCISE 13. Let $\mathbf{x}_1$ and $\mathbf{x}_2$ represent eigenvectors of an $n \times n$ matrix $\mathbf{A}$, and let c_1 and c_2 represent nonzero scalars. Under what circumstances is the vector $\mathbf{x} = c_1\mathbf{x}_1 + c_2\mathbf{x}_2$ an eigenvector of $\mathbf{A}$?

Solution. Let λ_1 and λ_2 represent the two (not-necessarily-distinct) eigenvalues to which $\mathbf{x}_1$ and $\mathbf{x}_2$ correspond. Then, by definition, $\mathbf{Ax}_1 = \lambda_1\mathbf{x}_1$ and $\mathbf{Ax}_2 = \lambda_2\mathbf{x}_2$, and consequently

$$\mathbf{Ax} = c_1\mathbf{Ax}_1 + c_2\mathbf{Ax}_2 = c_1\lambda_1\mathbf{x}_1 + c_2\lambda_2\mathbf{x}_2 = \lambda_1\mathbf{x} + (\lambda_2 - \lambda_1)c_2\mathbf{x}_2 .$$

Thus, if $\lambda_2 = \lambda_1$, then $\mathbf{x}$ is an eigenvector of $\mathbf{A}$ [unless $\mathbf{x}_2 = -(c_1/c_2)\mathbf{x}_1$, in which case $\mathbf{x} = \mathbf{0}$]. Alternatively, if $\lambda_2 \neq \lambda_1$, then, according to Theorem 21.4.1, $\mathbf{x}_1$ and $\mathbf{x}_2$ are linearly independent, implying that $\mathbf{x}$ and $\mathbf{x}_2$ are linearly independent (as can be easily verified) and hence that there does not exist any scalar c such that $\lambda_1\mathbf{x} + (\lambda_2 - \lambda_1)c_2\mathbf{x}_2 = c\mathbf{x}$. We conclude that if $\lambda_2 \neq \lambda_1$ then $\mathbf{x}$ is not an eigenvector of $\mathbf{A}$.

EXERCISE 14. Let $\mathbf{A}$ represent an $n \times n$ matrix, and suppose that there exists an $n \times n$ nonsingular matrix $\mathbf{Q}$ such that $\mathbf{Q}^{-1}\mathbf{AQ} = \mathbf{D}$ for some diagonal matrix $\mathbf{D} = \{d_i\}$. Further, for $i = 1, \ldots, n$, let $\mathbf{r}_i'$ represent the ith row of $\mathbf{Q}^{-1}$. Show (a) that $\mathbf{A}'$ is diagonalized by $(\mathbf{Q}^{-1})'$, (b) that the diagonal elements of $\mathbf{D}$ are the (not necessarily distinct) eigenvalues of $\mathbf{A}'$, and (c) that $\mathbf{r}_1, \ldots, \mathbf{r}_n$ are eigenvectors of $\mathbf{A}'$ (with $\mathbf{r}_i$ corresponding to the eigenvalue d_i).

Solution. The validity of Part (a) is evident upon observing that

$$\begin{aligned} \mathbf{D} = \mathbf{D}' = (\mathbf{Q}^{-1}\mathbf{AQ})' = \mathbf{Q}'\mathbf{A}'(\mathbf{Q}^{-1})' &= [(\mathbf{Q}')^{-1}]^{-1}\mathbf{A}'(\mathbf{Q}^{-1})' \\ &= [(\mathbf{Q}^{-1})']^{-1}\mathbf{A}'(\mathbf{Q}^{-1})'. \end{aligned}$$

And, observing also that $\mathbf{r}_i$ is the ith column of $(\mathbf{Q}^{-1})'$, the validity of Parts (b) and (c) follows from Parts (7) and (8) of Theorem 21.5.1.

EXERCISE 15. Show that if an $n \times n$ nonsingular matrix $\mathbf{A}$ is diagonalized by an $n \times n$ nonsingular matrix $\mathbf{Q}$, then $\mathbf{A}^{-1}$ is also diagonalized by $\mathbf{Q}$.

Solution. Suppose that $\mathbf{A}$ is diagonalized by $\mathbf{Q}$. Then, it follows from Theorem 21.5.1 that the columns of $\mathbf{Q}$ are eigenvectors of $\mathbf{A}$ and hence (in light of Lemma 21.1.3) of $\mathbf{A}^{-1}$, leading us to conclude (on the basis of Theorem 21.5.2) that $\mathbf{Q}$ diagonalizes $\mathbf{A}^{-1}$ as well as $\mathbf{A}$. [Another way to see that $\mathbf{Q}$ diagonalizes $\mathbf{A}^{-1}$ is to observe that $\mathbf{Q}^{-1}\mathbf{A}^{-1}\mathbf{Q} = (\mathbf{Q}^{-1}\mathbf{AQ})^{-1}$ and that (since $\mathbf{Q}^{-1}\mathbf{AQ}$ is a diagonal matrix) $(\mathbf{Q}^{-1}\mathbf{AQ})^{-1}$ is a diagonal matrix.]

EXERCISE 16. Let $\mathbf{A}$ represent an $n \times n$ matrix whose spectrum comprises k eigenvalues $\lambda_1, \ldots, \lambda_k$ with algebraic multiplicities $\gamma_1, \ldots, \gamma_k$, respectively, that sum to n. Show that $\mathbf{A}$ is diagonalizable if and only if, for $i = 1, \ldots, k$, $\text{rank}(\mathbf{A} - \lambda_i \mathbf{I}) = n - \gamma_i$.

Solution. Let $\nu_1, \ldots, \nu_k$ represent the geometric multiplicities of $\lambda_1, \ldots, \lambda_k$, respectively. Then, according to result (1.1), $\nu_i = n - \text{rank}(\mathbf{A} - \lambda_i \mathbf{I})$ $(i = 1, \ldots, k)$. Thus,

$$\text{rank}(\mathbf{A} - \lambda_i \mathbf{I}) = n - \gamma_i \quad \Leftrightarrow \quad \gamma_i = n - \text{rank}(\mathbf{A} - \lambda_i \mathbf{I}) \quad \Leftrightarrow \quad \nu_i = \gamma_i .$$

Moreover, it follows from Corollary 21.3.7 that $\nu_i = \gamma_i$ for $i = 1, \ldots, k$ if and only if $\sum_{i=1}^k \nu_i = \sum_{i=1}^k \gamma_i$ or equivalently (since $\sum_{i=1}^k \gamma_i = n$) if and only if $\sum_{i=1}^k \nu_i = n$. We conclude, on the basis of Corollary 21.5.4, that $\mathbf{A}$ is diagonalizable if and only if, for $i = 1, \ldots, k$, $\text{rank}(\mathbf{A} - \lambda_i \mathbf{I}) = n - \gamma_i$.

EXERCISE 17. Let $\mathbf{A}$ represent an $n \times n$ symmetric matrix with not-necessarily-distinct eigenvalues $d_1, \ldots, d_n$ that have been ordered so that $d_1 \leq d_2 \leq \cdots \leq d_n$. And, let $\mathbf{Q}$ represent an $n \times n$ orthogonal matrix such that $\mathbf{Q}'\mathbf{A}\mathbf{Q} = \text{diag}(d_1, \ldots, d_n)$ — the existence of which is guaranteed. Further, for $m = 2, \ldots, n-1$, define $S_m = \{\mathbf{x} \in \mathcal{R}^{n \times 1} : \mathbf{x} \neq \mathbf{0},\ \mathbf{Q}_m'\mathbf{x} = \mathbf{0}\}$ and $T_m = \{\mathbf{x} \in \mathcal{R}^{n \times 1} : \mathbf{x} \neq \mathbf{0},\ \mathbf{P}_m'\mathbf{x} = \mathbf{0}\}$, where $\mathbf{Q}_m = (\mathbf{q}_1, \ldots, \mathbf{q}_{m-1})$ and $\mathbf{P}_m = (\mathbf{q}_{m+1}, \ldots, \mathbf{q}_n)$. Show that, for $m = 2, \ldots, n-1$,

$$d_m = \min_{\mathbf{x} \in S_m} \frac{\mathbf{x}'\mathbf{A}\mathbf{x}}{\mathbf{x}'\mathbf{x}} = \max_{\mathbf{x} \in T_m} \frac{\mathbf{x}'\mathbf{A}\mathbf{x}}{\mathbf{x}'\mathbf{x}} .$$

Solution. Let $\mathbf{x}$ represent an arbitrary $n \times 1$ vector, and let $\mathbf{y} = \mathbf{Q}'\mathbf{x}$. Partition $\mathbf{Q}$ and $\mathbf{y}$ as $\mathbf{Q} = (\mathbf{Q}_m, \mathbf{R}_m)$ and $\mathbf{y} = \begin{pmatrix} \mathbf{y}_1 \\ \mathbf{y}_2 \end{pmatrix}$ (where $\mathbf{y}_1$ has $m-1$ elements). Then, $\mathbf{y}_1 = \mathbf{Q}_m'\mathbf{x}$, $\mathbf{y}_2 = \mathbf{R}_m'\mathbf{x}$, and

$$\mathbf{x} = \mathbf{Q}\mathbf{y} = \mathbf{Q}_m\mathbf{y}_1 + \mathbf{R}_m\mathbf{y}_2 .$$

Moreover, since the columns of $\mathbf{R}_m$ are linearly independent and the columns of $\mathbf{Q}_m$ are linearly independent, $\mathbf{y}_2 = \mathbf{0} \Leftrightarrow \mathbf{R}_m\mathbf{y}_2 = \mathbf{0}$ (or equivalently $\mathbf{y}_2 \neq \mathbf{0} \Leftrightarrow \mathbf{R}_m\mathbf{y}_2 \neq \mathbf{0}$) and $\mathbf{y}_1 = \mathbf{0} \Leftrightarrow \mathbf{Q}_m\mathbf{y}_1 = \mathbf{0}$. Thus,

$$\mathbf{Q}_m'\mathbf{x} = \mathbf{0} \quad \Leftrightarrow \quad \mathbf{y}_1 = \mathbf{0} \quad \Leftrightarrow \quad \mathbf{x} = \mathbf{R}_m\mathbf{y}_2 .$$

It follows that $\mathbf{x} \in S_m$ if and only if $\mathbf{x} = \mathbf{R}_m\mathbf{y}_2$ for some $(n-m+1) \times 1$ nonnull vector $\mathbf{y}_2$.

It is now clear (since $\mathbf{R}_m'\mathbf{R}_m = \mathbf{I}$) that

$$\min_{\mathbf{x} \in S_m} \frac{\mathbf{x}'\mathbf{A}\mathbf{x}}{\mathbf{x}'\mathbf{x}} = \min_{\mathbf{y}_2 \neq \mathbf{0}} \frac{(\mathbf{R}_m\mathbf{y}_2)'\mathbf{A}\mathbf{R}_m\mathbf{y}_2}{(\mathbf{R}_m\mathbf{y}_2)'\mathbf{R}_m\mathbf{y}_2} = \min_{\mathbf{y}_2 \neq \mathbf{0}} \frac{\mathbf{y}_2'(\mathbf{R}_m'\mathbf{A}\mathbf{R}_m)\mathbf{y}_2}{\mathbf{y}_2'\mathbf{y}_2} .$$

Moreover, $\mathbf{R}_m'\mathbf{A}\mathbf{R}_m = \text{diag}(d_m, d_{m+1}, \ldots, d_n)$, so that $\mathbf{R}_m'\mathbf{A}\mathbf{R}_m$ is a symmetric matrix whose smallest eigenvalue is d_m. Thus, as a consequence of Theorem

21.5.6, we have that

$$\min_{\mathbf{y}_2 \neq \mathbf{0}} \frac{\mathbf{y}_2'(\mathbf{R}_m'\mathbf{A}\mathbf{R}_m)\mathbf{y}_2}{\mathbf{y}_2'\mathbf{y}_2} = d_m .$$

And, we conclude that

$$\min_{\mathbf{x} \in S_m} \frac{\mathbf{x}'\mathbf{A}\mathbf{x}}{\mathbf{x}'\mathbf{x}} = d_m .$$

That $\max_{\mathbf{x} \in T_m} \mathbf{x}'\mathbf{A}\mathbf{x}/\mathbf{x}'\mathbf{x} = d_m$ follows from a similar argument.

EXERCISE 18. Let $\mathbf{A}$ represent an $n \times n$ symmetric matrix, and adopt the following notation: $d_1, \ldots, d_n$ are the not-necessarily-distinct eigenvalues of $\mathbf{A}$, $\mathbf{q}_1, \ldots, \mathbf{q}_n$ are orthonormal eigenvectors corresponding to $d_1, \ldots, d_n$, respectively, $\mathbf{Q} = (\mathbf{q}_1, \ldots, \mathbf{q}_n)$, $\{\lambda_1, \ldots, \lambda_k\}$ is the spectrum of $\mathbf{A}$; and, for $j = 1, \ldots, k$, $S_j = \{i : d_i = \lambda_j\}$, $\mathbf{E}_j = \sum_{i \in S_j} \mathbf{q}_i\mathbf{q}_i'$, and $\mathbf{Q}_j = (\mathbf{q}_{i_1}, \ldots, \mathbf{q}_{i_{\nu_j}})$, where $i_1, \ldots, i_{\nu_j}$ denote the elements of S_j.

(a) Show that the matrices $\mathbf{E}_1, \ldots, \mathbf{E}_k$, which appear in the spectral decomposition $\mathbf{A} = \sum_{j=1}^{k} \lambda_j \mathbf{E}_j$, have the following properties:

(1) $\mathbf{E}_1 + \cdots + \mathbf{E}_k = \mathbf{I}$;

(2) $\mathbf{E}_1, \ldots, \mathbf{E}_k$ are nonnull, symmetric, and idempotent;

(3) for $t \neq j = 1, \ldots, k$, $\mathbf{E}_t\mathbf{E}_j = \mathbf{0}$; and

(4) $\text{rank}(\mathbf{E}_1) + \cdots + \text{rank}(\mathbf{E}_k) = n$.

(b) Take $\mathbf{F}_1, \ldots, \mathbf{F}_r$ to be $n \times n$ nonnull idempotent matrices such that $\mathbf{F}_1 + \cdots + \mathbf{F}_r = \mathbf{I}$. And, suppose that, for some distinct scalars $\tau_1, \ldots, \tau_r$,

$$\mathbf{A} = \tau_1\mathbf{F}_1 + \cdots + \tau_r\mathbf{F}_r .$$

Show that $r = k$ and that there exists a permutation $t_1, \ldots, t_r$ of the first r positive integers such that (for $j = 1, \ldots, r$) $\tau_j = \lambda_{t_j}$ and $\mathbf{F}_j = \mathbf{E}_{t_j}$.

Solution. (a) (1)

$$\mathbf{E}_1 + \cdots + \mathbf{E}_k = \sum_{j=1}^{k} \sum_{i \in S_j} \mathbf{q}_i\mathbf{q}_i' = \sum_{i=1}^{n} \mathbf{q}_i\mathbf{q}_i' = \mathbf{Q}\mathbf{Q}' = \mathbf{I}.$$

(2) Observe that (for $j = 1, \ldots, k$) $\mathbf{E}_j = \mathbf{Q}_j\mathbf{Q}_j'$ and $\mathbf{Q}_j'\mathbf{Q}_j = \mathbf{I}$. Then, clearly, $\mathbf{E}_j$ is symmetric. And, $\mathbf{Q}_j$ is nonnull, implying (in light of Corollary 5.3.2) that $\mathbf{E}_j$ is nonnull. Further

$$\mathbf{E}_j^2 = \mathbf{Q}_j\mathbf{Q}_j'\mathbf{Q}_j\mathbf{Q}_j' = \mathbf{Q}_j\mathbf{I}\mathbf{Q}_j' = \mathbf{Q}_j\mathbf{Q}_j' = \mathbf{E}_j ,$$

so that $\mathbf{E}_j$ is idempotent.

(3) and (4) In light of Theorem 18.4.1, Parts (3) and (4) follow from Parts (1) and (2).

(b) Observe (in light of Theorem 18.4.1) that, for $t \neq j = 1, \dots, r$, $\mathbf{F}_t\mathbf{F}_j = \mathbf{0}$. Then, for $j = 1, \dots, r$, we find that

$$\mathbf{AF}_j = \tau_1\mathbf{F}_1\mathbf{F}_j + \cdots + \tau_r\mathbf{F}_r\mathbf{F}_j = \tau_j\mathbf{F}_j^2 = \tau_j\mathbf{F}_j,$$

implying that τ_j is an eigenvalue of $\mathbf{A}$ and that any nonnull column of $\mathbf{F}_j$ is an eigenvector of $\mathbf{A}$ corresponding to τ_j. Consequently, there exists some subset $T = \{t_1, \dots, t_r\}$ of the first k positive integers such that (for $j = 1, \dots, r$) $\tau_j = \lambda_{t_j}$ (so that $r \leq k$). Further,

$$\mathcal{C}(\mathbf{E}_{t_j}) = \mathcal{C}(\mathbf{Q}_{t_j}\mathbf{Q}'_{t_j}) = \mathcal{C}(\mathbf{Q}_{t_j}) = \mathcal{N}(\mathbf{A} - \lambda_{t_j}\mathbf{I}) = \mathcal{N}(\mathbf{A} - \tau_j\mathbf{I}),$$

so that $\mathbf{F}_j = \mathbf{E}_{t_j}\mathbf{L}_j$ for some $n \times n$ matrix $\mathbf{L}_j$.

We have that

$$\mathbf{A} = \lambda_{t_1}\mathbf{E}_{t_1}\mathbf{L}_1 + \cdots + \lambda_{t_r}\mathbf{E}_{t_r}\mathbf{L}_r,$$

implying [in light of Part (a) and equality (5.5)] that (for $j = 1, \dots, r$)

$$\lambda_{t_j}\mathbf{E}_{t_j} = \lambda_{t_j}\mathbf{E}_{t_j}^2 = \mathbf{E}_{t_j}\mathbf{A} = \lambda_{t_j}\mathbf{E}_{t_j}^2\mathbf{L}_j = \lambda_{t_j}\mathbf{E}_{t_j}\mathbf{L}_j = \lambda_{t_j}\mathbf{F}_j, \tag{S.3}$$

so that $\lambda_{t_j} = 0$ or $\mathbf{F}_j = \mathbf{E}_{t_j}$. Moreover, for $j \notin T$, we find that

$$\lambda_j\mathbf{E}_j = \lambda_j\mathbf{E}_j^2 = \mathbf{E}_j\mathbf{A} = \mathbf{0},$$

implying (since $\mathbf{E}_j \neq \mathbf{0}$) that $\lambda_j = 0$ (and hence that $k \leq r + 1$).

To complete the proof, it suffices to show that $r = k$ and that (for $j = 1, \dots, r$) $\mathbf{F}_j = \mathbf{E}_{t_j}$. Let us consider separately the following two cases: (1) $\lambda_{t_j} \neq 0$ for $j = 1, \dots, r$; and (2) $\lambda_{t_s} = 0$ for some integer s $(1 \leq s \leq r)$.

In Case (1), it follows from result (S.3) that (for $j = 1, \dots, r$) $\mathbf{F}_j = \mathbf{E}_{t_j}$. Moreover, $r = k$, since otherwise there would exist a positive integer $s \notin T$ such that $\lambda_s = 0$, and we would have [in light of Part (a)] that

$$\mathbf{E}_s = \mathbf{I} - \sum_{j=1}^{r}\mathbf{E}_{t_j} = \mathbf{I} - \sum_{j=1}^{r}\mathbf{F}_j = \mathbf{I} - \mathbf{I} = \mathbf{0},$$

which [since, according to Part (a), $\mathbf{E}_s \neq \mathbf{0}$] would lead to a contradiction.

In Case (2), it is clear that $r = k$ and also [in light of result (S.3) and Part (a)] that $\mathbf{F}_j = \mathbf{E}_{t_j}$ for $j \neq s$ and

$$\mathbf{F}_s = \mathbf{I} - \sum_{j \neq s}\mathbf{F}_j = \mathbf{I} - \sum_{j \neq s}\mathbf{E}_{t_j} = \mathbf{E}_{t_s}.$$

EXERCISE 19. Let $\mathbf{A}$ represent an $n \times n$ symmetric matrix, and let $d_1, \dots, d_n$ represent the (not-necessarily-distinct) eigenvalues of $\mathbf{A}$. Show that $\lim_{k\to\infty}\mathbf{A}^k = \mathbf{0}$ if and only if, for $i = 1, \dots, n$, $|d_i| < 1$.

Solution. Let $\mathbf{D} = \text{diag}(d_1, \ldots, d_n)$. Then, according to Corollary 21.5.9, there exists an $n \times n$ orthogonal matrix $\mathbf{Q}$ such that $\mathbf{Q}'\mathbf{A}\mathbf{Q} = \mathbf{D}$. And, it follows from result (5.6) that $\mathbf{A}^k = \mathbf{Q}\mathbf{D}^k\mathbf{Q}'$. Moreover, since $\mathbf{D}^k = \text{diag}(d_1^k, \ldots, d_n^k)$,

$$|d_i| < 1 \text{ for } i = 1, \ldots, n \quad \Leftrightarrow \quad \lim_{k\to\infty} d_i^k = 0 \text{ for } i = 1, \ldots, n$$
$$\Leftrightarrow \quad \lim_{k\to\infty} \mathbf{D}^k = \mathbf{0}.$$

Now, suppose that, for $i = 1, \ldots, n$, $|d_i| < 1$. Then,

$$\lim_{k\to\infty} \mathbf{A}^k = \mathbf{Q}(\lim_{k\to\infty} \mathbf{D}^k)\mathbf{Q}' = \mathbf{Q}\,\mathbf{0}\,\mathbf{Q}' = \mathbf{0}.$$

Conversely, suppose that $\lim_{k\to\infty} \mathbf{A}^k = \mathbf{0}$. Then, observing that $\mathbf{D}^k = \mathbf{Q}'\mathbf{A}^k\mathbf{Q}$, we find that

$$\lim_{k\to\infty} \mathbf{D}^k = \lim_{k\to\infty} \mathbf{Q}'\mathbf{A}^k\mathbf{Q} = \mathbf{Q}'(\lim_{k\to\infty} \mathbf{A}^k)\mathbf{Q} = \mathbf{Q}'\mathbf{0}\mathbf{Q} = \mathbf{0}.$$

And, it follows that, for $i = 1, \ldots, n$, $|d_i| < 1$.

EXERCISE 20. (a) Show that if 0 is an eigenvalue of an $n \times n$ not-necessarily-symmetric matrix $\mathbf{A}$, then it is also an eigenvalue of $\mathbf{A}^+$ and that the geometric multiplicity of 0 is the same when it is regarded as an eigenvalue of $\mathbf{A}^+$ as when it is regarded as an eigenvalue of $\mathbf{A}$.

(b) Show (via an example) that the reciprocals of the nonzero eigenvalues of a square nonsymmetric matrix $\mathbf{A}$ are not necessarily eigenvalues of $\mathbf{A}^+$.

Solution. (a) Suppose that 0 is an eigenvalue of $\mathbf{A}$. Then, according to Lemma 21.1.1, $\text{rank}(\mathbf{A}) < n$. And, since (according to Theorem 20.5.1) $\text{rank}(\mathbf{A}^+) = \text{rank}(\mathbf{A})$, it follows that $\text{rank}(\mathbf{A}^+) < n$, implying (in light of Lemma 21.1.1) that 0 is also an eigenvalue of $\mathbf{A}^+$. Further, making use of Lemma 11.3.1, we find that

$$\dim[\mathcal{N}(\mathbf{A}^+)] = n - \text{rank}(\mathbf{A}^+) = n - \text{rank}(\mathbf{A}) = \dim[\mathcal{N}(\mathbf{A})],$$

so that the geometric multiplicity of 0 is the same when it is regarded as an eigenvalue of $\mathbf{A}^+$ as when it is regarded as an eigenvalue of $\mathbf{A}$.

(b) Consider the $n \times n$ matrix $\mathbf{A} = (\mathbf{1}_n, \mathbf{0})$ (where $n \geq 2$). We find that $\mathbf{A}'\mathbf{A} = \text{diag}(n, 0, 0, \ldots, 0)$ and hence that $(\mathbf{A}'\mathbf{A})^+ = \text{diag}(1/n, 0, 0, \ldots, 0)$, implying (in light of Corollary 20.5.5) that

$$\mathbf{A}^+ = (\mathbf{A}'\mathbf{A})^+\mathbf{A}' = \begin{pmatrix} (1/n)\mathbf{1}_n' \\ \mathbf{0} \end{pmatrix}.$$

Since $\mathbf{A}$ and $\mathbf{A}^+$ are triangular, it is easy to see that the distinct eigenvalues of $\mathbf{A}$ are 0 and 1, while those of $\mathbf{A}^+$ are 0 and $1/n$. Obviously, $1/n$ is not (for $n \geq 2$) the reciprocal of 1.

EXERCISE 21. Show that, for any positive integer n that is divisible by 2, there exists an $n \times n$ orthogonal matrix that has no eigenvalues. [*Hint*. Find a 2×2 orthogonal matrix $\mathbf{Q}$ that has no eigenvalues, and then consider the block-diagonal matrix $\text{diag}(\mathbf{Q}, \mathbf{Q}, \ldots, \mathbf{Q})$.]

Solution. Let $\mathbf{Q} = \begin{pmatrix} 0 & 1 \\ -1 & 0 \end{pmatrix}$. Clearly, $\mathbf{Q}$ is orthogonal, and (as shown in Section 21.1) it has no eigenvalues. Now, consider the $n \times n$ block-diagonal matrix $\text{diag}(\mathbf{Q}, \mathbf{Q}, \ldots, \mathbf{Q})$ (having $n/2$ diagonal blocks). This matrix is orthogonal (as is easily verified), and, as a consequence of Part (2) of Lemma 21.2.1, it has no eigenvalues.

EXERCISE 22. Let $\mathbf{Q}$ represent an $n \times n$ orthogonal matrix, and let $p(\lambda)$ represent the characteristic polynomial of $\mathbf{Q}$. Show that (for $\lambda \neq 0$)

$$p(\lambda) = \pm\lambda^n p(1/\lambda).$$

Solution. Let λ represent an arbitrary nonzero scalar. Then,

$$\mathbf{Q} - \lambda\mathbf{I} = \mathbf{Q} - \lambda\mathbf{Q}\mathbf{Q}' = -\lambda\mathbf{Q}[\mathbf{Q}' - (1/\lambda)\mathbf{I}] = -\lambda\mathbf{Q}[\mathbf{Q} - (1/\lambda)\mathbf{I}]'.$$

Thus, making use of Theorem 13.3.4, Lemma 13.2.1, and Corollaries 13.2.4 and 13.3.6, we find that

$$\begin{aligned} p(\lambda) = |\mathbf{Q} - \lambda\mathbf{I}| &= |-\lambda\mathbf{Q}||[\mathbf{Q} - (1/\lambda)\mathbf{I}]'| \\ &= (-\lambda)^n|\mathbf{Q}||\mathbf{Q} - (1/\lambda)\mathbf{I}| \\ &= (-1)^n\lambda^n(\pm 1)p(1/\lambda) = \pm\lambda^n p(1/\lambda). \end{aligned}$$

EXERCISE 23. Let $\mathbf{A}$ represent an $n \times n$ matrix, and suppose that the scalar 1 is an eigenvalue of $\mathbf{A}$ of geometric multiplicity ν. Show that $\nu \leq \text{rank}(\mathbf{A})$ and that if $\nu = \text{rank}(\mathbf{A})$, then $\mathbf{A}$ is idempotent.

Solution. That $\nu \leq \text{rank}(\mathbf{A})$ is an immediate consequence of Corollary 21.3.8.

Now, suppose that $\nu = \text{rank}(\mathbf{A})$. Then, it follows from Corollary 21.3.8 that $\mathbf{A}$ has no nonzero eigenvalues other than 1, and it follows from Corollary 21.5.4 that $\mathbf{A}$ is diagonalizable. Thus, as a consequence of Theorem 21.8.3, we have that $\mathbf{A}$ is idempotent.

EXERCISE 24. Let $\mathbf{A}$ represent an $n \times n$ nonsingular matrix. And, let λ represent an eigenvalue of $\mathbf{A}$, and $\mathbf{x}$ represent an eigenvector of $\mathbf{A}$ corresponding to λ. Show that $|\mathbf{A}|/\lambda$ is an eigenvalue of $\text{adj}(\mathbf{A})$ and that $\mathbf{x}$ is an eigenvector of $\text{adj}(\mathbf{A})$ corresponding to $|\mathbf{A}|/\lambda$.

Solution. Note (in light of Lemma 21.1.1 and Theorem 13.3.7) that $\lambda \neq 0$ and $|\mathbf{A}| \neq 0$.

According to Lemma 21.1.3, $1/\lambda$ is an eigenvalue of $\mathbf{A}^{-1}$, and $\mathbf{x}$ is an eigenvector of $\mathbf{A}^{-1}$ corresponding to $1/\lambda$. And, since (according to Corollary 13.5.4) $\text{adj}(\mathbf{A}) =$

$|\mathbf{A}|\mathbf{A}^{-1}$, it follows from the results of Section 21.10 that $|\mathbf{A}|/\lambda$ is an eigenvalue of adj($\mathbf{A}$) and that $\mathbf{x}$ is an eigenvector of adj($\mathbf{A}$) corresponding to $|\mathbf{A}|/\lambda$.

EXERCISE 25. Let $\mathbf{A}$ represent an $n \times n$ matrix, and let $p(\lambda)$ represent the characteristic polynomial of $\mathbf{A}$. And, let $\lambda_1, \ldots, \lambda_k$ represent the distinct eigenvalues of $\mathbf{A}$, and $\gamma_1, \ldots, \gamma_k$ represent their respective algebraic multiplicities, so that (for all λ)

$$p(\lambda) = (-1)^n q(\lambda) \prod_{j=1}^{k} (\lambda - \lambda_j)^{\gamma_j}$$

for some polynomial $q(\lambda)$ (of degree $n - \sum_{j=1}^{k} \gamma_j$) that has no real roots. Further, define $\mathbf{B} = \mathbf{A} - \lambda_1 \mathbf{U}\mathbf{V}'$, where $\mathbf{U} = (\mathbf{u}_1, \ldots, \mathbf{u}_{\gamma_1})$ is an $n \times \gamma_1$ matrix whose columns $\mathbf{u}_1, \ldots, \mathbf{u}_{\gamma_1}$ are (not necessarily linearly independent) eigenvectors of $\mathbf{A}$ corresponding to λ_1 and where $\mathbf{V} = (\mathbf{v}_1, \ldots, \mathbf{v}_{\gamma_1})$ is an $n \times \gamma_1$ matrix such that $\mathbf{V}'\mathbf{U}$ is diagonal.

(a) Show that the characteristic polynomial, say $r(\lambda)$, of $\mathbf{B}$ is such that (for all λ)

$$r(\lambda) = (-1)^n q(\lambda) \prod_{i=1}^{\gamma_1} [\lambda - (1 - \mathbf{v}_i'\mathbf{u}_i)\lambda_1] \prod_{j=2}^{k} (\lambda - \lambda_j)^{\gamma_j}. \tag{E.1}$$

[*Hint*. Since the left and right sides of equality (E.1) are polynomials (in λ), it suffices to show that they are equal for all λ other than $\lambda_1, \ldots, \lambda_k$.]

(b) Show that in the special case where $\mathbf{U}'\mathbf{V} = c\mathbf{I}$ for some nonzero scalar c, the distinct eigenvalues of $\mathbf{B}$ are either $\lambda_2, \ldots, \lambda_{s-1}, \lambda_s, \lambda_{s+1}, \ldots, \lambda_k$ with algebraic multiplicities $\gamma_2, \ldots, \gamma_{s-1}, \gamma_s + \gamma_1, \gamma_{s+1}, \ldots, \gamma_k$, respectively, or $(1-c)\lambda_1, \lambda_2, \ldots, \lambda_k$ with algebraic multiplicities $\gamma_1, \gamma_2, \ldots, \gamma_k$, respectively [depending on whether or not $(1-c)\lambda_1 = \lambda_s$ for some s $(2 \le s \le k)$].

(c) Show that, in the special case where $\gamma_1 = 1$, (1) $\mathbf{u}_1$ is an eigenvector of $\mathbf{B}$ corresponding to the eigenvalue $(1 - \mathbf{v}_1'\mathbf{u}_1)\lambda_1$ and (2) for any eigenvector $\mathbf{x}$ of $\mathbf{B}$ corresponding to an eigenvalue λ [other than $(1 - \mathbf{v}_1'\mathbf{u}_1)\lambda_1$], the vector

$$\mathbf{x} - \lambda_1(\lambda_1 - \lambda)^{-1}(\mathbf{v}_1'\mathbf{x})\mathbf{u}_1$$

is an eigenvector of $\mathbf{A}$ corresponding to λ.

Solution. (a) Let λ represent any scalar other than $\lambda_1, \ldots, \lambda_k$. And, observe that

$$(\mathbf{A} - \lambda\mathbf{I})\mathbf{U} = \mathbf{A}\mathbf{U} - \lambda\mathbf{U} = (\lambda_1 - \lambda)\mathbf{U},$$

so that $(\mathbf{A} - \lambda\mathbf{I})^{-1}\mathbf{U} = -(\lambda - \lambda_1)^{-1}\mathbf{U}$. Then, making use of Corollary 18.1.2, we find that

$$\begin{aligned} r(\lambda) &= |\mathbf{A} - \lambda\mathbf{I} - \lambda_1\mathbf{U}\mathbf{V}'| \\ &= |\mathbf{A} - \lambda\mathbf{I}||\mathbf{I} - \lambda_1(\mathbf{A} - \lambda\mathbf{I})^{-1}\mathbf{U}\mathbf{V}'| \end{aligned}$$

$$
\begin{aligned}
&= p(\lambda)|\mathbf{I}_n + \lambda_1(\lambda - \lambda_1)^{-1}\mathbf{U}\mathbf{V}'| \\
&= p(\lambda)|\mathbf{I}_{\gamma_1} + \lambda_1(\lambda - \lambda_1)^{-1}\mathbf{V}'\mathbf{U}| \\
&= p(\lambda)\prod_{i=1}^{\gamma_1}[1 + \lambda_1(\lambda - \lambda_1)^{-1}\mathbf{v}_i'\mathbf{u}_i] \\
&= p(\lambda)(\lambda - \lambda_1)^{-\gamma_1}\prod_{i=1}^{\gamma_1}(\lambda - \lambda_1 + \lambda_1\mathbf{v}_i'\mathbf{u}_i) \\
&= (-1)^n q(\lambda)\prod_{i=1}^{\gamma_1}[\lambda - (1 - \mathbf{v}_i'\mathbf{u}_i)\lambda_1]\prod_{j=2}^{k}(\lambda - \lambda_j)^{\gamma_j}.
\end{aligned}
$$

(b) Part (b) follows from Part (a) upon observing that, in this special case,

$$\prod_{i=1}^{\gamma_1}[\lambda - (1 - \mathbf{v}_i'\mathbf{u}_i)\lambda_1] = [\lambda - (1 - c)\lambda_1]^{\gamma_1}.$$

(c) (1) In this special case,

$$\mathbf{B}\mathbf{u}_1 = \mathbf{A}\mathbf{u}_1 - \lambda_1\mathbf{u}_1\mathbf{v}_1'\mathbf{u}_1 = \lambda_1\mathbf{u}_1 - \lambda_1(\mathbf{v}_1'\mathbf{u}_1)\mathbf{u}_1 = (1 - \mathbf{v}_1'\mathbf{u}_1)\lambda_1\mathbf{u}_1.$$

(2) Suppose that $\gamma_1 = 1$, and let $d = \lambda_1(\lambda_1 - \lambda)^{-1}\mathbf{v}_1'\mathbf{x}$. Since (by definition) $\mathbf{B}\mathbf{x} = \lambda\mathbf{x}$, we have that

$$(\mathbf{A} - \mathbf{A}\mathbf{u}_1\mathbf{v}_1')\mathbf{x} = (\mathbf{A} - \lambda_1\mathbf{u}_1\mathbf{v}_1')\mathbf{x} = \lambda\mathbf{x}$$

and hence that

$$\mathbf{A}[\mathbf{x} - (\mathbf{v}_1'\mathbf{x})\mathbf{u}_1] = \lambda\mathbf{x}.$$

Thus,

$$\mathbf{A}(\mathbf{x} - d\mathbf{u}_1) = \mathbf{A}[\mathbf{x} - (\mathbf{v}_1'\mathbf{x})\mathbf{u}_1 + (\mathbf{v}_1'\mathbf{x})\mathbf{u}_1 - d\mathbf{u}_1] = \lambda\mathbf{x} - (d - \mathbf{v}_1'\mathbf{x})\lambda_1\mathbf{u}_1.$$

Moreover, $\lambda_1 d - \lambda d = (\lambda_1 - \lambda)d = \lambda_1\mathbf{v}_1'\mathbf{x}$, implying that $(d - \mathbf{v}_1'\mathbf{x})\lambda_1 = \lambda d$. And, it follows that

$$\mathbf{A}(\mathbf{x} - d\mathbf{u}_1) = \lambda(\mathbf{x} - d\mathbf{u}_1).$$

Since $\mathbf{x} - d\mathbf{u}_1 \neq \mathbf{0}$ (as is evident upon observing that $\mathbf{x}$ and $\mathbf{u}_1$ are eigenvectors of $\mathbf{B}$ that correspond to different eigenvalues and hence, in light of Theorem 21.4.1, that $\mathbf{x}$ and $\mathbf{u}_1$ are linearly independent), we conclude that $\mathbf{x} - d\mathbf{u}_1$ is an eigenvector of $\mathbf{A}$ corresponding to λ.

EXERCISE 26. Let $\mathbf{A}$ represent an $m \times n$ matrix of rank r. And, take $\mathbf{P}$ to be any $m \times m$ orthogonal matrix and $\mathbf{D}_1$ to be any $r \times r$ nonsingular diagonal matrix such that

$$\mathbf{P}'\mathbf{A}\mathbf{A}'\mathbf{P} = \begin{pmatrix} \mathbf{D}_1^2 & \mathbf{0} \\ \mathbf{0} & \mathbf{0} \end{pmatrix}.$$

Further, partition $\mathbf{P}$ as $\mathbf{P} = (\mathbf{P}_1, \mathbf{P}_2)$, where $\mathbf{P}_1$ has r columns, and let $\mathbf{Q} = (\mathbf{Q}_1, \mathbf{Q}_2)$, where $\mathbf{Q}_1 = \mathbf{A}'\mathbf{P}_1\mathbf{D}_1^{-1}$ and where $\mathbf{Q}_2$ is any $n \times (n-r)$ matrix such that $\mathbf{Q}_1'\mathbf{Q}_2 = \mathbf{0}$. Show that

$$\mathbf{P}'\mathbf{A}\mathbf{Q} = \begin{pmatrix} \mathbf{D}_1 & \mathbf{0} \\ \mathbf{0} & \mathbf{0} \end{pmatrix}.$$

Solution. Upon applying Theorem 21.12.1 with $\mathbf{A}'$ in place of $\mathbf{A}$ (and with n and m in place of m and n, respectively) and writing $\mathbf{P} = (\mathbf{P}_1, \mathbf{P}_2)$ for $\mathbf{Q} = (\mathbf{Q}_1, \mathbf{Q}_2)$ and $\mathbf{Q} = (\mathbf{Q}_1, \mathbf{Q}_2)$ for $\mathbf{P} = (\mathbf{P}_1, \mathbf{P}_2)$, we find that

$$\mathbf{Q}'\mathbf{A}'\mathbf{P} = \begin{pmatrix} \mathbf{D}_1 & \mathbf{0} \\ \mathbf{0} & \mathbf{0} \end{pmatrix}.$$

Thus,

$$\mathbf{P}'\mathbf{A}\mathbf{Q} = (\mathbf{Q}'\mathbf{A}'\mathbf{P})' = \begin{pmatrix} \mathbf{D}_1 & \mathbf{0} \\ \mathbf{0} & \mathbf{0} \end{pmatrix}' = \begin{pmatrix} \mathbf{D}_1 & \mathbf{0} \\ \mathbf{0} & \mathbf{0} \end{pmatrix}.$$

Or, alternatively, the equality $\mathbf{P}'\mathbf{A}\mathbf{Q} = \begin{pmatrix} \mathbf{D}_1 & \mathbf{0} \\ \mathbf{0} & \mathbf{0} \end{pmatrix}$ can be established via an argument paralleling the proof of Theorem 21.12.1.

EXERCISE 27. Let $\mathbf{A}$ represent an $m \times n$ matrix. And, let $\mathbf{P}$ represent an $m \times m$ orthogonal matrix, $\mathbf{Q}$ an $n \times n$ orthogonal matrix, and $\mathbf{D}_1 = \{s_i\}$ an $r \times r$ diagonal matrix with (strictly) positive diagonal elements such that $\mathbf{P}'\mathbf{A}\mathbf{Q} = \begin{pmatrix} \mathbf{D}_1 & \mathbf{0} \\ \mathbf{0} & \mathbf{0} \end{pmatrix}$; denote by $\mathbf{p}_1, \ldots, \mathbf{p}_r$ the first through rth columns of $\mathbf{P}$ and by $\mathbf{q}_1, \ldots, \mathbf{q}_r$ the first through rth columns of $\mathbf{Q}$; and define $\alpha_1, \ldots, \alpha_k$ to be the distinct values represented among $s_1, \ldots, s_r$. Then, the singular value decomposition of $\mathbf{A}$ is $\mathbf{A} = \sum_{j=1}^{k} \alpha_j \mathbf{U}_j$, where (for $j = 1, \ldots, k$) $\mathbf{U}_j = \sum_{i \in L_j} \mathbf{p}_i \mathbf{q}_i'$ with $L_j = \{i : s_i = \alpha_j\}$. Show that the matrices $\mathbf{U}_1, \ldots, \mathbf{U}_k$, which appear in the singular value decomposition, are such that $\mathbf{U}_j\mathbf{U}_j'\mathbf{U}_j = \mathbf{U}_j$ (for $j = 1, \ldots, k$) and $\mathbf{U}_t'\mathbf{U}_j = \mathbf{0}$ and $\mathbf{U}_t\mathbf{U}_j' = \mathbf{0}$ (for $t \neq j = 1, \ldots, k$).

Solution. By definition,

$$\mathbf{p}_v'\mathbf{p}_i = \mathbf{q}_v'\mathbf{q}_i = \begin{cases} 1, & \text{for } v = i \\ 0, & \text{for } v \neq i. \end{cases}$$

Thus,

$$\mathbf{U}_j'\mathbf{U}_j = \Big(\sum_{v \in L_j} \mathbf{p}_v\mathbf{q}_v'\Big)' \sum_{i \in L_j} \mathbf{p}_i\mathbf{q}_i' = \sum_{v \in L_j} \sum_{i \in L_j} \mathbf{q}_v\mathbf{p}_v'\mathbf{p}_i\mathbf{q}_i' = \sum_{i \in L_j} \mathbf{q}_i\mathbf{q}_i',$$

so that

$$\mathbf{U}_j\mathbf{U}_j'\mathbf{U}_j = \sum_{v \in L_j} \mathbf{p}_v\mathbf{q}_v' \sum_{i \in L_j} \mathbf{q}_i\mathbf{q}_i' = \sum_{v \in L_j} \sum_{i \in L_j} \mathbf{p}_v\mathbf{q}_v'\mathbf{q}_i\mathbf{q}_i' = \sum_{i \in L_j} \mathbf{p}_i\mathbf{q}_i' = \mathbf{U}_j.$$

And, for $t \neq j$,

$$\mathbf{U}_t'\mathbf{U}_j = (\sum_{v \in L_t} \mathbf{p}_v \mathbf{q}_v')' \sum_{i \in L_j} \mathbf{p}_i \mathbf{q}_i' = \sum_{v \in L_t} \sum_{i \in L_j} \mathbf{q}_v \mathbf{p}_v' \mathbf{p}_i \mathbf{q}_i' = \mathbf{0},$$

and

$$\mathbf{U}_t\mathbf{U}_j' = \sum_{v \in L_t} \mathbf{p}_v \mathbf{q}_v' (\sum_{i \in L_j} \mathbf{p}_i \mathbf{q}_i')' = \sum_{v \in L_t} \sum_{i \in L_j} \mathbf{p}_v \mathbf{q}_v' \mathbf{q}_i \mathbf{p}_i' = \mathbf{0}.$$

EXERCISE 28. Let $\mathbf{A}$ represent an $m \times n$ matrix. And, as in Exercise 27, take $\mathbf{P}$ to be an $m \times m$ orthogonal matrix, $\mathbf{Q}$ an $n \times n$ orthogonal matrix, and $\mathbf{D}_1$ an $r \times r$ nonsingular diagonal matrix such that $\mathbf{P}'\mathbf{A}\mathbf{Q} = \begin{pmatrix} \mathbf{D}_1 & \mathbf{0} \\ \mathbf{0} & \mathbf{0} \end{pmatrix}$. Further, partition $\mathbf{P}$ and $\mathbf{Q}$ as $\mathbf{P} = (\mathbf{P}_1, \mathbf{P}_2)$ and $\mathbf{Q} = (\mathbf{Q}_1, \mathbf{Q}_2)$, where each of the matrices $\mathbf{P}_1$ and $\mathbf{Q}_1$ has r columns. Show that $\mathcal{C}(\mathbf{A}) = \mathcal{C}(\mathbf{P}_1)$ and that $\mathcal{N}(\mathbf{A}) = \mathcal{C}(\mathbf{Q}_2)$.

Solution. Clearly,

$$\mathbf{A} = \mathbf{P}\begin{pmatrix} \mathbf{D}_1 & \mathbf{0} \\ \mathbf{0} & \mathbf{0} \end{pmatrix}\mathbf{Q}' = \mathbf{P}_1\mathbf{D}_1\mathbf{Q}_1',$$

implying that $\mathcal{C}(\mathbf{A}) \subset \mathcal{C}(\mathbf{P}_1)$. Moreover, as a consequence of result (12.12), $\mathcal{C}(\mathbf{P}_1) \subset \mathcal{C}(\mathbf{A})$. Thus, $\mathcal{C}(\mathbf{A}) = \mathcal{C}(\mathbf{P}_1)$.

We have that

$$\begin{pmatrix} \mathbf{Q}_1'\mathbf{Q}_1 & \mathbf{Q}_1'\mathbf{Q}_2 \\ \mathbf{Q}_2'\mathbf{Q}_1 & \mathbf{Q}_2'\mathbf{Q}_2 \end{pmatrix} = \mathbf{Q}'\mathbf{Q} = \mathbf{I}_n = \begin{pmatrix} \mathbf{I}_r & \mathbf{0} \\ \mathbf{0} & \mathbf{I}_{n-r} \end{pmatrix},$$

so that $\mathbf{Q}_1'\mathbf{Q}_2 = \mathbf{0}$. Thus,

$$\mathbf{A}\mathbf{Q}_2 = \mathbf{P}_1\mathbf{D}_1\mathbf{Q}_1'\mathbf{Q}_2 = \mathbf{0}.$$

Moreover, making use of result (12.9), we find that

$$\operatorname{rank}(\mathbf{Q}_2) = n - r = n - \operatorname{rank}(\mathbf{A}).$$

And, we conclude, on the basis of Lemma 11.4.1, that $\mathcal{N}(\mathbf{A}) = \mathcal{C}(\mathbf{Q}_2)$.

EXERCISE 29. Let $\mathbf{A}_1, \ldots, \mathbf{A}_k$ represent $n \times n$ not-necessarily-symmetric matrices, each of which is diagonalizable. Show that if $\mathbf{A}_1, \ldots, \mathbf{A}_k$ commute in pairs, then $\mathbf{A}_1, \ldots, \mathbf{A}_k$ are simultaneously diagonalizable.

Solution. The proof is a modified version of the mathematical induction argument employed in Section 13 (in showing that symmetric matrices that commute in pairs are simultaneously diagonalizable).

That one diagonalizable matrix $\mathbf{A}_1$ is "simultaneously" diagonalizable is obvious. Suppose now that any $k - 1$ diagonalizable matrices (of the same order) that commute in pairs can be simultaneously diagonalized (where $k \geq 2$). And, let $\mathbf{A}_1, \ldots, \mathbf{A}_k$ represent k diagonalizable matrices of arbitrary order n that commute in pairs. Then, to complete the induction argument, it suffices to show that $\mathbf{A}_1, \ldots, \mathbf{A}_k$ can be simultaneously diagonalized.

Let $\lambda_1, \ldots, \lambda_r$ represent the distinct eigenvalues of $\mathbf{A}_k$, and let $\nu_1, \ldots, \nu_r$ represent their respective geometric multiplicities. Take $\mathbf{Q}_j$ to be an $n \times \nu_j$ matrix whose columns are linearly independent eigenvectors of $\mathbf{A}_k$ corresponding to the eigenvalue λ_j or, equivalently, whose columns form a basis for the ν_j-dimensional linear space $\mathcal{N}(\mathbf{A}_k - \lambda_j \mathbf{I})$ $(j = 1, \ldots, r)$. And, define $\mathbf{Q} = (\mathbf{Q}_1, \ldots, \mathbf{Q}_r)$.

Since $\mathbf{A}_k$ is diagonalizable, it follows from Corollary 21.5.4 that $\sum_{j=1}^{r} \nu_j = n$, so that (in light of Theorem 21.4.2) $\mathbf{Q}$ has n linearly independent columns and hence is nonsingular. Further,

$$\mathbf{Q}^{-1}\mathbf{A}_k\mathbf{Q} = \operatorname{diag}(\lambda_1 \mathbf{I}_{\nu_1}, \ldots, \lambda_r \mathbf{I}_{\nu_r}) \tag{S.4}$$

(as is evident from Theorems 21.5.2 and 21.5.1).

As in the case where $\mathbf{A}_1, \ldots, \mathbf{A}_k$ are symmetric, there exists a $\nu_j \times \nu_j$ matrix $\mathbf{B}_{ij}$ such that

$$\mathbf{A}_i\mathbf{Q}_j = \mathbf{Q}_j\mathbf{B}_{ij} \tag{S.5}$$

$(i = 1, \ldots, k-1;\ j = 1, \ldots, r)$, and we find that (for $i = 1, \ldots, k-1$)

$$\mathbf{Q}^{-1}\mathbf{A}_i\mathbf{Q} = \operatorname{diag}(\mathbf{B}_{i1}, \ldots, \mathbf{B}_{ir}). \tag{S.6}$$

Now, partition $\mathbf{Q}^{-1}$ as

$$\mathbf{Q}^{-1} = \begin{pmatrix} \mathbf{T}_1 \\ \vdots \\ \mathbf{T}_r \end{pmatrix},$$

where (for $j = 1, \ldots, r$) $\mathbf{T}_j$ has ν_j rows. And, note that

$$\mathbf{T}_j\mathbf{Q}_m = \begin{cases} \mathbf{I}_{\nu_j}, & \text{for } m = j = 1, \ldots, r, \\ \mathbf{0}, & \text{for } m \neq j = 1, \ldots, r. \end{cases}$$

Then, using result (S.5), we find that, for $j = 1, \ldots, r$,

$$\mathbf{B}_{ij} = \mathbf{I}_{\nu_j}\mathbf{B}_{ij} = \mathbf{T}_j\mathbf{Q}_j\mathbf{B}_{ij} = \mathbf{T}_j\mathbf{A}_i\mathbf{Q}_j$$

$(i = 1, \ldots, k-1)$ and that

$$\begin{aligned} \mathbf{B}_{sj}\mathbf{B}_{ij} = \mathbf{T}_j\mathbf{A}_s\mathbf{Q}_j\mathbf{B}_{ij} &= \mathbf{T}_j\mathbf{A}_s\mathbf{A}_i\mathbf{Q}_j \\ &= \mathbf{T}_j\mathbf{A}_i\mathbf{A}_s\mathbf{Q}_j = \mathbf{T}_j\mathbf{A}_i\mathbf{Q}_j\mathbf{B}_{sj} = \mathbf{B}_{ij}\mathbf{B}_{sj} \end{aligned} \tag{S.7}$$

$(s > i = 1, \ldots, k-1)$.

Since (by definition) $\mathbf{A}_i$ is diagonalizable, there exists a nonsingular matrix $\mathbf{L}_i$ and a diagonal matrix $\mathbf{F}_i$ such that $\mathbf{L}_i^{-1}\mathbf{A}_i\mathbf{L}_i = \mathbf{F}_i$, or equivalently such that $\mathbf{A}_i\mathbf{L}_i = \mathbf{L}_i\mathbf{F}_i$, and hence such that

$$\mathbf{Q}^{-1}\mathbf{A}_i\mathbf{Q}(\mathbf{Q}^{-1}\mathbf{L}_i) = \mathbf{Q}^{-1}(\mathbf{L}_i\mathbf{F}_i) \tag{S.8}$$

$(i = 1, \ldots, k-1)$. Comparing (for $i = 1, \ldots, k-1$ and $j = 1, \ldots, r$) the jth group of rows of the left and right sides of equality (S.8) and using equality (S.6), we find that

$$(\mathbf{0}, \ldots, \mathbf{0}, \mathbf{B}_{ij}, \mathbf{0}, \ldots, \mathbf{0})\mathbf{Q}^{-1}\mathbf{L}_i = \mathbf{T}_j\mathbf{L}_i\mathbf{F}_i$$

and hence that

$$\mathbf{B}_{ij}(\mathbf{T}_j\mathbf{L}_i) = (\mathbf{T}_j\mathbf{L}_i)\mathbf{F}_i \,.$$

Thus, any nonnull column of the $\nu_j \times n$ matrix $\mathbf{T}_j\mathbf{L}_i$ is an eigenvector of $\mathbf{B}_{ij}$. Moreover, since $\mathbf{L}_i$ is nonsingular and since the rows of $\mathbf{Q}^{-1}$ are linearly independent,

$$\text{rank}(\mathbf{T}_j\mathbf{L}_i) = \text{rank}(\mathbf{T}_j) = \nu_j \,,$$

so that (according to Theorem 4.4.10) $\mathbf{T}_j\mathbf{L}_i$ contains ν_j linearly independent columns and it follows from Corollary 21.5.3 that $\mathbf{B}_{ij}$ is diagonalizable.

We have (for $j = 1, \ldots, r$) that each of the matrices $\mathbf{B}_{1j}, \ldots, \mathbf{B}_{k-1,j}$ is diagonalizable, and it follows from result (S.7) that $\mathbf{B}_{1j}, \ldots, \mathbf{B}_{k-1,j}$ commute in pairs. Thus, by supposition, $\mathbf{B}_{1j}, \ldots, \mathbf{B}_{k-1,j}$ can be simultaneously diagonalized; that is, there exists a $\nu_j \times \nu_j$ nonsingular matrix $\mathbf{S}_j$ and $\nu_j \times \nu_j$ diagonal matrices $\mathbf{D}_{1j}, \ldots, \mathbf{D}_{k-1,j}$ such that (for $i = 1, \ldots, k-1$)

$$\mathbf{S}_j^{-1}\mathbf{B}_{ij}\mathbf{S}_j = \mathbf{D}_{ij} \,. \tag{S.9}$$

Define $\mathbf{S} = \text{diag}(\mathbf{S}_1, \ldots, \mathbf{S}_r)$ and $\mathbf{P} = \mathbf{QS}$. Then, clearly, $\mathbf{S}$ is nonsingular, and hence $\mathbf{P}$ is also nonsingular. Further, using results (S.6), (S.9), and (S.4), we find that, for $i = 1, \ldots, k-1$,

$$\mathbf{P}^{-1}\mathbf{A}_i\mathbf{P} = \mathbf{S}^{-1}\mathbf{Q}^{-1}\mathbf{A}_i\mathbf{QS} = \text{diag}(\mathbf{D}_{i1}, \ldots, \mathbf{D}_{ir})$$

and that

$$\mathbf{P}^{-1}\mathbf{A}_k\mathbf{P} = \mathbf{S}^{-1}\mathbf{Q}^{-1}\mathbf{A}_k\mathbf{QS} = \text{diag}(\lambda_1\mathbf{I}_{\nu_1}, \ldots, \lambda_r\mathbf{I}_{\nu_r}),$$

so that all k of the matrices $\mathbf{A}_1, \ldots, \mathbf{A}_k$ are simultaneously diagonalized by the nonsingular matrix $\mathbf{P}$.

EXERCISE 30. Let $\mathbf{V}$ represent an $n \times n$ symmetric nonnegative definite matrix, $\mathbf{X}$ an $n \times p$ matrix of rank r, and $\mathbf{d}$ a $p \times 1$ vector. Using the results of Exercise 19.11 (or otherwise), show that each of the following three conditions is necessary and sufficient for the vector $\mathbf{X}(\mathbf{X}'\mathbf{X})^{-}\mathbf{d}$ to be a solution, for every $\mathbf{d} \in \mathcal{C}(\mathbf{X}')$, to the problem of minimizing the quadratic form $\mathbf{a}'\mathbf{V}\mathbf{a}$ (in $\mathbf{a}$) subject to $\mathbf{X}'\mathbf{a} = \mathbf{d}$:

(a) there exists an orthogonal matrix that simultaneously diagonalizes $\mathbf{V}$ and $\mathbf{P_X}$;

(b) there exists a subset of r orthonormal eigenvectors of $\mathbf{V}$ that is a basis for $\mathcal{C}(\mathbf{X})$;

(c) there exists a subset of r eigenvectors of $\mathbf{V}$ that is a basis for $\mathcal{C}(\mathbf{X})$.

Solution. (a) Recall from Part (3) of Theorem 12.3.4 that $\mathbf{P_X}$ is symmetric. Then, as a consequence of Corollary 21.13.2, there exists an orthogonal matrix that simultaneously diagonalizes $\mathbf{V}$ and $\mathbf{P_X}$ if and only if $\mathbf{P_X V} = \mathbf{V P_X}$. And, it follows from the results of Exercise 19.11 that the existence of an orthogonal matrix that simultaneously diagonalizes $\mathbf{V}$ and $\mathbf{P_X}$ is a necessary and sufficient condition for $\mathbf{X}(\mathbf{X}'\mathbf{X})^{-}\mathbf{d}$ to be a solution [for every $\mathbf{d} \in \mathcal{C}(\mathbf{X}')$] to the problem of minimizing $\mathbf{a}'\mathbf{V}\mathbf{a}$ subject to $\mathbf{X}'\mathbf{a} = \mathbf{d}$.

(b) and (c). Suppose that there exist r (possibly orthonormal) eigenvectors $\mathbf{u}_1, \ldots, \mathbf{u}_r$ of $\mathbf{V}$ that form a basis for $\mathcal{C}(\mathbf{X})$, and let $\mathbf{U} = (\mathbf{u}_1, \ldots, \mathbf{u}_r)$. Then, $\mathbf{VU} = \mathbf{UD}$ for some (diagonal) matrix $\mathbf{D}$. Moreover, since clearly $\mathcal{C}(\mathbf{U}) = \mathcal{C}(\mathbf{X})$, $\mathbf{X} = \mathbf{UT}$ and $\mathbf{U} = \mathbf{XS}$ for some matrices $\mathbf{T}$ and $\mathbf{S}$. Thus,

$$\mathbf{VX} = \mathbf{VUT} = \mathbf{UDT} = \mathbf{XSDT} = \mathbf{XQ}$$

for $\mathbf{Q} = \mathbf{SDT}$. And, it follows from the results of Exercise 19.11 that $\mathbf{X}(\mathbf{X}'\mathbf{X})^{-}\mathbf{d}$ is a solution [for every $\mathbf{d} \in \mathcal{C}(\mathbf{X}')$] to the problem of minimizing $\mathbf{a}'\mathbf{V}\mathbf{a}$ subject to $\mathbf{X}'\mathbf{a} = \mathbf{d}$.

Conversely, suppose that $\mathbf{X}(\mathbf{X}'\mathbf{X})^{-}\mathbf{d}$ is a solution [for every $\mathbf{d} \in \mathcal{C}(\mathbf{X}')$] to the problem of minimizing $\mathbf{a}'\mathbf{V}\mathbf{a}$ subject to $\mathbf{X}'\mathbf{a} = \mathbf{d}$. Then, it follows from Part (a) that there exists an $n \times n$ orthogonal matrix $\mathbf{Q}$ that simultaneously diagonalizes $\mathbf{P_X}$ and $\mathbf{V}$. That is, there exists an $n \times n$ orthogonal matrix $\mathbf{Q}$ such that $\mathbf{Q}'\mathbf{P_X}\mathbf{Q} = \mathrm{diag}(d_1, \ldots, d_n)$ and $\mathbf{Q}'\mathbf{V}\mathbf{Q} = \mathrm{diag}(f_1, \ldots, f_n)$ for some scalars $d_1, \ldots, d_n$ and $f_1, \ldots, f_n$.

Further, it follows from Theorem 21.5.1 that the (not necessarily distinct) eigenvalues of $\mathbf{P_X}$ are $d_1, \ldots, d_n$ and the (not necessarily distinct) eigenvalues of $\mathbf{V}$ are $f_1, \ldots, f_n$ and that the first, ..., nth columns of $\mathbf{Q}$ are eigenvectors of $\mathbf{P_X}$ corresponding to $d_1, \ldots, d_n$, respectively, and are also eigenvectors of $\mathbf{V}$ corresponding to $f_1, \ldots, f_n$, respectively. And, since [according to Part (8) of Theorem 12.3.4] $\mathrm{rank}(\mathbf{P_X}) = r$, $n - r$ of the eigenvalues of $\mathbf{P_X}$ are (in light of Lemma 21.1.1) equal to 0.

Now, let $\mathbf{Q}_1$ represent the $n \times r$ matrix obtained from $\mathbf{Q}$ by deleting those columns that are eigenvectors of $\mathbf{P_X}$ corresponding to 0. Then, in light of Theorem 21.4.3 and Part (7) of Theorem 12.3.4, $\mathcal{C}(\mathbf{Q}_1) = \mathcal{C}(\mathbf{P_X}) = \mathcal{C}(\mathbf{X})$. And, since the columns of $\mathbf{Q}_1$ are (orthonormal) eigenvectors of $\mathbf{V}$, there exist r orthonormal eigenvectors of $\mathbf{V}$ that form a basis for $\mathcal{C}(\mathbf{X})$.

EXERCISE 31. Let $\mathbf{A}$ represent an $n \times n$ symmetric matrix, and let $\mathbf{B}$ represent an $n \times n$ symmetric positive definite matrix. And, let $\lambda_{\max}$ and $\lambda_{\min}$ represent, respectively, the largest and smallest roots of $|\mathbf{A} - \lambda\mathbf{B}|$. Show that

$$\lambda_{\min} \leq \frac{\mathbf{x}'\mathbf{A}\mathbf{x}}{\mathbf{x}'\mathbf{B}\mathbf{x}} \leq \lambda_{\max}$$

for every nonnull vector $\mathbf{x}$ in $\mathcal{R}^n$.

Solution. Let $\mathbf{S}$ represent any $n \times n$ nonsingular matrix such that $\mathbf{B} = \mathbf{S}'\mathbf{S}$, let $\mathbf{R} = (\mathbf{S}^{-1})'$ (so that $\mathbf{B}^{-1} = \mathbf{R}'\mathbf{R}$), and let $\mathbf{C} = \mathbf{RAR}'$. Then, in light of result (14.7), $\lambda_{\max}$ and $\lambda_{\min}$ are, respectively, the largest and smallest eigenvalues of $\mathbf{C}$.

And, it follows from Theorem 21.5.6 that

$$\lambda_{\min} \leq \frac{\mathbf{y}'\mathbf{C}\mathbf{y}}{\mathbf{y}'\mathbf{y}} \leq \lambda_{\max}$$

for every nonnull vector $\mathbf{y}$ in $\mathcal{R}^n$.

Now, let $\mathbf{x}$ represent an arbitrary nonnull vector in $\mathcal{R}^n$, and let $\mathbf{y} = \mathbf{S}\mathbf{x}$. Then, $\mathbf{y}$ is nonnull, and

$$\frac{\mathbf{y}'\mathbf{C}\mathbf{y}}{\mathbf{y}'\mathbf{y}} = \frac{\mathbf{x}'\mathbf{S}'\mathbf{C}\mathbf{S}\mathbf{x}}{\mathbf{x}'\mathbf{S}'\mathbf{S}\mathbf{x}} = \frac{\mathbf{x}'\mathbf{A}\mathbf{x}}{\mathbf{x}'\mathbf{B}\mathbf{x}}.$$

Thus,

$$\lambda_{\min} \leq \frac{\mathbf{x}'\mathbf{A}\mathbf{x}}{\mathbf{x}'\mathbf{B}\mathbf{x}} \leq \lambda_{\max}.$$

EXERCISE 32. Let $\mathbf{A}$ represent an $n \times n$ symmetric matrix, and let $\mathbf{B}$ represent an $n \times n$ symmetric positive definite matrix. Show that $\mathbf{A} - \mathbf{B}$ is nonnegative definite if and only if all n (not necessarily distinct) roots of $|\mathbf{A} - \lambda\mathbf{B}|$ are greater than or equal to 1 and is positive definite if and only if all n roots are (strictly) greater than 1.

Solution. Let $d_1, \ldots, d_n$ represent the n (not necessarily distinct) roots of $|\mathbf{A} - \lambda\mathbf{B}|$. And, let $\mathbf{S}$ represent any $n \times n$ nonsingular matrix such that $\mathbf{B} = \mathbf{S}'\mathbf{S}$, let $\mathbf{R} = (\mathbf{S}^{-1})'$ (so that $\mathbf{B}^{-1} = \mathbf{R}'\mathbf{R}$), and let $\mathbf{C} = \mathbf{R}\mathbf{A}\mathbf{R}'$. Then, in light of result (14.7), the (not necessarily distinct) eigenvalues of $\mathbf{C}$ are $d_1, \ldots, d_n$, and it follows from Corollary 21.5.9 that there exists an $n \times n$ orthogonal matrix $\mathbf{P}$ such that $\mathbf{P}'\mathbf{C}\mathbf{P} = \mathbf{D}$, where $\mathbf{D} = \text{diag}(d_1, \ldots, d_n)$.

Now, take $\mathbf{Q} = \mathbf{S}'\mathbf{P}$. Then, according to results (14.2) and (14.1), $\mathbf{A} = \mathbf{Q}\mathbf{D}\mathbf{Q}'$ and $\mathbf{B} = \mathbf{Q}\mathbf{Q}'$. Thus,

$$\mathbf{A} - \mathbf{B} = \mathbf{Q}(\mathbf{D} - \mathbf{I}_n)\mathbf{Q}' = \mathbf{Q}\,\text{diag}(d_1 - 1, \ldots, d_n - 1)\mathbf{Q}'.$$

And, it follows from Corollary 14.2.15 that $\mathbf{A} - \mathbf{B}$ is nonnegative definite if and only if, for $i = 1, \ldots, n,\ d_i - 1 \geq 0$ and is positive definite if and only if, for $i = 1, \ldots, n,\ d_i - 1 > 0$. Or, equivalently, $\mathbf{A} - \mathbf{B}$ is nonnegative definite if and only if, for $i = 1, \ldots, n,\ d_i \geq 1$ and is positive definite if and only if, for $i = 1, \ldots, n,\ d_i > 1$.

22

Linear Transformations

EXERCISE 1. Let $\mathcal{U}$ represent a subspace of a linear space $\mathcal{V}$, and let S represent a linear transformation from $\mathcal{U}$ into a linear space $\mathcal{W}$. Show that there exists a linear transformation T from $\mathcal{V}$ into $\mathcal{W}$ such that S is the restriction of T to $\mathcal{U}$.

Solution. Let $\{\mathbf{X}_1, \ldots, \mathbf{X}_r\}$ represent a basis for $\mathcal{U}$. Then, it follows from Theorem 4.3.12 that there exist matrices $\mathbf{X}_{r+1}, \ldots, \mathbf{X}_{r+k}$ such that $\{\mathbf{X}_1, \ldots, \mathbf{X}_r, \mathbf{X}_{r+1}, \ldots, \mathbf{X}_{r+k}\}$ is a basis for $\mathcal{V}$.

Now, for $i = 1, \ldots, r$, define $\mathbf{Y}_i = S(\mathbf{X}_i)$; and, for $i = r+1, \ldots, r+k$, take $\mathbf{Y}_i$ to be any matrix in $\mathcal{W}$. And, letting $\mathbf{X}$ represent an arbitrary matrix in $\mathcal{V}$, take T to be the transformation from $\mathcal{V}$ into $\mathcal{W}$ defined by

$$T(\mathbf{X}) = c_1\mathbf{Y}_1 + \cdots + c_r\mathbf{Y}_r + c_{r+1}\mathbf{Y}_{r+1} + \cdots + c_{r+k}\mathbf{Y}_{r+k},$$

where $c_1, \ldots, c_r, c_{r+1}, \ldots, c_{r+k}$ are the (unique) scalars that satisfy $\mathbf{X} = c_1\mathbf{X}_1 + \cdots + c_r\mathbf{X}_r + c_{r+1}\mathbf{X}_{r+1} + \cdots + c_{r+k}\mathbf{X}_{r+k}$ — since $\mathbf{Y}_1, \ldots, \mathbf{Y}_r, \mathbf{Y}_{r+1}, \ldots, \mathbf{Y}_{r+k}$ are in the linear space $\mathcal{W}$, $T(\mathbf{X})$ is in $\mathcal{W}$. Clearly, if $\mathbf{X} \in \mathcal{U}$, then $c_{r+1} = \cdots = c_{r+k} = 0$, and hence

$$T(\mathbf{X}) = c_1\mathbf{Y}_1 + \cdots + c_r\mathbf{Y}_r = \sum_{i=1}^{r} c_i S(\mathbf{X}_i) = S\left(\sum_{i=1}^{r} c_i \mathbf{X}_i\right) = S(\mathbf{X}).$$

Moreover, it follows from Lemma 22.1.8 that T is linear. Thus, T is a linear transformation from $\mathcal{V}$ into $\mathcal{W}$ such that S is the restriction of T to $\mathcal{U}$.

EXERCISE 2. Let T represent a 1-1 linear transformation from a linear space $\mathcal{V}$ into a linear space $\mathcal{W}$. And, write $\mathbf{U}\bullet\mathbf{Y}$ for the inner product of arbitrary matrices

$\mathbf{U}$ and $\mathbf{Y}$ in $\mathcal{W}$. Further, define $\mathbf{X} * \mathbf{Z} = T(\mathbf{X}) \cdot T(\mathbf{Z})$ for all matrices $\mathbf{X}$ and $\mathbf{Z}$ in $\mathcal{V}$. Show that the "$*$-operation" satisfies the four properties required of an inner product for $\mathcal{V}$.

Solution. Observe (in light of Lemma 22.1.3) that $\mathcal{N}(T) = \{\mathbf{0}\}$ and hence that $T(\mathbf{X}) = \mathbf{0}$ if and only if $\mathbf{X} = \mathbf{0}$. Then, letting $\mathbf{X}$, $\mathbf{Z}$, and $\mathbf{Y}$ represent arbitrary matrices in $\mathcal{V}$ and letting k represent an arbitrary scalar, we find that

(1) $\mathbf{X} * \mathbf{Z} = T(\mathbf{X}) \cdot T(\mathbf{Z}) = T(\mathbf{Z}) \cdot T(\mathbf{X}) = \mathbf{Z} * \mathbf{X}$;

(2) $$\mathbf{X} * \mathbf{X} = T(\mathbf{X}) \cdot T(\mathbf{X}) \begin{cases} > 0, & \text{if } T(\mathbf{X}) \neq \mathbf{0} \text{ or, equivalently, if } \mathbf{X} \neq \mathbf{0}, \\ = 0, & \text{if } T(\mathbf{X}) = \mathbf{0} \text{ or, equivalently, if } \mathbf{X} = \mathbf{0}; \end{cases}$$

(3) $$\begin{aligned} (k\mathbf{X}) * \mathbf{Z} &= T(k\mathbf{X}) \cdot T(\mathbf{Z}) \\ &= [kT(\mathbf{X})] \cdot T(\mathbf{Z}) = k[T(\mathbf{X}) \cdot T(\mathbf{Z})] = k(\mathbf{X} * \mathbf{Z}); \text{ and} \end{aligned}$$

(4) $$\begin{aligned} (\mathbf{X} + \mathbf{Z}) * \mathbf{Y} &= T(\mathbf{X} + \mathbf{Z}) \cdot T(\mathbf{Y}) \\ &= [T(\mathbf{X}) + T(\mathbf{Z})] \cdot T(\mathbf{Y}) \\ &= [T(\mathbf{X}) \cdot T(\mathbf{Y})] + [T(\mathbf{Z}) \cdot T(\mathbf{Y})] = (\mathbf{X} * \mathbf{Y}) + (\mathbf{Z} * \mathbf{Y}). \end{aligned}$$

EXERCISE 3. Let T represent a linear transformation from a linear space $\mathcal{V}$ into a linear space $\mathcal{W}$, and let $\mathcal{U}$ represent any subspace of $\mathcal{V}$ such that $\mathcal{U}$ and $\mathcal{N}(T)$ are essentially disjoint. Further, let $\{\mathbf{X}_1, \ldots, \mathbf{X}_r\}$ represent a linearly independent set of r matrices in $\mathcal{U}$.

(a) Show that $T(\mathbf{X}_1), \ldots, T(\mathbf{X}_r)$ are linearly independent.

(b) Show that if $r = \dim(\mathcal{U})$ (or equivalently if $\mathbf{X}_1, \ldots, \mathbf{X}_r$ form a basis for $\mathcal{U}$) and if $\mathcal{U} \oplus \mathcal{N}(T) = \mathcal{V}$, then $T(\mathbf{X}_1), \ldots, T(\mathbf{X}_r)$ form a basis for $T(\mathcal{V})$.

Solution. (a) Let $c_1, \ldots, c_r$ represent any scalars such that $\sum_{i=1}^r c_i T(\mathbf{X}_i) = \mathbf{0}$. Then, $T(\sum_{i=1}^r c_i \mathbf{X}_i) = \sum_{i=1}^r c_i T(\mathbf{X}_i) = \mathbf{0}$, implying that $\sum_{i=1}^r c_i \mathbf{X}_i$ is in $\mathcal{N}(T)$ and hence (since clearly $\sum_{i=1}^r c_i \mathbf{X}_i \in \mathcal{U}$) that $\sum_{i=1}^r c_i \mathbf{X}_i$ is in $\mathcal{U} \cap \mathcal{N}(T)$. Thus, $\sum_{i=1}^r c_i \mathbf{X}_i = \mathbf{0}$, and (since $\mathbf{X}_1, \ldots, \mathbf{X}_r$ are linearly independent) it follows that $c_1 = \cdots = c_r = 0$. And, we conclude that $T(\mathbf{X}_1), \ldots, T(\mathbf{X}_r)$ are linearly independent.

(b) Suppose that $r = \dim(\mathcal{U})$ and that $\mathcal{U} \oplus \mathcal{N}(T) = \mathcal{V}$. Then, making use of Theorem 22.1.1 and of Corollary 17.1.6, we find that

$$\dim[T(\mathcal{V})] = \dim(\mathcal{V}) - \dim[\mathcal{N}(T)] = r.$$

And, in light of Theorem 4.3.9 and the result of Part (a), it follows that $T(\mathbf{X}_1)$, $\ldots, T(\mathbf{X}_r)$ form a basis for $T(\mathcal{V})$.

An alternative proof [of the result of Part (b)] can be obtained [in light of the result of Part (a)] by showing that the the set $\{T(\mathbf{X}_1), \ldots, T(\mathbf{X}_r)\}$ spans $T(\mathcal{V})$. Continue to suppose that $r = \dim(\mathcal{U})$ and that $\mathcal{U} \oplus \mathcal{N}(T) = \mathcal{V}$. And, let $\mathbf{Z}_1, \ldots, \mathbf{Z}_s$ represent any matrices that form a basis for $\mathcal{N}(T)$. Further, observe, in light of Theorem 17.1.5, that the $r + s$ matrices $\mathbf{X}_1, \ldots, \mathbf{X}_r, \mathbf{Z}_1, \ldots, \mathbf{Z}_s$ form a basis for $\mathcal{V}$.

Now, let $\mathbf{Y}$ represent an arbitrary matrix in $T(\mathcal{V})$. Then, $\mathbf{Y} = T(\mathbf{X})$ for some matrix $\mathbf{X}$ in $\mathcal{V}$ and $\mathbf{X} = \sum_{i=1}^{r} c_i \mathbf{X}_i + \sum_{j=1}^{s} k_j \mathbf{Z}_j$, so that

$$\mathbf{Y} = T\left(\sum_{i=1}^{r} c_i \mathbf{X}_i + \sum_{j=1}^{s} k_j \mathbf{Z}_j\right) = \sum_{i=1}^{r} c_i T(\mathbf{X}_i) + \sum_{j=1}^{s} k_j T(\mathbf{Z}_j) = \sum_{i=1}^{r} c_i T(\mathbf{X}_i).$$

Thus, $\{T(\mathbf{X}_1), \ldots, T(\mathbf{X}_r)\}$ spans $T(\mathcal{V})$.

EXERCISE 4. Let T and S represent linear transformations from a linear space $\mathcal{V}$ into a linear space $\mathcal{W}$, and let k represent an arbitrary scalar.

(a) Verify that the transformation kT is linear.

(b) Verify that the transformation $T + S$ is linear.

Solution. (a) For any matrices $\mathbf{X}$ and $\mathbf{Z}$ in $\mathcal{V}$ and for any scalar c,

$$\begin{aligned}(kT)(\mathbf{X}+\mathbf{Z}) = kT(\mathbf{X}+\mathbf{Z}) &= T[k(\mathbf{X}+\mathbf{Z})] \\ &= T(k\mathbf{X}+k\mathbf{Z}) \\ &= T(k\mathbf{X}) + T(k\mathbf{Z}) \\ &= kT(\mathbf{X}) + kT(\mathbf{Z}) = (kT)(\mathbf{X}) + (kT)(\mathbf{Z}),\end{aligned}$$

and

$$(kT)(c\mathbf{X}) = kT(c\mathbf{X}) = k[cT(\mathbf{X})] = c[kT(\mathbf{X})] = c(kT)(\mathbf{X}).$$

(b) For any matrices $\mathbf{X}$ and $\mathbf{Z}$ in $\mathcal{V}$ and for any scalar c,

$$\begin{aligned}(T+S)(\mathbf{X}+\mathbf{Z}) &= T(\mathbf{X}+\mathbf{Z}) + S(\mathbf{X}+\mathbf{Z}) \\ &= T(\mathbf{X}) + T(\mathbf{Z}) + S(\mathbf{X}) + S(\mathbf{Z}) \\ &= T(\mathbf{X}) + S(\mathbf{X}) + T(\mathbf{Z}) + S(\mathbf{Z}) \\ &= (T+S)(\mathbf{X}) + (T+S)(\mathbf{Z}),\end{aligned}$$

and

$$\begin{aligned}(T+S)(c\mathbf{X}) = T(c\mathbf{X}) + S(c\mathbf{X}) &= cT(\mathbf{X}) + cS(\mathbf{X}) \\ &= c[T(\mathbf{X}) + S(\mathbf{X})] = c(T+S)(\mathbf{X}).\end{aligned}$$

EXERCISE 5. Let S represent a linear transformation from a linear space $\mathcal{U}$ into a linear space $\mathcal{V}$, and let T represent a linear transformation from $\mathcal{V}$ into a linear space $\mathcal{W}$. Show that the transformation TS is linear.

Solution. For any matrices $\mathbf{X}$ and $\mathbf{Z}$ in $\mathcal{U}$ and for any scalar c,

$$\begin{aligned}(TS)(\mathbf{X}+\mathbf{Z}) = T[S(\mathbf{X}+\mathbf{Z})] &= T[S(\mathbf{X}) + S(\mathbf{Z})] \\ &= T[S(\mathbf{X})] + T[S(\mathbf{Z})] = (TS)(\mathbf{X}) + (TS)(\mathbf{Z}),\end{aligned}$$

and

$$(TS)(c\mathbf{X}) = T[S(c\mathbf{X})] = T[cS(\mathbf{X})] = cT[S(\mathbf{X})] = c(TS)(\mathbf{X}).$$

EXERCISE 6. Let T represent a linear transformation from a linear space $\mathcal{V}$ into a linear space $\mathcal{W}$, and let R represent a linear transformation from a linear space $\mathcal{U}$ into $\mathcal{W}$. Show that if $T(\mathcal{V}) \subset R(\mathcal{U})$, then there exists a linear transformation S from $\mathcal{V}$ into $\mathcal{U}$ such that $T = RS$.

Solution. Suppose that $T(\mathcal{V}) \subset R(\mathcal{U})$. And, let $\{\mathbf{X}_1, \dots, \mathbf{X}_r\}$ represent a basis for $\mathcal{V}$. Then, for $i = 1, \dots, r$, $T(\mathbf{X}_i) \in R(\mathcal{U})$, and consequently $T(\mathbf{X}_i) = R(\mathbf{Y}_i)$ for some matrix $\mathbf{Y}_i$ in $\mathcal{U}$.

Now, let $\mathbf{X}$ represent an arbitrary matrix in $\mathcal{V}$, and let $c_1, \dots, c_r$ represent the (unique) scalars that satisfy the equality $\mathbf{X} = \sum_{i=1}^r c_i \mathbf{X}_i$. And, take S to be the transformation from $\mathcal{V}$ into $\mathcal{U}$ defined by $S(\mathbf{X}) = \sum_{i=1}^r c_i \mathbf{Y}_i$. Then,

$$\begin{aligned} T(\mathbf{X}) = T\left(\sum_{i=1}^r c_i \mathbf{X}_i\right) &= \sum_{i=1}^r c_i T(\mathbf{X}_i) \\ &= \sum_{i=1}^r c_i R(\mathbf{Y}_i) \\ &= R\left(\sum_{i=1}^r c_i \mathbf{Y}_i\right) = R[S(\mathbf{X})] = (RS)(\mathbf{X}). \end{aligned}$$

Thus, $T = RS$. Moreover, it follows from Lemma 22.1.8 that S is linear.

EXERCISE 7. Let T represent a transformation from a linear space $\mathcal{V}$ into a linear space $\mathcal{W}$, and let S and R represent transformations from $\mathcal{W}$ into $\mathcal{V}$. And, suppose that $RT = I$ (where the identity transformation I is from $\mathcal{V}$ onto $\mathcal{V}$) and that $TS = I$ (where the identity transformation I is from $\mathcal{W}$ onto $\mathcal{W}$).

(a) Show that T is invertible.

(b) Show that $R = S = T^{-1}$.

Solution. (a) For any matrices $\mathbf{X}$ and $\mathbf{Z}$ in $\mathcal{V}$ such that $T(\mathbf{X}) = T(\mathbf{Z})$,

$$\mathbf{X} = I(\mathbf{X}) = (RT)(\mathbf{X}) = R[T(\mathbf{X})] = R[T(\mathbf{Z})] = (RT)(\mathbf{Z}) = I(\mathbf{Z}) = \mathbf{Z}.$$

Thus, T is 1-1. Further, for any matrix $\mathbf{Y}$ in $\mathcal{W}$,

$$\mathbf{Y} = I(\mathbf{Y}) = (TS)(\mathbf{Y}) = T(\mathbf{X}),$$

where $\mathbf{X} = S(\mathbf{Y})$. And, it follows that T is onto. Since T is both 1-1 and onto, we conclude that T is invertible.

(b) Using results (3.3) and (3.1), we find that

$$R = RI = R(TT^{-1}) = (RT)T^{-1} = IT^{-1} = T^{-1}$$

and

$$S = IS = (T^{-1}T)S = T^{-1}(TS) = T^{-1}I = T^{-1}.$$

EXERCISE 8. Let T represent an invertible transformation from a linear space $\mathcal{V}$ into a linear space $\mathcal{W}$, let S represent an invertible transformation from a linear space $\mathcal{U}$ into $\mathcal{V}$, and let k represent an arbitrary scalar. Using the results of Exercise 7 (or otherwise), show that

(a) kT is invertible and $(kT)^{-1} = (1/k)T^{-1}$ and that

(b) TS is invertible and $(TS)^{-1} = S^{-1}T^{-1}$.

Solution. (a) In light of the results of Exercise 7, it suffices to show that

$$((1/k)T^{-1})(kT) = I \quad \text{and} \quad (kT)((1/k)T^{-1}) = I.$$

Using results (2.12), (3.1), (3.3), (2.2), and (2.1), we find that

$$\begin{aligned}((1/k)T^{-1})(kT) &= (1/k)(T^{-1}(kT)) \\ &= (1/k)(k(T^{-1}T)) = (1/k)(kI) = [(1/k)k]I = 1I = I\end{aligned}$$

and similarly that

$$\begin{aligned}(kT)((1/k)T^{-1}) &= (1/k)((kT)T^{-1}) \\ &= (1/k)(k(TT^{-1})) = (1/k)(kI) = [(1/k)k]I = 1I = I.\end{aligned}$$

(b) In light of the results of Exercise 7, it suffices to show that

$$(S^{-1}T^{-1})(TS) = I \quad \text{and} \quad (TS)(S^{-1}T^{-1}) = I.$$

Using results (2.9), (3.1), (3.3), and (2.13), we find that

$$\begin{aligned}(S^{-1}T^{-1})(TS) &= ((S^{-1}T^{-1})T)S \\ &= (S^{-1}(T^{-1}T))S = (S^{-1}I)S = S^{-1}S = I\end{aligned}$$

and similarly that

$$\begin{aligned}(TS)(S^{-1}T^{-1}) &= ((TS)S^{-1})T^{-1} \\ &= (T(SS^{-1}))T^{-1} = (TI)T^{-1} = TT^{-1} = I.\end{aligned}$$

EXERCISE 9. Let T represent a linear transformation from an n-dimensional linear space $\mathcal{V}$ into an m-dimensional linear space $\mathcal{W}$. And, write $\mathbf{U} \bullet \mathbf{Y}$ for the inner product of arbitrary matrices $\mathbf{U}$ and $\mathbf{Y}$ in $\mathcal{W}$. Further, let B represent a set of matrices $\mathbf{V}_1, \ldots, \mathbf{V}_n$ (in $\mathcal{V}$) that form a basis for $\mathcal{V}$, and let C represent a set of matrices $\mathbf{W}_1, \ldots, \mathbf{W}_m$ (in $\mathcal{W}$) that form an orthonormal basis for $\mathcal{W}$. Show that the matrix representation of T with respect to B and C is the $m \times n$ matrix whose ijth element is $T(\mathbf{V}_j) \bullet \mathbf{W}_i$.

Solution. As a consequence of Theorem 6.4.4, we have that (for $j = 1, \ldots, n$)

$$T(\mathbf{V}_j) = \sum_{i=1}^{m} [T(\mathbf{V}_j) \bullet \mathbf{W}_i]\mathbf{W}_i.$$

And, upon comparing this expression for $T(\mathbf{V}_j)$ with expression (4.3), we find that the matrix representation of T with respect to B and C is the $m \times n$ matrix whose ijth element is $T(\mathbf{V}_j) \bullet \mathbf{W}_i$.

EXERCISE 10. Let T represent the linear transformation from $\mathcal{R}^{m\times n}$ into $\mathcal{R}^{n\times m}$ defined by $T(\mathbf{X}) = \mathbf{X}'$. And, let C represent the natural basis for $\mathcal{R}^{m\times n}$, comprising the mn matrices $\mathbf{U}_{11}, \mathbf{U}_{21}, \ldots, \mathbf{U}_{m1}, \ldots, \mathbf{U}_{1n}, \mathbf{U}_{2n}, \ldots, \mathbf{U}_{mn}$, where (for $i = 1, \ldots, m$ and $j = 1, \ldots n$) $\mathbf{U}_{ij}$ is the $m \times n$ matrix whose ijth element equals 1 and whose remaining $mn - 1$ elements equal 0; and similarly let D represent the natural basis for $\mathcal{R}^{n\times m}$. Show that the matrix representation for T with respect to the bases C and D is the vec-permutation matrix $\mathbf{K}_{mn}$.

Solution. Making use of results (4.11) and (16.3.1), we find that, for any $m \times n$ matrix $\mathbf{X}$,

$$(L_D^{-1} T L_C)(\text{vec}\,\mathbf{X}) = \text{vec}[T(\mathbf{X})] = \text{vec}(\mathbf{X}') = \mathbf{K}_{mn}\text{vec}\,\mathbf{X}.$$

And, in light of result (4.7), it follows that the matrix representation of T (with respect to C and D) equals $\mathbf{K}_{mn}$.

EXERCISE 11. Let $\mathcal{W}$ represent the linear space of all $p \times p$ symmetric matrices, and let T represent a linear transformation from $\mathcal{R}^{m\times n}$ into $\mathcal{W}$. Further, let B represent the natural basis for $\mathcal{R}^{m\times n}$, comprising the mn matrices $\mathbf{U}_{11}, \mathbf{U}_{21}, \ldots, \mathbf{U}_{m1}, \ldots, \mathbf{U}_{1n}, \mathbf{U}_{2n}, \ldots, \mathbf{U}_{mn}$, where (for $i = 1, \ldots, m$ and $j = 1, \ldots, n$) $\mathbf{U}_{ij}$ is the $m \times n$ matrix whose ijth element equals 1 and whose remaining $mn - 1$ elements equal 0. And, let C represent the usual basis for $\mathcal{W}$.

(a) Show that, for any $m \times n$ matrix $\mathbf{X}$,

$$(L_C^{-1} T L_B)(\text{vec}\,\mathbf{X}) = \text{vech}[T(\mathbf{X})].$$

(b) Show that the matrix representation of T (with respect to B and C) equals the $p(p+1)/2 \times mn$ matrix

$$[\text{vech}\,T(\mathbf{U}_{11}), \ldots, \text{vech}\,T(\mathbf{U}_{m1}), \ldots, \text{vech}\,T(\mathbf{U}_{1n}), \ldots, \text{vech}\,T(\mathbf{U}_{mn})].$$

(c) Suppose that $p = m = n$ and that (for every $n \times n$ matrix $\mathbf{X}$)

$$T(\mathbf{X}) = (1/2)(\mathbf{X} + \mathbf{X}').$$

Show that the matrix representation of T (with respect to B and C) equals

$$(\mathbf{G}_n'\mathbf{G}_n)^{-1}\mathbf{G}_n'$$

(where $\mathbf{G}_n$ is the duplication matrix).

Solution. (a) Making use of result (3.5), we find that, for any $mn \times 1$ vector $\mathbf{x}$,

$$\begin{aligned}(L_C^{-1} T L_B)(\mathbf{x}) &= (L_C^{-1}(T L_B))(\mathbf{x}) \\ &= L_C^{-1}[(T L_B)(\mathbf{x})] = \text{vech}[(T L_B)(\mathbf{x})] = \text{vech}\{T[L_B(\mathbf{x})]\}.\end{aligned}$$

And, in light of result (3.4), it follows that, for any $m \times n$ matrix $\mathbf{X}$,

$$\begin{aligned}(L_C^{-1}TL_B)(\text{vec }\mathbf{X}) &= \text{vech}\{T[L_B(\text{vec }\mathbf{X})]\} \\ &= \text{vech}(T\{L_B[L_B^{-1}(\mathbf{X})]\}) = \text{vech}[T(\mathbf{X})].\end{aligned}$$

(b) For any $m \times n$ matrix $\mathbf{X} = \{x_{ij}\}$, we find [using Part (a)] that

$$\begin{aligned}(L_C^{-1}TL_B)(\text{vec }\mathbf{X}) &= \text{vech}\left[T\left(\sum_{i,j} x_{ij}\mathbf{U}_{ij}\right)\right] \\ &= \text{vech}\left[\sum_{i,j} x_{ij}T(\mathbf{U}_{ij})\right] \\ &= \sum_{i,j} x_{ij}\text{vech}[T(\mathbf{U}_{ij})] \\ &= [\text{vech }T(\mathbf{U}_{11}), \ldots, \text{vech }T(\mathbf{U}_{m1}), \\ &\qquad \ldots, \text{vech }T(\mathbf{U}_{1n}), \ldots, \text{vech }T(\mathbf{U}_{mn})]\text{vec}(\mathbf{X}).\end{aligned}$$

And, in light of result (4.7), it follows that the matrix representation of T (with respect to B and C) equals the $p(p+1)/2 \times mn$ matrix

$$[\text{vech }T(\mathbf{U}_{11}), \ldots, \text{vech }T(\mathbf{U}_{m1}), \ldots, \text{vech }T(\mathbf{U}_{1n}), \ldots, \text{vech }T(\mathbf{U}_{mn})].$$

(c) Using Part (a) and results (16.4.6), (16.3.1), and (16.4.15), we find that, for any $n \times n$ matrix $\mathbf{X}$,

$$\begin{aligned}(L_C^{-1}TL_B)(\text{vec }\mathbf{X}) &= \text{vech}[(1/2)(\mathbf{X}+\mathbf{X}')] \\ &= (1/2)\text{vech}(\mathbf{X}+\mathbf{X}') \\ &= (1/2)(\mathbf{G}_n'\mathbf{G}_n)^{-1}\mathbf{G}_n'\text{vec}(\mathbf{X}+\mathbf{X}') \\ &= (1/2)(\mathbf{G}_n'\mathbf{G}_n)^{-1}\mathbf{G}_n'[\text{vec}(\mathbf{X})+\mathbf{K}_{nn}\text{vec}(\mathbf{X})] \\ &= (1/2)[(\mathbf{G}_n'\mathbf{G}_n)^{-1}\mathbf{G}_n' + (\mathbf{G}_n'\mathbf{G}_n)^{-1}\mathbf{G}_n'\mathbf{K}_{nn}](\text{vec }\mathbf{X}) \\ &= (1/2)[(\mathbf{G}_n'\mathbf{G}_n)^{-1}\mathbf{G}_n' + (\mathbf{G}_n'\mathbf{G}_n)^{-1}\mathbf{G}_n'](\text{vec }\mathbf{X}) \\ &= (\mathbf{G}_n'\mathbf{G}_n)^{-1}\mathbf{G}_n'(\text{vec }\mathbf{X}).\end{aligned}$$

And, in light of result (4.7), it follows that the matrix representation of T (with respect to B and C) equals $(\mathbf{G}_n'\mathbf{G}_n)^{-1}\mathbf{G}_n'$.

EXERCISE 12. Let $\mathcal{V}$ represent an n-dimensional linear space. Further, let $B = \{\mathbf{V}_1, \mathbf{V}_2, \ldots, \mathbf{V}_n\}$ represent a basis for $\mathcal{V}$, and let $\mathbf{A}$ represent an $n \times n$ nonsingular matrix. And, for $j = 1, \ldots, n$, let

$$\mathbf{W}_j = f_{1j}\mathbf{V}_1 + f_{2j}\mathbf{V}_2 + \cdots + f_{nj}\mathbf{V}_n\,,$$

where (for $i = 1, \ldots, n$) f_{ij} is the ijth element of $\mathbf{A}^{-1}$. Show that the set C comprising the matrices $\mathbf{W}_1, \mathbf{W}_2, \ldots \mathbf{W}_n$ is a basis for $\mathcal{V}$ and that $\mathbf{A}$ is the matrix representation of the identity transformation I (from $\mathcal{V}$ onto $\mathcal{V}$) with respect to B and C.

Solution. Lemma 3.2.4 implies that the set C is linearly independent and hence that C is a basis for $\mathcal{V}$. Then, clearly, $\mathbf{A}^{-1}$ is the matrix representation of the identity transformation I (from $\mathcal{V}$ onto $\mathcal{V}$) with respect to C and B. And, it follows from Corollary 22.4.3 that $(\mathbf{A}^{-1})^{-1}$ is the matrix representation of I^{-1} with respect to B and C and hence [since $(\mathbf{A}^{-1})^{-1} = \mathbf{A}$ and $I^{-1} = I$] that $\mathbf{A}$ is the matrix representation of I with respect to B and C.

EXERCISE 13. Let T represent the linear transformation from $\mathcal{R}^{4\times 1}$ into $\mathcal{R}^{3\times 1}$ defined by

$$T(\mathbf{x}) = (x_1 + x_2, \quad x_2 + x_3 - x_4, \quad x_1 - x_3 + x_4)',$$

where $\mathbf{x} = (x_1, x_2, x_3, x_4)'$. Further, let B represent the natural basis for $\mathcal{R}^{4\times 1}$ (comprising the columns of $\mathbf{I}_4$), and let E represent the basis (for $\mathcal{R}^{4\times 1}$) comprising the four vectors $(1, -1, 0, -1)'$, $(0, 0, 1, 1)'$, $(0, 0, 0, 1)'$, and $(1, 1, 0, 0)'$. And, let C represent the natural basis for $\mathcal{R}^{3\times 1}$ (comprising the columns of $\mathbf{I}_3$), and F represent the basis (for $\mathcal{R}^{3\times 1}$) comprising the three vectors $(1, 0, 1)'$, $(1, 1, 0)'$, and $(-1, 0, 0)'$.

(a) Find the matrix representation of T with respect to B and C.

(b) Find (1) the matrix representation of the identity transformation from $\mathcal{R}^{4\times 1}$ onto $\mathcal{R}^{4\times 1}$ with respect to E and B and (2) the matrix representation of the identity transformation from $\mathcal{R}^{3\times 1}$ onto $\mathcal{R}^{3\times 1}$ with respect to C and F.

(c) Find the matrix representation of T with respect to E and F via each of two approaches: (1) a direct approach, using the equality

$$\mathbf{W}\mathbf{A} = [T(\mathbf{V}_1), \ldots, T(\mathbf{V}_n)], \tag{$*$}$$

where $\mathbf{A}$ is the matrix of a linear transformation T from an n-dimensional linear space $\mathcal{V}$ into $\mathcal{R}^{m\times 1}$, where $\{\mathbf{V}_1, \ldots, \mathbf{V}_n\}$ is the basis for $\mathcal{V}$, and where $\mathbf{W}$ is an $m \times m$ matrix whose columns form the basis for $\mathcal{R}^{m\times 1}$; and (2) an indirect approach, using the results of Parts (a) and (b) in combination with the result that if $\mathbf{A}$ is the matrix representation of a linear transformation T from a linear space $\mathcal{V}$ into a linear space $\mathcal{W}$ with respect to bases B and C (for $\mathcal{V}$ and $\mathcal{W}$, respectively), then the matrix representation of T with respect to alternative bases E and F is $\mathbf{S}^{-1}\mathbf{A}\mathbf{R}$, where $\mathbf{R}$ is the matrix representation of the identity transformation from $\mathcal{V}$ onto $\mathcal{V}$ with respect to E and B and $\mathbf{S}$ is the matrix representation of the identity transformation from $\mathcal{W}$ onto $\mathcal{W}$ with respect to F and C.

(d) Find rank T and $\dim[\mathcal{N}(T)]$. Do so by, for instance, using the result that the rank of a linear transformation T from an n-dimensional linear space $\mathcal{V}$ into a linear space $\mathcal{W}$ equals the rank of its matrix representation (with respect to any bases B and C) and the result that $\dim[\mathcal{N}(T)] = n - \text{rank}(T)$.

Solution. (a) Let $\mathbf{A}$ represent the matrix representation of T with respect to B and C, and denote the first, ..., 4th columns of $\mathbf{I}_4$ by $\mathbf{e}_1, \ldots, \mathbf{e}_4$, respectively. Then, in light of the discussion in Part 1 of Section 4b, we find that

$$\mathbf{A} = \mathbf{I}_3\mathbf{A} = [T(\mathbf{e}_1), \ldots, T(\mathbf{e}_4)] = \begin{pmatrix} 1 & 1 & 0 & 0 \\ 0 & 1 & 1 & -1 \\ 1 & 0 & -1 & 1 \end{pmatrix}.$$

(b) (1) In light of the discussion in Part 3 of Section 4b, the matrix representation of the identity transformation from $\mathcal{R}^{4\times 1}$ onto $\mathcal{R}^{4\times 1}$ with respect to E and B is the 4×4 matrix whose first, ..., 4th columns are the vectors that form E, that is, the 4×4 matrix

$$\begin{pmatrix} 1 & 0 & 0 & 1 \\ -1 & 0 & 0 & 1 \\ 0 & 1 & 0 & 0 \\ -1 & 1 & 1 & 0 \end{pmatrix}.$$

(2) Let $\mathbf{S}$ represent the 3×3 nonsingular matrix whose inverse $\mathbf{S}^{-1}$ is the matrix representation of the identity transformation from $\mathcal{R}^{3\times 1}$ onto $\mathcal{R}^{3\times 1}$ with respect to C and F. Then, in light of Part 3 of Section 4b,

$$\begin{pmatrix} 1 & 1 & -1 \\ 0 & 1 & 0 \\ 1 & 0 & 0 \end{pmatrix} \mathbf{S}^{-1} = \mathbf{I}_3,$$

so that

$$\mathbf{S}^{-1} = \begin{pmatrix} 0 & 0 & 1 \\ 0 & 1 & 0 \\ -1 & 1 & 1 \end{pmatrix}.$$

(c) Let $\mathbf{H}$ represent the matrix representation of T with respect to E and F.

(1) Equality $(*)$ [or equivalently equality (4.8)] gives

$$\begin{pmatrix} 1 & 1 & -1 \\ 0 & 1 & 0 \\ 1 & 0 & 0 \end{pmatrix} \mathbf{H} = \begin{pmatrix} 0 & 0 & 0 & 2 \\ 0 & 0 & -1 & 1 \\ 0 & 0 & 1 & 1 \end{pmatrix}.$$

Thus,

$$\mathbf{H} = \begin{pmatrix} 0 & 0 & 1 & 1 \\ 0 & 0 & -1 & 1 \\ 0 & 0 & 0 & 0 \end{pmatrix}.$$

(2) The matrix S [from Part (b)] is (in light of Corollary 22.4.3) the matrix representation of the identity transformation from $\mathcal{R}^{3\times 1}$ onto $\mathcal{R}^{3\times 1}$ with respect to F and C. Thus, in light of the results of Parts (a) and (b), it follows from the

result cited (which is Theorem 22.4.4) that

$$\mathbf{H} = \begin{pmatrix} 0 & 0 & 1 \\ 0 & 1 & 0 \\ -1 & 1 & 1 \end{pmatrix} \begin{pmatrix} 1 & 1 & 0 & 0 \\ 0 & 1 & 1 & -1 \\ 1 & 0 & -1 & 1 \end{pmatrix} \begin{pmatrix} 1 & 0 & 0 & 1 \\ -1 & 0 & 0 & 1 \\ 0 & 1 & 0 & 0 \\ -1 & 1 & 1 & 0 \end{pmatrix}$$
$$= \begin{pmatrix} 0 & 0 & 1 & 1 \\ 0 & 0 & -1 & 1 \\ 0 & 0 & 0 & 0 \end{pmatrix}.$$

(d) It is clear from Part (c) that the rank of the matrix representation of T with respect to E and F equals 2. Thus, it follows from the first result cited (which is Theorem 22.5.2) that rank $T = 2$ and from the second result cited (which is part of Corollary 22.5.3) that $\dim[\mathcal{N}(T)] = 4 - 2 = 2$.

EXERCISE 14. Let T represent a linear transformation of rank k (where $k > 0$) from an n-dimensional linear space $\mathcal{V}$ into an m-dimensional linear space $\mathcal{W}$. Show that there exists a basis E for $\mathcal{V}$ and a basis F for $\mathcal{W}$ such that the matrix representation of T with respect to E and F is of the form $\begin{pmatrix} \mathbf{I}_k & \mathbf{0} \\ \mathbf{0} & \mathbf{0} \end{pmatrix}$.

Solution. Let B represent any basis for $\mathcal{V}$ and C any basis for $\mathcal{W}$. And, let $\mathbf{A}$ represent the matrix representation of T with respect to B and C. Then, as a consequence of Theorem 22.5.2, rank $\mathbf{A} = k$, and it follows from Theorem 4.4.9 that there exists an $n \times n$ nonsingular matrix $\mathbf{R}$ and an $m \times m$ nonsingular matrix $\mathbf{S}$ such that $\mathbf{A} = \mathbf{S}\begin{pmatrix} \mathbf{I}_k & \mathbf{0} \\ \mathbf{0} & \mathbf{0} \end{pmatrix}\mathbf{R}^{-1}$ or equivalently such that $\begin{pmatrix} \mathbf{I}_k & \mathbf{0} \\ \mathbf{0} & \mathbf{0} \end{pmatrix} = \mathbf{S}^{-1}\mathbf{A}\mathbf{R}$. And, based on Theorem 22.4.7, we conclude that $\begin{pmatrix} \mathbf{I}_k & \mathbf{0} \\ \mathbf{0} & \mathbf{0} \end{pmatrix}$ is the matrix representation of T with respect to some bases E and F.

EXERCISE 15. Let T represent a linear transformation from an n-dimensional linear space $\mathcal{V}$ into an m-dimensional linear space $\mathcal{W}$, and let $\mathbf{A}$ represent the matrix representation of T with respect to bases B and C (for $\mathcal{V}$ and $\mathcal{W}$, respectively). Use the result that an $n \times 1$ vector $\mathbf{x}$ is in $\mathcal{N}(\mathbf{A})$ if and only if the corresponding matrix $L_B(\mathbf{x})$ is in $\mathcal{N}(T)$ [or equivalently that a matrix $\mathbf{X}$ (in $\mathcal{V}$) is in $\mathcal{N}(T)$ if and only if the corresponding vector $L_B^{-1}(\mathbf{X})$ is in $\mathcal{N}(\mathbf{A})$] to devise a "direct" proof that $\dim[\mathcal{N}(T)] = \dim[\mathcal{N}(\mathbf{A})]$ (as opposed to deriving this equality as a corollary of the result that rank T = rank $\mathbf{A}$).

Solution. The result cited [which is Part (2) of Theorem 22.5.1] implies that $L_B[\mathcal{N}(\mathbf{A})] = \mathcal{N}(T)$ (as can be easily verified). Thus, there exists a 1-1 linear transformation from $\mathcal{N}(\mathbf{A})$ onto $\mathcal{N}(T)$, namely, the linear transformation R defined [for $\mathbf{x} \in \mathcal{N}(\mathbf{A})$] by $R(\mathbf{x}) = L_B(\mathbf{x})$. And, it follows that $\mathcal{N}(\mathbf{A})$ and $\mathcal{N}(T)$ are isomorphic. Based on Theorem 22.3.1, we conclude that $\dim[\mathcal{N}(T)] = \dim[\mathcal{N}(\mathbf{A})]$.

EXERCISE 16. Let T represent a linear transformation from an n-dimensional linear space $\mathcal{V}$ into $\mathcal{V}$.

(a) Let $\mathcal{U}$ represent an r-dimensional subspace of $\mathcal{V}$, and suppose that $\mathcal{U}$ is invariant relative to T. Show that there exists a basis B for $\mathcal{V}$ such that the matrix representation of T with respect to B and B is of the (upper block-triangular) form $\begin{pmatrix} \mathbf{E} & \mathbf{F} \\ \mathbf{0} & \mathbf{H} \end{pmatrix}$ (where $\mathbf{E}$ is of dimensions $r \times r$).

(b) Let $\mathcal{U}$ and $\mathcal{W}$ represent subspaces of $\mathcal{V}$ such that $\mathcal{U} \oplus \mathcal{W} = \mathcal{V}$ (i.e., essentially disjoint subspaces of $\mathcal{V}$ whose sum is $\mathcal{V}$). Suppose that both $\mathcal{U}$ and $\mathcal{W}$ are invariant relative to T. Show that there exists a basis B for $\mathcal{V}$ such that the matrix representation of T with respect to B and B is of the (block-diagonal) form diag($\mathbf{E}$, $\mathbf{H}$) [where the dimensions of $\mathbf{E}$ equal dim($\mathcal{W}$)].

Solution. (a) Let $\mathbf{X}_1, \ldots, \mathbf{X}_r$ represent any r matrices that form a basis for $\mathcal{U}$. And, take B to be any basis for $\mathcal{V}$ comprising $\mathbf{X}_1, \ldots, \mathbf{X}_r$ and $n - r$ additional matrices $\mathbf{X}_{r+1}, \ldots, \mathbf{X}_n$ — the existence of such a basis is guaranteed by Theorem 4.3.12. Further, let $\mathbf{A} = \{a_{ij}\}$ represent the matrix representation of T with respect to B and B. Then, for $j = 1, \ldots, r$,

$$a_{1j}\mathbf{X}_1 + \cdots + a_{rj}\mathbf{X}_r + a_{r+1,j}\mathbf{X}_{r+1} + \cdots + a_{nj}\mathbf{X}_n = T(\mathbf{X}_j) \in \mathcal{U},$$

implying (since any matrix in $\mathcal{U}$ can be expressed as a linear combination of $\mathbf{X}_1, \ldots, \mathbf{X}_r$) that $a_{r+1,j} = \cdots = a_{nj} = 0$. Thus, $a_{ij} = 0$ for $i = r+1, \ldots, n$ and $j = 1, \ldots, r$.

(b) Let $r = \dim(\mathcal{U})$ [in which case $\dim(\mathcal{W}) = n - r$]. Further, let $\mathbf{X}_1, \ldots, \mathbf{X}_r$ represent any r matrices that form a basis for $\mathcal{U}$ and $\mathbf{X}_{r+1}, \ldots, \mathbf{X}_n$ any $n - r$ matrices that form a basis for $\mathcal{W}$. And, take B to be the basis for $\mathcal{V}$ comprising $\mathbf{X}_1, \ldots, \mathbf{X}_r, \mathbf{X}_{r+1}, \ldots, \mathbf{X}_n$ — that $\mathbf{X}_1, \ldots, \mathbf{X}_r, \mathbf{X}_{r+1}, \ldots, \mathbf{X}_n$ form a basis for $\mathcal{V}$ is evident from Theorem 17.1.5.

Now, let $\mathbf{A} = \{a_{ij}\}$ represent the matrix representation of T with respect to B and B. Then, for $j = 1, \ldots, r, r+1, \ldots, n$,

$$a_{1j}\mathbf{X}_1 + \cdots + a_{rj}\mathbf{X}_r + a_{r+1,j}\mathbf{X}_{r+1} + \cdots + a_{nj}\mathbf{X}_n = T(\mathbf{X}_j).$$

And, for $j = 1, \ldots, r$, $T(\mathbf{X}_j) \in \mathcal{U}$, implying (since any matrix in $\mathcal{U}$ can be expressed as a linear combination of $\mathbf{X}_1, \ldots, \mathbf{X}_r$) that $a_{r+1,j} = \cdots = a_{nj} = 0$. Similarly, for $j = r+1, \ldots, n$, $T(\mathbf{X}_j) \in \mathcal{W}$, implying (since any matrix in $\mathcal{W}$ can be expressed as a linear combination of $\mathbf{X}_{r+1}, \ldots, \mathbf{X}_n$) that $a_{1j} = \cdots = a_{rj} = 0$. Thus, $a_{ij} = 0$ for $i = r+1, \ldots, n$ and $j = 1, \ldots, r$; and also $a_{ij} = 0$ for $i = 1, \ldots, r$ and $j = r+1, \ldots, n$.

EXERCISE 17. Let $\mathcal{V}$, $\mathcal{W}$, and $\mathcal{U}$ represent linear spaces.

(a) Show that the dual transformation of the identity transformation I from $\mathcal{V}$ onto $\mathcal{V}$ is I.

(b) Show that the dual transformation of the zero transformation 0_1 from $\mathcal{V}$ into $\mathcal{W}$ is the zero transformation 0_2 from $\mathcal{W}$ into $\mathcal{V}$.

(c) Let S represent the dual transformation of a linear transformation T from $\mathcal{V}$ into $\mathcal{W}$. Show that T is the dual transformation of S.

(d) Let k represent a scalar, and let S represent the dual transformation of a linear transformation T from $\mathcal{V}$ into $\mathcal{W}$. Show that kS is the dual transformation of kT.

(e) Let T_1 and T_2 represent linear transformations from $\mathcal{V}$ into $\mathcal{W}$, and let S_1 and S_2 represent the dual transformations of T_1 and T_2, respectively. Show that $S_1 + S_2$ is the dual transformation of $T_1 + T_2$.

(f) Let P represent the dual transformation of a linear transformation S from $\mathcal{U}$ into $\mathcal{V}$, and let Q represent the dual transformation of a linear transformation T from $\mathcal{V}$ into $\mathcal{W}$. Show that PQ is the dual transformation of TS.

Solution. Write $\mathbf{X} \bullet \mathbf{Z}$ for the inner product of arbitrary matrices $\mathbf{X}$ and $\mathbf{Z}$ in $\mathcal{V}$, $\mathbf{U} * \mathbf{Y}$ for the inner product of arbitrary matrices $\mathbf{U}$ and $\mathbf{Y}$ in $\mathcal{W}$, and $\mathbf{A} \star \mathbf{B}$ for the inner product of arbitrary matrices $\mathbf{A}$ and $\mathbf{B}$ in $\mathcal{U}$.

(a) For every matrix $\mathbf{X}$ in $\mathcal{V}$ and every matrix $\mathbf{Y}$ in $\mathcal{V}$,

$$\mathbf{X} \bullet I(\mathbf{Y}) = \mathbf{X} \bullet \mathbf{Y} = I(\mathbf{X}) \bullet \mathbf{Y}.$$

Thus, I is the dual transformation of I.

(b) For every matrix $\mathbf{X}$ in $\mathcal{V}$ and every matrix $\mathbf{Y}$ in $\mathcal{W}$,

$$\mathbf{X} \bullet 0_2(\mathbf{Y}) = \mathbf{X} \bullet \mathbf{0} = 0 = \mathbf{0} * \mathbf{Y} = 0_1(\mathbf{X}) * \mathbf{Y}.$$

Thus, 0_2 is the dual transformation of 0_1.

(c) For every matrix $\mathbf{Y}$ in $\mathcal{W}$ and every matrix $\mathbf{X}$ in $\mathcal{V}$,

$$\mathbf{Y} * T(\mathbf{X}) = T(\mathbf{X}) * \mathbf{Y} = \mathbf{X} \bullet S(\mathbf{Y}) = S(\mathbf{Y}) \bullet \mathbf{X}.$$

Thus, T is the dual transformation of S.

(d) For every matrix $\mathbf{X}$ in $\mathcal{V}$ and every matrix $\mathbf{Y}$ in $\mathcal{W}$,

$$\begin{aligned}\mathbf{X} \bullet (kS)(\mathbf{Y}) &= \mathbf{X} \bullet [kS(\mathbf{Y})] \\ &= k[\mathbf{X} \bullet S(\mathbf{Y})] = k[T(\mathbf{X}) * \mathbf{Y}] = [kT(\mathbf{X})] * \mathbf{Y} = (kT)(\mathbf{X}) * \mathbf{Y}.\end{aligned}$$

Thus, kS is the dual transformation of kT.

(e) For every matrix $\mathbf{X}$ in $\mathcal{V}$ and every matrix $\mathbf{Y}$ in $\mathcal{W}$,

$$\begin{aligned}\mathbf{X} \bullet (S_1 + S_2)(\mathbf{Y}) &= \mathbf{X} \bullet [S_1(\mathbf{Y}) + S_2(\mathbf{Y})] \\ &= \mathbf{X} \bullet S_1(\mathbf{Y}) + \mathbf{X} \bullet S_2(\mathbf{Y}) \\ &= T_1(\mathbf{X}) * \mathbf{Y} + T_2(\mathbf{X}) * \mathbf{Y} \\ &= [T_1(\mathbf{X}) + T_2(\mathbf{X})] * \mathbf{Y} = (T_1 + T_2)(\mathbf{X}) * \mathbf{Y}.\end{aligned}$$

Thus, $S_1 + S_2$ is the dual transformation of $T_1 + T_2$.

(f) For every matrix $\mathbf{X}$ in $\mathcal{U}$ and every matrix $\mathbf{Y}$ in $\mathcal{W}$,

$$\mathbf{X} \star (PQ)(\mathbf{Y}) = \mathbf{X} \star P[Q(\mathbf{Y})] = S(\mathbf{X}) \bullet Q(\mathbf{Y}) = T[S(\mathbf{X})] * \mathbf{Y} = (TS)(\mathbf{X}) * \mathbf{Y}.$$

Thus, PQ is the dual transformation of TS.

EXERCISE 18. Let S represent the dual transformation of a linear transformation T from an n-dimensional linear space $\mathcal{V}$ into an m-dimensional linear space $\mathcal{W}$. And, let $\mathbf{A} = \{a_{ij}\}$ represent the matrix representation of T with respect to orthonormal bases C and D, and $\mathbf{B} = \{b_{ij}\}$ represent the matrix representation of S with respect to D and C. Using the result of Exercise 9 (or otherwise), show that $\mathbf{B} = \mathbf{A}'$.

Solution. Write $\mathbf{X} \bullet \mathbf{Z}$ for the inner product of arbitrary matrices $\mathbf{X}$ and $\mathbf{Z}$ in $\mathcal{V}$, and $\mathbf{U} * \mathbf{Y}$ for the inner product of arbitrary matrices $\mathbf{U}$ and $\mathbf{Y}$ in $\mathcal{W}$. And, let $\mathbf{X}_1, \ldots, \mathbf{X}_n$ represent the matrices that form the orthonormal basis C and $\mathbf{Y}_1, \ldots, \mathbf{Y}_m$ the matrices that form the orthonormal basis D. Then, for $i = 1, \ldots, m$ and $j = 1, \ldots, n$, it follows from the result of Exercise 9 that

$$a_{ij} = T(\mathbf{X}_j) * \mathbf{Y}_i$$

and

$$b_{ji} = S(\mathbf{Y}_i) \bullet \mathbf{X}_j,$$

implying that

$$b_{ji} = \mathbf{X}_j \bullet S(\mathbf{Y}_i) = T(\mathbf{X}_j) * \mathbf{Y}_i = a_{ij}$$

and hence that the jith element of $\mathbf{B}$ equals the jith element of $\mathbf{A}'$. Thus, $\mathbf{B} = \mathbf{A}'$.

EXERCISE 19. Let $\mathbf{A}$ represent an $m \times n$ matrix, let $\mathbf{V}$ represent an $n \times n$ symmetric positive definite matrix, and let $\mathbf{W}$ represent an $m \times m$ symmetric positive definite matrix. And let S represent the dual transformation of the linear transformation T from $\mathcal{R}^{n \times 1}$ into $\mathcal{R}^{m \times 1}$ defined by $T(\mathbf{x}) = \mathbf{Ax}$ (where $\mathbf{x}$ is an arbitrary $n \times 1$ vector). Taking the inner product of arbitrary vectors $\mathbf{x}$ and $\mathbf{z}$ in $\mathcal{R}^{n \times 1}$ to be $\mathbf{x}'\mathbf{Vz}$ and the inner product of arbitrary vectors $\mathbf{u}$ and $\mathbf{y}$ in $\mathcal{R}^{m \times 1}$ to be $\mathbf{u}'\mathbf{Wy}$, obtain a formula for $S(\mathbf{y})$ that generalizes the formula $S(\mathbf{y}) = \mathbf{A}'\mathbf{y}$ $(\mathbf{y} \in \mathcal{R}^{m \times 1})$ obtained in the special case of the usual inner products (i.e., in the special case where $\mathbf{V} = \mathbf{I}_n$ and $\mathbf{W} = \mathbf{I}_m$).

Solution. For every vector $\mathbf{x}$ in $\mathcal{R}^{n \times 1}$ and every vector $\mathbf{y}$ in $\mathcal{R}^{m \times 1}$,

$$\mathbf{x}'\mathbf{V}S(\mathbf{y}) = (\mathbf{Ax})'\mathbf{Wy} = \mathbf{x}'\mathbf{A}'\mathbf{Wy}.$$

And, in light of the uniqueness of S,

$$S(\mathbf{y}) = \mathbf{V}^{-1}\mathbf{A}'\mathbf{Wy}.$$

EXERCISE 20. Let S represent the dual transformation of a linear transformation T from a linear space $\mathcal{V}$ into a linear space $\mathcal{W}$.

(a) Show that $[S(\mathcal{W})]^{\perp} = \mathcal{N}(T)$ (i.e., that the orthogonal complement of the range space of S equals the null space of T).

(b) Using the result of Part (c) of Exercise 17 (or otherwise), show that $[\mathcal{N}(S)]^{\perp} = T(\mathcal{V})$ (i.e., that the orthogonal complement of the null space of S equals the range space of T).

(c) Show that rank S = rank T.

Solution. (a) Write $\mathbf{X} \bullet \mathbf{Z}$ for the inner product of arbitrary matrices $\mathbf{X}$ and $\mathbf{Z}$ in $\mathcal{V}$ and $\mathbf{U} * \mathbf{Y}$ for the inner product of arbitrary matrices $\mathbf{U}$ and $\mathbf{Y}$ in $\mathcal{W}$.

Let $\mathbf{X}$ represent an arbitrary matrix in $\mathcal{V}$. Suppose that $\mathbf{X} \in \mathcal{N}(T)$. Then, for every matrix $\mathbf{Y}$ in $\mathcal{W}$,

$$\mathbf{X} \bullet S(\mathbf{Y}) = T(\mathbf{X}) * \mathbf{Y} = \mathbf{0} * \mathbf{Y} = 0.$$

Thus, $\mathbf{X} \in [S(\mathcal{W})]^{\perp}$.

Conversely, suppose that $\mathbf{X} \in [S(\mathcal{W})]^{\perp}$. Then, since $T(\mathbf{X}) \in \mathcal{W}$,

$$T(\mathbf{X}) * T(\mathbf{X}) = \mathbf{X} \bullet S[T(\mathbf{X})] = 0,$$

implying that $T(\mathbf{X}) = \mathbf{0}$ and hence that $\mathbf{X} \in \mathcal{N}(T)$.

Thus, $[S(\mathcal{W})]^{\perp} = \mathcal{N}(T)$.

(b) Since [according to Part (c) of Exercise 17] T is the dual transformation of S, it follows from Part (a) that $[T(\mathcal{V})]^{\perp} = \mathcal{N}(S)$. And, making use of Theorem 12.5.4, we find that

$$[\mathcal{N}(S)]^{\perp} = \{[T(\mathcal{V})]^{\perp}\}^{\perp} = T(\mathcal{V}).$$

(c) Making use of Corollary 22.5.3 and Theorem 12.5.12 [together with the result of Part (a)], we find that

$$\begin{aligned}
\text{rank}\, T &= \dim(\mathcal{V}) - \dim[\mathcal{N}(T)] \\
&= \dim(\mathcal{V}) - \dim\{[S(\mathcal{W})]^{\perp}\} \\
&= \dim(\mathcal{V}) - \{\dim(\mathcal{V}) - \dim[S(\mathcal{W})]\} \\
&= \dim[S(\mathcal{W})] \\
&= \text{rank}\, S.
\end{aligned}$$

References

References
Bartle, R. G. (1976), *The Elements of Real Analysis* (2nd ed.), New York: John Wiley.

Goodnight, J. H. (1979), "A Tutorial on the SWEEP Operator," *The American Statistician*, 33, 149–158.

Magnus, J. R., and Neudecker, H. (1980), "The Elimination Matrix: Some Lemmas and Applications," *SIAM Journal on Algebra and Discrete Mathematics*, 1, 422–449.

Meyer, C. D. (1973), "Generalized Inverses and Ranks of Block Matrices," *SIAM Journal on Applied Mathematics*, 25, 597–602.

Index

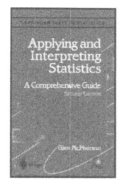

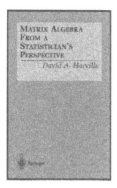

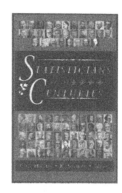